D1779534

ASTRONOMICAL DATA ANALYSIS SOFTWARE AND SYSTEMS II

A SERIES OF BOOKS ON RECENT DEVELOPMENTS IN ASTRONOMY AND ASTROPHYSICS

390 Ashton Avenue, San Francisco, California 94112

Printed by BookCrafters, Inc.

First published 1993

Library of Congress Catalog Card Number: 93-72266
ISBN 0-937707-71-6

D. Harold McNamara, Managing Editor of Conference Series
408 ESC Brigham Young University
Provo, UT 84602
801-378-2298

A SERIES OF BOOKS ON RECENT DEVELOPMENTS IN ASTRONOMY AND ASTROPHYSICS

Vol. 1-Progress and Opportunities in Southern Hemisphere Optical Astronomy: The CTIO 25th Anniversary Symposium
ed. V. M. Blanco and M. M. Phillips ISBN 0-937707-18-X

Vol. 2-Proceedings of a Workshop on Optical Surveys for Quasars
ed. P. S. Osmer, A. C. Porter, R. F. Green, and C. B. Foltz ISBN 0-937707-19-8

Vol. 3-Fiber Optics in Astronomy
ed. S. C. Barden ISBN 0-937707-20-1

Vol. 4-The Extragalactic Distance Scale: Proceedings of the ASP 100th Anniversary Symposium
ed. S. van den Bergh and C. J. Pritchet ISBN 0-937707-21-X

Vol. 5-The Minnesota Lectures on Clusters of Galaxies and Large-Scale Structure
ed. J. M. Dickey ISBN 0-937707-22-8

Vol. 6-Synthesis Imaging in Radio Astronomy: A Collection of Lectures from the Third NRAO Synthesis Imaging Summer School
ed. R. A. Perley, F. R. Schwab, and A. H. Bridle ISBN 0-937707-23-6

Vol. 7-Properties of Hot Luminous Stars: Boulder-Munich Workshop
ed. C. D. Garmany ISBN 0-937707-24-4

Vol. 8-CCDs in Astronomy
ed. G. H. Jacoby ISBN 0-937707-25-2

Vol. 9-Cool Stars, Stellar Systems, and the Sun. Sixth Cambridge Workshop
ed. G. Wallerstein ISBN 0-937707-27-9

Vol. 10-The Evolution of the Universe of Galaxies. The Edwin Hubble Centennial Symposium
ed. R. G. Kron ISBN 0-937707-28-7

Vol. 11-Confrontation Between Stellar Pulsation and Evolution
ed. C. Cacciari and G. Clementini ISBN 0-937707-30-9

Vol. 12-The Evolution of the Interstellar Medium
ed. L. Blitz ISBN 0-937707-31-7

Vol. 13-The Formation and Evolution of Star Clusters
ed. K. Janes ISBN 0-937707-32-5

Vol. 14-Astrophysics with Infrared Arrays
ed. R. Elston ISBN 0-937707-33-3

Vol. 15-Large-Scale Structures and Peculiar Motions in the Universe
ed. D. W. Latham and L. A. N. da Costa ISBN 0-937707-34-1

Vol. 16-Atoms, Ions and Molecules: New Results in Spectral Line Astrophysics
ed. A. D. Haschick and P. T. P. Ho ISBN 0-937707-35-X

Vol. 17-Light Pollution, Radio Interference, and Space Debris
ed. D. L. Crawford ISBN 0-937707-36-8

Vol. 18-The Interpretation of Modern Synthesis Observations of Spiral Galaxies
ed. N. Duric and P. C. Crane ISBN 0-937707-37-6

Vol. 19-Radio Interferometry: Theory, Techniques, and Applications, IAU Colloquium 131
ed. T. J. Cornwell and R. A. Perley ISBN 0-937707-38-4

Vol. 20-Frontiers of Stellar Evolution, celebrating the 50th Anniversary of McDonald Observatory
ed. D. L. Lambert ISBN 0-937707-39-2

Vol. 21-The Space Distribution of Quasars
ed. D. Crampton ISBN 0-937707-40-6

Vol. 22-Nonisotropic and Variable Outflows from Stars
ed. L. Drissen, C. Leitherer, and A. Nota ISBN 0-937707-41-4

Vol. 23-Astronomical CCD Observing and Reduction Techniques
ed. S. B. Howell ISBN 0-937707-42-4

Vol. 24-Cosmology and Large-Scale Structure in the Universe
ed. R. R. de Carvalho ISBN 0-937707-43-0

Vol. 25-Astronomical Data Analysis Software and Systems I
ed. D. M. Worrall, C. Biemesderfer, and J. Barnes ISBN 0-937707-44-9

Vol. 26-Cool Stars, Stellar Systems, and the Sun, Seventh Cambridge Workshop
ed. M. S. Giampapa and J. A. Bookbinder ISBN 0-937707-45-7

Vol. 27-The Solar Cycle
ed. K. L. Harvey ISBN 0-937707-46-5

Vol. 28-Automated Telescopes for Photometry and Imaging
ed. S. J. Adelman, R. J. Dukes, Jr., and C. J. Adelman ISBN 0-937707-47-3

Vol. 29-Workshop on Cataclysmic Variable Stars
ed. N. Vogt ISBN 0-937707-48-1

Vol. 30-Variable Stars and Galaxies, in honor of M. S. Feast on his retirement
ed. B. Warner ISBN 0-937707-49-X

Vol. 31-Relationships Between Active Galactic Nuclei and Starburst Galaxies
ed. A. V. Filippenko ISBN 0-937707-50-3

Vol. 32-Complementary Approaches to Double and Multiple Star Research, IAU Colloquium 135
ed. H. A. McAlister and W. I. Hartkopf ISBN 0-937707-51-1

Vol. 33-Research Amateur Astronomy
ed. S. J. Edberg ISBN 0-937707-52-X

Vol. 34-Robotic Telescopes in the 1990s
ed. A. V. Filippenko ISBN 0-937707-53-8

Vol. 35-Massive Stars: Their Lives in the Interstellar Medium
ed. J. P. Cassinelli and E. B. Churchwell ISBN 0-937707-54-6

Vol. 36-Planets and Pulsars
ed. J. A. Phillips, S. E. Thorsett, and S. R. Kulkarni ISBN 0-937707-55-4

Vol. 37-Fiber Optics in Astronomy II
ed. P. M. Gray ISBN 0-937707-56-2

Vol. 38-New Frontiers in Binary Star Research
ed. K. C. Leung and I. S. Nha ISBN 0-937707-57-0

Vol. 39-The Minnesota Lectures on the Structure and Dynamics of the Milky Way
ed. Roberta M. Humphreys ISBN 0-937707-58-9

Vol. 40-Inside the Stars, IAU Colloquium 137
ed. Werner W. Weiss and Annie Baglin ISBN 0-937707-59-7

Vol. 41-Astronomical Infrared Spectroscopy: Future Observational Directions
ed. Sun Kwok ISBN 0-937707-60-0

Vol. 42-GONG 1992: Seismic Investigation of the Sun and Stars
ed. Timothy M. Brown ISBN 0-937707-61-9

Vol. 43-Sky Surveys: Protostars to Protogalaxies
ed. B. T. Soifer ISBN 0-937707-62-7

Vol. 44-Peculiar Versus Normal Phenomena in A-Type and Related Stars
ed. M. M. Dworetsky, F. Castelli, and R. Faraggiana ISBN 0-937707-63-5

Vol. 45-Luminous High-Latitude Stars
ed. D. D. Sasselov ISBN 0-937707-64-3

Vol. 46-The Magnetic and Velocity Fields of Solar Active Regions, IAU Colloquium 141
ed. H. Zirin, G. Ai, and H. Wang ISBN 0-937707-65-1

Vol. 47-Third Decinnial US-USSR Conference on SETI
ed. G. Seth Shostak ISBN 0-937707-66-X

Vol. 48-The Globular Cluster-Galaxy Connection
ed. Graeme H. Smith and Jean P. Brodie ISBN 0-937707-67-8

Vol. 49-Galaxy Evolution: The Milky Way Perspective
ed. Steven R. Majewski ISBN 0-937707-68-6

Vol. 50-Structure and Dynamics of Globular Clusters
ed. S. G. Djorgovski and G. Meylan ISBN 0-937707-69-4

Vol. 51-Observational Cosmology
ed. G. Chincarini, A. Iovino, T. Maccacaro, and D. Maccagni ISBN 0-937707-70-8

Vol. 52-Astronomical Data Analysis Software and Systems II
ed. R. J. Hanisch, J. V. Brissenden, and Jeannette Barnes ISBN 0-937707-71-6

Inquiries concerning these volumes should be directed to the:
Astronomical Society of the Pacific
CONFERENCE SERIES
390 Ashton Avenue
San Francisco, CA 94112-1722
415-337-1100

ASTRONOMICAL SOCIETY OF THE PACIFIC
CONFERENCE SERIES

Volume 52

ASTRONOMICAL DATA ANALYSIS SOFTWARE AND SYSTEMS II

Edited by
R. J. Hanisch, R. J. V. Brissenden,
and Jeannette Barnes

Table of Contents

Section B. Space-Based Data

Section C. Software Tools and Data Structures

Section D. Electronic Publishing

Part 2. Data Analysis Systems

Section A. Next Generation Systems and Languages

Section B. Software Systems

Part 3. Data Acquisition

Section A. Real-Time Systems

Part 5. Data Analysis Applications

Section A. Imaging Algorithms and Techniques

Section B. Spectral Algorithms and Techniques

Section C. Other Algorithms and Techniques

Section D. Image Restoration

Preface

This volume contains the papers presented at the second conference on Astronomical Data Analysis Software and Systems—ADASS II—which was held in Boston, Massachusetts, on 2–4 November 1992. After only two years the ADASS conference series has established itself as the world's primary meeting on astronomical software issues. There were 302 registered participants at ADASS II, with 66 people from 16 countries or territories outside the United States.

The key topics for ADASS II were Next Generation Software Systems and Languages; Databases, Catalogs, and Archives; User Interfaces and Visualization; and Real-Time Data Acquisition and Scheduling. This volume is organized into sections and subsections according to these general categories, with the addition of a major section on Data Analysis Applications to accommodate the many contributed papers.

The ADASS conferences also include a number of BOF (Birds of a Feather) sessions on special topics. ADASS II had BOFs on User Interfaces, IDL, XANADU, World Coordinate Systems in FITS, and IRAF Site Management. In addition, an IRAF/STSDAS/PROS Users Group meeting was held. A further highlight of ADASS II was the special election night coverage arranged by the LOC. (Was it just coincidence that the Boston Democratic Party had its election celebration in the same hotel, complete with indoor fireworks, or had the LOC arranged for this as well?)

The conference was sponsored by the National Aeronautics and Space Administration, the National Optical Astronomy Observatories, the National Science Foundation, the Smithsonian Astrophysical Observatory, and the Space Telescope Science Institute. We thank these organizations for their continuing support of the ADASS conference series.

The conference Program Organizing Committee was comprised of Dennis Crabtree (Dominion Astrophysical Observatory), Carol Christian (Center for EUV Astronomy, U.C. Berkeley), Robert Hanisch (Space Telescope Science Institute), F. Rick Harnden, Jr. (Smithsonian Astrophysical Observatory), George Jacoby (National Optical Astronomy Observatories), Richard Shaw (Space Telescope Science Institute), Doug Tody (National Optical Astronomy Observatories), and Diana Worrall (Smithsonian Astrophysical Observatory).

The Local Organizing Committee was chaired by F. Rick Harnden, Jr., to whom we are most grateful for a job well done. Special thanks are also extended to Susan Tuttle, Patricia Buckley, and Sandra Terranova for their hard work. The LOC was rounded out by Nancy Adler, Jay Bookbinder, Deborah Brissenden, Roger Brissenden, James Cornell, Paul Grant, Dan Harris, Todd Karakashian, Wendy Roberts, and Diana Worrall, all of whom we thank for their efforts in supporting the conference. Technical support was ably provided by Doug Kline, David Meleedy, Sumitra Chary, Paul Rundans, and John McSweeney, with special help from John Curran of NEARnet.

We extend special thanks to Chris Biemesderfer of *ferberts associates* for preparation of the LaTEX style file and other editing tools used to produce this volume.

We look forward to the continuation of this conference series, with ADASS III scheduled for Victoria, British Columbia, in October 1993 and ADASS IV planned for Baltimore, Maryland, in November 1994.

Robert J. Hanisch
Space Telescope Science Institute

Roger J. V. Brissenden
Smithsonian Astrophysical Observatory

Jeannette Barnes
National Optical Astronomy Observatories

May 1993

Cover illustration: A plot produced by the IRAF **specfocus** task that estimates the dispersion width of spectral lines in sequences of arc spectra taken at different focus settings, provided by Francisco Valdes (see his paper, page 452).

Participant List

Alberto Accomazzi ⟨ CFA::ACCOMAZZI ⟩, Smithsonian Astrophysical Observatory, 60 Garden Street, Cambridge, MA 02138

Hans-Martin Adorf ⟨ adorf@eso.org ⟩, European Southern Observatory, Karl-Schwarzschild-Strasse 2, D-8046 Garching bei München, GERMANY

Miguel Albrecht ⟨ malbrech@eso.org ⟩, European Southern Observatory, Karl-Schwarzschild-Strasse 2, D-8046 Garching bei München, GERMANY

Rudolf Albrecht ⟨ ralbrech@eso.org ⟩, Space Telescope–European Coordinating Facility/European Space Agency, Karl-Schwarzschild-Strasse 2, D-8046 Garching bei München, GERMANY

David Alexander ⟨ dalex@cfa.harvard.edu ⟩, Smithsonian Astrophysical Observatory, 60 Garden Street, Cambridge, MA 02138

David J. Allan ⟨ dja@star.sr.bham.ac.uk ⟩, University of Birmingham, Department of Physics and Space Research, Edgbaston, Birmingham, UNITED KINGDOM

Peter M. Allan ⟨ pma@star.rl.ac.uk ⟩, Rutherford Appleton Laboratory, Chilton, Didcot OX11 0QX, ENGLAND

R. J. Allen ⟨ rjallen@stsci.edu ⟩, Space Telescope Science Institute, 3700 San Martin Drive, Baltimore, MD 21218

Roberta Allsman ⟨ robynallsman@llnl.gov ⟩, Lawrence Livermore National Laboratory, P.O. Box 808, Mail Stop 72, Livermore, CA 94550

Lorella Angelini ⟨ HEASRC::ANGELINI ⟩, NASA/Goddard Space Flight Center, Code 668, Greenbelt, MD 20771

Keith Arnaud ⟨ kaa@rosserv.gsfc.nasa.gov ⟩, NASA/Goddard Space Flight Center, Code 666, Greenbelt, MD 20771

Martina Belz Arndt ⟨ marndt@cfa.harvard.edu ⟩, Smithsonian Astrophysical Observatory, 60 Garden Street, Cambridge, MA 02138

Thomas R. Ayres ⟨ HYADES::AYRES ⟩, University of Colorado (CASA), Campus Box 389, Boulder, CO 80309-0389

Karin Loya Babst ⟨ kbabst@gsfcmail.gsfc.nasa.gov ⟩, NASA/Goddard Space Flight Center, CSC, 7926 Helmart Drive, Laurel, MD 20723

Roberto Baglioni ⟨ baglioni@arcetri.astro.it ⟩, Dipartimento Astronomia Firenze, Largo E. Fermi 5, I-50125 Firenze, ITALY

Jeremy Bailey ⟨ jab@aaoepp.oz.au ⟩, Anglo-Australian Observatory, P.O. Box 2121, Epping, N.S.W. 2121, AUSTRALIA

Pascal Ballester ⟨ pballest@eso.org ⟩, European Southern Observatory, Karl-Schwarzschild-Strasse 2, D-8046 Garching bei München, GERMANY

Klaus Banse ⟨ kbanse@eso.org ⟩, European Southern Observatory, Karl-Schwarzschild-Strasse 2, D-8046 Garching bei München, GERMANY

Irene Barg ⟨ ibarg@as.arizona.edu ⟩, Steward Observatory, 949 N. Cherry Ave, Room 260, Tucson, AZ 85721

Jeannette Barnes ⟨ jbarnes@noao.edu ⟩, National Optical Astronomy Observatories, P.O. Box 26732, Tucson, AZ 85726-6732

James W. Barrett ⟨ 44157::JBARRETT ⟩, SUNY at Stony Brook, ESS Department, Stony Brook, NY 11794-2100

Paul Barrett, NASA/Goddard Space Flight Center, USRA, Code 668, Greenbelt, MD 20771

Suzanne Bauer ⟨ sbauer@bdcrg.nrl.navy.mil ⟩, Naval Research Laboratory/Backgrounds Data Center, Hughes STX, Code 7604, 4555 Overlook Avenue SW, Washington, DC 20375

David Bazell ⟨ bazell@stsci.edu ⟩, Space Telescope Science Institute, 3700 San Martin Drive, Baltimore, MD 21218

Tomaso Belloni ⟨ 28773::TMB ⟩, Max-Planck-Institut für Extraterrestrische Physik, Karl-Schwarzschild-Strasse 1, D-8046 Garching bei München, GERMANY

Stephen P. Berczuk ⟨ berczuk@mit.edu ⟩, Massachusetts Institute of Technology – NE80-6015, 77 Massachusetts Avenue, Cambridge, MA 02138

Joel Berendzen ⟨ joelb@lanl.gov ⟩, Los Alamos National Laboratory, Biophysics Group, P-6, M715, Los Alamos, NM 87545

Chris Biemesderfer ⟨ cbiemes@noao.edu ⟩, ferberts associates, 1895 Mt. Lemmon Highway, P.O. Box 1180, Oracle, AZ 85623

Teresa Bippert-Plymate ⟨ teresa@sparky.as.arizona.edu ⟩, Steward Observatory, 949 N. Cherry Avenue, Room 276, Tucson, AZ 85721

Jeffrey J. Bloch ⟨ jbloch@alexis.lanl.gov ⟩, Los Alamos National Laboratory, Mail Stop D436, SST-9, Los Alamos, NM 87545

Rosalie Blum ⟨ blum@cfa.harvard.edu ⟩, Smithsonian Astrophysical Observatory, 60 Garden Street, Cambridge, MA 02138

Bruce Bohannan ⟨ bruce@noao.edu ⟩, Kitt Peak National Observatory, P.O. Box 26732, Tucson, AZ 85726-6732

Elizabeth Bohlen ⟨ edpo@cfa.harvard.edu ⟩, Smithsonian Astrophysical Observatory, 60 Garden Street, Cambridge, MA 02138

Jay Bookbinder ⟨ bookbind@cfa.harvard.edu ⟩, Smithsonian Astrophysical Observatory, 60 Garden Street, Cambridge, MA 02138

David Borden ⟨ borden@cfa.harvard.edu ⟩, Smithsonian Astrophysical Observatory, 60 Garden Street, Cambridge, MA 02138

Tye Brady ⟨ tye@space.mit.edu ⟩, Massachusetts Institute of Technology/Center for Space Research, 77 Massachusetts Avenue, Room 37-501, Cambridge, MA 02139

Jeffrey Breen ⟨ breen@cfa.harvard.edu ⟩, Smithsonian Astrophysical Observatory, 60 Garden Street, Cambridge, MA 02138

Alan Bridger ⟨ ab@jach.hawaii.edu ⟩, Joint Astronomy Centre, 660 N. Aohoku Place, University Park, Hilo, HI 96720

Deborah Brissenden ⟨ debbie@cfa.harvard.edu ⟩, Smithsonian Astrophysical Observatory, 60 Garden Street, Cambridge, MA 02138

Roger Brissenden ⟨ rjb@cfa.harvard.edu ⟩, Smithsonian Astrophysical Observatory, 60 Garden Street, Cambridge, MA 02138

Benjamin C. Bromley ⟨ bromley@mac.dartmouth.edu ⟩, Dartmouth College HB6127, Department of Physics and Astronomy, Hanover, NH 03755

Patricia A. Buckley ⟨ buckley@cfa.harvard.edu ⟩, Smithsonian Astrophysical Observatory, 60 Garden Street, Cambridge, MA 02138

Ivo Busko ⟨ busko@das.inpe.br ⟩, INPE, E.P. 515, 12200-Sao Jose Dos Campos-SP, BRAZIL

Jon Chappell ⟨ jhc@cfa.harvard.edu ⟩, Smithsonian Astrophysical Observatory, 60 Garden Street, Cambridge, MA 02138

Sumitra Chary ⟨ chary@cfa.harvard.edu ⟩, Smithsonian Astrophysical Observatory, 60 Garden Street, Cambridge, MA 02138

Bing Chen ⟨ CDSXB2::CHEN ⟩, Observatoire de Strasbourg, 11 rue de l'Universite, 67000 Strasbourg, FRANCE

Guido Chincarini ⟨ 39217::CHINCARINI ⟩, Universita Milano, Via E. Bianchi 46, I-22055 Merate, ITALY

Carol Christian ⟨ carolc@cea.berkeley.edu ⟩, Center for EUV Astrophysics, 2150 Kittredge Street, Berkeley, CA 04720

Peter Claes, Institue d'Astrophysique, Avenue de Cointe 5, B-4200 Liege, BELGIUM

Al Conrad ⟨ aconrad@keck.hawaii.edu ⟩, Keck Observatory, 65-1120 Mamalahoa Highway, Kamuela, HI 96743

Maureen Conroy ⟨ mo@cfa.harvard.edu ⟩, Smithsonian Astrophysical Observatory, 60 Garden Street, Cambridge, MA 02138

Mike Corcoran ⟨ HEASRC::CORCORAN ⟩, NASA/Goddard Space Flight Center, USRA, Code 666, Greenbelt, MD 20771

James Cornell, Smithsonian Astrophysical Observatory, 60 Garden Street, Cambridge, MA 02138

Mark E. Cornell ⟨ cornell@puck.as.utexas.edu ⟩, McDonald Observatory, University of Texas, RLM 15.308, Austin, TX 78712

Dennis Crabtree ⟨ crabtree@dao.nrc.ca ⟩, Dominion Astrophysical Observatory/Canadian Astronomy Data Centre, 5071 W. Saanich Road, Victoria, BC V8X 4M6, CANADA

Bill Craig ⟨ bill@physics.berkeley.edu ⟩, University of California, Department of Physics, Berkeley, CA 94720

David F. Crawford ⟨ crawford@suphys.su.oz.au ⟩, University of Sydney, School of Physics, N.S.W. 2006, AUSTRALIA

Geoffrey B. Crew ⟨ gbc@space.mit.edu ⟩, Massachusetts Institute of Technology/Center for Space Research, 77 Massachusetts Avenue, Room 37-275, Cambridge, MA 02139

Geoff Croes ⟨ gcroes@nrao.edu ⟩, National Radio Astronomy Observatory, 520 Edgemont Road, Charlottesville, VA 22093

Daniele Dal Fiume ⟨ daniele@botes1.bo.cnr.it ⟩, Instituto Tesre - CNR, Via de' Castagnoli 1, I-40126 Bologna, ITALY

Philip N. Daly ⟨ pnd@jach.hawaii.edu ⟩, Joint Astronomy Centre, 660 N. Aohoku Place, Hilo, HI 96720

Lindsey Davis ⟨ davis@noao.edu ⟩, National Optical Astronomy Observatories, P.O. Box 26732, Tucson, AZ 85726-6732

Michael M. Davis ⟨ mdavis@naic.edu ⟩, Arecibo Observatory, P.O. Box 995, Arecibo, Puerto Rico 00613-0995

Michele DeLaPena ⟨ IUEGTC::DELAPENA ⟩, IUE Observatory, 10,000-A Aerospace Road, Lanham-Seabrook, MD 20706

Janet DePonte ⟨ janet@cfa.harvard.edu ⟩, Smithsonian Astrophysical Observatory, 60 Garden Street, Cambridge, MA 02138

Erik Deul ⟨ deul@strw.leidenuniv.nl ⟩, Sterrewacht Leiden, P.O. Box G513, 2300 RA Leiden, THE NETHERLANDS

Daniel Dewey ⟨ dd@space.mit.edu ⟩, Massachusetts Institute of Technology/Center for Space Research, 77 Massachusetts Avenue, Room 37-635, Cambridge, MA 02139

Adam Dobrzycki ⟨ adam@cfa.harvard.edu ⟩, Smithsonian Astrophysical Observatory, 60 Garden Street, Cambridge, MA 02138

Bryan Dorland ⟨ dorland@bdcv8.nrl.navy.mil ⟩, Naval Research Laboratory, Code 7604, 4555 Overlook Avenue SW, Washington, DC 20375-5320

Kimberly Dow ⟨ dow@cfa.harvard.edu ⟩, Smithsonian Astrophysical Observatory, 60 Garden Street, Cambridge, MA 02138

Jeremy J. Drake ⟨ jdrake@cea.berkeley.edu ⟩, Center for EUV Astrophysics, 2150 Kittredge Street, Berkeley, CA 94720

Daniel Durand ⟨ durand@dao.nrc.ca ⟩, Dominion Astrophysical Observatory/Canadian Astronomy Data Centre, 5071 W. Saanich Road, Victoria, BC V8X 4M6, CANADA

Jerry D'Amico, Peak Technologies Group, 8990 Old Annapolis Road, Columbia, MD 21045-2179

Jo Ann Eder ⟨ eder@naic.edu ⟩, Arecibo Observatory, P.O. Box 995, Arecibo, Puerto Rico 00613-0995

Guenther Eichhorn ⟨ gei@cfa.harvard.edu ⟩, Smithsonian Astrophysical Observatory, 60 Garden Street, Cambridge, MA 02138

Jonathan Eisenhamer ⟨ eisenham@stsci.edu ⟩, Space Telescope Science Institute, 3700 San Martin Drive, Baltimore, MD 21218

Martin Elvis ⟨ elvis@cfa.harvard.edu ⟩, Smithsonian Astrophysical Observatory, 60 Garden Street, Cambridge, MA 02138

Brian Espey ⟨ espey@vms.cis.pitt.edu ⟩, University of Pittsburgh, Department of Physics and Astronomy, Pittsburgh, PA 15260

John Ewing ⟨ ewing@cobecl.dnet.nasa.gov ⟩, COBE/ARC, 7601 Ora Glen Drive, Greenbelt, MD 20770

Giuseppina Fabbiano ⟨ pepi@cfa.harvard.edu ⟩, Smithsonian Astrophysical Observatory, 60 Garden Street, Cambridge, MA 02138

Emilio E. Falco ⟨ falco@cfa.harvard.edu ⟩, Smithsonian Astrophysical Observatory, 60 Garden Street, Cambridge, MA 02138

Allen Farris ⟨ farris@stsci.edu ⟩, Space Telescope Science Institute, 3700 San Martin Drive, Baltimore, MD 21218

Christopher Fassnacht ⟨ fassnach@cfa.harvard.edu ⟩, Smithsonian Astrophysical Observatory, 60 Garden Street, Cambridge, MA 02138

Fabio Favata ⟨ ffavata@estsa2.estec.esa.nl ⟩, European Space Agency/ESTEC, P.O. Box 299, 2200 AG Noodwijk, THE NETHERLANDS

Mike Fitzpatrick ⟨ fitz@noao.edu ⟩, National Optical Astronomy Observatories, P.O. Box 26732, Tucson, AZ 85726-6732

Sharlene Ford ⟨ sharlene@cfa.harvard.edu ⟩, Smithsonian Astrophysical Observatory, 60 Garden Street, Cambridge, MA 02138

Brand Fortner, Spyglass, Inc., P.O. Box 6388, Champaign, IL 61826

James R. Fowler ⟨ jrf@galileo.apo.nmsu.edu ⟩, Apache Point Observatory, P.O. Box 59, Sunspot, NM 88374

Thomas W. Fuller ⟨ fuller@voodoo-physics.ucsb.edu ⟩, University of California, Physics Department, Santa Barbara, CA 93106-9530

Marc Gagne ⟨ gagne@jove.physast.uga.edu ⟩, University of Georgia, Department of Physics and Astronomy, Athens, GA 30602

Michael Garcia ⟨ garcia@cfa.harvard.edu ⟩, Smithsonian Astrophysical Observatory, 60 Garden Street, Cambridge, MA 02138

Severin Gaudet ⟨ gaudet@dao.nrc.ca ⟩, Dominion Astrophysical Observatory/Canadian Astronomy Data Centre, 5071 W. Saanich Road, Victoria, BC V8X 4M6, CANADA

Margaret J. Geller ⟨ mjg@cfa.harvard.edu ⟩, Smithsonian Astrophysical Observatory, 60 Garden Street, Cambridge, MA 02138

Jim Gettys ⟨ jg@crl.dec.com ⟩, Digital Equipment Corporation, Cambridge Research Lab, One Kendall Square, Cambridge, MA 02139

John Gibbons ⟨ gibbons@cfa.harvard.edu ⟩, Smithsonian Astrophysical Observatory, 60 Garden Street, Cambridge, MA 02138

Diane M. Gilmore ⟨ dgilmore@stsci.edu ⟩, Space Telescope Science Institute, 3700 San Martin Drive, Baltimore, MD 21218

Frank Giovane ⟨ 6646::FGIOVANE ⟩, NASA Headquarters, 600 Independence Avenue SW, Washington, DC 20546

Brian Glendenning ⟨ bglenden@nrao.edu ⟩, National Radio Astronomy Observatory, 520 Edgemont Road, Charlottesville, VA 22903

John C. Good ⟨ jcg@ipac.caltech.edu ⟩, IPAC, CalTech, Mail Stop 100-22, Pasadena, CA 91125

Paul Gorenstein ⟨ goren@cfa.harvard.edu ⟩, Smithsonian Astrophysical Observatory, 60 Garden Street, Cambridge, MA 02138

Paul Grant ⟨ paul@cfa.harvard.edu ⟩, Smithsonian Astrophysical Observatory, 60 Garden Street, Cambridge, MA 02138

Carolyn Stern Grant ⟨ stern@cfa.harvard.edu ⟩, Smithsonian Astrophysical Observatory, 60 Garden Street, Cambridge, MA 02138

William B. Green ⟨ bill_green@jplmail.jpl.nasa.edu ⟩, Jet Propulsion Laboratory, 4800 Oak Grove Drive, MS 168-527, Pasadena, CA 91109

Preben Grosbøl ⟨ pgrosbol@eso.org ⟩, European Southern Observatory, Karl-Schwarzschild-Strasse 2, D-8046 Garching bei München, GERMANY

Rainer Grüber ⟨ 28773::GRU ⟩, Max-Planck-Institut für Extraterrestrische Physik, Karl-Schwarzschild-Strasse 1, D-8046 Garching bei München, GERMANY

Stephen Guimond ⟨ guimond@cfa.harvard.edu ⟩, Smithsonian Astrophysical Observatory, 60 Garden Street, Cambridge, MA 02138

Shadia Habbal ⟨ habbal@cfa.harvard.edu ⟩, Smithsonian Astrophysical Observatory, 60 Garden Street, Cambridge, MA 02138

Mike Hall, Peak Technologies Group, Inc., 8990 Old Annapolis Road, Columbia, MD 21045-2179

Robert J. Hanisch ⟨ hanisch@stsci.edu ⟩, Space Telescope Science Institute, 3700 San Martin Drive, Baltimore, MD 21218

F. Rick Harnden, Jr. ⟨ frh@cfa.harvard.edu ⟩, Smithsonian Astrophysical Observatory, 60 Garden Street, Cambridge, MA 02138

Daniel E. Harris ⟨ harris@cfa.harvard.edu ⟩, Smithsonian Astrophysical Observatory, 60 Garden Street, Cambridge, MA 02138

R. Lee Hawkins ⟨ lhawkins@annie.wellesley.edu ⟩, Wellesley College, Department of Astronomy, Whitin Observatory, Wellesley, MA 02181

Susan Hazelton ⟨ swg@cfa.harvard.edu ⟩, Smithsonian Astrophysical Observatory, 60 Garden Street, Cambridge, MA 02138

Andre Heck ⟨ heck@ccsmvs.u-strasbg.fr ⟩, Observatoire Astronomique–Strasbourg, 5 rue Des Mesanges, F-67120 Duttlenheim, FRANCE

John Heise ⟨ johnj@sron.ruu.nl ⟩, SRO Netherlands, Sorbonnelaan 2, 4584 CA Utrecht, THE NETHERLANDS

Arne A. Henden ⟨ henden@mps.ohio-state.edu ⟩, Ohio State University, Astronomy Department, 174 W. 18th Avenue, Columbus, OH 43210

Rob Hewitt ⟨ hewitt@cfa.harvard.edu ⟩, Smithsonian Astrophysical Observatory, 60 Garden Street, Cambridge, MA 02138

James T. Himer ⟨ jthimer@iras.ucalgary.ca ⟩, University of Calgary/RAO, 339 Woodside Bay SW, Calgary, Alberta T2W 3K9, CANADA

Robert M. Hjellming ⟨ rhjellmi@nrao.edu ⟩, National Radio Astronomy Observatory, P.O. Box 0, Socorro, NM 87801-0379

Philip E. Hodge ⟨ hodge@stsci.edu ⟩, Space Telescope Science Institute, 3700 San Martin Drive, Baltimore, MD 21218

Richard Hook ⟨ rhook@eso.org ⟩, Space Telescope–European Coordinating Facility/European Southern Observatory, Karl-Schwarzschild-Strasse 2, D-8046 Garching bei München, GERMANY

Ulrich Hopp ⟨ hopp@mpia-hd.mpg.de ⟩, Max-Planck-Institut für Astronomie, Königstuhl 17, D-6900 Heidelberg, GERMANY

Allan Hornstrup ⟨ allan@dsri.dk ⟩, Danish Space Research Institute, G1. Lundtoftevej 7, 2800 Lyngby, DENMARK

Paul Hsieh ⟨ hsieh@cfa.harvard.edu ⟩, Smithsonian Astrophysical Observatory, 60 Garden Street, Cambridge, MA 02138

Jin-chung Hsu ⟨ hsu@stsci.edu ⟩, Space Telescope Science Institute, 3700 San Martin Drive, Baltimore, MD 21218

Maohai Huang ⟨ mhuang@bu-ast.bu.edu ⟩, Boston University, Department of Astronomy, 725 Commonwealth Avenue, Boston, MA 02215

Stephen Huber, Beaver College, Glenside, PA 19038

John Huchra ⟨ huchra@cfa.harvard.edu ⟩, Smithsonian Astrophysical Observatory, 60 Garden Street, Cambridge, MA 02138

David Huenemoerder ⟨ dph@space.mit.edu ⟩, Massachusetts Institute of Technology/Center for Space Research, 77 Massachusetts Avenue, Room 37-667, Cambridge, MA 02139

Steve Hulbert ⟨ hulbert@stsci.edu ⟩, Space Telescope Science Institute, 3700 San Martin Drive, Baltimore, MD 21218

Garth Hunt ⟨ ghunt@nrao.edu ⟩, National Radio Astronomy Observatory, 520 Edgemont Road, Charlottesville, VA 22903

Juhani Huovelin ⟨ juhani.huovelin@helsinki.fi ⟩, Observatory, University of Helsinki, Tahtitorninmai, SF-00130 Helsinki, FINLAND

Takashi Isobe ⟨ ti@space.mit.edu ⟩, Massachusetts Institute of Technology/Center for Space Research, 77 Massachusetts Avenue, Room 37-501, Cambridge, MA 02139

Carlo Izzo ⟨ 28773::IZZO ⟩, Max-Planck-Institut für Extraterrestrische Physik, Giessenbachstrasse, D-8046 Garching bei München, GERMANY

Paul Jacobs ⟨ jacobs@rosserv.gsfc.nasa.gov ⟩, NASA/Goddard Space Flight Center, Hughes-STX, Code 668, Greenbelt, MD 20771

George Jacoby ⟨ jacoby@noao.edu ⟩, Kitt Peak National Observatory, P.O. Box 26732, Tucson, AZ 85726-6732

Donald Jennings ⟨ jennings@antwrp.gsfc.nasa.gov ⟩, NASA/Goddard Space Flight Center, Code 668.1, Greenbelt, MD 20771

Kevin Jim ⟨ jim@uhifa.ifa.hawaii.edu ⟩, University of Hawaii, Institute for Astronomy, 2680 Woodlawn Drive, Honolulu, HI 96822

Paul Joachim ⟨ 28773::PAUL ⟩, Max-Planck-Institut für Extraterrestrische Physik, Giessenbachstrasse, D-8046 Garching bei München, GERMANY

Linda John ⟨ lindam@cea.berkeley.edu ⟩, Center for EUV Astrophysics, 2150 Kittredge Street, Berkeley, CA 94704

Jeff Johnson ⟨ jjohnson@stsci.edu ⟩, Space Telescope Science Institute, 3700 San Martin Drive, Baltimore, MD 21218

Mark D. Johnston ⟨ johnston@stsci.edu ⟩, Space Telescope Science Institute, 3700 San Martin Drive, Baltimore, MD 21218

James M. Jordan ⟨ jmj@grossc.gsfc.nasa.gov ⟩, NASA/Goddard Space Flight Center, Code 668.1, Greenbelt, MD 20771

Todd Karakashian ⟨ todd@cfa.harvard.edu ⟩, Smithsonian Astrophysical Observatory, 60 Garden Street, Cambridge, MA 02138

Ed Kellogg ⟨ emk@cfa.harvard.edu ⟩, Smithsonian Astrophysical Observatory, 60 Garden Street, Cambridge, MA 02138

B. D. Kelly ⟨ bdk@star.roe.ac.uk ⟩, Royal Observatory Edinburgh, Blackford Hill, Edinburgh EH9 1AN, UNITED KINGDOM

Bob Kibrick ⟨ kibrick@helios.ucsc.edu ⟩, UCO/Lick Observatory, Natural Sciences 2, University of California, Santa Cruz, CA 95064

Daniel A. Klinglesmith III ⟨ klinglesmith@stars.gsfc.nasa.gov ⟩, NASA/Goddard Space Flight Center, Code 684, Greenbelt, MD 20771

Fred Knight ⟨ knight@ll.mit.edu ⟩, Massachusetts Institute of Technology/Lincoln Laboratory, 244 Wood Street, C-483, Lexington, MA 02173

Michael Kopko, Jr. ⟨ kopko@psc.edu ⟩, Pittsburgh Supercomputing Center, 4400 Fifth Avenue, Pittsburgh, PA 15213

John Krist ⟨ krist@stsci.edu ⟩, Space Telescope Science Institute, 3700 San Martin Drive, Baltimore, MD 21218

Michael Kurtz ⟨ kurtz@cfa.harvard.edu ⟩, Smithsonian Astrophysical Observatory, 60 Garden Street, Cambridge, MA 02138

Wayne Landsman ⟨ landsman@stars.gsfc.nasa.gov ⟩, NASA/Goddard Space Flight Center, Hughes/STX, Code 681, Greenbelt, MD 20771

Nancy Laubenthal ⟨ laubenthal@lheavx.gsfc.nasa.gov ⟩, NASA/Goddard Space Flight Center, Code 664, Greenbelt, MD 20771

Susan Lauber ⟨ sml@cfa.harvard.edu ⟩, Smithsonian Astrophysical Observatory, 60 Garden Street, Cambridge, MA 02138

Soizick Lesteven ⟨ lesteven@simbad.u-strasbg.fr ⟩, Observatoire Astronomique (CDS), 11 rue de l'Universite, 67000 Strasbourg, FRANCE

Zoltan Levay ⟨ levay@stsci.edu ⟩, Space Telescope Science Institute, 3700 San Martin Drive, Baltimore, MD 21218

Ken Levenson ⟨ levenson@unhsmm.unh.edu ⟩, University of New Hampshire, Space Science Center, Durham, NH 03824

James Lewis ⟨ jlewis@cea.berkeley.edu ⟩, Center for EUV Astrophysics, 2150 Kittredge Street, Berkeley, CA 94720

J.R. Lewis ⟨ jrl@uk.ac.cam.ast-star ⟩, Royal Greenwich Observatory, Madingley Road, Cambridge CB3 0EZ, ENGLAND

Don J. Lindler, ACC, Inc., 11518 Gainsborough Road, Potomac, MD 20854

Bob Link ⟨link@cfht.hawaii.edu⟩, Canada-France-Hawaii Telescope, P.O. Box 1597, Kamuela, HI 96743

Harvey S. Liszt ⟨hliszt@nrao.edu⟩, National Radio Astronomy Observatory, 520 Edgemont Road, Charlottesville, VA 22903

Knox S. Long ⟨long@stsci.edu⟩, Space Telescope Science Institute, 3700 San Martin Drive, Baltimore, MD 21218

Caroline Lupfer, Smithsonian Astrophysical Observatory, 60 Garden Street, Cambridge, MA 02138

William Lupton ⟨wlupton@keck.hawaii.edu⟩, Keck Observatory, 65-1120 Mamalahoa Highway, Kamuela, HI 96743

Dyer M. Lytle ⟨lytle@noao.edu⟩, National Optical Astronomy Observatories, P.O. Box 26732, Tucson, AZ 85726-6732

Maria Cettina Maccarone ⟨40651::MACCARONE⟩, IFCAI/CNR, Piazza Verdi 6, I-90138 Palermo, ITALY

Barry F. Madore ⟨barry@ipac.caltech.edu⟩, IPAC, CalTech, Mail Stop 100-22, Pasadena, CA 91125

Eric Mandel ⟨eric@cfa.harvard.edu⟩, Smithsonian Astrophysical Observatory, 60 Garden Street, Cambridge, MA 02138

Kathleen R. Manning ⟨manning@cfa.harvard.edu⟩, Smithsonian Astrophysical Observatory, 60 Garden Street, Cambridge, MA 02138

Bob Marshall ⟨bob@noao.edu⟩, National Optical Astronomy Observatories/CCS, P.O. Box 26732, Tucson, AZ 85726-6732

Herman L. Marshall ⟨hermanm@cea.berkeley.edu⟩, Center for EUV Astrophysics, 2150 Kittredge Street, Berkeley, CA 94720

Paul L. Martenis ⟨leeds@cfa.harvard.edu⟩, Smithsonian Astrophysical Observatory, 60 Garden Street, Cambridge, MA 02138

Luis A. Martinez-Vazquez ⟨lamb@astroscu.unam.mx⟩, Inst. de Astronomia, UNAM, Apartado Postal 70-264, Cd. Univ., C.P. 04510, MEXICO, D.F.

Charles Maxson ⟨maxson@cfa.harvard.edu⟩, Smithsonian Astrophysical Observatory, 60 Garden Street, Cambridge, MA 02138

Steve Mazuk, The Aerospace Corporation, P.O. Box 92957, Mail Stop M2-255, Los Angeles, CA 90009

Laura M. McDonald ⟨laura@gamma.gsfc.nasa.gov⟩, NASA/Goddard Space Flight Center, Hughes-STX, Code 664, Greenbelt, MD 20771

Jonathan McDowell ⟨mcdowell@cfa.harvard.edu⟩, Smithsonian Astrophysical Observatory, 60 Garden Street, Cambridge, MA 02138

Thomas McGlynn ⟨mcglynn@grossc.gsfc.nasa.gov⟩, NASA/Goddard Space Flight Center, Code 668.1, Greenbelt, MD 20771

D. H. McNamara, Brigham Young University, Department of Physics and Astronomy, 408 ESC, Provo, UT 84602

Justin F. McNeill, Jr. ⟨jfme59@ipl.jpl.nasa.gov⟩, Jet Propulsion Laboratory, 4800 Oak Grove Drive, MS 168-414, Pasadena, CA 91109

David M. Meleedy ⟨ meleedy@cfa.harvard.edu ⟩, Smithsonian Astrophysical Observatory, 60 Garden Street, Cambridge, MA 02138

Karl R. Menten ⟨ menten@cfa.harvard.edu ⟩, Smithsonian Astrophysical Observatory, 60 Garden Street, Cambridge, MA 02138

Mark Metzger ⟨ metzger@alioth.mit.edu ⟩, Massachusetts Institute of Technology, 77 Massachusetts Avenue, Room 6-204, Cambridge, MA 02139

Thomas E. Milliman ⟨ tmilliman@unh.edu ⟩, University of New Hampshire, Space Science Center, Morse Hall, Durham, NH 03824

Douglas Mink ⟨ mink@cfa.harvard.edu ⟩, Smithsonian Astrophysical Observatory, 60 Garden Street, Cambridge, MA 02138

Bijoy Misra ⟨ CFA::MISRA ⟩, Smithsonian Astrophysical Observatory, 60 Garden Street, Cambridge, MA 02138

Phyllis Mitzman, Harvard University/OIT, 50 Church Street, Cambridge, MA 02138

Karen Modestino ⟨ karen@cfa.harvard.edu ⟩, Smithsonian Astrophysical Observatory, 60 Garden Street, Cambridge, MA 02138

Lorayne Mohammed ⟨ lorayne@cfa.harvard.edu ⟩, Smithsonian Astrophysical Observatory, 60 Garden Street, Cambridge, MA 02138

John Moran ⟨ jmoran@cfa.harvard.edu ⟩, Smithsonian Astrophysical Observatory, 60 Garden Street, Cambridge, MA 02138

Edward Morgan ⟨ ehm@space.mit.edu ⟩, Massachusetts Institute of Technology/Center for Space Research, 77 Massachusetts Avenue, Room 37-426, Cambridge, MA 02139

Stephen S. Murray ⟨ ssm@cfa.harvard.edu ⟩, Smithsonian Astrophysical Observatory, 60 Garden Street, Cambridge, MA 02138

Eiji Nishihara ⟨ eiji@c1.mtk.nao.ac.jp ⟩, NAO Japan, 2-21-1 Ohsawa, Mitaka, Tokyo 181, JAPAN

J.E. Noordam ⟨ noordam@nfra.nl ⟩, NFRA Dwingeloo, P.O. Box 2, 7990 AA Dwingeloo, THE NETHERLANDS

Stephen Odewahn ⟨ sco@apsl.spa.umn.edu ⟩, University of Minnesota, Department of Astronomy, 116 Church Street, Minneapolis, MN 55455

Eric Olson ⟨ ericco@cea.berkeley.edu ⟩, Center for EUV Astrophysics, 2150 Kittredge Street, Berkeley, CA 94720

Jeff Orszak ⟨ orszak@cfa.harvard.edu ⟩, Smithsonian Astrophysical Observatory, 60 Garden Street, Cambridge, MA 02138

Julian Osborne ⟨ 19838::JULO ⟩, Leicester University, Department of Physics and Astronomy, University Road, Leicester LE1 7RH, UNITED KINGDOM

Bruce O'Neel ⟨ oneel@heasrc.gsfc.nasa.gov ⟩, NASA/Goddard Space Flight Center, Hughes-STX, Code 668, Greenbelt, MD 20771

Earl J. O'Neil, Jr. ⟨ eoneil@noao.edu ⟩, Kitt Peak National Observatory, P.O. Box 26732, Tucson, AZ 85726-6732

Clive G. Page ⟨ 19838::CGP ⟩, University of Leicester, University Road, Leicester, LE1 7RH, UNITED KINGDOM

Laszlo Pasztor ⟨h2295pas@ella.hu⟩, MTA TAKI, H-1022 Budapest, Herman Otto Ut 15, HUNGARY

Lisa Paton ⟨lcp@cfa.harvard.edu⟩, Smithsonian Astrophysical Observatory, 60 Garden Street, Cambridge, MA 02138

Anita Pearson, Smithsonian Astrophysical Observatory, 60 Garden Street, Cambridge, MA 02138

Jose R. Munoz Peiro ⟨jmunoz@isosa4.estec.esa.nl⟩, ESTEC (SAI), Keplerlaan 1, 2200 AG Noordwijk, THE NETHERLANDS

William D. Pence ⟨pence@tetra.gsfc.nasa.gov⟩, NASA/Goddard Space Flight Center, Code 668, Greenbelt, MD 20771

Jeffrey W. Percival ⟨jwp@sal.wisc.edu⟩, University of Wisconsin, Space Astronomy Laboratory, 1150 University Avenue, Madison, WI 53706

Michele Peron ⟨mperon@eso.org⟩, European Southern Observatory, Karl-Schwarzschild-Strasse 2, D-8046 Garching bei München, GERMANY

Jeff Pier ⟨jrp@nofs.navy.mil⟩, U.S. Naval Observatory, P.O. Box 1149, Flagstaff, AZ 86002-1149

David Plummer ⟨plummer@cfa.harvard.edu⟩, Smithsonian Astrophysical Observatory, 60 Garden Street, Cambridge, MA 02138

Joseph Pollizzi ⟨pollizzi@stsci.edu⟩, Space Telescope Science Institute, 3700 San Martin Drive, Baltimore, MD 21218

Elisha Polomski ⟨elwood@cea.berkeley.edu⟩, Center for EUV Astrophysics, 2150 Kittredge Street, Berkeley, CA 94720

Andrea Prestwich ⟨prestwich@cfa.harvard.edu⟩, Smithsonian Astrophysical Observatory, 60 Garden Street, Cambridge, MA 02138

Andreas Priebe ⟨apr@babel.aip.de⟩, Astrophysical Institute Potsdam, An der Sternwarte 16, 0-1599 Potsdam, GERMANY

Francis Primini ⟨fap@cfa.harvard.edu⟩, Smithsonian Astrophysical Observatory, 60 Garden Street, Cambridge, MA 02138

Mauro Pucillo ⟨38439::PUCILLO⟩, Trieste Astronomical Observatory, P.O. Box Succ. 5, Via G.B. Tiepolo 11, I-34131 Trieste, ITALY

Bo Frese Rasmussen ⟨bfrasmus@nso.hq.eso.org⟩, European Southern Observatory, Karl-Schwarzschild-Strasse 2, D-8046 Garching bei München, GERMANY

Walter Rauh ⟨rauh@mpia-hd.mpg.de⟩, Max-Planck-Institut für Astronomie, Königstuhl 17, D-6900 Heidelberg, GERMANY

Alastair Reynolds ⟨areynold@estsaa.dnet.nasa.gov⟩, European Space Agency/ESTEC, Postbus 299, 2200 AG Noordwijk, THE NETHERLANDS

Sherri Reynolds ⟨reynolds@cfa.harvard.edu⟩, Smithsonian Astrophysical Observatory, 60 Garden Street, Cambridge, MA 02138

Wendy Roberts ⟨roberts@cfa.harvard.edu⟩, Smithsonian Astrophysical Observatory, 60 Garden Street, Cambridge, MA 02138

Peter Roelfsema ⟨pjotr@guspace.sron.rug.nl⟩, Space Research Organization, Postbus 800, 9700 AV Groningen, THE NETHERLANDS

John B. Roll, Jr. ⟨ john@cfa.harvard.edu ⟩, Smithsonian Astrophysical Observatory, 60 Garden Street, Cambridge, MA 02138

Robert S. Ronan ⟨ ronan@shiver.stanford.edu ⟩, Stanford University, CRL 328, Stanford, CA 94305-4055

Arnold Rots ⟨ arots@xebec.gsfc.nasa.gov ⟩, NASA/Goddard Space Flight Center, Code 666, Greenbelt, MD 20771

Diane Roussel-Dupre, Los Alamos National Laboratory, Mail Stop D436, SST-9, Los Alamos, NM 87545

Richard Saxton ⟨ 19838::RDS ⟩, Leicester University, Department of Physics and Astron., Leicester, LE1 7RH, UNITED KINGDOM

Jonathan F. Schachter ⟨ shaks@cfa.harvard.edu ⟩, Smithsonian Astrophysical Observatory, 60 Garden Street, Cambridge, MA 02138

Skip Schaller ⟨ skip@as.arizona.edu ⟩, Steward Observatory, University of Arizona, Tucson, AZ 85721

Eric M. Schlegel ⟨ eric@heasfs.gsfc.nasa.gov ⟩, NASA/Goddard Space Flight Center, USRA, Code 668, Greenbelt, MD 20771

Erwin Schmerling ⟨ eschmerling@nasamail ⟩, NASA Headquarters, Code SZ, Washington, DC 20546

Dennis Schmidt ⟨ dschmidt@cfa.harvard.edu ⟩, Smithsonian Astrophysical Observatory, 60 Garden Street, Cambridge, MA 02138

Daniel A. Schwartz ⟨ das@cfa.harvard.edu ⟩, Smithsonian Astrophysical Observatory, 60 Garden Street, Cambridge, MA 02138

Joseph Schwarz ⟨ joe@cfa.harvard.edu ⟩, Smithsonian Astrophysical Observatory, 60 Garden Street, Cambridge, MA 02138

Salvo Sciortino ⟨ sciorti@oapa.astropa.unipa.it ⟩, Osserv. Astronomico di Palermo, Piazza del Parlamento 1, I-90134 Palermo, ITALY

Robert L. Seaman ⟨ seaman@noao.edu ⟩, National Optical Astronomy Observatories, P.O. Box 26732, Tucson, AZ 85726-6732

Fred Seward ⟨ fds@cfa.harvard.edu ⟩, Smithsonian Astrophysical Observatory, 60 Garden Street, Cambridge, MA 02138

Richard A. Shaw ⟨ shaw@stsci.edu ⟩, Space Telescope Science Institute, 3700 San Martin Drive, Baltimore, MD 21218

Patrick Shopbell ⟨ pls@rice.edu ⟩, Rice University, Department of Space Physics, P.O. Box 1892, Houston, TX 77251

Keith Shortridge ⟨ ks@aaoepp.oz.au ⟩, Anglo-Australian Observatory, P.O. Box 296, Epping, N.S.W. 2121, AUSTRALIA

Andy Silber ⟨ andy@cfa.harvard.edu ⟩, Smithsonian Astrophysical Observatory, 60 Garden Street, Cambridge, MA 02138

David Silberberg ⟨ davids@stsci.edu ⟩, Space Telescope Science Institute, 3700 San Martin Drive, Baltimore, MD 21218

Oleg Smirnov ⟨ oms@airas.msk.su ⟩, Institute of Astronomy, Russian Academy of Sciences, 48 Pyatnitskaya St., Moscow 109017, RUSSIA

Barham W. Smith ⟨ barham@rainier.lanl.gov ⟩, Los Alamos National Laboratory, Mail Stop D436, SST-9, Los Alamos, NM 87545

Eric P. Smith ⟨ esmith@hubble.gsfc.nasa.gov ⟩, NASA/Goddard Space Flight Center, Code 681, Greenbelt, MD 20771

David M. Stern ⟨ davids@rsinc.com ⟩, Research Systems, Inc., 777 29th Street, Boulder, CO 80303

Julian R. Sternberg ⟨ 29738::JSTERNBE ⟩, ESTEC (SAI), Postbus 299, 2200 AG Noordwijk, THE NETHERLANDS

J. Malcolm Stewart ⟨ jms@star.roe.ac.uk ⟩, Royal Observatory Edinburgh, Blackford Hill, Edinburgh EH9 3HJ, UNITED KINGDOM

Elizabeth B. Stobie ⟨ stobie@stsci.edu ⟩, Space Telescope Science Institute, 3700 San Martin Drive, Baltimore, MD 21218

Mark Strickman ⟨ strickman@osse.nrl.navy.mil ⟩, Naval Research Laboratory, Code 7651.2, 4555 Overlook Avenue SW, Washington, DC 20375-5320

Karen M. Strom ⟨ kstrom@phast.umass.edu ⟩, University of Massachusetts, Astronomy Program, GRC 517B, Amherst, MA 01003

Sandra Terranova ⟨ teranova@cfa.harvard.edu ⟩, Smithsonian Astrophysical Observatory, 60 Garden Street, Cambridge, MA 02138

Robert Thomas ⟨ thomas@naic.edu ⟩, Arecibo Observatory, P.O. Box 995, Arecibo, PR 00613-0995

Doug Tody ⟨ tody@noao.edu ⟩, National Optical Astronomy Observatories, P.O. Box 26732, Tucson, AZ 85726-6732

Susan Tokarz ⟨ tokarz@cfa.harvard.edu ⟩, Smithsonian Astrophysical Observatory, 60 Garden Street, Cambridge, MA 02138

Jay Travisano ⟨ jay@stsci.edu ⟩, Space Telescope Science Institute, 3700 San Martin Drive, Baltimore, MD 21218

Francesca Tribioli ⟨ tribioli@arcetri.astro.it ⟩, Osserv. Astrofisico di Arcetri, Largo E. Fermi 5, I-50127 Firenze, ITALY

Massimo Trifoglio ⟨ 38045::MASSIMO ⟩, Instituto Tesre - CNR, Via de' Castagnoli 1, I-40126 Bologna, ITALY

Ginevra Trinchieri ⟨ ginevra@cfa.harvard.edu ⟩, Smithsonian Astrophysical Observatory, 60 Garden Street, Cambridge, MA 02138

Mario Tripiciano ⟨ 39405::TRIPICIANO ⟩, IFCAI/CNR, Piazza Verdi 6, I-90138 Palermo, ITALY

Benoit Turgeon ⟨ turgeon@nereid.sal.ists.ca ⟩, ISTS/Space Astrophysics Lab, York University, 202-2700 Steeles Avenue West, Concord, ON L4K 3C8, CANADA

Susan J. Tuttle ⟨ logan@cfa.harvard.edu ⟩, Smithsonian Astrophysical Observatory, 60 Garden Street, Cambridge, MA 02138

Michael F. Tyrrell ⟨ d6724@applelink.apple.com ⟩, Spyglass, Inc., One Kendall Square, Ste 2200, Cambridge, MA 02139

Frank Valdes ⟨ fvaldes@noao.edu ⟩, National Optical Astronomy Observatories/CCS, P.O. Box 26732, Tucson, AZ 85726-6732

Roland Vanderspek ⟨ roland@blitz.mit.edu ⟩, Massachusetts Institute of Technology/Center for Space Research, 77 Massachusetts Avenue, Room 37-527, Cambridge, MA 02139

Richard F. Vannelli ⟨ rfpvan@cfa.harvard.edu ⟩, Smithsonian Astrophysical Observatory, 60 Garden Street, Cambridge, MA 02138

Michael E. Van Steenberg ⟨ mev@ndadsa.gsfc.nasa.gov ⟩, NASA/Goddard Space Flight Center, Code 631, Greenbelt, MD 20771

David Van Stone ⟨ vanstone@cfa.harvard.edu ⟩, Smithsonian Astrophysical Observatory, 60 Garden Street, Cambridge, MA 02138

Frank Varosi ⟨ varosi@stars.gsfc.nasa.gov ⟩, NASA/Goddard Space Flight Center, Hughes-STX, Code 685, Greenbelt, MD 20771

Stephen A. Voels ⟨ voels@usm.uni-muenchen.de ⟩, Uni.-Sternwarte München, Scheinerstr. 1, D-8000 München, GERMANY

Wolfgang Voges ⟨ 28773::WHV ⟩, Max-Planck-Institut für Extraterrestrische Physik, Karl-Schwarzschild-Strasse 1, D-8046 Garching bei München, GERMANY

Patrick T. Wallace ⟨ ptw@star.rl.ac.uk ⟩, Rutherford Appleton Laboratory, Chilton, Didcot OX11 0QX, ENGLAND

Dennis Wang ⟨ wang@cedar.nrl.navy.mil ⟩, Naval Research Laboratory, Code 7660, 4555 Overlook Avenue SW, Washington, DC 20375-5000

Qingde Wang, JILA/CASA, University of Colorado, Campus Box 44, Boulder, CO 80309

Archibald Warnock III ⟨ warnock@hypatia.gsfc.nasa.gov ⟩, NASA/Goddard Space Flight Center, Hughes-STX, Code 631, Greenbelt, MD 20771

Rodney F. Warren-Smith ⟨ rfws@star.rl.ac.uk ⟩, Rutherford Appleton Laboratory, Chilton, Didcot OX11 0QX, ENGLAND

Joyce M. Watson ⟨ watson@cfa.harvard.edu ⟩, Smithsonian Astrophysical Observatory, 60 Garden Street, Cambridge, MA 02138

Len R. Watson ⟨ watson@selone.nrl.navy.mil ⟩, Naval Research Laboratory, Code 4160, 4555 Overlook Avenue SW, Washington, DC 20375-5000

Michael G. Watson ⟨ 19838::MGW ⟩, University of Leicester, Department of Physics and Astronomy, Leicester LE1 7RH, UNITED KINGDOM

Nicholas Weir ⟨ weir@phobos.caltech.edu ⟩, Caltech, 105-24, Pasadena, CA 91125

Don Wells ⟨ dwells@nrao.edu ⟩, National Radio Astronomy Observatory, 520 Edgemont Road, Charlottesville, VA 22903-2475

Rick Wenk ⟨ raw@physics.att.com ⟩, AT&T Bell Laboratories, 600 Mountain Avenue, Room 1D-316, Murray Hill, NJ 07974

Nick White ⟨ white@lheavx.gsfc.nasa.gov ⟩, NASA/Goddard Space Flight Center, Code 668, Greenbelt, MD 20771

Belinda Wilkes ⟨ belinda@cfa.harvard.edu ⟩, Smithsonian Astrophysical Observatory, 60 Garden Street, Cambridge, MA 02138

John Williams ⟨jwilliam@stsci.edu⟩, Space Telescope Science Institute, 3700 San Martin Drive, Baltimore, MD 21218

Winifred E. Williams ⟨wwilliams@noao.edu⟩, National Optical Astronomy Observatories/National Solar Observatory, P.O. Box 26732, Tucson, AZ 85726-6432

Ramon L. Williamson II ⟨williamson@stsci.edu⟩, Space Telescope Science Institute, 3700 San Martin Drive, Baltimore, MD 21218

George W. Wolf ⟨gww836f@vma.smsu.edu⟩, SW Missouri State, Physics and Astronomy Department, Springfield, MO 65804

Diana M. Worrall ⟨dmw@cfa.harvard.edu⟩, Smithsonian Astrophysical Observatory, 60 Garden Street, Cambridge, MA 02138

Nailong Wu ⟨nailong@stsci.edu⟩, Space Telescope Science Institute, 3700 San Martin Drive, Baltimore, MD 21218

William Wyatt ⟨wyatt@cfa.harvard.edu⟩, Smithsonian Astrophysical Observatory, 60 Garden Street, Cambridge, MA 02138

Hubert Yamada ⟨yamada@ifa.hawaii.edu⟩, University of Hawaii, Institute for Astronomy, 2680 Woodlawn Drive, Honolulu, HI 96822

Daryl Yentis ⟨yentis@xip.nrl.navy.mil⟩, Naval Research Laboratory, Code 7622, 4555 Overlook Avenue SW, Washington, DC 20375-5320

Nelson Zarate ⟨zarate@stsci.edu⟩, Space Telescope Science Institute, 3700 San Martin Drive, Baltimore, MD 21218

Alex Zepka ⟨zepka@astrosun.tn.cornell.edu⟩, Cornell University, 614 Space Sciences Building, Ithaca, NY 14853

Ping Zhao ⟨zhao@cfa.harvard.edu⟩, Smithsonian Astrophysical Observatory, 60 Garden Street, Cambridge, MA 02138

Hans-Ulrich Zimmermann ⟨28773::ZIM⟩, Max-Planck-Institut für Extraterrestrische Physik, Giessenbachstrasse, D-8046 Garching bei München, GERMANY

Martin V. Zombeck ⟨mvz@cfa.harvard.edu⟩, Smithsonian Astrophysical Observatory, 60 Garden Street, Cambridge, MA 02138

Saeid Zoonematkermani ⟨44902::SZOONEM⟩, SUNY at Stony Brook, Earth & Space Sciences, Stony Brook, NY 11794-2100

Second Annual Conference on Astronomical Data Analysis Software and Systems
November 2–4, 1992 Boston, MA

Part 1. Databases, Catalogs, and Archives

Section A. Ground-Based Data and Scanning Projects

Astronomical Data Analysis Software and Systems II
ASP Conference Series, Vol. 52, 1993
R. J. Hanisch, R. J. V. Brissenden, and J. Barnes, eds.

Archiving Data from Ground-Based Observatories

Miguel A Albrecht

European Southern Observatory, Karl–Schwarzschild–Str. 2, D–8046 Garching bei München, Germany

Abstract. Archiving data acquired with ground based telescopes is constrained by the fact that the life-time of instruments and detectors is considerably shorter than the expected life-time of the archive. This feature differentiates ground-based originated archives radically from their space-borne counterparts. The organisation of the observations catalogue becomes highly dependent on the capability of the archive to deal with new instrumental configurations. We introduce in this paper, the concept of an *archive database* as opposed to the static catalogue design currently in use in many archiving facilities, as a method to deal with this problem.

We present a brief review of activities currently in progress in this area and draw a perspective towards the archives and data systems of the new generation of observatories in the era of very large telescopes.

1. Introduction

Most ground-based observatories have not for many years taken any provision for data archiving other than a standard safety copy of observations. It has been felt that it may be easier to simply repeat an observation than having to deal with uncertainties associated to poorly catalogued archive data. In this paper we show that this situation is changing rapidly as two major factors start to play an important rôle in modern telescope operations: a) the growing demand for state-of-the-art observing facilities and the high cost involved in their development, and b) the growing awareness among astronomers of large amounts of unanalysed data. As happened in space missions, the cost-to-benefit ratio of ground-based observatory operations is becoming an important issue. The result being a number of data system activities—currently under development—with the goal to provide the astronomical community with long term archives and observation catalogues.

However, the task of building archives of observational data for facilities on the ground turns out to be of a high level of complexity because the inherent time scale of change of such facilities is by definition much shorter than the expected lifetime of the archive. The consequence being that the archive database must accommodate change of instrumental characteristics, such as pixel area of detectors, etc., as an integral part of the observations catalogue and therefore, the traditional observation log is replaced by an *archive database*. The importance of this database becomes evident when one considers that the number of instrumental elements subject to change on a particular telescope varies from complete instruments to detectors, filters, gratings, etc.

We will discuss in the next sections, first the current situation in greater detail with the objective of identifying the major problems to be solved when setting up an archive database. We then review some of the main activities in this area and summarise the most important features of the systems currently available. We then draw a perspective towards a possible data system design for the new generation of very large telescopes being built by a number of organisations: the *Great Observatories* of the ground. We conclude by discussing some of the major issues to be addressed by the community in relation to data archives; these being the central focus of an international workshop entitled *Handling & Archiving data from Ground-based Observatories* (held in Trieste, Italy, on April 21–23, 1993, see ESO Conf. Proc.).

2. Present Situation: Telescope Operations

The data handling aspects of telescope operations can be summarised in the following points.

Telescope Allocation Most telescopes on the ground are allocated on a night-by-night basis. During the period of time granted to a particular observer, it is left to the observer's discretion how to plan the observing sequence including target selection, exposure and calibration. Most large observatories will provide observers with assistance both during the day as well as during the night. This *modus operandi* represents the traditional way telescopes are operated world-wide. This set-up, however, results in a rather unpredictable and undocumented sequence of events during observing nights; in fact it is not unusual to find astronomers who bring their own 'private' set of filters to the telescope and request them to inserted into the optical path. Such events, which document the high degree of freedom that astronomers enjoy in the current usage of observing facilities, do have, however, a tremendous impact on any archiving efforts. If there is no record of the filter characteristics used in a particular observation, nobody will be able to analyse those frames other than the observer him/herself. The same being applicable to all configuration elements.

Changing Technical Environment Instrumental elements such as detectors, filters, gratings, etc., will in general be exchanged many times during the life time of the archive. This is a direct consequence of one the great advantages of telescopes on the ground: direct access to these elements allows the observatory to keep pace with state-of-the-art technology and to deploy the best available equipment at any given time. In addition to configuration changes, detectors will degrade their performance with time. Both facts impose on the archive database the need to keep track of instrumental characteristics for the complete archive history. Also, each observation needs to be associated to the parametric description of the configuration and technical performance in place at the time of the exposure, possibly by including this information in the FITS header.

Data and Operation Records Most observatories keep data records, however, only as a safety copy, i.e., they are catalogued for the purpose of retrieving by time or observer name only. Observation logs, if available at all, are mostly kept off-line. Also, since observatories mostly rely on observers to fill in these records, they are often incomplete. The telescope operations log is an essential part of the archive database. In fact this is the only way to keep track of volatile events such as weather conditions, special alarms, etc. In general, since such events are likely to happen during exposures, it is essential that the archive database contains the operations log in full for a given night and that each frame header includes all events occurred during the exposure. The night log allows for an overview of the overall performance while the exposure log gives hints to any anomalies during the exposure time.

Data Format In general, most observatories will deliver users with FITS formatted data. The basic FITS standard (see Grosbøl 1991, for an overview) covers both physical and syntactical format specifications; i.e., the standard prescribes the length of each block (2880 bytes), the existence of at least one header block and the syntax (`keyword = value`) of the header. Further, it defines a set of minimal keywords that describe the physical characteristics of the data set following the header. This set of descriptions has provided the community with a successful mean to transport data across many data analysis packages, computers systems and recording media. However, when it comes to archiving data, the headers need to include much more information than the mere physical structure of the frame. Because of data consistency and integrity considerations it is essential that the description of the *observation* is included in the header of the file. This poses, however, the problem of expressing a number of instrumental features (up to hundreds in the case of sophisticated instruments) into the FITS header, constrained by 8-char-long keyword names. The argument that these keywords will normally be read only by software and that therefore they can be name-coded efficiently, does not apply to long term archives where the expected life-time of the archive is much longer than the maintained life-time of the analysis software for a particular instrument. One has to foresee the case in which, e.g., 15 years after the exposure was done, someone needs to read the header. In general terms, one needs the means by which both static characteristics and dynamic configuration parameters of telescopes, instruments, detectors, adapters are coded in the observation header in a human readable form.

A second problem is given by the need to establish links between scientific observations and calibration frames or data. In space missions this is done by the the project operations team, because they are in charge of performing calibration exposures. At the telescopes of the ground, dark/bias and flat field frames are taken by the observer him/herself. In order for the archive database to properly associate the corresponding sets of scientific and calibration exposures, it is mandatory that these are tagged, and that the time of the observation is coded correctly.

3. Current Situation: Archiving Facilities

We will summarize in this section major features of some of the archiving facilities that are currently operational. There is no attempt to be exhaustive, however. We refer the reader for further detail, to the various archives and data systems described in Albrecht & Egret, 1991. In all cases, archives store the raw observations and the associated calibration exposures; there is no definition and implementation of standard reduction (calibration) procedures.

La Palma (ING) and Westerbork (WSRT) Archives The archives of the Isaac Newton Group of telescopes (La Palma) and of the Westerbork Observatories (Raimond 1991) contain many years of astronomical observations: 5 years of optical data obtained with the ING telescopes and more than 20 years now of radio observations acquired with the WSRT. The reason why these archives are usually described together lies in the fact that both have been set-up and are operated with the same software system. The La Palma archive holds data in FITS files which have been organised in such a way as to group sets of descriptions (telescope, instrument, detector, etc.), the data being associated to the detector group. WSRT data is stored in an internal format, but can be dearchived in FITS. The observations catalog of both archives can be queried on-line via the ARCQUERY retrieval program. The main features of ARCQUERY can be summarised as follows:

- *search-by-cone:* select all observations of objects within a projected cone on the sky
- *search-by-object-class:* select observations of objects of a particular class, e.g., elliptical galaxies, QSO's, etc.
- *qualified search:* select observations which match a search criteria expressed in terms of constraints on observation parameters.

The 'proprietary period' for the La Palma archive is one year, for the Westerbork archive 18 months.

ESO Archive (NTT EMMI/SUSI) ESO started routine archiving of it's New Technology Telescope (NTT) in April 1991. Two out of three NTT instruments (EMMI and SUSI) produce data descriptions that comply with the ESO Archive Data Interface specification and are both entered into the archive database and observations catalog. This specification makes use of hierarchical FITS keywords to describe instrumental and configuration parameters. As of November 1992, the archive holds around 11,000 scientific observations. Observers hold a one year data rights on their frames The observations catalog is currently under development; a first version is on-line via the STARCAT program (Pirenne et al. 1992). In addition to *search-by-cone* and *qualified searches*, STARCAT offers the possibility to visualise the selected observations either as lists (summaries) or in full detail one by one. The catalog itself, integrates information from a variety of sources including the ESO observation schedule, meteo and seeing night records and the instrumental static characteristics database. It is planned to cross-correlate the catalog with SIMBAD in order to obtain an approximated object classification. Also, it is planned to generate on-line 'quick-looks' of images, via *lossy* data compression; these can be displayed through the STARCAT `preview` command.

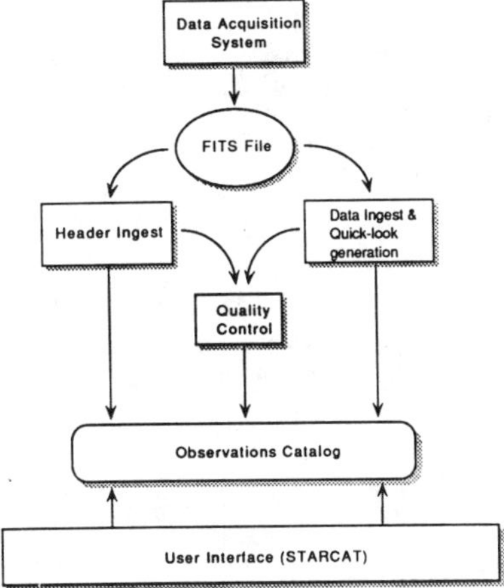

Figure 1. Data flow common to both the ESO and CFHT archives.

CFHT Archive at CADC The Canadian Astronomical data Center has only recently started (September 1992) with routine archiving of most of the instruments operated at the CFHT, Hawaii. Even though, the first public observations will only be available in the next fall (1993), we mention this activity because of the great momentum behind it. The observations catalog is available on-line via STARCAT.

As an example data flow, we show in Figure 1 the data flow common to both the CFHT and ESO archives[1]. The telescope data acquisition system provides the archive with FITS files. The archive system analyses the header in order to: a) make a consistency check of the frame in question, b) to control whether or not observation parameters have been filled in properly and c) to determine whether or not a quick-look of the frame will be generated. The file is catalogued in the archive database and the observation keywords ingested into the observations catalog. The user has access to this information via the common user interface (STARCAT).

It is worthwhile noting that STARCAT is the common user interface now to three major archiving facilities: The Space Telescope European Coordinating Facility (ST-ECF, HST archive), and the ESO and CFHT archives.

4. A Data System Design for Ground-Based Observatories

In the era of the new generation of telescopes, few facilities worldwide will be in the position of offering access to either single or combined focus 8m class telescopes. It is anticipated, therefore, that these telescopes will be largely oversubscribed. The operation of such facilities will possibly resemble more a space mission than the classical telescope operation outlined in section 2. On the

[1] Based on a figure by D. Crabtree, 1992

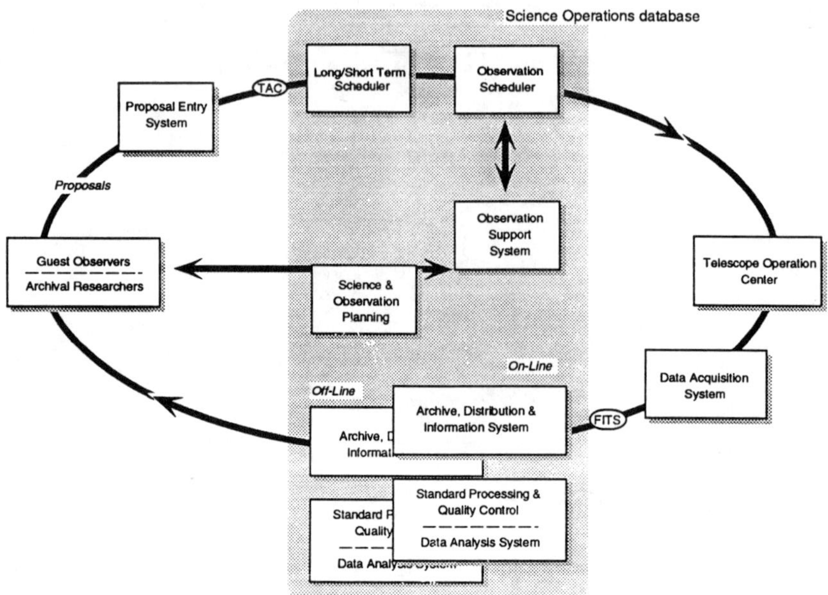

Figure 2. Science operations scenario for a data system of a ground-based observatory.

other hand, observatories on the ground do have advantages over those in space that should be fully exploited. In the following we propose a sketch of a data system for a ground-based observatory that should fulfill these requirements.

The end-to-end scenario shown in Figure 2 differentiates in two major aspects from that of a space mission (e.g., HST): a real-time *observation scheduler* (or very short term scheduler) and the existence of a *science operations database* (see Figure 3). The real-time scheduler originates from the need to adapt telescope operations to unforeseen events like bad weather conditions. The science operations database is required to keep track of change and to provide and information platform on which science decisions can be made. The scenario can be described as follows: Guest observers prepare observing time proposals possibly making use of the science and observation planning system (SOPS). Through the SOPS they have access to all information services of the science operations database. Proposals are submitted electronically to the proposal entry system which both validates syntax aspects and preliminary assesses the technical feasibility of the observation. The time allocation committee (TAC) peer reviews all proposals to determine the list of accepted observing projects. The long term scheduler establishes the telescope allocation to individual proposals according to constraints arising from moon, season of the year and technical conditions. The short term scheduler is used by the observer of a successful proposal to prepare the sequence of observations during a particular night. During the observing run, observers interact with the observation support system (OSS) to prepare the instrumental set-up and configuration needed for each exposure and

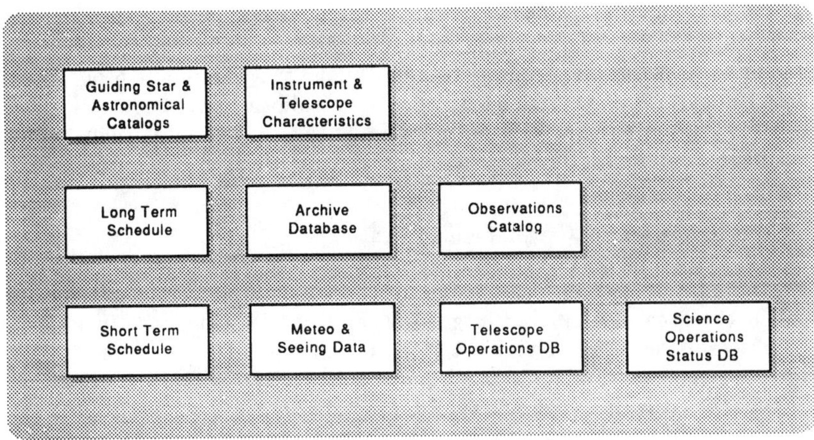

Figure 3. The Science Operations Database.

to control the output of each operation. Through the OSS, the observer queues all operation stacks onto the observation scheduler which, in term, passes telescope commands to the operations base for implementation. This scenario accommodates a variety of operation modes, like service observing (observations performed by the observatory on behalf of an absent user), remote observing and classical observing, i.e., with the observer present at the base. Upon detector exposure, the data acquisition system formats the frame in FITS and includes in the header the complete description of all configuration parameters. This file is passed onto the telescope on-line elements of the archive and distribution system which performs a consistency check of the frame. At regular intervals, the on-line data analysis system executes a set of procedures aimed at assessing the performance of all elements involved in the observation. This system is also possibly used by observers to visualize and perform a preliminary reduction of the data. The archive and information system stores the data files on the archive media and ingests all relevant parameters and events into the SciOps DB. This system also delivers guest observers the data acquired during the run as well as servicing dearchive requests from archival researchers. The general instrument reduction software included in the data analysis system is used to calibrate and reduce each of the observations.

Figure 3 depicts the elements of the science operations database. They are ordered by frequency of update, increasing frequency from top to bottom The long term schedule includes all information associated with observing programmes, like the programme title, it's abstract, PI and Co-I's, etc., in addition to the actual long term allocation. The short term schedule includes only target sequences: is used by the OSS to possibly optimize telescope operations and as the master plan for a given observing night. Meteorological and seeing measurements are performed regularly during the night and ingested into the DB.

These data are used both by the archive system to associate quality flags to individual frames as well as by the user to monitor weather and seeing conditions when conducting remote or service observing. The telescope operations database records all relevant event during the exposure such as slew, computer boot operations, etc. This information is included in the file header. The science operations status database, allows to monitor the status of night programme. It records the exposures performed by observers, calibration exposures performed by the observatory and any information related to the availability of instruments, detectors, optical disks and other elements.

5. Conclusions and Open Issues

Before attempting to summarise some of the most important issues still to be solved in the area of archiving in general, and of archiving data from ground-based observatories in particular, we note that, in spite of the numerous problems outlined in sections 1 and 2, the number of observatories that are pursuing archiving activities is growing rapidly. In fact many new telescope projects conceive the data archive as an integral part of the overall undertaking. This development is encouraging and represents, without doubt, a big challenge to the community at large.

Some selected issues which need yet to be discussed are:

- Database implementation: the commercial vs. ad hoc approach. Database Management systems (DBMS's) when bought commercially tend to be expensive, and in most of the cases they will not deliver astronomical functionality (e.g., search by coordinates) and user friendliness at the same time. Ad hoc developments, on the other hand, will probably lack the performance of their commercial counterparts. Hybrid solutions (commercial engines with 'astronomical' UIF's) have proven to offer the best of both sides but require to be maintained at relative high cost (manpower).
- How to express semantics in FITS: the domain name issue. When describing observations and the associated instrumental parameters, the basic FITS standard proves to be insufficient because of the need to 'code' complex names ito 8-bytes keyword names. A possible solution has been attempted by ESO in taking the hierarchical keyword approach, i.e., for instance the filter wheel position in the red side of an instrument can be coded `HIERARCH INS REDARM FILTER = '#601'`, which increases readability. This is important when analysing data with software other than the instrument custom package, or when retrieving old archive data.
- How to link scientific and calibration data: the rôle of data 'packages' or tables in FITS. When distributing archive data, archival researchers are confronted with the need to organise, possibly hundreds of files, in order to be able to reduce the data. Should archiving facilities deliver data 'packed' together (scientific frames plus calibration data)? Or is it more reasonable to provide the files and an accompanying table?
- Standard reduction procedures: the rôle of data analysis packages. One of the hottest issues is whether ground based observatories should set-up

standard reduction procedures for highly requested instruments. Looking at the experience with space missions, where it is customary that the project assumes the responsibility to deliver calibrated data with the best available calibration frames, it seems that observers have little interest in making use of the standard calibration: most of them prefer to perform the calibration by themselves, when possible. On the other hand, one of the most successful archives ever, the IUE archive, is largely used *because* it contains calibrated data. To find a balanced trade-off between costs and utilisation is difficult because of lacking precedents.

- Are Quick-looks to be considered data?: the rôle of lossy data compression. With growing data set sizes the issue of data compression becomes quite important. In addition, it is very desirable, while browsing through an observation catalog to have the possibility to first have a quick look at a frame in question before requesting it and reducing it. Some facilities have planned to generate quick-looks automatically as part of the ingest process. The question arises, whether these data, usually highly compressed and with substantial loss of information, should be made available, other than as screen dumps, to the community. In fact, some may be ideally suited for preliminary studies. On the other hand, the archiving facilities may hesitate to give out data which was never meant to be analysed scientifically.

Acknowledgments. The author would like to thank P. Grosbøl for valuable discussions regarding the future data system scenarios.

References

Albrecht, & Egret (eds.) 1991, Databases & On-line Data in Astronomy (Kluwer Acad. Publishers, Dordrecht)

Raimond, E. 1991, in Databases & On-line Data in Astronomy (Kluwer Acad. Publishers, Dordrecht)

Grosbøl, P. 1991, in Databases & On-line Data in Astronomy (Kluwer Acad. Publishers, Dordrecht)

Pirenne, et al. 1992, in Astronomy from Large Databases II, Conf. Proc., ESO

Crabtree, D. 1992, CDS Council, Strasburg

Discussion

Madore: Are any of the ground-based archives, that you have discussed, soliciting or even accepting data that have been processed and reduced by the end-user?

Albrecht: The overhead involved in the collection and quality control at large of end-user data would be much too large for the level of resources currently available. However, efforts are underway to archive data reduced by the end-user in the framework of 'key programmes' (observation programmes that span many years) where PI's assume the responsibility to deliver complete data sets of homogeneous quality.

Astronomical Data Analysis Software and Systems II
ASP Conference Series, Vol. 52, 1993
R. J. Hanisch, R. J. V. Brissenden, and J. Barnes, eds.

The Digital Archive of the International Halley Watch

D. A. Klinglesmith III, M. B. Niedner, Jr.

Goddard Space Flight Center, Code 684, Greenbelt MD 20771

E. Grayzeck

Small Bodies Node of the Planetary Data System, University of Maryland, Astronomy Department, College Park MD 20742

M. Aronsson, R. L. Newburn

Jet Propulsion Laboratory, 4800 Oak Grove Dr., MS 169-237, Pasadena CA 91109

A. Warnock III

Hughes STX, Goddard Space Flight Center, Code 631, Greenbelt MD 20771

1. Introduction

Before its formal existence as a funded project, the International Halley Watch, IHW, was an idea which was studied and advocated by a NASA study panel in 1979–1980 (Brandt, Newburn & Friedman 1980). The core notion of IHW was that proper study of a highly temporal and probably complex body like comet P/Halley would require detailed and voluminous observations obtained on a global, coordinated basis. Beyond the mere act of independent astronomers obtaining data was the obvious requirement that the observations be archived in a standardized, centralized way. This would give researchers an unprecedented opportunity to characterize a comet's behavior using complete and diverse datasets. More on the history of and concepts behind the IHW can be found in the article by Newburn and Rahe in *The Comet Halley Archive Summary Volume* (1991).

For this paper it is important to point out that the 26 CD-ROMs comprising the IHW digital archive form the principal component of the complete archive, and the first 24 disks are now complete. The data have indeed travelled a long path, beginning with the observers taking the original observations, the submission of the data to the nine IHW Discipline Specialist Teams which were formed to receive and standardize global data of different observational types, the central archiving of all the data at the IHW Lead Center at JPL, and finally the preparation (including pre-mastering) of the data for deposit on CD-ROM at the NASA/Goddard Space Flight Center. All along the way was an extensive verification process, the writing of documentation, the creation of various metadata and software tools, and the design and creation of the disks themselves. The disks were manufactured by the Digital Audio Disc Corporation (DADC) under contract to NASA-GSFC.

The eventual IHW CD-ROM Archive will consist of 26 disks. Volumes 1–23 of P/Halley, and Volume 24 of IHW "Trial Run" target comets P/Crommelin and P/Giacobini-Zinner, contain remote data exclusively and are currently ready for initial distribution. To be issued within the next year or so will be two additional disks containing *in situ* data returned by the spacecraft which visited P/Halley in 1986 March and P/Giacobini-Zinner in 1985 September.

2. Choice of Digital Medium

With an anticipated total volume of data of the order of 30 gigabytes, the IHW was forced to give careful consideration to the choice of digital medium. At the time that many of the decisions had to be made, in the early-to-mid 1980s, there was little feasibility for creating an on-line archive that would be readily accessible to the international scientific community, let alone contain 30 gigabytes of data. Accordingly, at a very early date the concept of a *distributed archive* became the central principle under which the IHW operated. As Newburn and Rahe have mentioned in the **preface.txt** file of Volumes 19–23 of the CD-ROM archive, the earliest expectations were that the IHW Archive would be distributed on paper, but given the need to present researchers with complete data, a *digital archive* quickly became a necessity.

The use of laser disks was considered but rejected due to lack of digital standards. CD-ROMs became the only available medium that had both international standard formats and widespread availability. However, even given CD-ROM's high data storage capacity (600–700 MB per disk), it was apparent by 1987 that > 40 disks would be required to hold the data, this large number being strongly driven by the Large-Scale Phenomena (L-SP) Discipline's 1,000+ wide-field images (15–30 MB typical file size). In order to both contain costs and create a more manageable CD-ROM archive, the decision was made to reduce the size of the L-SP dataset by compressing the initial 2-byte data into files containing successive 1-byte differences. This "previous pixel" compression technique saved 18 disks and brought the total number of CD-ROMs containing ground-based data down to 23 (for P/Halley).

3. Efforts to Define the Digital Archive: SAM

During the early years of the IHW it became clear not only that the volume of archivable data would be substantial, but also that data format standards needed to be developed. In 1983, the first Software Archive Meeting (or "SAM") was held. SAM I, chaired by D. A. Klinglesmith III, brought together both the scientists and the computer/archive personnel from each of the IHW Discipline Specialist Teams, to discuss a range of topics relevant to archiving details. As a result of SAM meetings between 1983 and 1987, many issues were resolved which were vital to fabrication of the final disks. Specifically, the use of FITS (Wells, Greisen & Harten 1981) and the standardization of the IHW FITS keywords were agreed upon at SAM. Discussions about how to order the data, what kind of directory structure to use, how to name the individual files, what kind of documentation would be supplied and by whom, what types of index and software support tools needed to be supplied, etc., were all issues either discussed or resolved at SAM meetings. It is important to note that activity of this type continued after the last SAM meeting, much of it at the NASA-GSFC. The CD-ROMs were prepared and pre-mastered at GSFC where the Planetary Data System labels were added as an alternate description of the data.

4. Content and Size of the CD-ROM Archive

The P/Halley remote-data disks come in two subsets. Volumes 1–18 hold wide-field digital images archived by the Large-Scale Phenomena Discipline, and Volumes 19–23 hold data from all the IHW Disciplines: Astrometry, Infrared Studies, Large-Scale Phenomena (subsampled images), Meteor Studies, Near-Nucleus Studies, Photometry and Polarimetry, Radio Studies, Spectroscopy and Spectrophotometry, and Amateur Observations. Volumes 19–23 are sometimes described as "mixed" as a result of their multi-Discipline content. The primary discriminator between disks in each subset is the time interval spanned by the data; i.e., the volumes are in chronological sequence.

The mixed disks (Vols. 19–23) have an enormous number of small files, while the L-SP disks (Vols. 1–18) have a relatively small number of large files. This basic distinction between the two subsets of remote-data disks is well drawn in Table 1. A complete description of the file sizes and counts can be found in the **volinfo.txt** file located in the **/document** directory on each of the "mixed disks".

Table 1. File and Directory Sizes and Counts

CD	Directories		Files		dates
NUM	#	Kb	#	MB	dd/mm/yy-dd/mm/yy
1-18	351	1,516	16,712	11,110	22/12/84-26/04/87
19	695	2,499	32,002	629	01/12/81-08/12/85
20	468	1,880	25,719	563	09/12/85-09/02/86
21	485	1,870	24,889	686	10/02/86-13/04/86
22	644	2,136	24,519	682	14/04/86-03/04/87
23	133	438	4,828	639	04/04/87-12/04/89
Totals	2,776	10,339	128,669	14,309	

The large number of files on the mixed disks has resulted in a rather complex directory structure given the need to keep "reasonable" the number files within any given directory. The data subdirectories have names based on the beginning time of the data contained in each, and there are from 400 to 700 such directories on each of the mixed disks. The directory tree structure has been broken out in two text files, **cdtree.txt** and **datatree.txt** that can be found in the **/document** subdirectory on each of the mixed disks.

It is extremely difficult to summarize the contents of an archive of this size and complexity in a brief manner. The previous paragraph dealt with the total number of data files, which is important to the database and/or computer systems manager who needs to find space or the means to access a formidable number of files. On the other hand the comet scientist would want to know how many and what kind of information is available within the archive, and Table 2 gives the breakdown of the data in terms of IHW Disciplines and sub-Disciplines.

Table 2. Number of Individual Observations

DISCIPLINE/subnet	#obs.	DISCIPLINE/subnet	#obs.
Astrometry		Radio Studies	
- 1835-36	158	- Continuum	97
- 1910-11	1696	- Occultation	6
- 1982-89	6475	- OH (molecule)	1,657
Infrared Studies		- Radar	1
- Image	95	- Spectral Line	189
- Photometry	2,204	Amateur Observation	
- Polarimetry	137	- Drawing	1,294
- Spectroscopy	84	- Photography	2,170
Large-Scale Phenomena	3,383	- Spectroscopy	45
Near Nucleus Studies	3,523	- Visual	11,641
Photometry		Meteor Studies	
- Broad Band	3,318	- Radar	6,962
- Narrow Band	18,495	- Visual	1,624
- Polarimetry	752	Spectroscopy	3,368
- Stokes Parameter	164		
		Grand Total	69,538

5. CD-ROM Support: Metadata Indices, Documentation, Software

An archive of this size and complexity demands that a variety of tools be created which allow the researcher to search for data of interest. Toward this end a number of "index files" have been constructed and deposited on the disks (this discussion is mostly about the mixed disks, Volumes 19–23). All of the indices contain values for some fraction of the FITS keywords, and the associated filename. The indices are presented as delimited tables which can be read directly into a database program, e.g., dBASE IV. Metadata for these tables is expressed as FITS headers with the table extension and PDS labels.

The IHW Disciplines are very different in terms of data type and FITS keyword content. As a result, there are three basic types of index files: global indices which contain IHW mandatory keywords and/or full directory paths, net-specific indices which contain a large fraction of all FITS keywords, including the COMMENT keywords, for each of several Disciplines, and indices for the so-called "Printed Archive," the latter files being constructed at the level of the sub-Disciplines.

The global indices are **quik_0nn.idx** (nn=19–23) and **pathtabl.idx**. Both provide the time of observation, the filename, and the full directory path; in addition the **quik_0nn** files contain the System code, Observer(s), and remaining mandatory IHW FITS keywords. The global indices are, by definition, inclusive of all the IHW Disciplines (except Meteor Studies, data for which reside in a relative handful of large FITS table files on Volume 23), and the arrangement is chronological. They are located in the **/index** directory of each mixed disk.

The net-specific and printed archive index tables are located in the **/index/netables** subdirectories of the mixed disks. These tables provide the detailed metadata about, and in some cases the actual data for, individual observations.

Another important aspect of an archive of this size is documentation which describes the contents in detail. Each of the CD-ROMs has a **/document** directory containing a wide assortment of individual files addressing different topics. Of particular importance are the "Discipline Appendix" files, written by each of the IHW Discipline Specialists and deposited in the **/document/appendix** directory of each of the mixed disks. Software for manipulating data and metadata is included in the **/software** directory.

6. Distribution of the Complete Archive

The IHW has produced three products: the CD-ROM archive discussed here, the *The Comet Halley Archive Summary Volume* (Sekanina 1991), and the *International Halley Watch Atlas of Large-Scale Phenomena* (Brandt, Niedner & Rahe 1992). The publishing of a "Printed Archive" is under consideration.

The initial distribution of all three archive components will be carried out by JPL from a mailing listed generated via responses to a questionnaire sent out in the late-1980s. Generally speaking, these initial recipients of the archive are either scientists who obtained the data (including space data) or research institutions and libraries. As the result of a memorandum of understanding worked out between NASA's National Space Sciences Data Center, NASA-NSSDC, the IHW, and the Small Bodies Node of the Planetary Data System, PDS-SBN, it has been agreed that additional individual copies of the digital Archive will be distributed by PDS-SBN and that bulk mailings will be handled by NASA-NSSDC.

Acknowledgments. The generation of this digital archive was clearly an international effort. It could not have been done without the cooperation of many people and institutions too numerous to mention here. Refer to the **acknwldg.txt** file in the root directory and the **preface.txt** file in the **/document/appendix** directory on Volume 23 of the digital archive for a complete list of credits.

References

Brandt, J.C., Newburn, R.L., & Friedman, L.D. 1980, The International Halley Watch, Report of the Science Working Group, NASA TM 82181

Brandt, J.C., Niedner, M.B., & Rahe, J. 1992, The International Halley Watch Atlas of Large-Scale Phenomena, ISBN 1-880768-00-3 (Johnson Publishing Co., Boulder, CO)

Sekanina, Z. 1991, The Comet Halley Archive Summary Volume, ed. Z. Sekanina, NASA/JPL 400-450 8/91

Wells, D.C., Greisen, E.W., & Harten, R.H. 1981, A&AS, 44, 363

Astronomical Data Analysis Software and Systems II
ASP Conference Series, Vol. 52, 1993
R. J. Hanisch, R. J. V. Brissenden, and J. Barnes, eds.

The NSO FTS Database Program and Archive (FTSDBM)

D. M. Lytle

National Optical Astronomy Observatories, Tucson, AZ 85719

Abstract. Data from the NSO Fourier transform spectrometer is being rearchived from half inch tape onto write-once compact disk. In the process, information about each spectrum and a low resolution copy of each spectrum is being saved into an on-line database. FTSDBM is a simple database management program in the NSO external package for IRAF. A command language allows the FTSDBM user to add entries to the database, delete entries, select subsets from the database based on keyword values including ranges of values, create new database files based on these subsets, make keyword lists, examine low resolution spectra graphically, and make disk number/file number lists. Once the archive is complete, FTSDBM will allow the database to be efficiently searched for data of interest to the user and the compact disk format will allow random access to these data.

1. Background

The old FTS archive at NOAO is, physically, approximately one thousand 2400 foot half inch tapes and access to the database is via searching through a large text file containing one line of text per spectrum. These data are written in an old, arcane, Cyber computer format that requires special software to read.

The new FTS archive will reside on approximately 50 compact disks, access to the database will be via a simple, efficient, database program, and the data will be written in an international standard format (FITS).

Write-once compact disks were chosen for their predicted archival quality, their compact format, and the ability to access the data randomly rather than sequentially as with other media (e.g., Exabyte). This is, admittedly, very new technology and there is some risk involved since the true archival properties are unknown and the future popularity of the CD format may change drastically.

2. Foreground

The four key pieces of software to be used for this project are *rearcfts, wfits, ftsdbm,* and *mkdsc.* Rearcfts reads FTS archive tapes and produces IRAF images. Wfits will write these IRAF images into FITS format images. Ftsdbm creates and keeps track of the database associated with this project. Mkdsc formats and writes the data onto compact disk. The first three programs are contained in IRAF, rearcfts and ftsdbm are in the nso/fts package and wfits is in the dataio package. Mkdsc is a stand-alone program that has an X interface window and sends commands to another stand-alone program called *makedisk*. These two were supplied by Young Minds Inc. along with the CD writing equipment purchased from them.

3. Reading the Old Tapes with REARCFTS

Rearcfts was written solely to ingest old archive tapes and return IRAF images containing the data found on those tapes. FTS data are generally written on the archive tapes in three parts for each spectrum. First there is a 4096 point low resolution image of the entire spectrum the first 2048 points of which are the real amplitude and the last 2048 points of which are the phase information. (Remember that these data are taken as an interferogram and only after being Fourier transformed is the spectrum visible.) The second part is the real or amplitude version of the spectrum. This can be any length between 4096 points and over a million points. The third part of the spectrum is the full, complex, spectrum which ranges up to a length of three million points.

Currently the processing we are doing saves the low resolution image in an on-line database, writes the real image onto compact disk in FITS format, and saves the complex image on Exabyte tape.

On some of the early tapes (and perhaps on some of the later tapes) the low resolution spectrum associated with any given spectrum may not exist on the tape. In this case, since we need a low resolution spectrum to enter into the database, we use the IRAF program *blkavg* to block average the real spectrum down to a length of 2048 to create a low resolution spectrum.

4. The Database and FTSDBM

Ftsdbm is an IRAF task in the FTS package that allows the user to create and access databases containing FTS data headers and low resolution spectra.

The basic list of commands for *ftsdbm* are listed in Table 1.

Table 1. FTSDBM Command Summary.

Command	Arguments	Description
create	filename	create a new database
insert	imagename	insert record from an image
delete	date scan	delete a record by date/scan
update	date scan filename	update a record
database	filename/register	set database
select	select-string	add query to query list
result	register	filter data, result in reg
clist		concise list of database
vlist		multi list tool
plot		plot low res data
write	filename	write new database from reg
tapeindex	filename	write tapeindex to file
slist		list select query list
sclear		clear select query list
help		detailed help
quit		exit ftsdbm

Select strings are of the form *keyword* = *value* and may contain wildcards. Currently, allowed keywords are *datescan, bsplit, wrange, airmass, resolve, observer,* and *source.*

5. Storing the Data, WFITS and MakeDisk

As the rearchiving project progresses, the larger part of the data, the complex spectra, are written in FITS format on Exabyte tape. These complex spectra are little used and can, without a great amount of effort, be reproduced from the raw interferograms. Thus the decision was made to store them on tape rather than on compact disk. The IRAF task *wfits* is used to write these spectra to tape.

The low resolution spectra are entered into a database file. Eventually this file will contain all of the spectra. Copies of this file will be kept in various places both to add to its usefulness and to protect the file from loss. Users may choose to generate various other database files containing subsets of this universal file. This is possible to do in FTSDBM. These low resolution spectra will also be stored on compact disk at intervals although they too can be reproduced from the full real spectra.

The real spectra are transformed from IRAF images on disk to FITS format files on disk by the IRAF program wfits. A program called *MakeDisk*, published by Young Minds Incorporated, is used to format these FITS files into ISO 9660 standard format and the result is written to compact disk using a unit made by Young Minds called the CD-Studio which includes a hardware disk writer made by Philips. This unit allows us to write 600 Mbytes of data to a blank 12 cm diameter compact disk in 30 minutes. These disks can then be mounted in a CD-reader as a High Sierra File System (HSFS) and accessed just like a hard magnetic disk (although somewhat slower).

Later in this project, the raw interferograms will also be archived to compact disk.

6. Archive Management

At NOAO, the database file will be kept on disk at all times so that it will be available for anyone that wants to look through it. The disks containing the real spectra will be available in a central location so that users can mount any disk they want to read (CD readers will be provided on most machines). A notes file will be kept with the database describing the database and its current state, last update, persons to take inquiries to, etc.

Astronomical Data Analysis Software and Systems II
ASP Conference Series, Vol. 52, 1993
R. J. Hanisch, R. J. V. Brissenden, and J. Barnes, eds.

DENIS—DEep Near Infrared Survey of the Southern Sky

Erik R. Deul

Sterrewacht Leiden, P.O.Box 9513 - 2300 RA Leiden, The Netherlands

Abstract. DENIS (Epchtein 1992) will be a complete deep near infrared survey of the southern sky, made with the objective of providing full coverage in 2 near infrared bands (J at 1.25μm and K at 2.2μm) and one optical band (I at 0.8μm), using a ground-based telescope and digital array detectors. The products of this survey will be databases of calibrated images, extended sources, and small objects. In addition, catalogs of small and extended sources will be produced. We expect the survey to be completed within five years; restricted access to the databases is possible during the second half of the survey. The production of catalogs (to be distributed) will take a some additional time.

1. The DENIS Survey

DENIS (DEep Near Infrared Survey of the Southern Sky) is a joint project of 18 European and South American Institutes, aiming to provide digitized maps of the southern sky in a spectral region, similar to that of the TMSS (Neugebauer & Leighton 1969) carried out some 25 years ago, but providing a sensitivity 10 magnitudes deeper. Using the 1-m photometric telescope of ESO, on La Silla, the southern celestial hemisphere will be fully mapped simultaneously in three bands, namely J (1.25μm), K$'$ (2.2μm) and I (0.8μm), using CCD and NICMOS array detectors. The spatial resolutions and anticipated limiting magnitudes will be $3''$, $3''$ and $1\farcs5$, and $16^{\rm m}$, $14\fm5$ and $18^{\rm m}$, respectively. The survey will start in october '93, with an expected duration of at least four years.

The observations will be performed in step-and-stare mode, using "strips" as basic units. Each strip is $12'$(one frame) wide in RA, and $30°$ long in declination, consisting of 180 overlapping frames. For telescope pointing and brightness calibration purposes additional frames will accompany each unit. DENIS will yield a huge quantity of data - one million survey fields will be observed, providing about 4 TBs of data.

ESO has granted this project the Key Program status, and provided up to 66% observing time on the 1 m telescope during the planned data acquisition period. The EEC granted 350 kECUs for the project through its Science Program.

2. Data Analysis Centers

The aspect of computer analysis starts at the data acquisition site, La Silla, of the ESO, where the enormous amount of raw data (4Gb per night) will be recorded on DAT tape. There data will be transferred to the two European Data Analysis Centers, at the Sterrewacht Leiden and at the Institute d'Astrophysique de Paris, for further processing.

The Leiden Data Analysis Center (LDAC) will extract objects, ranging from point sources to small extended sources, parameterize them, and archive them into a source catalog. The Paris Data Analysis Center (PDAC) is responsible for archiving and preprocessing the raw data to provide a homogeneous set of images suitable for both the Leiden and Paris data analysis streams. The PDAC will also extract and archive images for the sources flagged by the LDAC as extended, and possibly also creating a catalog of galaxies. Both DAC's are working in close collaboration, supplementing each other, to perform a coherent and complete data reduction and analysis task.

Final archiving on permanent media (e.g., CD-ROM), primarily for the catalogs, will be achieved after the completion of the survey, other aspects of the archiving are under discussion with ESO.
The products that will be delivered are:

- A Bright Source Catalog extracted in real-time at La Silla for preliminary investigation and to prompt objects for astronomical interest.
- Small Object Database (SODA) that can be remotely interrogated. A condensed form of this database is the Small Source Catalog.
- Large Object Database (LODA) in the form of an image database and catalog (of galaxies) that can be remotely interrogated.
- Processed Frame Database and Data Tracking Catalog (FOURBI), again remotely accessible. Possibly an additional database of selected areas will be available containing mosaiced images, mosaicing of frame sets will also be possible.

2.1. The Leiden Data Analysis Center (LDAC)

The LDAC will concentrate on the extraction of "small objects" from the DENIS images at all three proposed wavelength bands. Production of a catalog will be done during the survey data acquisition and should result in a first-order data product at the end of the observing period. The derivation of object parameters will be based solely on image properties; no *a priori* astronomical information will enter this catalog.

All the objects above the 5σ noise level will be extracted. The objects will be de-blended and their positions calculated using the frame centers computed with the help of cross-identifications with Input Catalog sources. In order to improve the positional accuracy, strip-wide astrometric solutions will be computed. The small objects will be photometrically calibrated using a set of DENIS standards distributed around the southern sky.

Basic information to be stored in the catalog consists of astrometric position, photometric intensity, image classification, and geometry. These parameters will be derived using the moments, up to second order, of each objects pixel distribution. As the extraction pattern analyzer recognizes the objects, cumulative sums of pixel intensity (at all three passbands) are calculated, and intensity-weighted and unweighted pixel positions are stored.

All this will result in the creation of the Small Object DAtabase (SODA). The SODA will grow continuously during the survey period. The SODA will be a relational database storing four main entities: Objects, Frames, Strips, and The Survey. The latter entity will dominate the view of future users. Each entity will keep the following information in the form of attributes:

- Objects: x, y, x', y', xy', I, J, K, flags, local S/N quality, local background level, where x, y are the pixel positions within a frame (first moments), x', y', xy' are the second moments of the intensity distribution. I, J, and K are the zeroth moments (fluxes). Many flags are kept to store conditions of the source extraction.
- Frames: center, exp-time, grey-filter, time, etc., where center stores the derived Ra and Dec of the frame, exp-time gives the total exposure time for the frame (addition of sub-frame exposures), and grey-filter is a flag denoting the use of a grey filter during exposure.
- Strips: date, weather, observer, etc., where date is the date of the observations, weather stores information on the weather conditions, such as seeing, distance to the moon, etc.
- Survey: none.

The known relations in the database are:

- Survey-strip: List of all strips, includes strip id's and info whether they are normal survey strips, or additional observations.
- Strip-frame: A set of lists of frame id's belonging to this particular strip.
- Frame-object: A large set of tables describing which objects belong to this particular frame.
- Object-object: This table contains information on links between the individual objects that were decomposed in a de-blending operation.
- Frame-frame: This table stores the links between objects observed in different frames but found to be identical on positional arguments.
- Strip-strip: This table stores the links between objects observed in different (consecutive) strips, but found to be identical on positional arguments.
- Object-IC: A table storing the connections between the extracted objects and the Input Catalog.

Output from searches through the SODA may be used as input to the tools for visualization, basic statistics, etc., such as plotting packages (e.g., Mongo, GNUplot) or statistical analysis packages (e.g., SAS, BLISS). During and after the data-acquisition period the SODA will be available for the scientific community through computer networks. The LDAC will process the database further, merging the objects on the overlapping areas and possibly cross-correlating them with existing astronomical catalogs. We call the final product the Small Source Catalog.

Statistical studies will be carried out to assess the quality of the catalog. These studies will quantify the completeness, reliability, source confusion (Beichman et al. 1988) of the catalog, and provide measures of the variability of sources, and of their positional and photometric accuracy. For those studies a small fraction of the southern sky (about 2%) will be observed repeatedly (about five times during the mini-survey).

2.2. The Paris Data Analysis Center (PDAC)

The PDAC will perform a number of processing steps to prepare the data for further processing/analysis. First a de-glitching of the frames at all wavelengths needs to be performed. The second processing step is the flat-fielding of the strip images using flat-fields derived by the real-time processing on the mountain. The third step is shuffling the frames around to obtain color-grouped sets of data for each pointing position (the raw tapes are single channel "isochromatic" data tapes).

Then the processed frames (strips) will be stored in the Processed Frame Database (PROFDA) and also sent to the LDAC for further processing. The PROFDA will later allow mosaicing the individual images. The raw data will also be kept on DAT tapes in the PDAC. Another database produced and maintained at the PDAC will be the Data Tracking - Survey Information Database (FOURBI). This database allows verification of the observing strategy goal, full mapping of the southern sky, and also keeps track of the data quality and storage administration.

The return flow of data from the LDAC, that most likely will be through international networks, will involve astrometric calibration and the extracted source information. At the PDAC a version of the SODA is kept; the astrometric information will be stored in the FOURBI. From the SODA the extended sources are recognized and using data-tracking information the extended source parameters and maps are derived. This will include, in many cases, doing mosaicing. The results are stored in a final Extended Source Database.

3. The Input Catalog

DENIS needs an Input Catalog to get positional information "from the sky" (supplementing the information "from the telescope"), and to give some *a priori* indication on the presence of very bright sources or the absence of sources in the field. The only existing catalog deep enough to contain at least one source per frame is the HST Guide Star Catalog (Lasker et al. 1990; Russel et al. 1990; Jenkner et al. 1990). (One can expect 7 GSC stars per frame, compared to the expected average value of 100 sources per frame for DENIS.) For the purpose of the survey the CDS will combine the multiple entries of the same GSC object, and supply better coordinates for the sources present in the TYCHO catalog from the HIPPARCOS mission. In order to enable real-time access to the Input Catalog the GSC will be re-structured using a simple file-structure based on $10'$ constant declination strips and about $12'$ wide fields in each strip. In this way a query for Input Catalog sources in a frame would need to access six catalog fields (files), each containing less than 10 sources (average).

4. The Small Source Catalog (SSC)

4.1. Description

The SSC is expected to contain some 10^8 sources (in the K band), out of which more than 10^6 are galaxies. Stars brighter than 7^m above the limiting magnitude will give a saturated signal - this range can be somewhat extended using a grey filter. The SSC will be extracted from the SODA, removing astronomically less interesting information, and merging multiply observed objects.

4.2. Quality Estimates

The LDAC has been testing its source extraction algorithm using synthetic data of the form produced by the actual data acquisition equipment, and preprocessed at the PDAC. Sets of synthetic images at I-band and K′-band wavelength have been produced using a dynamical range of 10^6. We adopted a uniform spatial distribution function, and a magnitude distribution function according to the Soneira distribution. The micro-scanning technique was mimiced through the production of sub-images shifted with respect to the original by the micro-scanning step (0.33 pixel) and then combining these subimages in the same way the micro-scanned images will be combined during real-time processing. Images with different numbers of stars (200 – 5000) were made.

Knowing the input parameters: stellar position and magnitude, we tested the quality of the object extraction algorithm by determining an artificial calibration curve and an estimate of the completeness/reliability/confusion measures.

The photometric accuracy can be assessed from a synthetic calibration curve that shows the regime of noise-dominated extractions ($M_I > 16\overset{m}{.}5$) and that of the overexposure effects ($M_I < 9\overset{m}{.}5$). The actual standard deviation in the photometric parameter derived from this analysis shows high that accuracy is achieved over a large range in magnitudes. The photometric accuracy of the object extraction (rms error) will be about $0\overset{m}{.}01$ just below the saturation limit, and $0\overset{m}{.}2$ at the faint cut-off.

The positional accuracy will be influenced mainly by the systematic errors of the GSC (Taff et al. 1990) which on average are about $1''$. The relative rms errors (within a DENIS frame) will be in the order of $0\overset{''}{.}1$. After the TYCHO catalog from the HIPPARCOS mission becomes available (we expect one TYCHO star per DENIS frame), the positional errors can be reduced to the order of a few tenths of arcseconds.

References

Beichman, C.A., et al. (eds.) 1988, IRAS Catalogs and Atlases, Explanatory Supplement NASA RP-1190 VIII

Epchtein, N. (ed.) 1992, A Deep Near Infrared Survey of the Southern Sky

Jenkner, H., et al. 1990, AJ, 99, 2082

Lasker, B.M., et al. 1990, AJ, 99, 2019

Neugebauer, G., & Leighton, R.B. 1969, Two Micron Sky Survey, NASA SP-3047

Russel, J.L., et al. 1990, AJ, 99, 2059

Taff, L.G., et al. 1990, ApJ, 353, L45

Astronomical Data Analysis Software and Systems II
ASP Conference Series, Vol. 52, 1993
R. J. Hanisch, R. J. V. Brissenden, and J. Barnes, eds.

Miyun 232 MHz Survey I Fields Centred at: $\alpha : 00^h41^m, \delta : 41°12'$ and $\alpha : 07^h00^m, \delta : 35°00'$

Zhang Xizhen, Zheng Yijia, Chen Hongsheng, and Wang Shouguan

Beijing Astronomical Observatory, CAS, Beijing 100080, China

Abstract. A new meter-wave survey of the sky region north of declination +30° has been carried out with the Miyun 232 MHz Synthesis Radio Telescope (MSRT). The instrument, observation and method of data reduction are briefly described in this paper. As a first result, two 8° × 8° regions centred respectively at $\alpha : 00^h41^m, \delta : 41°12'$ and $\alpha : 07^h00^m, \delta : 35°00'$ were observed and reduced. On the average 4–5 sources per square degree were recorded with position accuracy of 5" / S(Jy). The BGPW scale is adopted for the flux density calibration (Baars et al. 1977). The accuracy of flux determination is limited by the background fluctuation which is about 30 mJy. The catalogue is complete for sources with flux larger than 0.25 Jy. The total number of sources listed in the paper amounts to 687.

1. Instrument and Observation

The Miyun 232 MHz Survey is a moderately deep meter-wave survey. Most of the sky north of Dec 30 degrees has been observed. The survey has its working frequency located between the 6C (Baldwin et al. 1985) and B3 surveys (Ficarra et al. 1985), and has a resolution and sensitivity similar to them.

The Miyun aperture synthesis system (Wang 1984), working at 232 MHz and 327 MHz, consists of an E-W array of 28 dishes each of 9 m in diameter. The characteristics of the array are summarized in Table 1. For details about the telescope, refer to Beijing Observatory Meter-Wave Radio Group (1985).

All the combinations $A_i \times B_j$ (i=1–16, j=1–12) are used to form 192 interferometers with baselines incrementing by 6 meters from 18 meters to 1164 meters. In a complete set of Earth Rotation Synthesis, when weighted in a natural manner, the MSRT gives an overall resolution of $3.8' \times 3.8' csc\delta$ and a thermal noise limited sensitivity of 0.01 Jy/beam for the 232 MHz system.

The observations were made with MSRT for the two fields respectively on January 1985 for the 7h field and October same year for the Andromeda field. Two sets of observations were carried out for the second field and four for the first. Gain and phase calibrations for each array element were made by observing Cyg A for 10 minutes before or after the observation. To check the stability of the receiving system, the calibration observation of Cyg A was made with long duration about once a month.

Table 1. The characteristics of the MSRT

observing frequency	232 MHz
aerials	9 m parabolic
number of aerials	28
primary beam	$10° \times 12°$
baseline	1164 m East-West
spacing interval	6 m
number of baselines	192
min. and max. baselines	18 m - 1164 m
synthesised beam width(232)	$3.8' \times 3.8' csc\delta$
trans. frontend noise	100° K
band width	1.5 MHz
sampling interval	10 sec.
rms noise/spacing/sampling	10 Jy(232)
path compensation	digital
correlator	96 dig.(1 bit)

2. Data-Editing

This is a procedure employed to remove the effect of interference and compensate for the zero-offset errors. Some interference may come from either inside or outside of the instrument. To overcome the effect of the former, a double modulation has been added to the hardware, but the interference received by the dishes still exist.

To cope with the interference, two software techniques have been developed. The first one, to remove some strong spikes and subtract low level interference of slow variation from each baseline data, is designed to reject the distorted data by iteration and running mean algorithm when there is no source stronger than 10 Jy, otherwise subtract the source by modeling first. The power level of the interference is usually much higher than the signal level of a 20 Jy source when received by a single interferometer, and the rms of the system noise is just about 10 Jy. In a field of $10° \times 10°$, few sources exceed 20 Jy. This method is very effective for the sky area near the north pole.

The second method, utilizing the so-called frequency domain filtering technique (actually it does a Fourier analysis for each baseline data), is much better for the elimination of interference from long duration and low level. The main advantage of this method is that it retains almost all the observation data. The interference can be separated from the normal record by the examination of the spatial frequency spectrum and can thus be subtracted from the initial data. The method for taking off the zero-offset error is based on the same idea.

3. Mapping, CLEAN and Self-Calibration

As mentioned above, preliminary calibration with Cyg A gives the phase and gain differences between the array elements. But Cyg A is not a simple point source for our telescope. Referring to the data at other frequencies, we model it as a double point-source, and calculate the variations of phase and amplitude with the hour angle. Daily calibrations are carried out with the help of the calculated-variations. After pre-processing, the "dirty" map is usually good enough for forming a model to be used in the self-calibration procedures. A method of one dimension correlation-analysis had been developed by Mingzhi Wei (1988). In late 1985, with the help of Dr. E. B. Fomalont, the AIPS software package was installed in Beijing Astronomical Observatory. Since then, AIPS has been used for the mapping, CLEANing and some self-calibration of the MSRT maps.

To reduce the effects of short-time fluctuation of the ionosphere, two 12 hour UV-data sets were added together for each field before imaging. Usually a 512×512 map is made by MX, an AIPS task, and two or three self-calibration loops of phase-cal. only were done by ASCAL. For collecting total flux densities, the input-model for self-cal. is made with the sources inside the full primary beam (much wider than the FWHM).

For the field centred at $\alpha : 0^h41^m, \delta : 41°12'$, the strong source Cas A is not far beyond the boundary. We had to subtract the effect of Cas A first, before making a model from the sources in the main field. Since the primary beam is so wide that many sources can be recorded, we modified the CC-table of AIPS to ensure that most sources were included, rather than just take the components before the first negative component. We used several hundred iterations for the 7h field and a few thousand for the Andromeda field since there is a extended source in the field.

4. Coordinate Conversion and Calibration of Flux Density

After an ordinary coordinate conversion from X, Y to R.A. and Dec, large scale gradients in the ionosphere give rise to systematic shifts in the apparent positions of sources. These shifts can be as large as tens of arcsec. To remove the shifts, a number of reference sources were chosen from the B3 Survey and the Texas Survey for the correction of the two field of view respectively.

Research aimed at establishing flux density standards covering a wide range of wavelengths has been done by many authors. The absolute flux density system BGPW and RBC (Roger et al. 1973) are among the most widely used. Laing et al. (1980) give the calibrated flux density spectra of 165 3CR sources at frequency ranges 10–178 MHz and 750 MHz – 15 GHz. Riley (1988) presents the flux density at 408 MHz of a number of sources.

We rely on the reference sources to determine the flux density scale in terms of the recorded values. It is necessary to establish a homogeneous reference system over the whole sky. For the flux scale reference of the Miyun Survey, we used the BGPW, absolute flux scale. The flux densities of some sources at 178 MHz were supplemented with that in the LP table and RBC system. For the maps presented in this paper, two reference sources, 3C13 and 3C173, were used. Their flux densities were taken to be 10.4 Jy and 8.4 Jy respectively.

The primary antenna pattern was corrected before proceeding to the flux density calibration. As the calibration of the flux density is done in the map-plane, the AGC system has no affect on the calibration.

5. Estimation of Errors

Errors in position and flux were examined by three different methods: (a) the Monte-Carlo method, which serves to estimate the degradation of position and flux accuracy by background noise; (b) method of direct comparison, i.e., comparing the observed data with well-chosen references; (c) experimental formulas.

Monte Carlo experiments were used on four different ranks, i.e., 0.7 Jy, 1.0 Jy, 1.5 Jy and 2.0 Jy. From the results, the uncertainty of flux was found to be independent of the artificial sources. These results showed that about 60% of the sources had apparent flux densities within ±30 mJy of their "true" flux densities.

By means of these three methods, the rms of position errors can be expressed as

$$\sigma_\alpha(\text{arcsec}) \simeq 5'' / S(\text{Jy}) \tag{1}$$

$$\sigma_\delta = \sigma_\alpha \times csc\delta \tag{2}$$

Based on cross identifications with other catalogues, it is suggested that sources with 232 MHz flux density ≥ 0.25 Jy are complete and reliable in this survey.

6. Results

Figure 1 shows one of the two maps reported in this paper. Due to the page limitation, some results are summarized.

In the source list of the two fields, the following information was listed for each source. In total 687 sources with flux densities larger than 0.20 Jy were listed.

Items (1) to (8) in the list are: (1) name of the sources, (2) $\alpha(1950.0)$, (3) $\delta(1950.0)$, (4) flux density (Jy)(estimated from the peak value), (5) integrated flux density(Jy), (6) local zero level (Jy), i.e., rms of background fluctuation around the source., (7) name in 87GB catalogue, and (8) estimated spectral index (232 MHz to 4850 MHz).

Several interesting sources were found such as extended sources, steep spectra sources, and sources with convex spectra.

Acknowledgments. The authors wish to acknowledge contributions of all colleagues of the Miyun Metre-wave Radio Astronomy Group. It is the collective effort that have made the operation of MSRT in proper condition. We would also like to thank Dr. E. B. Fomalont for his kind help in installing AIPS, and say many special thanks to Drs. W. M. Goss, J. P. Ge, T. Speostral and P. Warner for their observations of some interesting sources found from the survey.

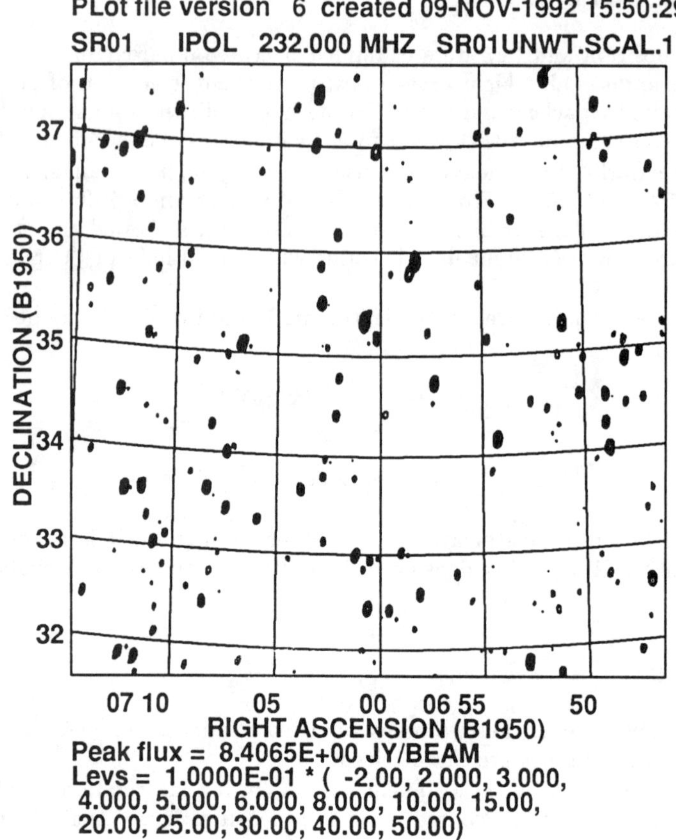

Figure 1. Map of the field centred at $\alpha : 07^h00^m, \delta : 35°00'$

References

Baars, J.W.M., Genzel, R., Pauliny-Toth, I.I.K., & Witzel, A. 1977, A&A, 61, 99

Baldwin, J.E., Boysen, R.C., Hales, S.E.G., Jenning, J.E., Waggett, P.C., Warner, P.J., & Wilson, D.M.A. 1985, MNRAS, 217, 717

Beijing Observatory Metre-Wave Radio Astronomy Group 1985, ACTA ASTROPHYSICA SINICA, 5, 245

Ficarra, A., Grueff, G., & Tomassetti, G. 1985, A&AS, 59, 255

Laing, R.A., & Peacock, J.A. 1980, MNRAS, 190, 903

Riley, J.M. 1988, MNRAS, 233, 225

Roger, R.S., Bridle, A.H., & Costain, C.H. 1973, AJ, 78, 1030

Wang Shouguan 1984, Publications of Beijing Astronomical Observatory, 1

Astronomical Data Analysis Software and Systems II
ASP Conference Series, Vol. 52, 1993
R. J. Hanisch, R. J. V. Brissenden, and J. Barnes, eds.

Wide-Field Direct CCD Observations Supporting the Astro-1 Ultraviolet Imaging Telescope

Eric P. Smith[1], Paul Hintzen[1,2], K.-P. Cheng[1,3]

Laboratory for Astronomy & Solar Physics, NASA-Goddard Space Flight Center, Greenbelt, MD 20771

Ronald Angione, Freddie Talbert

Department of Astronomy, San Diego State University, San Diego, CA 92182

1. Summary

We are obtaining wide field direct CCD observations to complement the ultraviolet (1200 Å$< \lambda <$ 3000 Å) images provided by Astro's Ultraviolet Imaging Telescope (UIT) during a Space Shuttle flight in December 1990. By the end of 1992, ground-based observations should have been acquired for virtually all of the Astro-1 UIT fields. Because of the wide variety of projects addressed by UIT, the fields observed include Galactic supernova remnants and globular clusters, the Magellanic Clouds, M33, M81, and other galaxies in the Local Group, and rich clusters of galaxies, principally the Perseus cluster and Abell 1367. The optical images allow identification of individual UV sources in each field and provide the long baseline in wavelength necessary for accurate analysis of UV-bright sources. To facilitate use of our optical images for analysis of UIT data and other projects, we plan to archive them, with the UIT images, at the National Space Science Data Center (NSSDC), where they will be universally accessible via anonymous FTP.

2. The Ultraviolet Imaging Telescope

The UIT, one of three telescopes comprising the Astro spacecraft, is a 38-cm f/9 Ritchey-Chretien telescope on which high quantum efficiency, solar-blind image tubes are used to record UV images on photographic film (OII-A). Five filters with passbands centered between 1250Å and 2500Å provide both UV colors and a measurement of extinction via the 2200Å dust feature. The resulting calibrated UV pictures are 40 arcminutes in diameter at 3.5 arcseconds resolution. The

[1]Visiting Astronomer, Cerro Tololo Inter-American Observatory and Kitt Peak National Observatory. CTIO and KPNO are operated by AURA, Inc. under contract to the National Science Foundation.

[2]Department of Physics, University of Nevada, Las Vegas, NV

[3]National Research Council Postdoctoral Fellow

capabilities of UIT therefore complement HST's WFPC: the latter has 40 times greater collecting area, while UIT's usable field has 170 times WFPC's field area.

3. Mission Science

The UIT obtained data for a wide variety of scientific projects. Low redshift galaxies were major targets for UIT, as study of their OB associations should yield fundamental information on the initial mass function for massive star formation, the resulting supernova rates, and ionizing radiation fields. At larger distances, UIT surface brightness observations are similarly being used to study the initial mass functions for star formation in a sample of nearby galaxies, since integrated UV colors and color profiles provide the most sensitive available measure of the formation rate of massive stars. The blue stages later in stellar evolution are also being studied (e.g., UIT observations at 1500Å detected 1300 hot horizontal branch stars (HHB) in Omega Cen). Such a large statistical sample is necessary to place constraints on mass loss during the red giant phase, which is probably the dominant factor determining the future evolution of these stars as well as a central issue in the chemical enrichment of the interstellar medium. These data will also allow more accurate determination of the helium abundances and metallicities of HHB stars.

In all of these projects, supporting ground-based data are critical. Depending on the type of object, optical images have been obtained in some combination of filters which may include U, B, V, R and/or narrow-band Hα and [OIII] λ5007Å (Table 1 is a current listing of the ground database). A mere three years ago, only photographic plates would have allowed coverage of the 40 arcminute diameter UIT fields. However, the fortuitous development of a new generation of large-format CCDs has allowed us to acquire optical observations with seeing-limited resolution and linear flux response while still covering a substantial fraction of each UIT frame with one to four integrations. Objects accessible from the Northern Hemisphere were observed from Kitt Peak National Observatory or San Diego State University's Mount Laguna Observatory; their complementary observing seasons allow year-round coverage. Objects accessible only from the Southern Hemisphere, including the Magellanic Clouds, were observed from Cerro Tololo InterAmerican Observatory. Data from the National Observatories were taken with Tek 2048 CCDs on 0.9-meter telescopes, yielding seeing-limited resolution over fields of 20+ arcminutes (KPNO) and 14 arcminutes (CTIO). At Mount Laguna Observatory, a focal reducer was used in conjunction with a TI 800 CCD, providing fields of 11 to 18 arcminutes, depending on the field lens used. All data were reduced using standard IRAF routines. The reduced, photometrically calibrated data, in FITS format, will be distributed by the NSSDC to interested investigators upon request. This effort marks the first time that NASA will provide a complementary ground-based data set matched to its space-based observations. This combined data set should allow a wider variety of scientific problems to be addressed with the UIT observations.

Table 1. Ground Support Database

Object Name	Observatory	Observing Run Date	Filters
3C273	KPNO/ CTIO	30-May-90 / 31-Mar-91	R
Abell 1367	KPNO	9-Mar-91	B, R
Abell 665	KPNO	8-Mar-91	B, V
Crab Nebula	KPNO	11-Mar-91	U, B, V, Hα, 5007Å
Cygnus Loop	MLO	10-Jun-91	R, 5007Å
G191B2B	KPNO	9-Mar-91	B, R
G70D8247 GRW	KPNO	12-Mar-91	B
HZ43	KPNO	1-Jun-90	R
M3	MLO	19-Feb-91	R
M3	KPNO	9-Mar-91	B, V, R
M13	KPNO / MLO	1-Jun-90 / 18-Feb-91	B, R
M31 nucleus	MLO	7-Oct-91	B
M31 (8' E 8' N)	MLO	7-Oct-91	B
M31 (8' W 8' N)	MLO	7-Oct-91	B, V
M31 (8' S 8' W)	MLO	7-Oct-91	B, V
M31 (8' S 8' E)	MLO	8-Oct-91	B, V
M31 - NE	KPNO	2-Jan-92	B, R, Hα
M31 - SW	KPNO	2-Jan-92	U, B, R, Hα
M33 - NE & NW	KPNO	1-Jan-92	R, Hα
M33 - SE & SW	KPNO	1-Jan-92	R, Hα
M74	MLO/ CTIO	8-Oct-91 / 6-Dec-91	B, V, R, Hα
M79	MLO	19-Feb-91	B, R
M79	KPNO	8-Mar-91	B, V, R
M81	MLO	20-Feb-91	B, V, R
M81	KPNO	7-Mar-91	B, V, R, Hα
M82	MLO	19-Feb-91	B, V, R,
M82	KPNO	8-Mar-91	B, V, R, Hα
M82	MLO	12-May-91	V, R, Hα
M82	KPNO	2-Jan-92	U, B, R, Hα
M87	MLO	18-Feb-91 & 10-May-91	B, V, R Hα
M100	MLO	19-Feb-91	B, R
M100	KPNO	11-Mar-91	B, V, R, Hα
M100	MLO	12-May-91	V, R, Hα
M100	KPNO	2-Jan-92	U, R, Hα
NGC 891	MLO	23-Aug-92	R
NGC 1068	MLO / KPNO	20-Feb-91 / 11-Mar-91	B, R, Hα
NGC 1316/17	CTIO	9-Dec-91	B V, R, 6600Å
NGC 1399	MLO	19-Feb-91	B, R,
NGC 1850	CTIO	8-Dec-91	U, B, V
NGC 1851	CTIO	6-Dec-91	U, B, V, R
NGC 1851 160" East	CTIO	9-Dec-91	B, V
NGC 1851 320" East	CTIO	9-Dec-91	B, V
NGC 2146	KPNO	8-Mar-91	B, R
NGC 2146	KPNO	31-Dec-91	B, R, Hα
NGC 2992	KPNO	1-Jan-92	B, R, 6619Å
NGC 2992/3	CTIO	8-Dec-91	B, V, R, 6600Å
NGC 3486	MLO	19-Feb-91	B, V, R
NGC 4151	KPNO	1-Jun-90	R
NGC 4151	MLO	19-Feb-91	B, R
NGC 4151	KPNO	7-Mar-91	B, V, R, Hα
NGC 4639 sn	MLO	18-Feb-91	B, V, R
NGC 4639 sn	MLO	11-May-91	B
NGC 5474	MLO	11-May-91	B, V, R, I
NGC 7023	MLO	23-Aug-92	B,V,R
Omega Centauri	CTIO	31-Mar-91	U, Strömgren u
Perseus Cluster	KPNO	7-Mar-91	U, B, V, R
Perseus - East & West	KPNO	1-Jan-92	B, R, Hα
Q1700+642	KPNO	8-Mar-91	B, R
Q1821+642	KPNO / MLO	7-Mar-91 / 9-Jun-91	B, R
SN 1987a	CTIO	31-Mar-91	B, R
U-GEM HD64511	KPNO	8-Mar-91	B, R
UGC 6697	KPNO	1-Jan-92	U, B, R
UX-UMA	KPNO / MLO	7-Mar-91 / 10-Jun-91	B, R
Z-CAM	KPNO	8-Mar-91	B, R

Astronomical Data Analysis Software and Systems II
ASP Conference Series, Vol. 52, 1993
R. J. Hanisch, R. J. V. Brissenden, and J. Barnes, eds.

STARBASE: Database Software for the Automated Plate Scanner

S. C. Odewahn, R. M. Humphreys, and P. Thurmes

Astronomy Department, University of Minnesota, 116 Church St., Minneapolis, MN 55455

Abstract. The Automated Plate Scanner (APS) of the University of Minnesota is being used to scan and digitize the first epoch Palomar Sky Survey. The resultant database will be used to produce a catalog of approximately a billion stars and several million galaxies. We describe the ongoing development of a dedicated APS database management system which will be made available to the astronomical community via Internet.

1. Introduction

The Automated Plate Scanner (APS) of the University of Minnesota, a unique high speed "flying spot" laser scanner, is currently being used to scan and digitize the 936 O and E plate pairs of the first epoch Palomar Sky Survey. The resultant database will be used to produce a catalog of approximately a billion stars and several million galaxies. A specialized DBMS called STARBASE has been written to provide fast access to the hundreds of millions of images collected by the APS. This system provides an initial reduction mode for parameterizing APS images and classifying image types using a novel set of neural network image classifiers. A second analysis mode, which will be that commonly used by the general user, provides for searches of the database which may be constrained by any combination of physical and positional parameters. Through the use of two dimensional pointer hash tables, the system has been optimized for extremely fast positional searches. In addition to fast data retrieval, the system provides a graphical interface for displaying scatter plots or histograms of the collected data. In addition, a specialized image display system is being developed to allow the user to view densitometric data for all objects classified as extended by the neural network system. Finally, STARBASE will have a flexible programmable interface which allows other programs to access information in the database. This will allow users to write applications suited to their particular needs to process APS data.

2. A Summary of the APS POSS Scanning Project

Originally designed as a proper motion measuring machine, the APS has been redeveloped using modern electronics and computers to produce a very flexible scientific instrument. This device uses a rotating prism in combination with a beam splitter to produce a pair of spots which scan across a pair of plates.

Because of this unique design, the APS is able to scan and digitize two 14 inch square Schmidt fields simultaneously in about 6 hours. Using a series of low resolution background scans and high resolution "forward" scans, the device utilizes a thresholding mode to detect objects which lie above some user specified plate transmittance level. Pixel data are collected for all regions of the scan detected above this threshold level. All images are reconstructed and parameterized during the scanning process by software running on both a Sun Sparc 10 and a Silicon Graphics R-4D220S.

The APS group has just completed scanning the glass copies of the blue and red (O and E) plates in the Palomar Observatory–National Geographic Sky Survey (POSS I). All of the detected images are being cataloged with positions, isophotal shapes, isophotal diameters, magnitudes and colors. The stellar and non-stellar objects are separated using a set of neural network image classifiers (Odewahn et al. 1992). Densitometric data will be collected for all extended objects. We estimate that there will be nearly a billion stellar objects and image data for several million galaxies when the project is completed. **The final goal of this work is a database, including the cataloged objects and the pixel data for the non stellar images, which will be made available to the astronomical community on-line via Internet.** The APS Catalog of the POSS I will complement other existing or planned sky-survey size catalogs such as the Guide Star Catalog (Lasker et al. 1990). We summarize the properties of our catalog in Table 1.

Table 1. APS POSS I Catalog Characteristics

Stars Catalog:	
	10^9 records per color
	Magnitude range: 12-22 in O, 12-20.7 in E
	Matched (O and E) images only
	Astrometry accurate to 0.4 arcseconds
	Photometry accurate to 0.1 to 0.2 mag
Galaxies Catalog:	
	Nx10^6 records per color
	Integrated isophotal magnitudes: 12-20 in O, 12-19.5 in E
	Matched (O and E) images only
	Photometry accurate to 0.3 to 0.5 mag

3. A Summary of STARBASE

STARBASE is a custom database engine which is being uniquely tuned to the requirements of the APS catalog. The package was designed and written at the University of Minnesota by E. B. Stockwell under the supervision of R. L. Pennington. Currently, the code is being maintained and expanded by N. Kavuri and R. Mavuduru. The three major requirements that this engine had to fulfill were portability, minimized storage requirements, and the ability to treat the full set of plates as a single entity. It is portable to any 32 bit integer machine and is written to conform to the ANSI C standard. Each record in the database (1 record per object per color) is bit-packed into 45 bytes. This contains all of the parameterized information from the APS for an image, transmittance information, indices to the same object on other plates, and a pointer to an extended field that may contain other information (references to other catalogs, etc.). An image which has been classified as extended by the neural network classifier will also include pixel information in the extended field.

Each color in the survey will comprise a database of 45 GB. Searching this by brute force will require prohibitive amounts of time. Search times have been greatly reduced by producing hash lists for each searchable field and two-dimensional search trees for the positional information. Search times are generally seconds or minutes for reasonably sized searches.

STARBASE may be used interactively in two different ways. First, the user may enter direct database access commands to produce ASCII tables of data for records that meet the user imposed search criteria. In the second mode, the data are accessed and manipulated through display and graphics programs which present the data visually. We are currently developing a graphical user interface to allow the user to view the densitometric data for extended objects selected during a STARBASE search. A fast two-dimensional graphics package will also be provided for producing scatter plot diagrams and histograms. The user will be able to use a cursor to graphically select any of the plotted data and produce subset files for more detailed analysis. Using these Xwindows-based graphics tools, the astronomer is able to get a quick look at the characteristics of the selected data before generating the final data file(s) to be returned to the home institution by ftp.

4. Operation of STARBASE

The variables presently operated on by STARBASE are the following:

xc, yc This is the (x,y) position of the image center on the plate, measured in eres (1 ere=0.366 microns).

ra,dec Right-ascension and declination are computed, if the necessary astrometry has been done when the database is constructed.

theta Angle of rotation of the ellipse that is fitted to the image. Theta ranges from 0.0 to 179.9.

dia Image diameter, measured in eres, of a circle with the same area as the ellipse that fits the image.

mag Integrated magnitude derived through diameter relation.

ell Ellipticity of the image $(1 - b/a)$.

fuz RMS error in the fit of the ellipse to the raw transit endpoints.

jitter A measure of the error in the X,Y centroid position.

avg_in Average pixel transmittance in the image.

cen_in Transmittance value of the central pixel.

class Image classes derived by the neural network image classifier. This field will have one of the following values: GALAXY, OTHER, STAR, UNKNOWN. OTHER is for images that are not stars or galaxies, but that have been classified.

starnum Plate relative absolute starnumber.

Other fields, such as **color index** will be available in the near future. The user specifies the type of output to be generated using the **report** command:

report *report_type report_options* [[file=]*filename*]

This command specifies how the output will be formatted. The actual reports are generated with the **get** command. Using the file option enables the output to be direct to a disk file.

A tabular output is specified using the **column** type:

report column [*switch fieldname ...*] [[file=]*filename*]

This generates a column formatted report, suitable for manipulation with Unix text processing tools, such as awk, and sort.

A plot is generated with the plot tool using the same command syntax:

report plot [*fieldname fieldname*] [[file=]*filename*]

The basic command used in STARBASE to apply user selection criteria is the **get** command. This does the real work. It searches the database for all records that match the given expression. A report is generated in the format set by the most recent report command. The following are examples of expressions using get:

```
get dia > 34.9
get ell < 0.5 and ell > 0.45
get class = galaxy
get (class = unknown and ell > 0.5) or class = galaxy
get dia > 200 and ell < 0.5 and ell > 0.45
get plate = O_1393 and (stripe = 3 or stripe = 2)
get stripe != 4 and stripe != 5
get not plate = O_1393
get ((ell > 0.3) or ((dia > 0.9) and (dia < 0.96)))
```

Values may be compared with the operators: =, >, <, !=. Expressions may be joined in any combination with logical operators **and**, **or**, **not**. When constructing complex expressions, parentheses are used to make the order of evaluation explicit. Any of the fields listed above may be used in these expressions.

References

Odewahn, S.C., Stockwell, E.B., Pennington, R.M., Humphreys, R.M., & Zumach, W.A. 1992, AJ, 103, 318

Lasker, B.M., Sturch, C.R., McLean, B.J., Russell, J.L., Jenkner, H., & Shara, M.M. 1990, AJ, 99, 2019

Astronomical Data Analysis Software and Systems II
ASP Conference Series, Vol. 52, 1993
R. J. Hanisch, R. J. V. Brissenden, and J. Barnes, eds.

SKICAT: A Cataloging and Analysis Tool for Wide Field Imaging Surveys

Nicholas Weir

Palomar Observatory, California Institute of Technology, 105-24, Pasadena, CA 91125

Usama M. Fayyad

Jet Propulsion Laboratory, California Institute of Technology, 525-3660, Pasadena, CA 91109

S. Djorgovski

Palomar Observatory, California Institute of Technology, 105-24, Pasadena, CA 91125

Joseph C. Roden, Nicolas Rouquette

Jet Propulsion Laboratory, California Institute of Technology, 525-3660, Pasadena, CA 91109

Abstract. We describe the design and applications of a software system for analyzing the digital scans of the Second Palomar Observatory Sky Survey (POSS-II). The system (SKICAT) integrates new and existing packages for performing the full spectrum of tasks from raw pixel processing, to object classification, to the photometric matching of multiple Schmidt plates. The system also provides a variety of tools for interactively querying and analyzing the resulting object catalogs. Portions of SKICAT are directly applicable to digital sky surveys of the near future and large-format CCD imagery of today.

1. Introduction

The Palomar – ST ScI Digital Sky Survey, a digitization of the photographic Second Palomar Observatory Sky Survey (POSS-II), will amount to nearly three terabytes of imagery data. The traditional means of extracting useful information from such surveys is through the construction of object catalogs. Thanks to developments in the fields of pattern recognition and machine learning, it is possible to reliably construct such catalogs objectively and automatically.

Caltech Astronomy and the JPL Artificial Intelligence Group are engaged in an effort to integrate state-of-the-art computing methods for the scientific utilization of the digitized POSS-II. In this paper, we describe the system we are producing (SKICAT, Sky Image Cataloging and Analysis Tool) for constructing and analyzing classified object catalogs of uniform calibration and quality down to the faintest limits of the survey.

The POSS-II will consist of almost 3,000 photographic plates covering the entire northern sky in the green, red, and near-infrared passbands (J, F, and N). The limiting magnitudes for point sources are typically $B_J \sim 22.5$, $R_F \sim 20.8$, and $I_N \sim 19.5$ (Reid et al. 1991). They are being digitized at ST ScI with 1-arcsec sampling, producing $23{,}040 \times 23{,}040$ pixel images at 2 bytes/pixel; i.e., 1 GB/plate (Reid & Djorgovski 1992). ST ScI also provides an astrometric solution with approximately one arcsec rms accuracy. The quality of the plates and scans is such that we expect to be able to accurately identify over 2×10^7 galaxies in the survey, and over 2×10^8 stars, including over 10^5 quasars.

2. The SKICAT System

The SKICAT system, currently running on a Sparcstation II under SunOS, integrates four essential tasks: image processing, object classification, catalog merging, and catalog analysis. Individual plate scans, as well as CCD calibration sequences, are processed from pixels into catalog form fully automatically. The system applies a modified version of the FOCAS package (Jarvis & Tyson 1979) for object detection and parameter measurement, and the SAS database library for catalog maintenance. Database management will soon be accomplished using Sybase. Objects are classified as stars, galaxies, and artifacts using a machine-trained decision tree method described below. SKICAT merges plate and CCD catalogs into a master (whole sky) catalog in which multiple images of the same source are matched together. The CCD catalogs, in addition to the overlap regions between plates, are in turn used for calibrating both the photometry and object classifications within the master catalog. The same query and output tool used in performing these calibrations may be used for a variety of scientific analyses of the catalogs.

2.1. Image Processing

The most domain specific aspects of SKICAT, like any survey analysis system, are those which manipulate the data in its most primitive form. A digitized POSS-II Schmidt plate image is provided as a 23×23 matrix of sub-rasters. These data are readily converted from densities to intensities using a polynomial fit to the sensitometry spots in the Southwest corner of each plate. To improve processing efficiency and map out field effects in the point spread function (PSF), we analyze each plate as a set of 13×13 overlapping "footprint" images. Each footprint is 2048^2 in size, with a minimum overlap between adjacent footprints of 4.5 arcmin. This large overlap allows all but the largest objects to be reliably measured in this piecemeal fashion, while providing accurate statistics on certain measurement errors. Our compilation of the matching statistics using a variety of consistency checks confirms that the errors introduced by the footprint gridding process are far below the measurement errors resulting from noise in the data.

The footprint sky is initially estimated by binning the image into blocks of 64^2 pixels each, accumulating the median and quartile sigma for each block, then accumulating the median and quartile sigma for all of the block measurements. This process allows for relatively accurate initial sky and sky sigma estimation even in the presence of bright sources. The footprint image is divided by a bilinearly interpolated version of this median sky to help correct for differential sensitivity due to vignetting and emulsion variations. Image quality control checks, which test for scanning artifacts, are also applied at this stage.

Next, FOCAS routines provide a more refined sky estimate used for detecting objects. Objects are defined as contiguous pixels a specified threshold above the sky. Subsequent FOCAS routines estimate the sky value for each individual object by sampling an annulus around its border, measuring a variety of parameters for each object, and splitting the objects into constituent components. The result of these steps is a catalog of primary object parameters for virtually every detectable object in the footprint.

An additional set of object parameters is derived by automatically estimating and subtracting the "stellar locus" from the parameters m_{core}, the magnitude of the brightest 3×3 pixel region, of total intensity L_{core}; $\log Area$, the log of the isophotal area; $ir1$, the intensity weighted first moment radius; and S, where $S = Area/\log[L_{core}/(9 \times ispht)]$. The stellar locus is the attribute value as a function of magnitude around which point sources are fairly narrowly distributed, at least at brighter magnitudes. We have found that the resulting revised attributes are relatively insensitive to footprint-to-footprint, and even plate-to-plate, variations, and are thus robust parameters for use in object classification.

In order to derive even more powerful classification parameters, we must form an empirical estimate of the PSF for each footprint. Along with magnitude and ellipticity, the four revised attributes described above are combined to form a six dimensional parameter space in which star/galaxy separation is automatically performed. We have applied a decision tree algorithm, described in the following section, to this task. In the magnitude range B_J = 16–20, our automatic star-selection classifier has demonstrated an error rate of less that two percent, providing a very accurate composite estimate of the PSF for the footprint. Using the PSF template, the FOCAS resolution routine (Valdes 1982) determines the best-fitting 'scale' and 'fraction' values, representing the optimal parametric fit of the PSF to each object. Along with the revised attributes, these parameters form a very powerful space in which to perform object classification.

On a row by row basis, the individual footprint catalogs are merged into the plate catalog. Numerous overlap statistics concerning matches and parameter measurements are automatically monitored and saved to assure the quality of the final catalog. A list of unmeasured 'border' objects, which are not fully measured within any single footprint, is also maintained.

2.2. Object Classification

The goal in computer-based object classification is, given a list of attribute values, automatically provide the probability of that object's membership in a list of classes. The purpose of machine learning in this task is, given a training set of object attributes and classes, automatically infer a near-optimal set of classification rules from the data. By using historical, human generated, classification methods to provide derived object attributes, the computer can benefit from past experience before embarking on its own search for new classification rules. Encoding these methods in the form of the revised object attributes described above, as well as computing the FOCAS scale and fraction values, was therefore a very important part of our effort to automate POSS-II object classification.

The machine learning techniques we have employed are the GID3* and O-BTree decision tree induction algorithms developed by Fayyad (1991, and see Fayyad et al. 1992). Unlike neural networks, which we have also experimented with, these algorithms produce output in the form of decision trees, which are readily understood lists of classification rules, rather than obscure weights. As we have found the two methods to provide virtually the same error rate, we have opted for the decision tree approach. We have also used a technique for statistically combining sets of independently produced trees to form even more robust and accurate lists of classification rules.

Our first classification problem, described in the image processing section above, involved developing a robust means of automatically selecting a set of "sure thing" stars within a footprint. These were needed to form an empirical estimate of the PSF, which is in turn used by FOCAS to derive an object's scale and fraction parameters. For this task, we trained the machine learning algorithms on a data set of objects from multiple plates, each classified by eye. The induction algorithms produced decision trees which used magnitude, ellipticity, and the four revised attributes to select virtually all relatively bright stars from a footprint catalog.

The final classification of all objects is accomplished after the footprint catalogs have been merged into a plate catalog. The same six attributes used above, plus scale and fraction, are employed for this task. For the preliminary tests reported here, the decision tree algorithms were trained using a data set of over a thousand objects which one of us (NW) classified by eye from the digitized scans of two J plates. Subsequent comparison with several hundred exact classifications provided by CCD data indicated an error rate of $< 5\%$ in the list classified by eye. By applying our derived classification rules to a comparably large set of test data from the same two plates, we obtained an error rate of less than 7% down to a B_J limit of 21.0. The object classes were restricted simply to stars and galaxies. These results indicate our ability to achieve very high rates of reliable classification down to between 1.0 and 1.5 magnitudes of our estimated point source detection limit. In comparison, the APM catalog, derived from UKST J plates, achieved comparable levels of classification accuracy a magnitude brighter than our limit, with galaxy incompleteness and contamination sharply increasing thereafter (Maddox et al. 1990). We therefore expect to obtain at least twice the density of classified galaxies in our catalog relative to theirs.

2.3. Catalog Merging

Plate and CCD catalogs are matched and merged into a master object catalog using the third major component of SKICAT. A master catalog may be created and appended to using any number of input catalogs. Objects are matched according to their RA and Dec by finding the nearest non-matched object within some maximum radius. Separate entries are maintained for matched objects appearing in multiple catalogs, with indices pointing to their input source and other objects to which they were matched. The user chooses which subset of object parameters are loaded into the master catalog for later retrieval. In the future, when each individual plate and CCD catalog may also be stored on-line, the master catalog will primarily serve as an index to these constituent data sets, where the full list of object attributes will always be available.

2.4. Catalog Analysis

The fourth SKICAT component facilitates analysis of the various catalogs. Using an *X* window tool, the user can graphically construct complicated queries of individual or master catalogs, outputting any number of original or derived object parameters to a text file for subsequent processing, plotting, etc. The same query/output mechanism may be driven using a text-based specification file for batch processing. For example, the user, or a user-defined program, can readily select all objects within the overlap of multiple plates and/or CCDs and derive offsets for calibrating the master catalog's photometry. A database of magnitude corrections for each plate may thus be created and, in turn, used to produce a new object parameter, such as a calibrated magnitude. Such derived parameters are defined within SKICAT using user-supplied text files containing standard C code with references to database attributes as variables.

The same query/output mechanism may be used not just for calibration and quality assessment, but for extracting information from the survey catalog for most scientific objectives imaginable. The user simply employs SKICAT to filter the data set appropriately, then uses her own tools for manipulating the output.

3. Future Work

While SKICAT was designed expressly for analyzing POSS-II, every effort was made to design the system as generally as possible. In turn, all but the most basic image processing tasks should be readily adaptable to a variety of image analysis needs. In the short term, we plan to extend SKICAT's catalog analysis capabilities to provide the end-user with a variety of statistical analysis tools, including a user-friendly interface to the machine learning algorithms we have applied, as well as a selection of multivariate analysis and unsupervised classification methods.

Acknowledgments. The POSS-II is funded by grants from the Eastman Kodak Company, The National Geographic Society, The Samuel Oschin Foundation, NSF Grants AST 84-08225 and AST 87-19465, and NASA Grants NGL 05002140 and NAGW 1710. This work was supported in part by a NSF graduate fellowship (NW), the Caltech President's Fund, NASA Contract NAS5-31348, and the NSF PYI Award AST-9157412 (SGD). The work described in this paper was carried out in part by the Jet Propulsion Laboratory, California Institute of Technology, under a contract with the National Aeronautics and Space Administration.

References

Fayyad, U. 1991, PhD Dissertation, EECS Dept. The University of Michigan

Fayyad, U., Doyle, R., Weir, N., & Djorgovski, S. 1992, in Proceedings of the Machine Discovery Workshop, Ninth International Conference on Machine Learning, Aberdeen, Scotland

Jarvis, J.F., & Tyson, A.J. 1979, in SPIE Proc. Instrumentation in Astronomy III, 172, 422

Maddox, S.J., Sutherland, W.J., Efstathiou, G., & Loveday, J. 1990, MNRAS, 243, 692

Reid, I. N., et al. 1991, PASP, 103, 661

Reid, I. N., & Djorgovski, S. 1993, in Sky Surveys: Protostars to Protogalaxies, A.S.P. Conf. Ser., ed. B.T. Soifer, in press.

Valdes, F. 1982, in SPIE Proc. Instrumentation in Astronomy IV, 331, 465

Discussion

Shaw: Do you find cases where an object's classification changes with color?

Weir: We have not yet looked for systematic variations in object classification as a function of image passband. Eventually, we intend to take measurements of matched objects on difference color plates to train a classifier which uses the color information and redundancy to provide even more accurate results.

Kurtz: What is the performance advantage of using your method versus using just the standard FOCAS resolution classifier and FINDSTARS?

Weir: The FOCAS FINDSTARS script is unable to produce uniformly adequate lists of PSF star candidates because of large deviations in the stellar locus within and between plates. The use of multiple revised attributes and the machine trained classifier solves this problem. While we have found that the FOCAS fraction and scale parameters carry the majority of the weight in determining final object classifications, we estimate that the use of additional attributes and machine-learned classification rules extend our 90% completeness/10% contamination limit by 0.25 to 0.5 magnitude relative to the standard FOCAS resolution classifier.

M. Davis: Will you include information in the final catalogue on the relative reliability of the star/galaxy classification (solid ... very uncertain)?

Weir: We intend to include such an estimate.

Pier: When does the machine learning stop? You want to use a uniform set of criteria so that your catalog is homogeneous, but there are tremendous difference (background fog, seeing, zenith distance/differential refraction, etc.) and plates.

Weir: A great deal of our research effort involved determining a parameter space within which stars and galaxies are homogeneously distributed for all plates and for all positions within plates. We have found that the four revised attributes described in the text, combined with ellipticity, quasi-calibrated magnitude, and FOCAS resolution and scale comprise such an attribute space. Previous tests have shown that when we train our classifier using these measurements for objects from different locations on a set of plates, the classifier performs with nearly identical accuracy on test objects from the same plates or completely independent ones.

Astronomical Data Analysis Software and Systems II
ASP Conference Series, Vol. 52, 1993
R. J. Hanisch, R. J. V. Brissenden, and J. Barnes, eds.

SIMBAD Quality-Control

Soizick Lesteven
Centre de données astronomiques de Strasbourg, Observatoire astronomique, F-67000 Strasbourg (lesteven@simbad.u-strasbg.fr)

Abstract. Taking into consideration the amount and the complexity of SIMBAD data, it is necessary to use automatic methods to control and assure the quality of the SIMBAD database. One possibility is to apply multivariate data analysis to the content of documents related to astronomical data. The method and first results are presented.

1. Introduction

The SIMBAD astronomical database, developed at the Centre de Données astronomiques de Strasbourg, presently contains 1,000,000 objects (stellar and non stellar). It has the unique characteristic of being specifically structured for astronomical objects. All types of heterogeneous data (bibliographic references, measurements and sets of identification) are connected with each object. The attributes that define the quality of the database include:

- Reliability – cross-identification should not rely upon only exact values for object coordinates. It also means that information attached to one single object should be consistent. We have to control the existing data in order to start with a reliable base and to cross-identify new data ensuring the quality as the database grows.

- Exhaustivity – delays between publications of new information and their inclusion in the database should be as short as possible. We have to maintain the integrity of the database as data accumulate.

Taking into consideration the amount of data and the rate of new data production, it is necessary to use automatic methods. The data are complex and heterogeneous in structure, content and in relation to other components of the database. To analyze the astronomical data (identifiers, measurements), it is possible to use an expert system. We need to define a set of suitable data representations in order to structure the astronomer's knowledge and to set up efficient algorithmic methods to analyze the database in the best possible way. To analyze the bibliography related to astronomical data, multivariate data analysis seems to be more appropriate.

2. Methods

One of the possibilities in determining the content of a document is to use multivariate data analysis. Factor Space (Ossorio 1965) is an n-dimensional relevancy space described by n-axes representing a set of n subject matter headings. Words and phrases can be used to scale the axes, and documents are then a vector average of the terms within them. These relevancy scores may be obtained either directly, using human judgement (Ossorio 1965), or via automated evaluation of classified collections of documents using statistical analysis (Kurtz 1992).

Currently, we simplify the Factor Space by using keywords instead of subject matter headings. Each document is defined by a list of keywords and can then be located within the n-dimensional space. On this set of data, we can apply a Principal Component Analysis (PCA). The object of our PCA is to take the keywords as variables and find combinations of these to produce new variables that are uncorrelated. The lack of correlation means that the variables measure different dimensions in the data. The meaning of the axes are given by the combination of keywords. The new variables are also ordered, the first one displays the largest amount of variation. This is a way to reduce the large number of variables to a small number of transformed variables without losing information.

Our research is presently based on the NASA-STI bibliographic database. The selected data concern astronomy, astrophysics and space radiation (102,963 references from 1975 to 1991 including 8070 keywords). References are described by the title, the authors, the journal, a list of keywords, the publication year, the abstract and other information. Our Factor Space is built from these bibliographic data.

By computing the distances between documents in the n-space, we have a tool for information retrieval. Given a position in the space (or a single document) we can find other documents clustered around it.

3. Results

For our first application, we selected a small data set consisting of the astronomy and astrophysics references from 1989. This sample contains 5915 documents and 463 keywords.

Applying the PCA on this sample, we obtained new variables that defined the n-axes of our space. This application is not straightforward, the first principal component only accounts for about 2% of the variation in the data. 200 components are needed to account for 82% of the variation.

- The first component is rather meaningful in terms of the object types. Most of the keywords concerning stars and their environment are grouped with low values, then we have a group of keywords describing astrophysics in general, then the keywords concerning galaxies, universe and cosmology.
- The second component discriminates stellar features (negative values) from interstellar matter (positive values).
- The third component represents methods. It sets observational methods against modelisation.

Each document can be located in the n-dimensional space. We computed distances between documents in the 30 first dimensions and the clusters around a point (or a document) are meaningful. These documents had common subject matter but not necessarily the same keywords.

4. Future Work

Future work will concentrate on the following issues.

From 1983 to 1991 there is approximately 40% overlap between the SIMBAD bibliography and the NASA-STI bibliography. Furthermore, searching the SIMBAD objects associated with a bibliographic reference, it is possible to position each physical measurement of these SIMBAD objects in the space defined by the NASA-STI keywords. The physical properties of stars (e.g., their UBV colors) are not randomly distributed. Different types of stars have measures which cluster in different regions of a physical parameter space. We will show that there is a relation between these clusters in a space of physical measurements and the literature concerning these objects clustered in a Factor Space. We will investigate the nature of the relationship between the physical and the literature spaces.

By using a thesaurus and words extracted from abstracts instead of keywords to build the Factor Space, we minimize subjective influences thereby improving this information retrieval method.

By comparing the Factor Space positions obtained from the NASA-STI keywords with the Factor Space positions obtained from the SIMBAD titles, we will be able to check whether it is possible to retrieve information using a restricted set of words only. If the comparison is valid, this will be a way to use bibliographic information in the SIMBAD quality control process. This comparison could perhaps differentiate between information from titles and abstracts.

Acknowledgments. It is a pleasure to thank Dr. M. Crézé and Dr. M. Kurtz for valuable discussions, Dr. M. Kurtz for providing access to the NASA-STI bibliographic data, and the ADASS '92 Organizing committee. This work is supported by the CNES.

References

Egret, D., Wenger, M., & Dubois, P. 1991, in Databases and On-line Data in Astronomy, eds. D. Egret & M. Albrecht, 77

Kurtz, M.J. 1992, Content Based Document Indexing, On-line Astronomy Documentation and literature, eds. F. Giovane & C. Pilachowski, NASA conference, proc., in press.

Ossorio, P.G. 1965, Classification space : a multivariate procedure for automatic document indexing and retrieval, J. Multivariate Behavioral Research 2, 479

Part 1. Databases, Catalogs, and Archives

Section B. Space-Based Data

Astronomical Data Analysis Software and Systems II
ASP Conference Series, Vol. 52, 1993
R. J. Hanisch, R. J. V. Brissenden, and J. Barnes, eds.

The Cool-Star Spectral Catalog: A Uniform Collection of IUE SWP-LOs

T. Ayres, D. Lenz, R. Burton

Center for Astrophysics & Space Astronomy, University of Colorado, Boulder, CO 80309

J. Bennett

Astrophysics Division, National Aeronautics & Space Administration, Washington, DC 20546

Abstract. We have assembled an extensive electronically-accessible catalog of low-dispersion far-ultraviolet spectra of chromospheric emission-line stars observed with the *International Ultraviolet Explorer.*

1. Introduction

Over the past decade and a half of its operations, the IUE satellite has recorded low-dispersion (5 Å resolution) spectrograms in the 1150–2000 Å far-ultraviolet band of more than 800 stars of late spectral type (F–M). The FUV contains a number of emission lines—like O I λ1304, C II λ1335, and C IV λ1549—that are key diagnostics of physical conditions in the high-excitation chromospheres and subcoronal "transition zones" of such stars. Many of the sources have been observed a number of times, and the available collection of SWP-LO exposures in the IUE Archives exceeds 4,000.

With support from the Astrophysics Data Program, we have assembled a subset of the archival material into a catalog. Our goal was to process and measure the FUV spectra in as uniform (and automated) a way as possible, and provide the results in a form amenable to electronic access through the Astrophysics Data System. A brief summary of our efforts follows. A more thorough report will be presented elsewhere (Ayres et al. 1993a).

2. The Sample

We focused on a group of approximately 600 relatively normal F–K stars, that were detected in at least the 12μm survey band of the *Infrared Astronomical Satellite.* The brightness-limited sample is represented by approximately 3,500 SWP-LO images in the IUE archives. We ordered the SWP raw images from the National Space Sciences Data Center using the automated Mail-Retrieval system, typically a hundred images at a time.

3. Image Processing

3.1. Photometric Linearization, Spectral Extraction, and Calibration

The configuration of the IUE image processing system (IUESIPS) has evolved significantly over the 15-year duration of the mission. In addition, a bug in the system (fully appreciated only recently) has prevented it from properly removing the pixel-to-pixel fixed sensitivity pattern, with the result that production-processed IUE spectra are much noisier than they should be. Thus, in the spirit of applying a uniform reduction to the entire collection of spectra, we devised an independent method for processing the raw vidicon images.

In particular, we first rotate the raw image so that the low-dispersion footprint is parallel to the image "SAMPLE" axis. We then photometrically-linearize the 512×48 subimage containing the large-aperture spectrum using an Intensity Transfer Function based on the 1985 recalibration of the SWP camera. The ITF was derived in the rotated frame: the individual pixel sensitivity curves were represented as 4th-order polynomials. We identified, and removed, point-like cosmic-ray hits and other transient defects using an automated procedure. We then filtered the subimage with a 1.5-pixel FHWM Gaussian to suppress "misregistration" noise. Next, we evaluated, and subtracted, the off-spectrum background by spatially-filtering and heavily smoothing the fluxes in reference bands above and below the spectral swath.

We extracted the stellar spectrum using an "Optimal" weighted slit, like that described by Kinney, Bohlin, and Neill (1991), based on the local cross-dispersion profile of the spectral trace and a "noise model" that assigns a photometric uncertainty to each pixel flux according to its intensity. The noise model was derived from the same epoch-1985 UV-Flood flatfield images that were utilized to construct the polynomial ITF (see Ayres 1993). The Optimal extraction explicitly accounts for spatially-extended spectral traces, like trailed or multiple exposures, and is designed to track the small-scale "wiggles" of the spectrum which result from shears in the camera's fiber-optic coupler.

Finally, we applied an absolute calibration by reference to a series of SWP-LO point-source and trailed spectra of the hot-WD G191-B2B—a fundamental flux standard for the IUE—which were compared with theoretical energy distributions. Prior to applying the absolute calibration, we subtracted a "scattered light" level determined by assessing the apparent fluxes shortward of 1150 Å (where the window of the SWP camera is opaque, so any detected photons must be scattered from longer wavelengths). The wavelength scale was determined by fitting a quadratic relation to the apparent positions of prominent emission lines in the rich spectrum of the bright RS CVn-binary λ Andromedae.

3.2. Measurement of Emission Lines and Continuum Bands

We measured the calibrated SWP-LO spectra using a semi-autonomous algorithm developed originally by Bennett (1987). It establishes a smooth continuum via a specialized numerical filter, and then fits the significant emissions (or absorptions) by means of a constrained Bevington-type multiple-Gaussian procedure. Nineteen separate emission lines and nine continuum bands are measured. The algorithm assigns uncertainties to the fitted fluxes—or upper limits in the

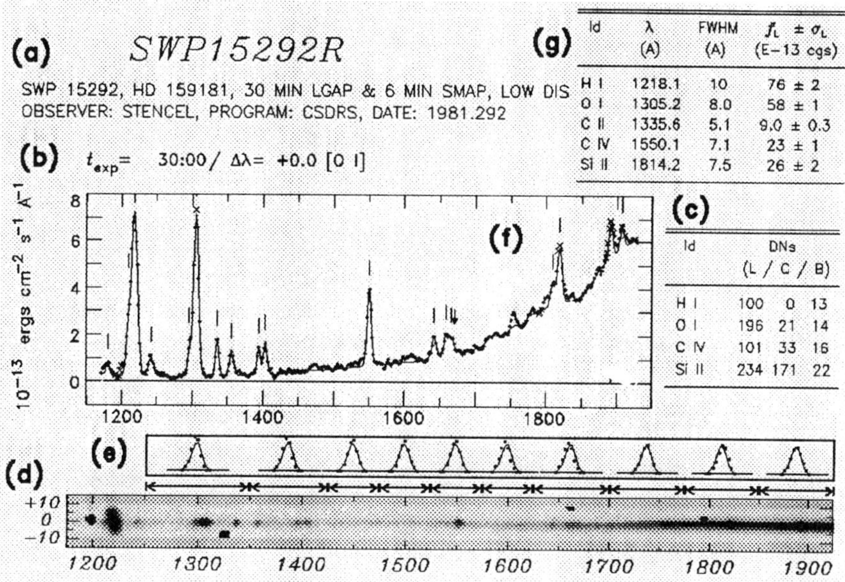

Figure 1. Visualization of SWP-LO reduction scheme. Panel (d) illustrates a stretched version of the photometrically-corrected image: the stellar continuum is visible, particularly at $\lambda > 1700$ Å, and a number of prominent emission lines (including diffuse Lyα sky emission through IUE's large aperture). The sharp-edged dark features are reseau marks. Panel (e) depicts the cross-dispersion profiles of the Optimal extraction: points refer to local profiles; solid curves are the global average. In panel (f), the points are the calibrated fluxes; the solid curve is the fitted spectrum.

absence of a significant detection—according to error prescriptions developed for Gaussian profiles by Lenz and Ayres (1992). The processed data—spectrum, flux errors, continuum fit, tables of line and continuum flux parameters—are stored in a simple ASCII file. The storage file also contains a compressed (one character per pixel) version of the photometrically-corrected, background-subtracted 2-D spectral image, which has been stretched to reveal faint structure. A simple procedure allows one to visualize the contents of the storage files. Figure 1 is an example.

The unified processing and measuring code runs 1 CPU minute per image on a VAXstation 3100/M38 under VMS. The code itself is written in IDL V2, and has been ported to other platforms (e.g., a SUN SPARCstation). Each ASCII storage file occupies 20 kbytes: the entire collection of $\approx$ 3500 spectra is maintained online. A subsample of the images—500 SWP-LOs of about 50 stars—was processed completely many times until the code—and the underlying numerical strategies—stabilized. As a result, we have had to reprocess the full sample only twice (so far!): each reprocessing required roughly a month of

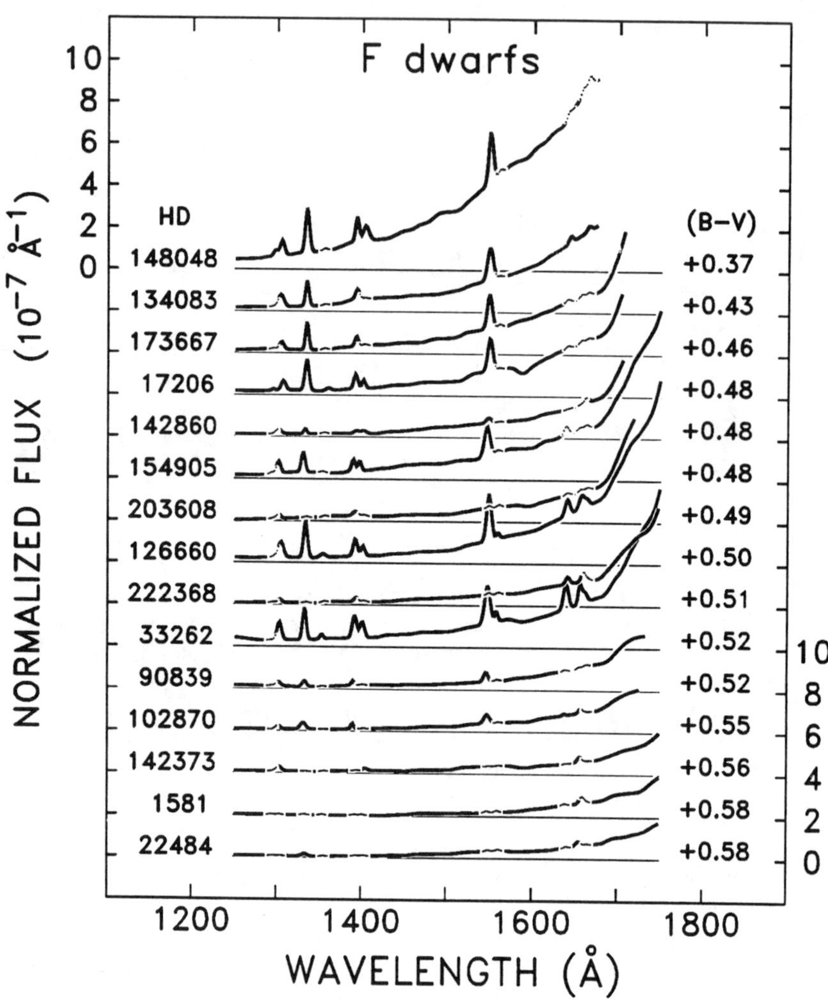

Figure 2. Examples of (mostly coadded) spectra from the test sample. Each calibrated spectrum was divided by the bolometric flux, f_{bol}, of the star for display purposes.

background time at a rate of a few-hundred images per night (limited more by available storage than CPU speed: A SPARCstation with 4 Gbytes of disk could reprocess the entire collection overnight).

4. Present Status of the Catalog

We presently are constructing auxiliary tables required by the Astrophysics Data System to permit ready access to the catalog. We also are coadding spectra for those stars for which several suitable SWP-LOs are available (about 20% of the sample), so that we can provide high-quality average spectra. Examples are illustrated in Figure 2 for F-type stars from the "test sample".

5. For the Future

We are developing low-dispersion processing codes for the IUE's two longwavelength cameras. Specialized high-dispersion software has been devised as well (see Ayres et al. 1993b). Our approach is complementary to that of the IUE Project in its long-term "NEWSIPS" development. In particular, our strategies are devised primarily for late-type emission-line sources, and with the intent to rapidly make available highly-distilled versions of the archival spectra suitable for quick-look analysis and other catalog-style purposes. We feel that our photometric corrections and spectral extractions are comparable in quality to those which will be provided by the NEWSIPS for the majority of IUE images, except perhaps for observations taken at extremes of camera temperature (see discussion in Ayres 1993). However, some types of in-depth analysis projects—particularly those for which high-quality *spatial* information perpendicular to the dispersion is desired—likely will turn to the Final Archive for the ultimate in S/N and geometrical fidelity, using catalogs like the present one for the initial reconnaissance of the available spectra.

Acknowledgments. We gratefully acknowledge support from NASA's Astrophysics Data Program. We also thank the participants in the Final Archive Definition Committee and the staff of the IUE Project for helpful advice. The raw SWP frames were provided by the National Space Sciences Data Center.

References

Ayres, T.R. 1993, PASP, submitted

Ayres, T.R., Bennett, J.O., Lenz, D.D., & Burton, R.B. 1993a, PASP, submitted

Ayres, T.R., Brown, A., Gayley, K.G., & Linsky, J.L. 1993b, ApJ, 402, in press: January 10

Bennett, J.O. 1987, Ph.D. thesis, University of Colorado

Kinney, A.L., Bohlin, R.C., & Neill, J.D. 1991, PASP, 103, 694

Lenz, D.D., & Ayres, T.R. 1992, PASP, 104, in press: November

Astronomical Data Analysis Software and Systems II
ASP Conference Series, Vol. 52, 1993
R. J. Hanisch, R. J. V. Brissenden, and J. Barnes, eds.

The EUVE Proposal Database

C. A. Christian and E. C. Olson

Center for EUV Astrophysics, 2150 Kittredge Street, University of California, Berkeley, CA 94720

Abstract. The proposal database and scheduling system for the Extreme Ultraviolet Explorer is described. The proposal database has been implemented to take input for approved observations selected by the EUVE Peer Review Panel and output target information suitable for the scheduling system to digest. The scheduling system is a hybrid of the SPIKE program and EUVE software which checks spacecraft constraints, produces a proposed schedule and selects spacecraft orientations with optimal configurations for acquiring star trackers, etc. This system is used to schedule the In Orbit Calibration activities that took place this summer, following the EUVE launch in early June 1992. The strategy we have implemented has implications for the selection of approved targets, which have impacted the Peer Review process. In addition, we will discuss how the proposal database, founded on Sybase, controls the processing of EUVE Guest Observer data.

1. Introduction

The Extreme Ultraviolet Explorer (EUVE) Satellite was successfully launched on June 7, 1992. This satellite was built under the auspices of the NASA explorer class missions and is devoted to acquiring astronomical data in the wavelength range from 70 to 760 Å. The science payload, fabricated and tested at the University of California, Berkeley, incorporates first, a collection of three grazing incidence imaging telescopes used to conduct an all-sky survey in the EUVE and second, the deep survey/spectrometer (DS/S) telescope. The deep survey is an imaging instrument used to obtain deep exposures along a 5 degree wide strip along the ecliptic while the all sky survey is taking place. Three spectrometer instruments which share the primary and secondary mirrors with the DS are primarily used to acquire pointed observations for Guest Observers sponsored by a NASA Research Opportunity Program. An overview of the mission is described by Bowyer, Malina and Marshall (1988) and Bowyer and Malina (1991).

During the pointed phase of the mission it is desirable to track the data on each target, from the time the observation is scheduled, through data acquisition and processing. For EUVE the data is tracked with the Proposal Database which interacts with the scheduler, obtains information about data taken, retrieves data for processing and logs various pieces of information relevant to each observation. The database also provides mild protection for proprietary data and can serve as the access port for network retrieval of EUVE data through the EUVE Astrophysics Data System node.

2. EUVE Proposal Database Implementation

2.1. Rationale

Each year, 150–200 individual pointings will be conducted with the EUVE satellite. While this number is not large, it is most practical to track each observation within a database. The Proposal Database for EUVE pointed observations is designed to provide information about not only the target, the observation and the Principal Investigator for the data, but also information about the observation schedule, the location of the acquired data and various history information such as the software version used to process the data. The processing information is particularly useful for organizing the appropriate calibration data relevant to each observation, especially if any attributes of the instrumentation change over time.

2.2. Scheduling

The EUVE Proposal Database is the centerpiece for controlling the flow of data associated with each target, as shown in Figure 1. The database itself is implemented in Sybase, and the list of targets, positions, exposure times, Principal Investigator names, addresses and other ancillary information are entered usually through an interface that reads ascii observing forms.

The list of targets is subsequently provided to the scheduler who computes the feasibility (or suitability) of each observation as limited by spacecraft constraints. The targets are then scheduled using the SPIKE software from the Space Telescope Science Institute. The scheduler reports back to the database which targets have been scheduled, an attribute that can then be queried, just as any other attribute concerning the observation.

2.3. Observation Information

Since EUVE observations are long, it is typical for an observation to be divided into several parts. As each of these observation periods are completed, a report is sent to the database. This report includes information about the duration of the observation, target position, and a complete observation log.

When an observation is completed, an EUVE Guest Observer (EGO) Center scientist staff member decides which calibration data set is relevant to the completed observation. Note that proposals with multiple observations of the same target may require different calibration data sets because the calibration data set reflects the state of the instrument during a particular observation. Over long periods of time, the calibration data set will change. There are other factors that can effect the calibration data set used. For example, in November 1992 the EUVE guide star catalog was expanded which improved the accuracy of the spacecraft aspect solution. Observations completed before this date may require a different aspect calibration data set than observations completed after this date.

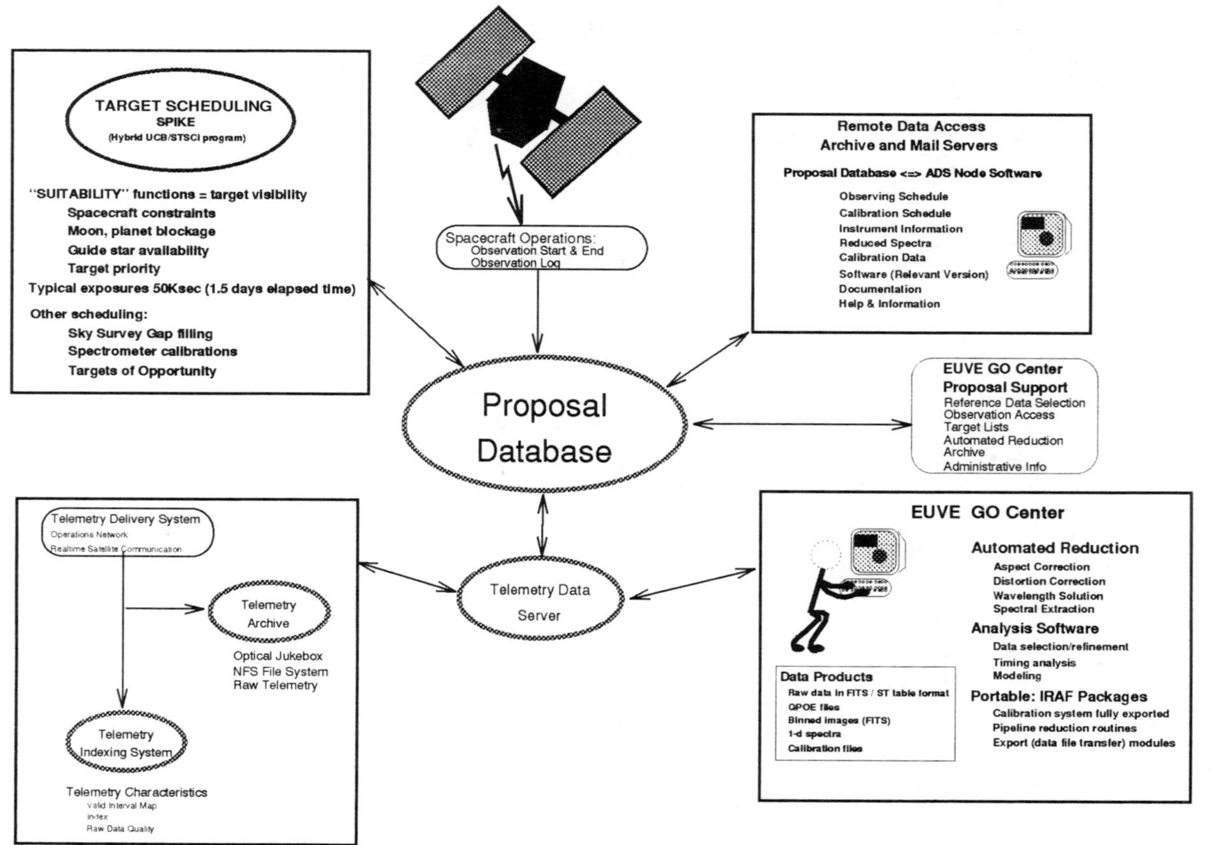

Figure 1. EUVE Proposal Database and Processing Flow

2.4. Processing Flow

The proposal database essentially controls the processing flow for EUVE pointed spectrometer data. Data for each observation is proprietary and so access to the data is provided through the database. A restricted set of users (staff) may request proprietary data. The general user, for example an archival researcher, will retrieve data through the database once the key for the data is unlocked at the expiration of the proprietary period.

When the user requests data through this interface, the proposal database verifies the observation id and access code before delivering all the telemetry associated with the observation, to the user. Currently there is only one interface to the telemetry delivery system. This interface is the IRAF/EUV layered package in the IRAF. In this interface, all telemetry data are restructured into tabular format using ST Tables. The IRAF/EUV layered package provides several packages of IRAF tasks to assist in the analysis of EUVE data. These tasks supplement existing IRAF tasks, as well as provide new functionality.

Preliminary processing of GO data generates several standard data products including fully reduce 1-D (counts *vs.* wavelength) spectra. These data products are stored in FITS format and stored in a TAR file. When this processing is completed, the location of the TAR file is stored in the proposal database.

2.5. EUVE GO Center Staff

The proposal database provides an interface for the EGO Center staff. The staff provides information about the proposals, the associated targets, and calibration data sets. These features mean that the database must have an interactive feel where separate pieces of information are added to the database during the various processing steps. Additionally, the staff ensures that the nominal processing of GO data is done properly, and that information in the database is up-to-date and correct. Several tools have been built to make simple adjustments to the proposal database without resorting to SQL. However, for making significant adjustments to the proposal database one must rely on SQL.

3. Conclusion

The EUVE GO Proposal database provides targeting information to the SPIKE scheduling system, records observation information, and controls the preliminary processing of GO data. The Sybase DBMS provides all the required functionality needed to implement this database. However, the cost of developing and maintaining this system can grow large. A complete understanding of SQL and the DBMS is needed in order to make an effective implementation. Less efficient, more familiar, existing tools might have been a more cost effective and quicker solution in the near term, but in the longer term it is hoped that the DBMS will pay off for archival researchers and for future projects. Groups that already have an expert SQL programmer, will no doubt be able to implement a similar system more quickly.

Acknowledgments. We wish to thank the EUVE staff research assistants and programming staff who assisted in the performing the analysis and writing the software which contributed to this paper. We wish to thank the principal investigators of the EUVE project, Prof. Stuart Bowyer, and Dr. Roger Malina and the EUVE science team for their advice and support. This work has been supported by NASA contract No. NAS5–30180 to UCB.

References

Bowyer, S., & Malina, R.F. 1991, in Extreme Ultraviolet Astronomy, eds. R.F. Malina & S. Bowyer (New York: Pergamon), 397

Bowyer, S., Malina, R.F., & Marshall, H.L. 1988, JBIS, 41, 357

Astronomical Data Analysis Software and Systems II
ASP Conference Series, Vol. 52, 1993
R. J. Hanisch, R. J. V. Brissenden, and J. Barnes, eds.

The Extreme Ultraviolet Explorer Archive

Jeremy J. Drake, Carl Dobson, and Elisha Polomski

Center for EUV Astrophysics, University of California, 2150 Kittredge Street, Berkeley CA 94720.

Abstract. The Extreme Ultraviolet Explorer (EUVE) public archive was commissioned to handle the storage, maintenance and distribution of EUVE data and ancillary documentation, information and software. It was initiated on the 17 July 1992, 40 days after the launch of the satellite itself. This paper gives a brief overview of the Archive and its anticipated directions of development over the next two years.

1. EUVE

1.1. Mission Overview

The Extreme Ultraviolet Explorer (EUVE) was launched on June 7, 1992. The EUVE science payload includes four telescopes, which will support two mission phases: a six-month all-sky-survey; and a subsequent Guest Observer (GO) program.

EUVE entered survey mode on 24 July 1992, following approximately 1 month of initial in-orbit calibration (IOC) observations. The all-sky survey is being carried out by three scanning telescopes, which are observing in the four bandpasses 50–180 Å, 160–240 Å, 345–605 Å, and 500–740 Å. Simultaneously, the fourth telescope, which feeds Deep Survey and Spectrometer instruments, is surveying the ecliptic in a strip approximately 2 degrees wide, and will cover approximately 180 degrees during the 6 month survey. During the GO phase, pointed observations of individual EUV sources will be undertaken, with typical exposure times of 40,000 seconds or more.

A detailed overview of the EUVE mission is presented in Bowyer and Malina (1990). For complete descriptions of the EUVE instruments and their calibration, see Welsh et al. (1989), and Malina et al. (1992).

1.2. Science Data Flow

By nature, the EUVE science data flow is akin to that of earlier astronomical space missions. An end-to-end system (EES) has been devised (Marshall 1989) which converts unwieldy raw data into more scientifically useful information. In the case of survey data, this might be photon positions in the sky as a function of time, and additionally, in the case of GO spectroscopic observations, dispersed in wavelength space.

All of the data have proprietary periods. Sky survey data remain the property of Berkeley for a period of one year following survey gap filling. Guest

Observers will retain the proprietary rights to their approved pointed spectroscopic data for a period of one year following the data delivery. Survey gap filling is anticipated to be completed within the first year of the Guest Observer program.

Minimum storage requirements for the EUVE mission data are quite large. The mission is currently expected to extend for at least 3.5 years, three of these being devoted to GO pointed observations, during which it will generate in excess of 1 TByte of data.

2. The Archive

Analogous to previous missions, one TByte of raw data being rather cumbersome for the average astronomer equipped with workstation, there is a salient need for palatable means of access to the scientifically interesting EUVE data products. Such a requisite inevitably spawns an "archive". Descriptions of many of the new and existing astrophysical archive and database systems can be found in Albrecht and Egret (1991). The goals of the EUVE Archive are identical to those of other mission archives: the dissemination of data and information in the most expedient manner technology and budget constraints permit.

2.1. Contents

The data products to be archived will be EUV skymaps, catalogues, spectra, relevant telemetry, and any pertinent data reduction software, documentation and related scientifically useful information. All data will be "released" as soon as data rights expire: the first GO spectral data will become publicly available in January 1994; while all-sky survey data is likely to become available towards the end of 1994. FITS format will generally be supported in the case of images and spectral data.

At the time of writing, only 6 months after launch, 14 spectra have already been on limited release from preliminary reductions of IOC data. As an example, we illustrate in Figure 1 observations of AU Mic during flare outburst and quiescence taken by the short wavelength spectrometer during IOC. Other data currently available through the Archive include a revised catalogue of past and future EUVE calibration observations, past editions of the EUVE Electronic Newsletter, a bibliography of EUVE science team publications, and C routines to calculate interstellar medium transmission at EUV wavelengths.

2.2. Interfaces

Interfaces through which astronomers currently access remote sites and data vary in complexity from simple ftp-sites and mailservers, to complex interactive environments which offer a variety of analytical tools and database manipulation and retrieval functions. Examples of the latter are described elsewhere in this volume.

While the EUVE Archive is currently committed to the development of such a remotely accessible, interactive research tool and data retrieval system, it will also differ significantly from other space mission archives to-date in that a substantial part of the distribution of data products will be through the release

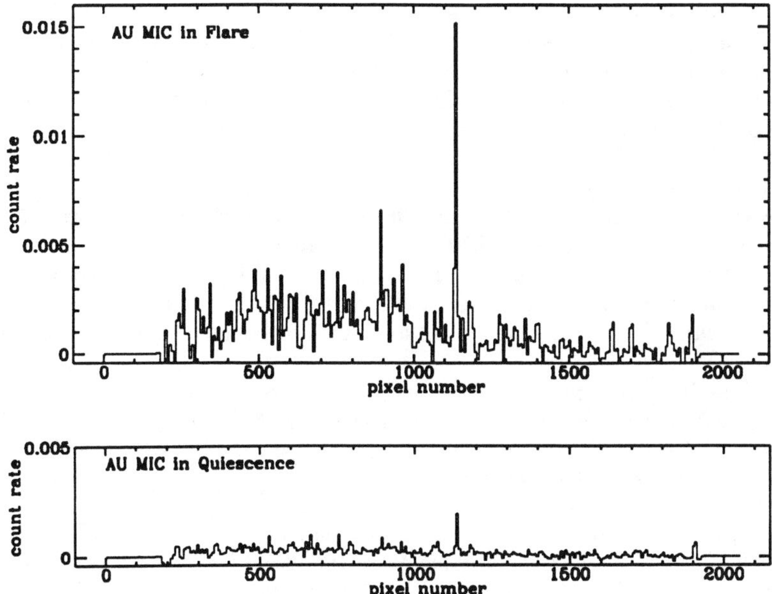

Figure 1. AU Mic as observed by the short wavelength spectrometer during IOC. The wavelength coverage is approximately 70–170 Å. Note that the y-axes of both quiescent and flare spectra are the same.

of a large number (perhaps up to 30 a year, for the duration of the mission) of CD ROMs. In this way, the EUVE archive will spread from a remotely accessible site, to office or library shelves of astronomers worldwide.

Interfaces to EUVE data might thus be grouped under the following headings:

1. Mailserver and FTP site
2. The Astrophysics Data System and related interactive interfaces
3. CD ROM distributions

We give a brief précis of these below. Since all EUVE data undergoes proprietary periods, all methods of access have to incorporate the appropriate data protection protocols. More sophisticated interfaces, such as the ADS, can incorporate access restrictions readily into the host user agent software.

Mailserver and FTP Site An FTP site perhaps defines the minimum requirements for a remotely accessible on-line archive. While limited in scope, FTP sites have the distinct advantage of uniformity and transparent ease of use. The simple command `mget *` can initiate the transfer of the entire contents of an archive at typical rates of 20 KBytes/s or so. The EUVE Archive will continue to make all publicly released data available through "anonymous ftp." To reach the FTP site, connect to cea-ftp.cea.berkeley.edu, enter the username `anonymous`, and your e-mail address as a password.

While more limited for data transfer, mailserver programs are becoming increasingly common, and can offer services which a simple FTP site cannot. The queuing facility intrinsic to electronic mail systems also allows a user to file requests for data and information while lines of communication are slow, or even broken. The EUVE Archive mailserver has been slowly evolving since its inception in July 1992, and now has the capability of passing data supplied by the user, within a mail message, into useful programs, and returning the results back to the user via e-mail, automatically. An example of this is a code called *ism*, which folds user-supplied fluxes through an interstellar medium transmission function calculated for supplied values of H and He column densities. The system is generic, such that adding new programs for multifarious tasks is as trivial as updating a source code index.

The mailserver contains further information on the contents, and use of, the EUVE Archive; a message containing the word `help` on the first line of the main body of the mail text, with no preceding spaces, suffices to generate an automatic response containing the Archive "help" file.

Astrophysics Data System and Related Interactive Interfaces NASA's Astrophysics Data System (see for example, Weiss and Good 1991 for a recent overview, and elsewhere in this volume) is a distributed, database information system which employs a client/server architecture. EUVE is now an ADS primary node and is actively participating in the ADS development program. The panoply of research tools and facilities planned for ADS would appear to make it a promising archival resource. We anticipate that within the next two years the ADS will become one of the primary archival interfaces with EUVE data.

Similar directions have been taken by White and co-workers of the the HEASARC group (this volume). The HEASARC interface has been in place for some time and provides an excellent data perusal and analysis tool. Indeed, we are likely to adopt some of this perusal software, and, if demand exists, HEASARC analysis packages suitable for EUVE data.

CD ROMs CD ROMs enjoy the advantage of greater durability and more rapid access times compared with magnetic tape media. Longevity of CD ROMs appears to be related to quality of manufacture, and although CD lifetimes are currently not expected to exceed much more than 5 years, it is very likely that improved manufacturing techniques and standards will extend lifetimes indefinitely (Fox 1992).

The first EUVE CD release is planned for the 181st AAS meeting in June 1993—less than one year after launch! The contents of this CD will include spectroscopic data acquired during the IOC phase of the mission, the first release of the EUVE IRAF data reduction software package for spectral extraction, interstellar medium data useful for planning observations at EUV wavelengths compiled by researchers at Berkeley, and miscellaneous pertinent tabular data.

3. Present and Future

While the more complex interfaces to, and aspects of, the EUVE archive remain in the planning and development stages, the more basic are already in place and are handling up to 50 requests and queries each week.

The youth of the EUVE mission at the time of writing necessitates a somewhat skeletal overview of the Archive. However, we have endeavoured to give interested astronomers a taste of how they might access the wealth of data EUVE promises to deliver over the next 2–3 years. While we are entering the prosopographical near-future of White and Giommi (1991)—"It can be anticipated that in the near future astronomers will access vast amounts of astronomical data, both new and archival, from terminals/workstations located thousands of kilometers from where the data are collected and stored"—the same astronomers will also be looking at the EUVE spectrum of AU Mic as a function of time, using a portable CD ROM drive and a laptop computer, in the airplane on the way to the Tenth Cambridge Workshop on Cool Stars Stellar Systems and the Sun.

Acknowledgments. This work was supported by NASA contract NAS5-30810, administered by the Center for EUV Astrophysics at UC Berkeley. The authors thank the Principal Investigators, Stuart Bowyer and Roger Malina, and the EUVE science team for their advice and support.

References

Bowyer, S., & Malina, R. 1991, in Extreme Ultraviolet Astronomy, eds. R.F. Malina & S. Bowyer (New York: Pergamon Press), 397

Fox, B. 1992, New Scientist, 134(1815), 19

Malina, R.F., Jelinsky, P., Finley, D.S., Vallerga, J.V., & Vedder, P.W. 1993, ApJ, submitted

Marshall, H.L. 1989, in Data Analysis in Astronomy III, eds. V. di Gesu et al. (New York: Plenum), 169

Pasian, F., & Richmond, A. 1991, in Databases and On-line Data in Astronomy, eds. M.A. Albrecht & D. Egret (Dordrecht: Kluwer), 235

Weiss, J.R., & Good, J.C. 1991, in Databases and On-line Data in Astronomy, eds. M.A. Albrecht & D. Egret (Dordrecht: Kluwer), 139

Welsh, B.Y., Vallerga, J.V., Jelinsky, P, Vedder, P.W., Bowyer, S., & Malina, R.F. 1989, Opt. Eng., 29, 752

White, N.E., & Giommi, P. 1991, in Databases and On-line Data in Astronomy, eds. M.A. Albrecht & D. Egret (Dordrecht: Kluwer), 11

Astronomical Data Analysis Software and Systems II
ASP Conference Series, Vol. 52, 1993
R. J. Hanisch, R. J. V. Brissenden, and J. Barnes, eds.

Quality Control of EUVE Databases

Linda M. John
Center for EUV Astrophysics, 2150 Kittredge Street, University of California, Berkeley, CA 94720

Abstract. The publicly accessible databases for the Extreme Ultraviolet Explorer (EUVE) include: the EUVE Archive Mailserver, the Center for EUV Astrophysics ftp site, the EUVE Guest Observer Mailserver, and the Astronomical Data System node. The EUVE Performance Assurance team is responsible for verifying that these public databases are working properly and that the public availability of EUVE data contained therein does not infringe any data rights which may have been assigned. In this paper, we describe the quality assurance (QA) procedures we have developed from approaching QA as a service organization; this approach reflects the overall EUVE philosophy of QA integrated into normal operating procedures, rather than imposed as an external, post-facto, control mechanism.

1. Introduction

Performance assurance (PA) activities can be seen either as external managerial approval and certification of procedures and operations, or instead as internal services provided to support generation and improvement of procedures and operations. We have chosen the latter approach in performance assurance of the EUVE databases and have found this approach very successful both in maintaining the desired levels of quality and in allocating our limited PA resources to the areas most in need of attention. In particular, our attention is directed to consulting on procedures early in order to prevent errors from occurring, and to incorporating quality control activities into these procedures from the start so that PA activities do not become interruptive and tedious, as can occur when they are added later.

The EUVE archive databases include the EUVE Archive Mailserver, the EUVE Guest Observer Mailserver, the Center for EUV Astrophysics (CEA) ftp site, and the Astronomical Data System (ADS) node.

The Archive Mailserver (archive@cea.berkeley.edu on Internet) provides e-mail access to publicly released spectra, lists of detected sources, and lists of calibration targets. The Guest Observer Mailserver (egoinfo@cea.berkeley.edu) provides e-mail access to the IRAF/euv software package, the Guest Observer handbook and proposal instructions, associated tables of EUVE instrument characteristics, and EUVE bibliographic information. The CEA ftp site (ftp cea-ftp) provides access to these same two sets of data, archive and egoinfo. Finally, the ADS node provides access via ADS to the list of calibration targets and, in the future, will include a wide variety of publicly released EUVE data products.

2. Consulting on Archive Procedures and Quality Control

Rather than waiting until after the system was in place, the PA team has participated in the creation and review of the Archive procedures. We began by generating a list of questions similar to those that would be used if we were performing a PA audit of the system. The Archive procedure questions included how data are selected for placement in various archival databases; how public data rights are confirmed for the data and by whom; what steps must be performed to correctly add the data to each archive database; how errors are reported, corrected and tracked; and how data are removed, when necessary.

This pre-audit set of questions was then reviewed with the Archive administrator and procedure writer. During the review, some questions were answered, some new ones were generated, and some others were found to be inappropriate or not applicable. Because this pre-audit was done while the procedures were still in draft form, there was ample opportunity to modify each procedure to accommodate the changes desired for quality control, database security, and data rights protection. Because this process is similar to an actual audit, information and statistics needed for future audits have been identified, their value assessed, and procedures for gathering the information are being generated (and can even be automated). The procedure writers received a clear understanding of what needs to be addressed, resulting in clear procedures that have not had to undergo numerous revisions.

The individuals who must follow the procedure are the ones writing it, and they decide how to meet the quality control requirements determined during the pre-audit. Since the writers are the most knowledgeable about the system, they will be able to find the most efficient methods for meeting these requirements. Moreover, since they have created these procedures, they are willing to follow them.

3. Archive Problem Reporting and Tracking

Another service that the EUVE PA team provides to the Archive team is a single, projectwide EUVE problem reporting (EPR) system. This automated system provides problem reporting and tracking, while minimizing the paperwork and tedious meetings and virtually eliminating the possibility of permanently lost problem reports. EPRs are generated electronically (on-line) by any EUVE user and are automatically e-mailed to the Software Assurance and Test (SWAT) committee, which is headed by the Performance Assurance Manager. When outside users send problems via e-mail, the appropriate Archive or Mailserver administrator originates the EPR. EPRs are prioritized and assigned within a week. When an EPR is assigned, it is automatically e-mailed to the assignee, and the originator is also notified via e-mail.

This reliance on e-mail reduces the printing needs incredibly, allows for on-line response to the EPR, provides a forum for discussion of problems that warrant special attention, and maintains a complete history of all EUVE problems reported. With an average of 46 EPRs generated every month projectwide, this automated system has fulfilled the PA needs, reduced the time required for processing problem reports, and freed the PA team to perform the consulting services described above.

In support of the EUVE Management team, the EUVE PA team also acts as a central collection point for metrics charts tracking the activity levels and performance of both the archive and egoinfo Mailservers. The number of requests and number of customers are charted monthly, as are the number of complaints received and the number of personal responses made.

4. Independent Testing of the Archive Software System

A third service provided by the EUVE PA team is independent testing of software systems. In the case of the Archive databases, this is realized by the monthly tests and audits done on each of the CEA Database/Archive sources: ADS, the CEA ftp site, and the egoinfo and archive Mailservers.

The SWAT team accesses the files as an outside user would. When necessary, we have novice users perform the data accessing to confirm the ease of use of the system and of our procedures. We access and print each of our new CEA Database/Archive files monthly, making sure the formats are as expected. The Performance Assurance manager confirms with the Archive scientist that all new files are approved for release.

From inside the ftp site, we verify file ownership and protection monthly. Protection of subdirectories is also checked, and we confirm that simple attempts to access other areas of the file server are unsuccessful. These simple security checks are designed to protect against damage by inadvertent or malicious but ignorant attempts at unauthorized file access.

By making these activities a PA responsibility, the burden of repetitive or routine tests and audits has been removed from the Archive team, freeing them to concentrate on new data and resolving known problems, and giving the PA team experience and understanding of the Archive system and the information needed to determine if any projectwide problem trends are occurring.

5. Future Audits of the Archive User Interface and the Data Protection

When all the procedures are finally in place and have been used for several months, an audit will be scheduled to verify compliance and to determine necessary improvements in the procedures. Since the Archive administrator is generating the procedures, compliance discrepancies found several months from now are expected to result from inadequate or incorrect procedures, rather than lack of knowledge of the procedures or laxness in following them. The resultant report will identify errors, give specific recommendations for rectifying errors, and then allow the administrator to make necessary corrections to the procedures. Wherever possible, automation opportunities and removal of unnecessary steps will also be identified.

6. Conclusions

Performance Assurance as a service organization has proved to be an excellent choice for maintaining the quality of the EUVE database. Not only has there been less duplication of effort in testing, more effective procedure generation, and a reduction of problem reporting paperwork, we have experienced a very high level of cooperation from the Archive team and throughout the project in problem reporting and resolution and during audits. There is a projectwide team approach to quality and quality control, resulting in a positive "get the job done" attitude and reducing dramatically the distrust of the PA team as an unnecessary interruption.

Acknowledgments. I am grateful to Carl Dobson, Jeremy Drake, Jim Lewis, and Alex Wiercigroch for assistance in preparing and reviewing this material. I thank the principal investigators, Stuart Bowyer and Roger F. Malina, and the EUVE science team for their advice and support. This research has been supported by NASA contracts NAS5–30180 and NAS5–29298.

Astronomical Data Analysis Software and Systems II
ASP Conference Series, Vol. 52, 1993
R. J. Hanisch, R. J. V. Brissenden, and J. Barnes, eds.

The EXOSAT Database and Archive

A. P. Reynolds and Arvind Parmar

Astrophysics Division, Space Science Department of ESA, ESTEC, Keplerlann 1, Postbus 299, 2200 AG Noordwijk, The Netherlands

1. The EXOSAT Observatory

- The European X-ray Astronomy Satellite EXOSAT was operational from May 1983 to April 1986. It made 1780 observations of most classes of astronomical object including:
  - Active galactic nuclei
  - Cataclysmic variables
  - Stellar coronae
  - X-ray binaries
  - Supernova remnants
  - Clusters of galaxies
  - Isolated white dwarfs and neutron stars
- Three complementary instruments:
  - Low energy imaging telescopes (LE). X-ray images plus broad band filter spectroscopy.
  - Medium energy proportional counter array (ME). High time resolution light curves and spectra.
  - Gas scintillation proportional counter (GS). Smaller area than the ME, but better spectral resolution.

2. The EXOSAT Data

- All the data are now in the public domain
- Analysis requires specialised software and expertise
- Some (large) users have their own analysis systems
- Others use the interactive analysis system at ESTEC
- Often impossible for non X-ray astronomers or users who belong to small groups to invest the manpower/effort required to analyze EXOSAT data.

2.1. The EXOSAT Database

- Is designed to:
  - Ensure that the data and results are not lost for posterity
  - Be a prime tool for archival research

- Make the data more accessible to the non-specialist
- Be available over computer networks

- On-line since April 1989
- Distributed at several sites
- Very much a collaborative effort

2.2. Four Levels of Access to the Database

- An observation log.
  - See if a source has been observed
  - details about the observer and past requests for the data
  - search the EXOSAT bibliography
- A summary of the results from each EXOSAT observation.
  - from a specific observation
  - from all observations of a particular object
  - from a class of objects
- The final products, i.e., spectra, images, lightcurves and calibration data.
  - utilities to view and extract the products
  - application programs to allow further analysis
  - combine EXOSAT data with that from other telescopes
- If necessary return to the raw data for a more detailed analysis.
  - only for the more specialized cases, unexpected discoveries, etc.
  - Interactive analysis system will still be supported at ESTEC. Works in a Unix environment. Intend to make available over the the networks.

3. Software Sites

- ESTEC, EXOSAT Observatory, NL
- Institute of Astronomy, Cambridge, UK
- Astronomical Institute, Amsterdam, NL
- Observertoire Haute-Provence, F
- Physics Dept, Univ of Leicester, UK
- MPI für Extraterrestrische Physik, D
- Lancashire Polytechnic, UK
- Geneva Observatory, CH
- DSR, University of Birmingham, UK
- TESRE, Bologna
- IFCAI/CNR Palermo
- Instituto di Radioastronomia, Bologna, I

- Tata Institute, Bombay, India
- Dip. Fisica, Univ. Milano, I
- Astronomical Observatory, Brera, I
- IFCTR/CNR Milano, I
- Astronomical Observatory Pino Torinese, I
- NASA/GSFC, USA
- NASA/MSFC, USA
- Space Telescope Science Inst, USA
- Dept of Physics and Astronomy, Univ of Iowa, USA
- Instituto di Cosmo-Geofisica, Torino, I
- High Energy Astrophysics Division, CfA, Cambridge, USA
- Dept of Physics, MIT, Cambridge, USA
- Dept of Physics, Nagoya Univ, Japan
- Dept of Astronomy, Penn state, USA
- SSL, Berkeley, USA
- Inst. Study Earth Ocean and Space, Univ of New Hampshire, USA
- Dept of Physics, Korean Inst of Technology, Korea
- ICRA, Univ of Rome, I
- IOA, Univ of Rome, I
- Astronomical Institute, Univ of Tubigen, D
- CDS, Strasbourg, F
- ESRIN, Frascatti, I
- SRON, Utrecht, NL
- HEAD, Chinese Academy of Sciences, Beijing, China
- CfA, Univ of Science and Technology, Hefei, China

4. Available Databases

- EXOSAT Observation Log
- EXOSAT LE database
- EXOSAT ME database
- EXOSAT GS database
- EXOSAT grating database
- Various Einstein HRI databases
- Einstein EMSS and SSS databases
- Various Einstein IPC databases
- HEAO-1 A2, A3, A4 Catalogs
- IUE ULDA database of X-ray Sources

- Cos-B database
- Hipparcos input catalog
- Merged X-ray catalogs
- ROSAT AO1 pointings and log
- IRAS Point source and faint source catalogues
- IRAS FSC associations catalogue
- Merged optical catalogues
- Merged radio catalogs
- HD, SAO and variable star catalogues
- HST Guide star catalog

5. The EXOSAT Database Software Packages

- BROWSE Database Management
  - Catalog/database search
  - SQL interface
  - Data retrieval
  - Statistical analysis
- XSPEC Spectral analysis
  - Supports EXOSAT, Einstein, IUE
  - Simulations
  - User defined models
- XRONOS Timing analysis
  - General statistical analysis
  - Period searches (FFTs, folding)
  - Correlations
- XIMAGE Image analysis
  - Image display
  - Source detection
  - Image Algebra
- QDP General plotting and function fitting
  - ASCII files as input
  - Publication quality output
  - Data fitting
- Mission Independent

  EXOSAT, Einstein, Rosat, IUE, Cos-B, XMM, SAX, Astro-D, AXAF, Jet-X
- Frequency Independent, but best for X-rays

  Support ranges from radio to gamma ray data.

- Operating System Independent

  VMS and Unix versions already available for XSPEC, XRONOS and QDP. XIMAGE and BROWSE run under VMS but are currently being ported to SUN-OS.

- Terminal Independent

  Supports most graphic terminals, workstations, etc.

- Interfaced with other S/W Packages

  SAO-IMAGE
  IDL
  IRAF (via FITS files)
  IRAF-PROS (via FITS A3D files)
  MIDAS (via FITS files)
  Other packages (via FITS)

5.1. Institutes Involved

- ESTEC/EXOSAT Observatory, NL
- Goddard SFC/HEASARC, USA
- Brera Observatory - Milan I
- Marshall SFC, USA
- University of Leicester, UK
- ESRIN/Frascatti, I

5.2. Future Development

- SSD

  - Standard for SAX
  - Maintenance of Current System
  - FOT Optical Archive
  - Current Support: 1 man/year

- HEASARC

  - Port to Unix + Further Development
  - Standard for Astro-D
  - Standard for GRO science center
  - Alternative to PROS for Rosat
  - Already part of ADS
  - Current Support: 20 man/year

- ESRIN/Frascatti

  - Prototype ESIS Correlation Environment
  - FOT Optical Archive
  - Current Support: 3 man/year

6. More Information?

e-mail EXOSA0::ALASTAIR or EXOSA0::ARVIND

Part 1. Databases, Catalogs, and Archives

Section C. Software Tools and Data Structures

Astronomical Data Analysis Software and Systems II
ASP Conference Series, Vol. 52, 1993
R. J. Hanisch, R. J. V. Brissenden, and J. Barnes, eds.

Evaluation of Relational Database Packages for use in Astronomy

C. G. Page and A. C. Davenhall

Department of Physics & Astronomy, University of Leicester, UK

Abstract. A number of relational database management systems (RDBMS) have been evaluated for use in astronomical data analysis. Several areas are identified, however, in which commercial database packages do not provide all the facilities needed by astronomers. The possibilities of building a package based on FITS binary tables are considered.

1. Introduction

Tabular datasets arise in many areas of astronomy. Catalogs of celestial objects and photon-event lists are important examples of large tabular datasets, while datasets in tabular form but usually of smaller size are often used or generated during data analysis. The operations to be performed on these tables are, mostly, just those provided by relational database management systems.

The Starlink software collection includes a relational database package called SCAR (Starlink Catalogue Access and Reporting), but a replacement was desirable because of its inflexibility, poor user interface, and dependency on VMS. Rather than attempt major modifications to the existing code, we hoped to find a commercial RDBMS that could be adapted for general astronomical use.

Our evaluation criteria (Davenhall 1992) outlined the overall properties needed for scientific use. It was accepted that certain specifically astronomical features (such as sexagesimal formatting, great-circle distance function, producing all-sky plots, etc.) would have to be provided by writing additional software, but one of the important considerations was how easily each package could be extended in this way.

The commercial packages in our initial list included: Dataflex, dBASE-IV, Empress, Gembase, Ingres, MicroRim, Oracle, SAS, Sybase, System-1032, and Tekbase. For more detailed evaluation and benchmarking we selected Ingres as representative of the major commercial systems, and Tekbase, which seemed the most promising of the few packages specifically designed to handle technical data. Although we did not find a package that was entirely suitable, it was possible to draw a number of general conclusions about the suitability of commercial RDBMS for handling astronomical data.

2. Some Common Limitations in Commercial Systems

We note a number of areas in which most commercially available database systems seem poorly matched to the requirements of astronomers. These are, of course, generalizations, and all of them do not apply to every package that we examined.

2.1. Table Modularity

Most astronomical tables are modular and self-describing (e.g., FITS tables). Astronomers expect to be able to move tables from one directory to another, and one machine to another over the network. This is much easier if *table≡file.* Commercial systems nearly all use a *system catalog* or *common data dictionary* to hold table and column descriptors, with (at most) only the row-and-column data held in the user's own file-space. As a result, tables have to be registered and deregistered explicitly, and a database administrator may be required to handle the access control.

2.2. Name-Space

We need to have a set of (system-owned) star catalogs that everyone can read, and allow users to have as many private tables as their disk quota allows. Joins may be required between private and public tables in any combination. In contrast, many commercial systems have the concept of number of separate *databases* each containing a set of closely-related tables. Within a database there is a single name-space for tables, and often also for columns. Several systems only permit *joins* between tables in the same database. These features can cause severe problems in an unstructured multi-user environment.

2.3. Data Exploration

It seems important to encourage data exploration, for example with a *browse* operation, or by allowing further selections on a subset previously selected. Commercial systems are mostly designed for transaction processing (TP), which means permitting multiple concurrent updates to a single table. As a consequence, the state of a table may change unpredictably between one operation and the next. Indeed the reason that SQL has such complicated syntax is that all the operations in the sequence (select, project, join, sort, group, etc.) must be completed while the relevant parts of the table are locked against other updates. This makes the user-interface less suitable for data exploration. The overheads of TP systems (record-locking, two-phase commit, audit-trails, etc.) can also be a disadvantage in an astronomical research context, as most astronomical tables are updated infrequently and then only by one user at a time.

2.4. Data Types

Astronomical tables mostly contain floating point data, while character strings and integers are the main-stay of the commercial world. We really need both 32-bit and 64-bit reals (to conserve disk space) which not all systems support; vector types are rare, and scaled integers (as used in many FITS tables) are quite unknown. Display formats for floating-point data are often limited: we need to have at the very least both E and F formats and some way of displaying angles in sexagesimal notation. Fixed length vectors of the basic data types are also desirable, e.g., to hold the magnitude of a source in different wavebands. These are supported only by a few commercial systems.

2.5. Relational Joins

In commercial systems the only join operation provided is normally the equi-join, i.e., requiring exact match. When using imprecise floating-point data a *fuzzy-join* is required. And the important special case of joining two star catalogs on celestial position requires a join based on the great-circle distance function. We have found no way of doing this efficiently via SQL where only an equi-join is provided.

2.6. Export and Import

Because of the need to exchange tabular data with many other programs, flexible import and export arrangements are essential. Most commercial systems can only import and export fixed-format text files, which means the operations are slow and wasteful of disk space, and often extremely awkward to use.

2.7. Procedure Interfaces

We are certain to write additional database applications in Fortran or C which make use of the database engine, so a good procedure interface is required. In many systems the input interface just passes an SQL command. But there seems to be no standard for output from these embedded-SQL systems, and the arrangements for returning tabular data to the calling program are system-dependent and awkward to use.

2.8. Keyed Access

Keyed access is essential as astronomical tables can be very large. Home-grown systems like SCAR usually sort celestial catalogs on a coordinate such as declination and use a binary search to locate a range of records efficiently. Sorting seems not be used in commercial systems, which instead index via hash tables, ISAM files, or B-trees. Although these indexes are fully dynamic, sorting uses no extra disk space. For efficient access to really large celestial catalogs a 2-dimensional indexing scheme would be desirable, but none of the systems examined so far provide this.

2.9. Meta-data

It is often necessary to attach parameters to astronomical tables, for example to record the equinox, epoch, author's name, date, or processing history of the table. Commercial systems seem to make no provision for this type of information, nor do they permit columns to be labelled with their physical units or descriptive comments.

3. Non-commercial Packages

Our preliminary conclusion is that the typical commercial RDBMS is not very well suited to the task of handling astronomical data. The amount of work needed to modify its behavior and provide the missing functions looks not much different in magnitude from that of writing a suitable system from scratch. We therefore considered in addition a number of "free" packages that were designed for astronomical or geophysical data. These included:

- DIRA2 (Astronet Database Group, CNR Bologna),
- EXOSAT database system (originally from ESOC/ESTEC, now supported by the HEASARC at GSFC),
- MIDAS table file system (ESO),
- R-EXEC (Scientific Databases Section, Rutherford Appleton Laboratory).

Each of these packages contains a wealth of good features especially designed for scientific data and was carefully designed for its original purpose. At present, however, none of these packages appears to have all the features that we think are needed to cover the whole range of astronomical applications. It should be pointed out, however, that all of these packages are still under active development. Some recent enhancements to the MIDAS table system are described by Peron and Grosbøl (1993).

4. FITS Tables

Many tabular datasets are already being distributed to the astronomical community as FITS tables, in either the ASCII or binary variants. This prompted us to consider the FITS binary table as a possible file format for internal use in a database package. FITS binary table format seems quite suitable for the job: it allows the encapsulation in a single file of the tabular data, the column descriptors, and an unlimited number of table parameters. The format is well defined and machine-independent, and FITS binary tables can include all the required data types, and they support features such as scaled integers, vectors, and null values.

In parallel with the evaluation exercise, we started to examine in more detail the feasibility of a package based on FITS tables. This exercise resulted in the development of a basic prototype called PANTHER (Portable Astronomical Table Handler). This program has a simple command-line interface and allows users to manipulate FITS binary tables. The present prototype includes only limited functionality, but enough to demonstrate the general feasibility of many of the concepts involved. Some of its main features are:

- Versions have been tested on VMS, SunOS, Ultrix, Irix, and MS-DOS.
- It can handle FITS binary and ASCII tables, and access SCAR binary tables, and those of the EXOSAT DBMS. Other table formats such as MIDAS, dBASE, R-EXEC, could easily be added.
- Keyed access is provided in two ways: for sorted static tables the binary search can be used; B+ trees can be created as indexes to dynamic tables.
- All FITS data types are supported, including scaled types, and vectors can be used.
- Missing (null) values are supported throughout using 3-valued logic.

- The expression compiler (used in selections and projections) includes a range of mathematical and trigonometric functions, and can easily be extended to include all the required astronomical functions (such as precession).
- Data exploration is supported by arranging that each selection produces a subset of rows which is pushed on a stack of subsets. The initial selection works on the whole table, subsequent ones work on the latest subset. Other operations such as listing, sorting, computing statistics, etc., can be performed at any point, and also operate on the latest subset, if there is one, or else on the whole table.
- Display options include Fortran-90 formats (such as B, O, Z, EN and ES), there are also facilities for displaying angles in sexagesimal notation, and for displaying Modified Julian Dates as calendar date and time.

No decision has been taken on whether to develop this prototype further, but at least it suggests that a viable database engine can be developed on the basis of FITS tables. The hard part is designing a good user interface.

Users of FITS tables should be aware of the FITSIO library, produced at GSFC (Pence et al. 1993) which provides a convenient and powerful procedure interface to FITS tables. GSFC are also developing a set of utilities for handling FITS files called `ftools`; many of the underlying ideas are clearly similar to those outlined above, but the `ftools` have a different style of user-interface, being modules that run independently (or under IRAF). An important principle in the development of PANTHER was the preservation of context between one command and the next.

References

Davenhall, A.C. 1992, Starlink internal report: Evaluation Criteria for a Starlink DBMS

Pence, W.D., Blackburn, J.K., & Greene, E. 1993, this volume

Peron, M., & Grosbøl, P, 1993, this volume

Astronomical Data Analysis Software and Systems II
ASP Conference Series, Vol. 52, 1993
R. J. Hanisch, R. J. V. Brissenden, and J. Barnes, eds.

A Generic Archive Protocol and an Implementation

J. M. Jordan, D. G. Jennings, T. A. McGlynn, N. G. Ruggiero, and T. A. Serlemitsos

COSSC/NASA-GSFC/CSC, Code 668.1, NASA/GSFC, Greenbelt, MD, 20771

Abstract. Archiving vast amounts of data has become a major part of every scientific space mission today. GRASP, the Generic Retrieval/Archive Services Protocol, addresses the question of how to archive the data collected in an environment where the underlying hardware archives and computer hosts may be rapidly changing.

1. Introduction

GRASP, the Generic Retrieval/Archive Services Protocol, addresses the problem of creating a consistent front-end to an archive system when the archive device and the host computer are changing regularly. GRASP insulates the archive user from the details of the archiving devices being used. The archive system is designed in two layers, a static top layer for the user to interface to, and, a bottom layer maintaining a consistent interface to the top layer while changing to interact with any archive device.

Read, write, deletion and status functions are provided in the GRASP system. Access to all of the GRASP functions is specified at both the command line and the functional level. The current implementation, written in C for Unix and VMS systems, provides access to a GRASP system at both levels and assumes a non-specific archive device accessible as a file system appropriate to the operating system.

Data at the Compton Observatory Science Support Center (COSSC) is currently being stored in a GRASP driven archive. Experience with the archive has already led to one update of the implementation, and more enhancements are on the way. This implementation is operational for the COSSC archive. Modifications are underway to cover the full scope of the COSSC archive system.

2. Design Concept

GRASP is designed as a two part system. The Transfer Interface is computer/archive device independent code while the Action interface contains code which is dedicated to each computer/archive device addressed.

A typical system using the GRASP system for archiving would have a structure resembling the following outline:

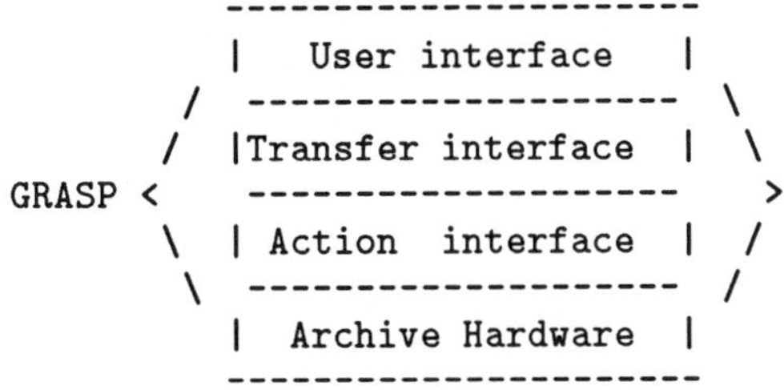

The Transfer Interface is based on a set of atomic functions implemented in the Action Interface. Functions at the Transfer Interface level act as the interpreters of complex user commands which may involve multiple files and implicitly create sets of files.

Filesets are an extremely useful feature of GRASP used for associating groups of data. Filesets are divided into two types, regular and derived. Regular filesets consist of groups of files which are stored together in the archive during the archiving process. Derived filesets are used to associate data already within the archive and make related data easier to retrieve with a single operation. A derived fileset appears to the user to be the same as a regular fileset, but may span the contents of many single files or other filesets within an archive. The association of a calibration file with all of the daily readings of an instrument is an example of the use for derived filesets. Files associated with each day would be grouped in regular filesets. Then derived filesets linking the separately stored calibration data can be created.

Operations at the Transfer Interface level use only three items of information, file names, tag names, and archive names. Separation of the transfer interface from the operating system is easily maintained using these names. The archive and tag names are ascii strings. Individual archives are addressed by the user assigned archive name. Once a file or fileset has been archived, it is referenced only by a tagname specified by the user. Only the file name is system dependent as it may contain a path name, but, subfunctions within the action interface are used to parse these strings for dependencies.

At the Action Interface level, atomic functions are defined for the GRASP system which perform archiving procedures on single files. Additionally, functions are defined which return information on individual files in the archive.

For maximum flexibility, the Action Interface works with files in two parts, the data and the context. The context of the file consists of the specification of the type of file. For example, under the VMS operating system, files may have many formats, from fixed length sequential to varying length random access. The context of the VMS file is the format of the file. The data of the VMS file is the stream of bytes left in the file after format information has been stripped out. At the other extreme, Unix systems represent all files as streams of bytes, so there is no context information on a Unix system.

The advantage of context and data separation becomes evident when files archived in a VMS archive are required on a Unix system. As the data in the files on the VMS archive is stored as a stream of bytes, the context of the file is ignored and the stream of bytes is retrieved to the Unix system. As another case, if VMS files are archived in a Unix archive, but are needed on a VMS system, then the context information stored in the archive can be used to restore the archived file to its proper VMS format.

3. Functionality

GRASP covers four areas of archive operation; reading, writing, deletion, and status information. Reading, writing, and deletion work on either a single file or a group of files.

Groups of files are operated on as file sets and may be specified in two ways, explicit and implicit. For write operations implicit filesets are generated by using wildcards in the filename specification. If more than one file matches the wildcard specification, then a file set is created. Explicit filesets are created by giving the write command the name of a file which is a list of files to be archived in the fileset. Read and delete operations automatically detect whether tag names indicate single or multiple files.

Status commands in the GRASP system are specified to determine the state of the archive and the state of tags within the archive. The ability to store files and the remaining space in a filesystem can be queried, as may the association of a tag. If desired, a valid, unused tag name can be requested.

Functions of the GRASP system are specified to work both from an operating system command line and as function calls within a program. The commands in GRASP, as they might be specified in a function call, are as follows.

```
status  = GrWrite(archive, filename, tagname)
```

This function writes a file `filename` to `archive` under the name `tagname`. If `filename` is preceded by a '@' it is the name of a file with a list of files to be archived in an explicit fileset.

```
status  = GrRead(archive, tagname, filename)
```

This function reads a tag `tagname` from `archive` and stores it in `filename`. If `tagname` refers to a fileset, then `filename` is considered to be a location, typically a directory, to put the fileset files.

```
status  = GrDelete(archive, tagname)
```

GrDelete deletes `tagname` from `archive` and handles deletion of single files, regular filesets, and derived filesets. Note that the deletion of a derived fileset removes no actual files, just the record of the derived fileset.

```
status  = GrMakeFileSet(archive, filename, tagname)
```

Derived filesets are created with this function. Here, `filename` refers to the file which contains the list of tagnames to be included in the derived fileset.

```
status  = GrCanStore(archive, filename)
```

GrCanStore queries `archive` to see if it has enough space to store the file(s) specified by `filename`.

```
size = GrSpace(archive)
```

The amount of space (in bytes) available in the archive is returned by this function.

```
status  = GrStat(archive)
```

This function checks whether `archive` is available.

```
status  = GrTag(archive, tagname)
```

A valid, unused `tagname` is returned by this function if possible.

```
status  = GrTagStat(archive, tagname)
```

The association of `tagname` with a single file or fileset is checked and returned by this function.

```
message = GrPrStat(status)
```

This function returns a string explaining the meaning of the numeric `status` codes returned by the other GRASP functions. All of the `status` returns are integers which may be plugged into GrPrStat to find out the string description of the status return.

4. Implementation

GRASP has been implemented as a C program on Unix and VMS. The Transfer Interface is written in strict ANSI standard C so as to be as transportable as possible. The Transfer Interface currently runs, with no modifications, on DECstations, SPARCstations, and VAX/VMS systems.

The Action Interface runs on the same three systems. The SPARCstation and DECstation versions of the Action Interface are identical, while the VAX/VMS version has minor variations which are placed, using C precompiler statements, in the same implementation.

The GRASP implementation uses the inherent file system structure of each operating system as the database for the archive. This approach greatly simplifies the program implementation from a data storage point of view, but must deal with the path naming schemes of individual operating systems. To address the complexity of working with the different path naming schemes, a number of small functions modularize the path interpretation/construction processes. This modularization was also designed in view of the necessity of reprogramming the Action Interface for a variety of archive devices. With these modules, the same model of programming for the Action Interface can be used, with the small modules providing an interpretation cogent to the current device.

With the file system devices currently targeted by the GRASP implementation, archives are built as a directory. Each separate archive has its own directory. Single files are stored in the top level of the directory. Filesets are stored as subdirectories of the main archive directory. Derived filesets are implemented as a set of pointers which take up minimal space. In the Unix version, the pointers are implemented as soft links. The VMS version of GRASP uses path names within files in the derived fileset to point to the actual file being referenced.

Context and data separation is not currently implemented for the VMS version of GRASP, but the software to perform the separation exists and will be integrated into the GRASP system.

Currently, the GRASP system manages the COSSC archive on a hard disk. Soon, the GRASP driven archive will be transferred to a 90 GB jukebox of magneto-optical disks. A secondary archive of less used data will be maintained on an optical jukebox of WORM disks to which a GRASP driven system will provide read access.

5. Future Directions

COSSC is currently using GRASP to archive data from the Compton Gamma Ray Observatory (CGRO). The experience of dealing with archiving the CGRO data has provided insight into the implementation requirements of the GRASP software, and the second version of the GRASP implementation is currently in use. A third version is in work. It will address the requirement for multidisk spanning which is driven by the need for more storage space on the current COSSC hard disk archive. Archive management functions will also be added as to provide a consistent basis for archive handling.

Also on the horizon is GRASPnet, which will link archives between various machines. The GRASPnet subsystem will rely heavily on the context and data separation which is part of the GRASP specification.

6. For More Information

More information can be obtained from Jim Jordan at:

```
jmj@enemy.gsfc.nasa.gov (Internet)
grossc::jmj (SPAN)
```

The source code for version 2 of the implementation, the specification paper and this paper (both in LaTex), may be obtained via anonymous ftp from enemy.gsfc.nasa.gov in the grasp directory.

Astronomical Data Analysis Software and Systems II
ASP Conference Series, Vol. 52, 1993
R. J. Hanisch, R. J. V. Brissenden, and J. Barnes, eds.

Some Practicable Applications of Quadtree Data Structures/Representation in Astronomy

László Pásztor[1]

MTA TAKI, Budapest Herman Ottó út 15., H-1025, Hungary

Abstract. Beyond the wide applicability of quadtrees in image processing and spatial information analysis there may be numerous further applications in astronomy. Some of these practicable applications based on quadtree representation of astronomical data are presented. Statistics of nodes at different levels may provide useful information on spatial structure of astronomical data in question. A sampling method based on quadtree representation of an image is proposed, which may prove to be efficient where observations were carried out previously either with different resolution or/and in bands.

1. Introduction

Briefly, the quadtree is a class of hierarchical data structuring techniques which is based on the recursive partition of a square region into quadrants and subquadrants until a predefined limit. Development of quadtree as hierarchical data structuring technique for representing spatial data has been motivated to a large extent by storage requirements of images and maps. For many spatial algorithms the time-efficiency of quadtrees in terms of execution may be as important as their space-efficiency. According to the principle guiding the partition process, the spatial resolution of a quadtree represented data set is generally variable. The term 'quadtree' is due to the fact that the decomposition process is represented by a tree of degree four (for details see Samet 1990).

Quadtree data structures are becoming more and more common in astronomy, thus it can prove to be useful to develop and adopt techniques for analysis of this type of data representation to exploit information on the represented data without transforming the quadtree structure into other, well-known formats.

2. Revealing the Spatial Structure of Point Patterns by Quadtree Statistics

Point patterns may be classified into three basic categories: aggregated, random and regular (Diggle 1983). An aggregated pattern shows dominant density irregularities, a regular pattern is roughly evenly distributed, in the case of a random

[1]Department of Astronomy of the Eötvös University, Budapest Ludovika tér 2., H-1083, Hungary

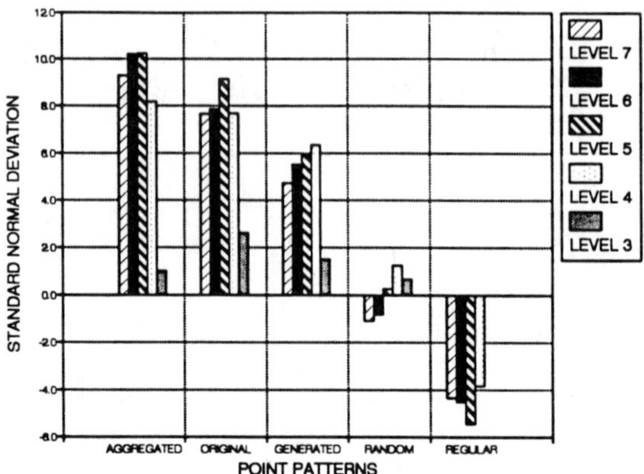

Figure 1. Join count (Moran) statistic (BB joins)

pattern there is no obvious structure. In terms of spatial autocorrelation aggregated and regular point patterns show positive and negative autocorrelation respectively, while there is no spatial autocorrelation in a random pattern. In course of the spatial analysis of a real sample, the first step is the classification of the pattern into one of the former categories based upon certain statistics.

The spatial behavior of a filtered sample of IRAS point sources (# = 272; $110^o < l < 130^o$, $5^o < b < 25^o$) was thoroughly studied previously (Pásztor et al. 1992). 10 groups were identified in the sample. In the course of this earlier work, as reference, Monte Carlo simulated samples were used which possessed the same large scale spatial structure as the original one but without its small scale fluctuations. In the present work the original IRAS sample, one of these generated samples and three reference samples representing the basic pattern classes are compared.

2.1. Join Count (Moran) Statistics

One of the simplest measures of spatial autocorrelation, join count statistics, was introduced by Moran (1948) for testing nominal scale data. In this context spatially structured data sets are treated as mosaics of areas with different colors. Elements with a positive nonzero common boundary are said to be linked by a join. The basis of join count statistics is the following: the distributions of joins between areas of two different and same colors respectively under the null hypothesis, H_0, of no spatial autocorrelation in the sample are asymptotically normal with moments determinable by the geometry of the data set. The first two moments can be used to test whether the number of various joins departs significantly from random expectations.

In the case of quadtree representation of point patterns binary coding seemed to be applicable. The criterion of partition of a quadrant into four subquadrants was based on the population of the quadrant: it was subdivided, if it contained more than one sample point. Black and white color was assigned to a leaf with surface density equal to or greater and smaller than $T/64$ re-

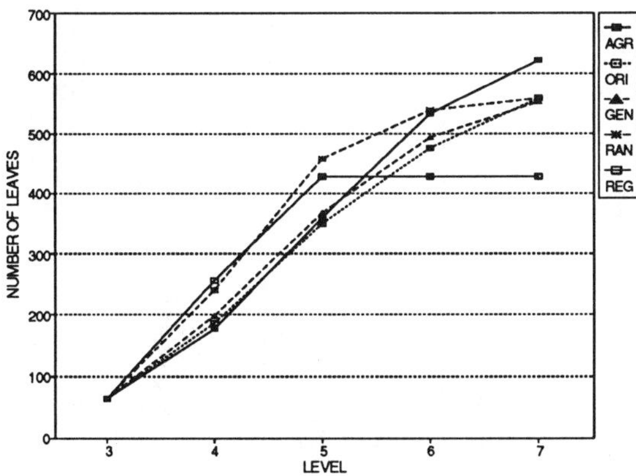

Figure 2. Leaf statistics

spectively (T is the area of the whole sample; black and white are traditional terms and referred to as B and W hereafter). In the binary case the first two moments of the BB join count are as follows: $\mu_1'(BB) = \frac{Cn_B(n_B-1)}{n(n-1)}$, $\mu_2(BB) = \mu_1'(BB) + \frac{Dn_B(n_B-1)(n_B-2)}{n(n-1)(n-2)} + \frac{[C(C-1)-D]n_B(n_B-1)(n_B-2)(n_B-3)}{n(n-1)(n-2)(n-3)} - [\mu_1'(BB)]^2$, where $C = \sum_{i=1}^{n} L_i$, $D = \sum_{i=1}^{n} L_i(L_{i-1})$, L_i is the number of joins of the ith leaf, n is the number of all leaves and n_B is number of B leaves.

According to Figure 1 the standard normal deviation of the BB join count statistic provides an efficient tool for discrimination among point patterns. The random point pattern behaves randomly from the point of view of Moran statistics as well. The regular sample shows negative spatial autocorrelation, while there is positive autocorrelation in the remaining three patterns, but to different extents, just according to their clustering characteristics.

2.2. Leaf Statistics

An additional informative statistic of quadtree represented point patterns is the cumulative number of leaves at different levels and the variation in number of leaves along the recursive partition. This kind of analysis may prove to be helpful in the design of storage requirements of different point patterns.

According to former results, the random point pattern can be considered as reference. As it can be seen in Figure 2, at lower levels the clustered samples are represented with fewer leaves than the random, while the regular uses more room. On the other hand, continuing the subdivision, storage conditions turn to the contrary. Consequently there is no a priori rule how to derive room requirements for different patterns because it is level depending.

However the section of a graph and the graph of reference (random) sample may provide certain parameters of spatial patterns. According to Figure 2 the section of graphs representing the regular and the random sample is situated between level 4 and 5 (closer to 4), which corresponds to a resolution between $T/256$ and $T/1024$, but closer to the former value. Since there are 272 points

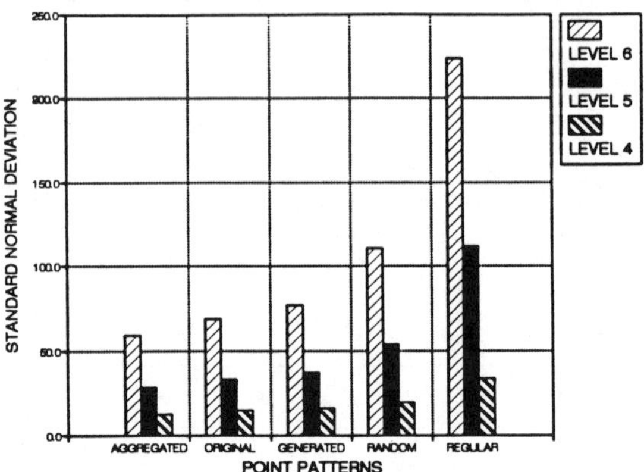

Figure 3. Leaf-pixel statistics (white pixels)

regularly scattered over the region, the average area assigned to each point is $T/272$, that is the characteristic dimension of the regular sample is reflected in the leaf statistics. The parameter of the aggregated sample may be derived similarly.

2.3. Leaf-Pixel Statistics

The leaf distribution of a quadtree represented data set may be transformed to a color distribution knowing the maximum resolution. Let's consider two subsequent quadtree decompositions of the same image descending until level $n-1$ and n respectively. Transition from level n to level $n-1$ involves merging leaves at level n into their father nodes, and the color of a pixel at level $n-1$ is determined by the colors of its four subpixels at level n according to a coloring rule.

If the pixels at level n are colored as B or W independently with probabilities $p_n(B)$ and $p_n(W)$ respectively, the probability of painting the merged pixel B or W can be computed by the aid of the coloring rule. In our case: $p_{n-1}(B) = p_n^4(B) + 4p_n^3(B)p_n(W) + 6p_n^2(B)p_n^2(W) + 4p_n(B)p_n^3(W), p_{n-1}(W) = p_n^4(W)$, at a regular transition. As the image is binary, the number of pixels with the same color has a binomial distribution, so both the expected number of B and W pixels and variance can be given. $E_{n-1}(C) = N_{n-1}p_{n-1}(C)$, $var_{n-1}(C) = N_{n-1}p_{n-1}(C)(1 - p_{n-1}(C))$; $\{C = B, W\}$, where N_{n-1} is the number of pixels at the given level. The resulting expected numbers and variances can be used to test randomness and significance respectively. According to Figure 3 leaf-pixel statistics may also be used for discrimination among different point patterns. However, from this point of view the random sample does not behave 'randomly', so it cannot be considered as reference. This is due to the definition of the guiding principle and coloring rule, which makes leaf-pixel statistics unsuitable for prediction of the necessary number of nodes to store the quadtree at a predefined level.

3. Elaboration of Sampling Strategy by Quadtree Decomposition

Prior to expounding how a sampling design may be elaborated based on the linkage of information theory and quadtrees, some basic guidelines should be emphasized. (i) A good sampling pattern is expected to provide information efficiently and economically. (ii) A priori knowledge may always prove to be useful in the design of a suitable sampling strategy. (iii) The guiding principle of the decomposition process is really flexible, so it may be defined based on some information theoretic criteria as well. (iv) The quadtree decomposition of a spatial data set may also be stopped between two subsequent levels.

There are a lot of definitions for measuring the information theoretical dissimilarities of an original image and its quadtree represented form: $I-$ and $J-$divergence, $J - M$ distance, entropy, etc. The information content of the quadtree represented image may also be expressed by its (maximum) entropy, Shannon-Wiener index, Simpson-index, etc. If the given quadtree decomposition is altered, the measures listed above change their values as well. To get a suitable sample pattern, the overall information measure should be reduced among certain predefined constraints.

Let's consider the following recursive algorithm. In the course of an elementary step the number of quadtree leaves (L) may remain constant, increase or decrease. In the first case a node is divided and four subquadrants are merged simultaneously, in the second case a leaf is subdivided and in the last case four sons are merged. At subdivision the leaf with maximum contribution to the overall information measure as well as that whose partition reduces the overall information measure in the greatest extent may be decomposed. At merging, similar criteria may be defined. To finish the algorithm, a stopping rule is required. In constant L case a limit should be given for the change in information measure, in altering L case, if the number of leaves reaches a predefined value, the algorithm may be stopped. So there are a great variety of techniques useful for the design of efficient sampling with variable resolution based on quadtree representations of an image, in which the places with great information content are focussed on due to the usage of information theory based guiding principles. However, the appropriate method should be selected by the user (Kummert et al. 1992).

Acknowledgments. I would like to thank the Organizing Committee and Jeannette Barnes for their assistance in making attendance to the Conference possible. I also acknowledge the useful conversations with M. Kertész, Á. Kummert and S. Kabos.

References

Diggle, P.J. 1983, Statistical Analysis of Spatial Point Patterns (Academic Press, London)

Kummert, Á., Csillag, F., & Kertész, M. 1992, in Proc. of EGIS '92, eds. J. Harts, H.F.L. Ottens & H.J. Scholten, 722

Moran P.A.P. 1948, J. Roy. Stat. Soc. ser. B 10, 243

Pásztor, L., Tóth, L.V. & Balázs, L.G. 1992, A&A, in press

Samet, H. 1990, Applications of Spatial Data Structures (Addison-Wesley)

Astronomical Data Analysis Software and Systems II
ASP Conference Series, Vol. 52, 1993
R. J. Hanisch, R. J. V. Brissenden, and J. Barnes, eds.

Data Indexing Techniques for the EUVE All-Sky Survey

James W. Lewis, Carl A. Dobson, and Vince Saba

Center for EUV Astrophysics, University of California, Berkeley

Abstract. This paper describes the use of quadrilateralized spherical cube sky addresses by the EUVE map library and catalog manipulation utilities.

1. EUVE Full Sky Maps

The Extreme Ultraviolet Explorer (EUVE) satellite, launched in June 1992, is nearing completion of its primary mission, an all-sky survey over the entire EUV band (Bowyer & Malina 1991). The EUVE end-to-end software system (Vedder et al. 1992) will use sky survey data to produce high-resolution full sky maps of raw photon counts, exposure time, and exposure-corrected count rates. To organize our maps, we have used the quadrilateralized spherical cube (also known as quad cube, quad tree, or cubic pixel) indexing scheme. This technique divides the sky into six faces, and subdivides each face into pixels having equal areas. A complete description of the forward and inverse transformations is beyond the scope of this paper; the details are explained in the original publication (Chan & O'Neill 1975). The EUVE libraries are an extension of the implementation used by the COBE project, and their overview of the technique (White & Stemwedel 1992) will be useful to prospective users.

In the EUVE implementation, the face number and pixel coordinates within that face are packed into a 32-bit quantity which can be used as an index into a map. The pixel coordinates are interleaved bit by bit in such a way that rebinning and subsampling can be done with simple shifting and bitwise masking operations. Another attractive property of this implementation is the locality of the addresses. Except for rare cases (such as adjacent pixels spanning a cube face boundary), pixels that are close to one another on the sky will have addresses that are close together. This is a key element for obtaining high-performance disk access in support of typical map operations. The forward and inverse transformations between pixel addresses and celestial coordinates are well-behaved, invertible, and efficient enough so that operations such as full-sky convolutions are feasible to perform directly in cubic pixel address space.

2. Map Layout and Buffering

Although our full sky maps are each conceptually a single data structure, they are implemented by splitting each map into a group of files. For example, the finest resolution possible with 32-bit pixel addresses is about 20 seconds of arc per pixel. A full sky map at this resolution would be split into 1536 files, each file containing about 2 megabytes of data (assuming the pixel values are stored as 16-bit integers). Owing to disk space constraints, we are usually forced to work at a coarser resolution for full sky maps. At 80 arcsecond resolution, a fully scanned map would be split into 96 files of 2 Mb each (assuming the same data type), for a total of 192 Mb. At this resolution, the maps are reasonably sized, while still being fine enough for tasks such as source detection.

The EUVE map library implements a fine-grained buffering scheme to minimize disk traffic. Each file is subdivided into blocks of 1024 bytes, and the map library maintains a cache of map blocks that have recently been accessed. Since nearly all operations on maps result in access patterns with excellent locality, a modest in-memory cache reduces disk traffic to nearly optimal levels.

The routines in our map library all take a map identifier (used in much the same way that Unix uses file pointers) to distinguish between different maps opened by the same application. The maps are indexed by cubic pixel addresses, which are reinterpreted by the map library as a set of bitfields specifying a file number, block number, and pixel offset. The data block (loaded from disk, if necessary) is retrieved from a hash table (using the map identifier, file number, and block number as a hash key), and the pixel offset is used to select the data of interest.

3. Why Not use Tangent Plane Maps?

We have found that the benefits of the EUVE mapping scheme outweigh many of the disadvantages, such as the current lack of standardization and the tendency for pixel shapes to vary across the sky. For example, while the maps are implemented as sets of distinct files, our map library provides a seamless, non-overlapping abstract data structure. Such a structure cannot be achieved with a set of tangent plane images, which must overlap in order to cover the entire sky without gaps. The equal area property of cubic pixel addresses protects EUVE skymaps from the distortions in area that occur in tangent plane maps as one moves away from the tangent point. (While it is easy to convert an EUVE skymap into a set of tangent plane images, the inverse conversion would be quite difficult because of overlaps and tangent plane distortions.) Our buffering scheme allows us to operate efficiently on the entire sky at once or to zoom in on a tiny region while only reading the parts of the map that are actually in use.

4. FITS Structure for Large Maps

The large storage requirements of high resolution full sky maps present serious difficulties for proposed FITS structures (White & Stemwedel 1992), which would store an entire map in a single FITS file. The idea of storing the map as a data cube in FACE, X, and Y coordinates does not scale up when the maps may each contain hundreds of megabytes of data, as they do for the EUVE survey. It may be more appropriate, in the case of high resolution maps, to treat a map as a one-dimensional array indexed by cubic pixel addresses. A FITS structure which includes the map resolution, data type, start address, and end address would then be able to represent an entire low-resolution map, or one piece of a high-resolution map.

5. Pixel Addresses as Position-Dependent Hash Keys

We have found another, novel use for cubic pixel addresses. Consider the problem of comparing two catalogs to find all the objects which are common (within a given positional tolerance) to both. For example, one might want to compare the output of a source detection algorithm to a master catalog of suspected EUV sources to see which ones were detected. The naive approach, a brute-force comparison of every item in the first catalog against every item in the master catalog, would be quite slow if the catalogs involved were very large. (Even if the catalogs are sorted by RA, one cannot necessarily restrict the search to a small range of coordinate values for each input object. Near the poles, two objects with completely arbitrary RA values might still match within the given tolerance.) Cubic pixel addresses provide an elegant solution: they can be thought of as a hashing scheme where the hash keys are directly related to sky position. We can make a table indexed by cubic pixel addresses at some fairly coarse resolution (tens of arc minutes) and treat each entry as a bucket containing catalog entries from that small part of the sky. This allows a huge reduction in the number of entries from the master catalog which need to be tested for each item in the input catalog; one need look at only the catalog entries for a single bucket and its neighbors, rather than scan the entire master catalog for each input entry.

Acknowledgments. We thank the principal investigators, Roger F. Malina and Stuart Bowyer, and the EUVE science team for their advice and support. We also thank Mike Lampton for reviewing this paper and giving us many helpful suggestions.

This research has been supported by NASA contracts NAS5-29298 and NAS5-30180.

References

Bowyer, S., & Malina, R.F. 1991, The EUVE Mission, in Extreme Ultraviolet Astronomy, eds. R.F. Malina & S. Bowyer (New York: Pergamon Press), 397

Chan, F.K., & O'Neill 1975, Feasibility Study of a Quadrilateralized Spherical Cube Earth Data Base, EPRF Tech. Rep., 2-75 (CSC)

Vedder, P.W., et al. 1992, in Astronomical Data Analysis Software and Systems I, A.S.P. Conf. Ser., Vol. 25, eds. D.M. Worrall, C. Biemesderfer, & J. Barnes, 496

White, R.A., & Stemwedel, S.W. 1992, in Astronomical Data Analysis Software and Systems I, A.S.P. Conf. Ser., Vol. 25, eds. D.M. Worrall, C. Biemesderfer, & J. Barnes, 379

Astronomical Data Analysis Software and Systems II
ASP Conference Series, Vol. 52, 1993
R. J. Hanisch, R. J. V. Brissenden, and J. Barnes, eds.

A Distributed Clients/Distributed Servers Model for STARCAT

B. Pirenne

Space Telescope – European Coordinating Facility, European Southern Observatory, Karl-Schwarzschild-Str., 2, Garching bei München, Germany

M. Albrecht

European Southern Observatory, Karl-Schwarzschild-Str., 2, Garching bei München, Germany

D. Durand, S. Gaudet

Canadian Astronomy Data Centre, Dominion Astrophysical Observatory, 5071 W. Saanich Road, R.R. 5, Victoria, B.C. V8X 4M6, Canada

Abstract. STARCAT, the Space Telescope ARchive and CATalogue user interface, has been around for a number of years already. During this time it has been enhanced and augmented in a number of different areas. This paper is about a new capability allowing geographically distributed STARCAT interfaces to connect to geographically distributed data servers. Moving towards a window-based STARCAT is another goal being pursued: a graphic/image server, a help/doc server and a cross-correlation tool are currently being added to it. They should further enhance the functionality, ease of use and access transparency.

1. Introduction

The development of the Space Telescope Archive and Catalogue (STARCAT) started some 6 years ago at the Space Telescope – European Coordinating Facility with collaborations from the STScI and later from the Canadian Astronomy Data Centre at DAO. The intention was to provide an ASCII-based interface to browse through the Hubble Space Telescope catalogue, expected to be made available immediately after launch.

The HST launch only took place 2 1/2 years ago and in the mean time STARCAT developed useful catalogues search capabilities. A number of important astronomical catalogues were introduced and internal users at ESO/ST-ECF and STScI quickly became familiar with the interface.

If six years ago, Unix and window-based user interfaces, client-server models and high-speed networks were available, they were not really introduced in astronomical observatories where the future users of the system were supposedly located. This historical fact explains why the system was developed chiefly for ASCII terminals. Early on, however, the underlying concepts were those of windowing systems (emulation of windows on ASCII terminal screens), of client-servers orientation (the database server can be reached by STARCAT anywhere on the network). STARCAT is also relying on an operating system interface which makes the software independent of the underlying computer architecture and of the operating system. All these reasons explain why STARCAT is still around and improving.

The next sections will explain what these improvements are and how they take advantage of the high speed networks and of the now popular graphical user interfaces, while remaining useful for remote-access ascii-terminal users.

2. The Current Client-Server Model

The mechanism used by STARCAT to provide users access to catalogue information is through a form interface that helps build relational database queries in the SQL language. Users need not know anything about the underlying system, for the language sentences are invisible to them. The generated queries are passed via local or wide area network to the database server machine which analyses them and sends answers using the same channel. STARCAT upon receipt of the results translates them into a form displayable to users and split the information across the various fields on the screen. See Figure 1 for an example of the STARCAT form interface. As mentioned above, STARCAT can be located anywhere on the TCP/IP network (or DECNET network, given a special interface). The transparency of the socket interface makes the connection possible. The only restriction remains the performance of the line. Upon installation at any given site, it suffices to tell STARCAT the address of the database server it has to connect to. In other words, one STARCAT installed anywhere on the network can only get the information from one data server. If for one reason or another that computer is not reachable the STARCAT session will fail.

3. The Future Multi-clients/Multi-servers Model

Given these considerations, it became evident to us that a simple improvement in our STARCAT code could lead to a tremendously powerful tool for users. As a matter of fact, if the STARCAT server connections could be made dynamic rather than fixed, we could automatically give any user immediate access to the data stored in any server with a minimum of requirements from the part of these data centers:

- grant access permission to this particular client,
- share a minimum of data structures with all other servers so that STARCAT can have access to the local data dictionary,
- have a database management system that is supported by STARCAT.

The advantages of such a system would be extremely appealing:

- transparent access to several data centers (the multi-server concept),

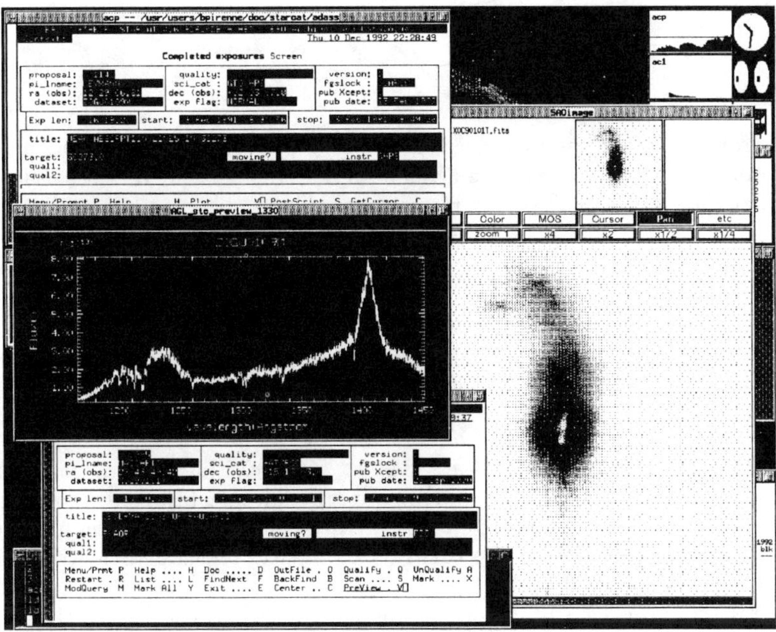

Figure 1. The present STARCAT with its form interface designed to facilitate the interrogation of catalogues. Note that already now, STARCAT takes advantage of the bitmap screens and can retrieve and display on-line quick-look images and spectra.

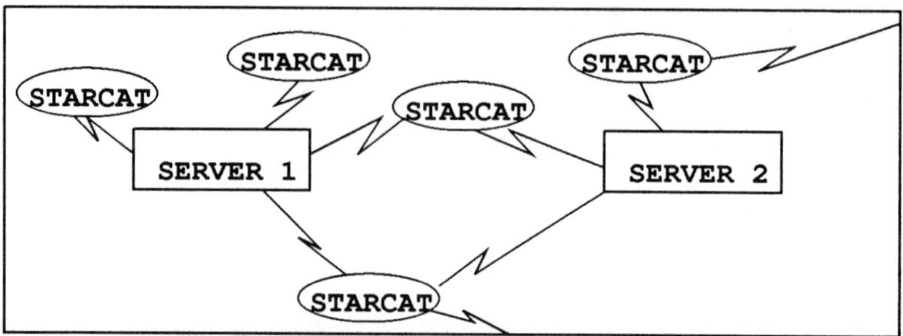

Figure 2. The multi-server concept or "STARCAT *village*". Users are transparently connected to a given data center depending on the service that they chose or that is currently available.

- access to a replication of the same information, if the default server for this information is temporarily not accessible (the backup data server),
- complementary parts of one catalogue could be available from two distinct servers (e.g., catalogue available in server A and B and associated quick look data only available from server B).

This concept which we call the STARCAT "*village*" is depicted in Figure 2. In the first implementation of the system, we are planning to involve 2 or 3 sites. The first two would be the ESO/ECF archive and the Canadian Astronomy Data Centre while the third one could be the Space Telescope Science Institute with which we are collaborating.

The system will be based on a configuration file which defines the various *services* available at all *data centers* participating in STARCAT. Some services available at more than one site will be called *redundant* and will be used to back each other up. Table 1 shows an example of such a configuration for an STARCAT client installed somewhere in Europe.

4. A Window-Based STARCAT

Bitmapped terminals and workstations are now familiar terms in many scientific institutions around the world. The advantages of such systems over the standard ASCII-terminals are enormous: multi-tasking capabilities, pointing device capabilities in conjunctions with easy-to-understand icons and pull-down menus. However, adapting existing complex user interfaces to a windowing system while keeping backward compatibility with previous ASCII terminal capabilities is not a simple enterprise. This is nevertheless the commitment that the STARCAT development teams have taken. To achieve this, we are planning to add more and more capabilities *outside* STARCAT, in close connection with it. The idea is to add future new functionality as external, window-based client applications to the existing STARCAT which would remain the query server for those external

Table 1. Example of a STARCAT services configuration table

Service	Service Type	Server	Database	Visit order	Report-Mail
astrocat	catalog	ESOECF	astrocat	1	bpirenne@eso.org
astrocat	catalog	CADC	astrocat	2	bpirenne@eso.org
astrocat	preview	ESOECF	preview	1	bpirenne@eso.org
astrocat	preview	CADC	preview	2	durand@dao.nrc.ca
hst	catalog	ESOECF	hst	1	bpirenne@eso.org
hst	catalog	CADC	hst	2	durand@dao.nrc.ca
hst	catalog	STSCI	dmf	3	dsimonton@stsci.edu
hst	request	ESOECF	archive	1	bpirenne@eso.org
hst	request	CADC	archive	2	gaudet@dao.nrc.ca
hst	request	STSCI	archive	3	dsimonton@stsci.edu
hst	preview	ESOECF	preview	1	bpirenne@eso.org
hst	preview	CADC	preview	2	durand@dao.nrc.ca
eso	catalog	ESOECF	archeso	1	malbrech@eso.org
eso	preview	ESOECF	preview	1	malbrech@eso.org
eso	request	ESOECF	archive	1	malbrech@eso.org
cfht	catalog	CADC	cfhtcat	1	durand@dao.nrc.ca
cfht	preview	CADC	cfhtcat	1	durand@dao.nrc.ca
cfht	request	CADC	archive	1	durand@dao.nrc.ca

applications. This conversion operation was already started in 1992 with the delivery of very popular features like the PreView quick look facility and the Stella finding charts generator. Future external applications to be released in 1993 also include a generalized image/graphic display, a documentation utility and a catalogue cross-correlation environment. Understandably, some of these functions simply cannot be made available to ASCII-based terminals users, but the basic STARCAT query capabilities will be maintained in any case.

Future "X" additions to the STARCAT query server itself will include in the first stage of implementation pointing device interaction capabilities to facilitate cursor positioning and command choice through the menus.

5. Conclusions

From a form-based database query interface for ASCII data, STARCAT has evolved to become a generalized data access interface supporting retrieval and display of ASCII data; 1-d and 2-d datasets from on-line data archives. The multi-server access capabilities will enhance the existing features and give users simple and transparent access to some of nowadays most active and dynamic astronomical observatories' data. We also think that the addition of long-awaited tools like the cross-correlation environment and the progressive compliance to today's graphical user interface standards will bring STARCAT up to date with the technology and more importantly will satisfy most archive access requirements.

Astronomical Data Analysis Software and Systems II
ASP Conference Series, Vol. 52, 1993
R. J. Hanisch, R. J. V. Brissenden, and J. Barnes, eds.

StarView: The Object Oriented Design of the ST DADS User Interface

J. Williams

Science & Engineering Systems Division, Space Telescope Science Institute, Baltimore, MD 21218

Abstract. Astronomical archives can hold vast amounts of data. Finding your way through this forest of data can be a daunting problem. StarView is the user interface for the Space Telescope Data Archive and Distribution Service (ST DADS). It was designed to aid scientists in finding their way through the enormous amount of data found in ST DADS. It was designed using object oriented methodology and implemented in C++. The object oriented approach has supported the inclusion of design features needed to manipulate and understand these data.

1. Background

Astronomical archives can hold vast amounts of data. Finding your way through this forest of data can be a daunting problem. The problem is even more complex when many different attributes are needed to describe this data in meaningful ways. The Space Telescope Data Archive and Distribution Service (ST DADS) provides a good example of this complexity. The catalog describing data has over 200 tables and 1500 attributes. Providing a method for flexibly and meaningfully viewing this volume of data is beyond the scope of existing user interfaces. The frameworks of interfaces such as STARCAT and ADS are oriented towards smaller databases. While adequate for smaller databases, they lack the fundamental capabilities needed to flexibly handle large amounts of data.

What is needed is a user interface that can provide the scientist an intuitive method of finding his way through the morass. This interface must provide support for the scientists' desire to quickly find the data they wish to study. It must be flexible enough to let them search for information in ways that are meaningful. It needs the power to support complex indeterminate queries. It should also be adaptable to other archives stored as relational databases. StarView was designed to accomplish these goals.

2. StarView as a Solution

StarView was designed from the beginning to handle the complexity of ST DADS. It has the fundamental underpinnings necessary to overcome the inadequacies of current systems. It accomplishes this by providing a variety of powerful tools to support the researcher. The AdHoc query generator lets the scientist define queries on the fly. He does not need to understand the structure of the database. This removes the necessity of the researchers becoming database or SQL gurus. It also frees them from the restriction of using only pre-defined forms.

StarView also expands their capabilities for viewing and combining information in new ways. The complete set of query results are kept in a local database. The scientists can scan back and forth through these results at will. No information will be lost. They can view them on a standard form or in a table-row (spreadsheet) format. There is no limit on the number of columns that can be displayed. The entire table-row format is scrollable both vertically and horizontally. If information within a column is longer than can be displayed at once, the field itself can be scrolled.

The local database is an extended relational database. It provides the scientists with a new range of possibilities. Information can be read in from other sources and correlated with the results of a query. If the scientists wish to view the query results outside of StarView, they can be saved in a FITS or ASCII format file. The scientists can also use imported information as the basis of queries. This capability would, for example, allow them to lead in a list of favorite supernovas and then do a search for these targets in the DADS catalog.

The framework of StarView was designed to accommodate change. Useful systems expand and grow. The object oriented approach to system design creates a design that is more amiable to change. Most changes to systems are changes in function. This can destroy a functional design. Object oriented systems are easily extended by adding new capabilities to well encapsulated existing objects or adding additional objects as necessary. Neither of these requires fundamental changes to an existing design.

3. StarView Design Features

StarView includes several features to meet the needs defined earlier. The internal relational database provides the researcher with new ways to manipulate and understand his query results. A form definition language was created which allows pre-defined forms to work on both CRTs and X window interfaces. StarView also incorporates its own data definition language which allows it to work with any relational database.

With the AdHoc generator, StarView supports the arbitrary selection of an indeterminate number of attributes to construct a query. The generator determines the joins automatically and generates the SQL query. StarView also lets the user create a single query from single or multiple forms. By merging the selected attributes into a single query, the user can create a meaningful query from selected portions of pre-existing forms.

The scientists can flexibly navigate query results. Single or multiple predefined forms can be used to show all or part of a returned result. Additionally, the user can select a completely scrollable table-row format. This is similar to a spreadsheet display.

4. Key StarView Object Oriented Abstractions

Object oriented analysis and design techniques were used to support the mechanisms and concepts required to implement these features. The object oriented approach provides several concepts that support the development of robust modifiable systems.

Information hiding is provided by encapsulation. Internal data structures and methods can be hidden within an object. This lets interfaces between objects remain stable while allowing internal class changes to occur. The effects of changes are localized within an object. Code reuse is supported by inheritance. This allows code functionality to be reused without being duplicated everywhere. Polymorphism supports late binding and allows decisions to be made at run time rather than compile time. Data abstraction is supported through the use of classes. With classes, the system can be designed and implemented in terms of the problem domain. Low level implementation issues can be hidden.

Another powerful concept is that of mechanism. A mechanism is the interaction of objects to produce a higher level behavior. A mechanism used in StarView is the Model-View-Controller (MVC). The MVC is defined by the interaction of three basic class types. The Model is the underlying system or data. In StarView this is the query and the results of that query. This is the "data" or model that is the heart of the system. The Viewer is the set of classes that display the model to the user. The Controller is the class which monitors and processes user input. The interaction of these three classes defines the visible user interface. The structure of the Model is independent of the view of the data. The model knows how to display itself. The Viewer provides a particular perspective on the model without having to know the defaults of data being displayed. The Controller interacts with the user to let him change his view of the model or change the model itself.

Within StarView, the MVC allows the user to have multiple views of the underlying database while providing a consistent mechanism for interaction regardless of the view. This approach supports both CRT and X windows interfaces while providing a common mode of user interaction. It supports both individual forms and table-row formats without affecting the underlying model. In the X version, multiple simultaneous views of the model are possible.

Another powerful abstraction in StarView is the Query Model. This is the set of classes that abstract the notion of a query. It allows a query to be built from single or multiple forms. Supporting this model is the AdHoc query generator. Both the model and generator are designed to work with any relational database system.

System connectivity has benefited from the object oriented approach. Connectivity is both isolated and interchangeable. StarView supports both direct connection and the client/server approach. Alternative connection approaches are feasible as needs change. StarView has abstracted the fundamental aspects of requesting services without unduly restricting specific approaches. Additional services can be easily added.

5. How the Object Oriented Approach Worked

What advantages have we actually seen using object oriented development? Code reuse through inheritance has reduced the number of classes we needed to develop. It reduced the amount of code we needed to write for classes that can inherit from others. The use of polymorphism has aided the development of a flexible user interface. Models for connection to different views can be determined by the user's actions at runtime. We do not need to be restrictive in the freedoms or flexibility we allow the user. In developing presentation classes, we have seen the value of encapsulation. In our effort to support both CRT and X interfaces, we have been able to isolate unique display issues.

The object oriented approach also makes extensive use of ongoing integration to create a system that continually evolves in functionality. We were able to establish milestones that showed this increasing functionality. This means that there is no big bang integration step at the end of the development process. By the end of development, we will already have significant experience in building and testing StarView. There are no major surprises or crises. The final integration step becomes one of thorough systemic testing.

Are there problems with the object oriented approach? Certainly. The development tools are not yet stable and are not available for all development platforms. UNIX systems have the best support. VMS is weak in comparison. We have also found that interface packages developed for C don't work well with C++. At best they simply need changes to the header files. At their worst, there are subtle interactions which cause the system to crash in inexplicable ways. Finally, C++ is a subtle language. While the basic semantics can be learned in a short time, the subtleties of object creation and destruction can vary from complier to complier. C++ requires a very thorough understanding of both an object's and a compiler's behavior. Object oriented development is not just a language issue. You can't simply send your team off for C++ training and expect them to create a robust system. Just learning how to create classes will get you a collection of bad classes. You need to understand what makes a good class and what makes for good relationships between them.

References

Booch, Grady 1991, Object Oriented Design with Applications (Benjamin Cummings)

Rumbaugh, James, et al. 1991, Object-Oriented Modeling and Design (Prentice Hall)

Astronomical Data Analysis Software and Systems II
ASP Conference Series, Vol. 52, 1993
R. J. Hanisch, R. J. V. Brissenden, and J. Barnes, eds.

Integrating a Local Database into the StarView Distributed User Interface

D. Silberberg

Space Telescope Science Institute, 3700 San Martin Drive, Baltimore, Md 21218

Abstract. StarView is a user interface onto the Space Telescope Data Archive and Distribution Service (ST-DADS). Scientists use StarView to browse the DADS catalog for archived files they wish to retrieve. Generally, user interfaces are limited in functionality by the database access medium and are limited in scope by the contents of the database. StarView incorporates a local database which gives users capabilities beyond that of the database medium and provides users with a faster response time. Most importantly, however, it broadens the view of the data by providing methods for manipulating derived attributes, and it permits scientists to understand the catalog data in a context far beyond that of just the DADS catalog.

1. Introduction

StarView is a user interface onto the Space Telescope Data Archive and Distribution Service (ST-DADS), the repository for all scientific and engineering data sets produced by the Hubble Space Telescope (HST) and ground system. DADS consists of the archive subsystem, which manages the archiving and retrieving of the data sets from optical disk files, the catalog subsystem, which is a relational database describing the archived data, and the host computers, which manage access to the archive and catalog. Scientists and engineers use StarView to browse the catalog data, select the records that describe the archived data they want to retrieve from DADS, and submit a request for the data.

StarView screens are windows onto the DADS Catalog database. The catalog database is comprised of almost 1500 descriptive fields distributed over approximately 40 relational tables. The catalog database is highly normalized. Each table is a meaningful grouping of fields, but are only informative when "joined" with fields from other tables. Therefore, typical meaningful StarView screens include fields from 4 to 10 tables. Users browse the catalog by entering field qualifications on the StarView screen and submitting the request. StarView builds a corresponding complex SQL query and transmits it to the DADS Catalog for evaluation. The result records are returned to StarView row-by-row. The data are displayed on the screen and also stored in a database table local to the StarView application. This paper is devoted to explaining why the local database provides more flexibility to manipulate and view the data, and permits scientists to understand the catalog data in much greater context than a standard user interface.

2. The Problem With Most User Interfaces

The world of user interfaces has been limited to fancy SQL interfaces onto catalog data. Generally, they can not provide more data access functionality than the database medium provides. If the database medium provides a "get next row" function, the user interface also provides a "get next row" function. If the database medium provides a "get previous row" function, the user interface also provides a "get previous row" function. The only view of the data is that provided by the database medium. The only operations on the data are those provided by the database medium. Since most of the scientific databases reside on relational databases, the access to and manipulation of the data is limited by SQL. Clearly, we want the ability to understand and manipulate data without limitations imposed by the database access methods.

SQL is limited by the lack of a "get previous row" command. Relational databases do not provide this because tables do not inherently order data. Scientists, on the other hand, retrieve data on a row-by-row basis and certainly want to review previously retrieved records via the user interface. Most user interface systems do not provide a way to review previously retrieved data without resubmitting the query or saving a limited amount of retrieved data. Scientists need to view all previously retrieved records in a lossless manner, without resubmitting the query.

Most queries of the DADS catalog are composed of complex, multi-table joins. Often, scientists need to refine query results. This is accomplished by resubmitting the complex query with a more restricted set of field qualifications. Executing this query requires significant database overhead due to the number of tables involved in the join and the volume of data in the catalog. However, this query should not be necessary. The data to be refined were already returned by the first query in the form of a single table. Certainly, refining a single table of limited data is more efficient than resubmitting the complex query to the DADS catalog. Also, users would not be burdened waiting for the query to complete.

Both "get previous row" and refinement operate on previously retrieved data. If these are implemented as queries issued to the DADS Catalog, network overhead and catalog overhead are incurred unnecessarily. It is more efficient to save the results of the initial query and operate on it during subsequent queries.

Finally, and most importantly, the user interface is limited to displaying and manipulating data from the catalog on which it is operating. While, some user interfaces have the ability to display derived attributes, they certainly do not have the capability of saving them, much less operating on them. It is desirable to cross reference other catalogs with the results of DADS queries, including the derived fields. Also, it is beneficial to provide the facility for exporting the results in ASCII and FITS format to be used by other analysis systems.

StarView addresses all of the aforementioned problems by incorporating a local database in the user interface.

3. The Local Database

StarView incorporates the Requiem relational database (Papazaglou & Valder 1989) into the user interface. Requiem is the database of choice for multiple reasons. It is public-domain software requiring no fees for distribution. StarView must be distributed free of charge to all interested scientists and engineers. Also, since it is in the public domain and is documented in a 500+ page book, it is extensible by the StarView developers. StarView must be supported on the Unix, Ultrix and VMS platforms—Requiem runs on any platform with a C compiler. Finally, Requiem is callable from StarView since StarView is written in C++ and Requiem is written in C. No other database system examined met all of these criteria.

When a DADS query is submitted, a corresponding table is created in the local database. As rows are returned, they are displayed on the StarView screen and placed in the local table. The Requiem software was modified to support a "get previous row" command on a single table. Therefore, the query does not have to be resubmitted to get the previous row. Nor are previously retrieved records lost since the local table captures the entire query result. the "get previous row" command retrieves data from the local table to the screen instantaneously.

When the user refines a query result, no longer does a query as complex as the first need to be submitted to the DADS catalog. Since a single local table of the query results exists, only a simple query operating on one table need be submitted to the local database. The query is executed quicker, thus giving a faster response time.

Each local database query lessens the load on the catalog database. This is significant when multiple users submit queries to the DADS Catalog. furthermore, queries submitted to the local database do not incur the overhead of network communication between the session and the host computers. Therefore, querying the local database increases the user interface response time and lessens the load on the catalog database.

Most importantly, scientists can store query results in tabular form, import other catalogs, and cross correlate them. This expands the scientists view of the data beyond the DADS catalog. It affords the opportunity to view the HST data in the broader context of other scientific catalogs. Furthermore, not only is the raw catalog data available for analysis, but the derived data also can be used integrally in the analysis. In the future, StarView will serve as the interface to other catalogs. Although distributed database technology is not generally available, StarView will provide the capability of creating local tables from multiple catalogs and cross correlating them. Local tables and cross correlation results can be exported in ASCII and FITS format for analysis by other systems.

4. Summary

Most database user interfaces are no more than a visual interface for database access methods. If database access methods are available, the user interface provides the corresponding functionality. If the access methods are not available, neither will they be available on the user interface—even if it is appropriate for the user interface. By integrating a local database, StarView broadens its view of the DADS catalog as well as exceeds the capabilities of the database access methods. The local database allows a "get previous row" operation, streamlines query refinement, and lessens the resource load on the network and DADS catalog. More importantly, query results, including derived fields, can be saved as local tables. External catalogs can be imported for cross correlation with the query results. Query results and cross correlation results can be exported in FITS and ASCII format for use by other analysis systems. In all, the use of a local database expands the horizons of scientists beyond that provided by standard database user interfaces.

References

Papazaglou, M., & Valder, W. 1989, Relational Database Management - A Systems Programming Approach (Prentice Hall International, UK)

Astronomical Data Analysis Software and Systems II
ASP Conference Series, Vol. 52, 1993
R. J. Hanisch, R. J. V. Brissenden, and J. Barnes, eds.

Recommendations for a Service Framework to Access Astronomical Archives

J. J. Travisano

Computer Sciences Corporation, Space Telescope Science Institute, Baltimore, MD 21218

J. A. Pollizzi

Space Telescope Science Institute, Baltimore, MD 21218

Abstract. This paper examines many of the issues involved in supporting distributed access to astronomical catalogs and archives.

1. Introduction

With the move towards client-server systems, the astronomical community would benefit by having a standard framework for distributed access to catalog and archive services. With a standard framework and a set of software interfaces, users would have access to more data from within their favorite user interfaces and data analysis systems.

In the sections below, many of the protocol and software issues of a client-server framework are discussed. We considered these issues during our work with StarView, the user interface to the Space Telescope Data Archive and Distribution Service (ST DADS).

2. Networking Protocols

The *de facto* standard networking protocols are those in the Internet Protocol (IP) suite: Transmission Control Protocol (TCP/IP) for reliable, connection-oriented service and User Datagram Protocol (UDP/IP) for connectionless, datagram service. DECnet is used in the VMS world, although many VMS sites support TCP/IP and UDP/IP as well.

The recommended approach for client-server systems today is to use TCP/IP and/or UDP/IP, depending on the need for reliability and performance. DECnet support would be beneficial for many sites but is not crucial.

3. Distributed Computing Packages

3.1. Interfaces

There are a few well-known software packages for distributed computing, specifically oriented towards the Remote Procedure Call (RPC) paradigm. These packages support a means to define remote procedures and their parameters, functions for client-server code to invoke these procedures, and supporting software for automatic data translation, networking input/output, etc. Automatic data translation—byte swapping, floating point conversion, and support for pointer-based data structures such as linked lists—is especially important in a heterogeneous computing environment.

Sun RPC, now part of Sun's Open Network Computing (ONC) architecture, has been around for many years. The source is freely available (with some restrictions on redistribution) and has been ported to a wide range of systems.

The Open Software Foundation (OSF) has developed the Distributed Computing Environment (DCE) using contributed software from HP/Apollo, DEC, and others. It is available on a few platforms now, with plans to support many more. Cost and licensing remain open issues, although DCE may soon be bundled with some operating systems.

ANSAware, based on the European ESPRIT project's Advanced Networked Systems Architecture (ANSA), is another package that supports distributed computing and the RPC paradigm. It is a commercial product with licensing and cost limitations.

These packages support the basic features needed to establish interfaces for catalog and archive access. Sun RPC has the advantages of being free and widely available on a number of systems. DCE has more advanced features and looks promising but is not yet widely available, so experience with it has been limited.

We recommend the use of Sun RPC now, due to its wide availability. Nonetheless, we look forward to DCE if the cost, licensing, and availability issues are resolved for the major platforms used in the community.

3.2. Servers

There are two modes of operation for servers: permanent and transient. Permanent servers are always running. They receive a client request, process it, and then wait for the next request or other connections. Transient servers start each time a client makes a connection, process the client's requests, and then exit when the client is done. This second mode is supported by the Unix Internet services daemon (inetd).

Permanent servers are generally preferred since only one server process need be running at a time. There may be a problem, however, if the server needs to block for any reason. For example, if the server blocks to access a database, then all other clients would also be blocked. The use of transient servers overcomes this problem by having separate copies of the server handle individual clients. This simplifies the software at the expense of additional cpu and memory usage, since there is the potential for a large number of such server processes.

Another strategy for supporting simultaneous clients is to have a multithreaded server. The server creates a processing thread for each client connection. Each thread can execute (and block) independently. Many operating systems and RPC packages now support this type of server concurrency.

Servers must also provide for security. It is often easier to have a transient server process start or later transform itself into the user id of the client, thus providing security implicitly through the operating system login mechanism.

There is no specific recommendation for the types of servers to use. It depends largely on the nature of the service. The client application usually does not know or care how the server is invoked.

3.3. Directory Services

A client needs to know how to connect to a server or a service on the network. In simple cases, this may be hard-coded or parameterized in an application. With more distributed applications, directory services are now prevalent.

Given the name or other unique identifier for a service, a client sends a lookup request to a directory or name server. The name server returns the information needed for the client to connect to the appropriate server. In Sun RPC, this name server is known as the *portmapper*. It translates an RPC program identifier (a unique number) into a TCP/IP port number on a given host. DCE incorporates a naming server which is compatible with the CCITT standard X.500 Global Directory Service.

Since we are recommending Sun RPC, the portmapper is used. Well-known services could have standard program ids and network port numbers.

4. Security

Various methods exist for supporting security in distributed computing. Authentication information can be passed from client to server as part of RPC connection packets, which the server uses to verify the user. DCE incorporates the Kerberos system from MIT's Project Athena. Kerberos supports a number of security features related to authentication with encryption. While authentication identifies a user to the server, the server is then responsible for granting access to specific resources. The server controls this access with support from the operating system or RPC package features such as access control lists.

On a higher level, there are different user registration requirements and security needs for each catalog and archive supplier. There may be sites where guest accounts are acceptable for all services, where others require full registration. In active astronomical missions such as HST, there are also proprietary data restrictions which must be enforced.

We recommend that only non-proprietary data be made available for distributed access now. Any protocols implemented for data access should be extensible to support evolving security features. The fuller use of security can be revisited once a common approach has been embraced by the community.

5. Catalog Access

5.1. Database Systems

We must consider the Data Base Management System (DBMS) used, its query language, and its programming interface (if any), in supporting an open client path to a catalog. Many sites use a commercial relational DBMS using Structured Query Language (SQL) and a proprietary programming interface. Other catalogs are implemented as custom database systems, flat files, or CD-ROM files, with specialized interfaces and query languages. It is often straightforward to add support for the basic SQL *select* clause to these non-relational systems.

We recommend an SQL-based interface into database systems.

5.2. Data Formats

We can retrieve data in ASCII assuming that the user will want to see it on a screen. This is the easiest and most portable approach. On the other hand, retrieving the data in binary may be preferable if there are stringent requirements for floating point precision or special formatting needs for data types such as positional coordinates (RA–Dec). RA–Dec values may also require coordinate system conversion and precession. The client application must determine the data format and type of an item in order to perform such translations. A Data Definition Language (DDL) is useful to detail these data formats and types.

We recommend the use of a Data Definition Language to describe catalog data for a client. This requires extra work on the part of the data supplier to fully describe the data as well as work for the application developer to support data conversions, but then everyone knows exactly what they are getting.

5.3. Mode of Operation

Are the data records retrieved one at a time from the database or are they all written to a staging area and then sent back to the client one (or N) at a time? Synchronous versus asynchronous operations further complicate these two cases. Getting records one at a time sends the data back to the user more immediately, but it requires the use of a programming interface and some loss in flexibility. Retrieving everything to a file first requires the user to wait if there are many records, and a lot of temporary disk space could be used. Ordinary SQL commands can be used, however, to achieve this without extensive programming.

For databases of known size and performance characteristics, it is quite reasonable to pre-fetch all the data for a given query and then feed it back to the client application from a staging area. For large and dynamic databases, such as the HST catalogs, this is not feasible. Instead, the preferred approach is to retrieve the records directly from the database and return them to the user as soon as possible. By recommending that the client receive database records one (or N) at a time, the client application should not know or care which approach is being used to retrieve the data.

6. Archive Access

For basic access to data sets in a small archive, the simplest method is to use FTP and download the files. For large archives, where the data sets may not be immediately accessible, the approach is often to browse a catalog with a user interface program, select the data sets of interest, and request that these be delivered. The delivery could be via the network in which case security issues need to be worked out in getting the data back to the user's machine. Does the data get pushed to the user's machine or must the user pull it from some staging area on the archive system?

We recommend that data be staged on the archive system and copied by users to their home site, at least until the security issues can be resolved.

Many archives have other services to offer—data browsing, analysis functions, name services, etc. Once a standard framework is adopted, adding new services is straightforward.

7. Current Systems

StarView (the ST DADS user interface) is designed using the client-server model. We plan to distribute the client portion of the software to interested users to access the ST DADS catalog or modify for use with other catalogs. Data values are retrieved a record at a time in binary through a standard database interface called STDB, which is portable across Unix and VMS and has been used with different relational DBMSs. A distributed version of STDB has been developed, called RemoteSTDB, which is built upon Sun RPC. Other remote services are planned, specifically file access and data set retrieval.

The Astrophysics Data System (ADS) also uses a client-server approach for accessing astronomical catalogs. Client software (a user interface program) is distributed to user sites, and server software is offered to data suppliers to make their catalogs available to ADS users. For database access, an SQL interface is used on top of the ANSAware package. Other remote services are being implemented or planned. The client-server software is proprietary, due to the use of ANSAware for networking and the Knowledge Dictionary System (KDS) for the core system. Currently, one can only use the ADS user interface client to get data through ADS servers.

8. Conclusion

The basic types of catalog and archive services for astronomy are well-known. There is agreement throughout the community on the need to have open, distributed access to these services. Below are our recommendations towards achieving this goal:

- Standardize on the networking mechanisms. Sun RPC over TCP/IP or UDP/IP is recommended now, while monitoring DCE developments.
- Support catalog access through SQL interfaces, with data type and format information defined via a Data Definition Language.
- Provide access to public data now. Reconsider proprietary data once all of the security issues are resolved.
- Publish archive services for data set retrieval.

There are separate efforts towards accomplishing these goals. We need to work towards combining these efforts to achieve truly open access to the many astronomical catalog and archive systems.

Astronomical Data Analysis Software and Systems II
ASP Conference Series, Vol. 52, 1993
R. J. Hanisch, R. J. V. Brissenden, and J. Barnes, eds.

Managing an Archive of Weather Satellite Images

Rob Seaman

National Optical Astronomy Observatories, Tucson, Arizona 85726

Abstract. The author's experiences are described of building and maintaining an archive of hourly weather satellite pictures at NOAO. This archive has proven very popular with visiting and staff astronomers, especially on windy days and cloudy nights. Given access to a source of such pictures, a suite of simple shell and IRAF CL scripts can provide a great deal of robust functionality with little effort.

These pictures and associated data products such as surface analysis (radar) maps and National Weather Service forecasts are updated hourly at anonymous ftp sites on the Internet, although your local atmospheric sciences department may prove to be a more reliable source. Contact the author for other suggestions. The raw image formats are unfamiliar to most astronomers, but reading them into IRAF is straightforward. Techniques for performing this format conversion at the host computer level are described which may prove useful for other chores.

Pointers are given to sources of data and of software, including a package of example tools. These tools include shell and Perl scripts for downloading pictures, maps, and forecasts, as well as IRAF scripts and host level programs for translating the images into IRAF and GIF formats and for slicing and dicing the resulting images. The author gives hints for displaying the images and for making hardcopies.

1. Introduction

Since the Summer of 1991, NOAO/Tucson has had continuously improving access to a local archive of hourly geosynchronous weather satellite pictures of the continental United States. These are similar to pictures that are available over the Internet, but are acquired in a different format from a source that is closer to the U.S. Government agency responsible for distributing them.

Three times an hour, 24 hours a day, a new picture is downloaded in one of three different wavelength bands: visible, infrared, and water vapor (narrow band IR sensitive to atmospheric moisture). These are first processed to translate the obscure native image format into IRAF. The IRAF images are then registered to the nearest pixel and overlaid with the North American geopolitical boundaries as solid yellow lines. The pictures are also translated into the popular GIF format.

We archive the last 24 hours worth of pictures for each of the three bandpasses in a directory on one of our downtown servers. The archive also includes GIFs and IRAF images of the Internet surface analysis maps indicating radar echoes, atmospheric pressure, weather fronts, and local conditions at cities around the United States and Canada. The downtown archive is mirrored over a T1 link to a machine on our Kitt Peak network which is accessible from the Sun workstations at each of the telescopes.

Many readers will be familiar with sources of GOES pictures accessible via anonymous ftp over the Internet. Our approach of managing the weather pictures within a local archive is superior due both to the control it gives us over the formatting and handling of the data, and also to the much greater and more reliable bandwidth through which the pictures are delivered to the user. This latter is especially critical for a large site that may potentially have dozens of users simultaneously viewing the pictures.

2. Acquiring the Data

We implemented our local weather picture archive after some amount of experience with retrieving the pictures via anonymous ftp. This would typically occur at the particular moment that the user wanted to display the picture. This is an awkward process, even when facilitated by a host shell script that keeps track of the bookkeeping details. Of necessity an intermediate file transfer to the local disk is required. It then becomes at least a three step process, first download the picture, second display it, third delete the picture.

By way of contrast, NOAO's weather archive performs the first and last of these steps automatically in a robust manner. Robust in the sense that data that are missed due to an intervening network outage will be recognized and downloaded as soon as the network comes back up. Only the second step is left to the user and that step can be easily parametrized with a CL script.

Nothing in this scheme of mirroring a remote archive on our local network places any great restrictions on the nature of that remote archive. Indeed we automatically download and locally archive the "surface analysis" maps from an anonymous ftp archive that will be mentioned in a moment. However, by relying on a well known network resource you will find yourself in competition for bandwidth with thousands of other users. One bit of fairly frequent maintenance is to move the moment of the anonymous download forward or backward by a few minutes to avoid the network blizzard that occurs when hundreds or thousands of automatic scripts login to the remote archive simultaneously.

This argues for locating a nearby, less well known source of weather data. An obvious place to start looking is the atmospheric sciences department of some neighboring university. One benefit of working in the astronomical community is that observatories carry a great currency of good will among the larger academic community (or so the author has found). Folks are glad to help when approached politely. Astronomers will also benefit from working together on these issues. The author welcomes correspondence on any weather topic, including pointers on locating sources of data.

Nothing has been said so far of the specifics of how the archive works. Suffice it to say that a small number of shell scripts reference a variety of familiar Unix utilities, including `ftp` and `rsh`. The scripts are invoked hourly by `crontab`. The key idea is to always generate complete lists of candidate image files and to perform matching operations between the corresponding remote and local lists to determine what subset should be downloaded and alternately, what other subset of the pictures has expired and should be deleted. This ensures that the archive is robust in the face of those network vagaries that we all face every day with no small amount of trepidation.

Note that while we have chosen to seek another source for weather satellite pictures, the Internet archives do provide an extremely valuable service to the community. You should investigate what they have to offer since they will undoubtedly come in handy some day. A good place to start is the file *sources.doc* in the directory *wx* on the machine *vmd.cso.uiuc.edu.*

3. Translating Formats

Network sources provide pictures in any of a number of formats (although GIF is certainly the most prevalent). These formats can be manipulated with varying degrees of convenience by several publically available packages of translation tools. I would recommend the San Diego Supercomputer Center's *Image Tools* package that is available from *sdsc.edu*, and also the *pbmplus* package that is distributed with the *X windows* release.

Since astronomy is very rich in image processing resources, it makes sense to apply these resources to the weather images. With that in mind, the NOAO solution is to translate the images into a format palatable to IRAF. This is currently the native IRAF image format, which is not the best fit to the characteristics of the weather pictures due to a current restriction to short integer pixels or larger if full IRAF access is desired. The GOES weather pictures as distributed by Unidata are only 8 bits deep, and therefore there is a factor of two wasted space. For local use the benefits of easy access overcome the penalties of the inflated size.

IRAF v2.10.3 should have support for FITS built right into the Image I/O interface. The weather archive will move to 8 bit FITS at that point. Another interesting development on the horizon is the application of (lossy or lossless) data compression techniques to astronomical (and weather) pictures. The weather pictures are quite compressible.

After the initial acquisition of the pictures, many of the chores associated with translating the data formats and manipulating the images are performed with CL scripts. The weather archive uses an experimental version of the CL that supports host level access to these scripts as Unix "interpreter" files, i.e., via a line such as *#!/iraf/bin.sparc/cl.e -f* at the top of the script that directs Unix to execute the named interpreter rather than the script itself, while passing the script's pathname as the first argument to that interpreter.

4. Distributing the Pictures

The group of people who are most likely to be rabidly interested in the weather at 3 o'clock in the morning are observational astronomers. In order to get the pictures to the observers, we have chosen to maintain not just our downtown master weather archive, but also a mirror archive on the other side of our mountain T1 link. The Unix `rdist` utility provides an ideal tool for effecting this transfer. This is a very reliable way to ensure that the downtown and mountain archives do not drift apart.

Note that a centralized distribution mechanism such as `rdist` avoids the periodic network overloads that a more diffuse mechanism like anonymous ftp is prone to. Since the master archive is responsible for all transfers, each mirror archive can be updated in turn. In this way the entire available bandwidth can be utilized if need be, rather than there being a requirement for significant headroom to cushion the network blizzards.

This is not to imply that the weather archive comes at a totally negligible cost. Receiving and retransmitting weather pictures is the single largest regular consumer of bandwidth on our network surpassing, for instance, the usenet news by a large margin. As new sources of data come on line this will only increase, placing ever greater emphasis on data compression techniques.

Of course the final step in distributing the pictures occurs when a user actually displays them. NOAO uses IRAF networking to access the archived pictures. Currently this relies on the normal per user association with remote kernel server processes—each user independently spawns kernel servers as needed on the archive machine. One restriction is that each user must have an account on the archive machine. This is somewhat problematic since our archive machine on Kitt Peak is currently a private scientist's workstation.

An alternative that is supported by IRAF v2.10.2 is to have the users connect to the archive machine using a kind of "anonymous IRAF networking". Version 2.10 of IRAF introduced a new networking driver that supports the notion of a *kernel server daemon.* The kernel server daemon is responsible for firing up the individual kernel server processes as they are needed by various networked IRAF sessions. Any IRAF session with knowledge of the correct port ID and authorization code can connect to a daemon and thus be granted access to that remote system with the permissions owned by the daemon. By arranging to have this daemon be managed by `inetd`, for instance, with proper process permissions and a restricted view of the filesystem supplied by `chroot`, a new type of anonymous networking can be implemented.

5. Displaying the Pictures

A local IRAF task, `wdisplay`, can be used to view the pictures in a variety of revealing ways using the *imtool* or *saoimage* display servers.

- Display the latest visible light picture: `wdisplay vis`
- Display the last four IR pictures in successive frames: `wdisp ir four+`
- Display the latest picture from each of the three bands, as well as the latest surface analysis map: `wdisp all`

There are also any number of *X windows* and other windowing system clients that can be used to view the GIF versions of the pictures. At NOAO, these pictures are most frequently viewed by astronomers who are observing at the various telescopes. The observing environment at NOAO relies heavily on IRAF (via the ICE software), thus NOAO's reliance on IRAF tools for viewing the data. It is well integrated with the observer workstations.

6. Weather Forecasts

An allied facility is supplied by the Unix `nws` command, which will retrieve the current weather forecast and conditions for many cities around the U.S. and Canada. This command is implemented as a `perl` script that is optimized for the particular syntax of another Internet resource, the *Weather Underground.*

- Retrieve the forecast for Tucson: `nws tus` (or simply `nws`)
- Retrieve the current conditions for Arizona: `nws az`
- Retrieve the national weather roundup: `nws usa`

X windows users can access the same information using the `xforecast` client, which will display a North American map and allow the desired city to be selected with the mouse.

7. Future Plans

Many incremental improvements have been made to our support for the weather archive over the past year, and these should continue at a steady pace. Future additions may include:

- New sources of data: (Europe, Australia, and the Pacific basin seem likely)
- New image formats: (compressed FITS)
- New data products: (bandpass principle components?)
- New display options: (movies, graphics overlays, new clients)
- Improved networking support: (anonymous IRAF networking)
- Extended distribution: (contact the author)

Suggestions and inquiries should be directed to *rseaman@noao.edu*, and are welcome from outside organizations as well as from NOAO. Note that only some of the current capabilities have been described here. A package of example tools is contained in the file *weather.tar.Z* in the directory */pub* on the machine *gemini.tuc.noao.edu.* These tools are meant to serve only as examples that should help you to avoid duplicated effort, but that are certainly guaranteed not to work "right out of the box".

Acknowledgments. The National Weather Service data are provided courtesy of the NSF-funded Unidata Project, the University of Arizona, the University of Michigan, and the University of Illinois, Urbana-Champaign.

Part 1. Databases, Catalogs, and Archives

Section D. Electronic Publishing

Astronomical Data Analysis Software and Systems II
ASP Conference Series, Vol. 52, 1993
R. J. Hanisch, R. J. V. Brissenden, and J. Barnes, eds.

Electronic Publishing & Advanced Information Retrieval

André Heck
Observatoire Astronomique, 11 rue de l'Université, F-67000 Strasbourg, France

Abstract. The present status of electronic publishing is reviewed, as well as its possible contribution to advanced information retrieval.

1. Introduction

If you want to go somewhere (to obtain scientific results), you certainly need the brain of a driver (a researcher). But other persons are also indispensable: those designing and assembling the cars, those laying out the highways, building the bridges and drilling the tunnels, those printing the maps and producing the travel guides, and so on. The better understanding of driving have all these people, the most efficiently they will bring the driver where he wants to go.

Electronic publishing (EP) and *information retrieval* (IR) have become a couple of bricks of the research edifice. EP is sometimes confused with *desktop publishing* (DTP). These concepts do not cover quite exactly the same things, even if they are intimately linked. The latter can be understood as the local production, through relatively sophisticated software packages and laser printers, of high-quality material ready for reproduction by a publisher with the traditional camera-ready copy (CRC) technique.

The former term would concern the electronic submission of material (a *compuscript*) from its originator (author, scientific editor) to a possible intermediary (scientific editor, journal editor, referee) and, in any case, to the publisher who will work directly on the electronic files and get the paper, journal, or book ready for printing through a succession of computer-assisted steps.

These techniques are widespread nowadays and a number of packages are used by astronomers, space scientists, engineers and technicians for producing their papers, reports, etc., as well as their everyday mail. The motivations behind the choice of a given package are various and not always rational (local availability, inertia, software/hardware dependence, financial constraints, word-of-mouth recommendations, adequate training, institutional pressure, journal policies, and so on).

It is obvious that the combination of EP/DTP and networking will profoundly reshape not only our ways of publishing, but also our procedures of communicating and retrieving information. The ultimate aim of EP might be considered as what is called *advanced information retrieval* (AIR). We are talking here not only of bibliographical data, but also of the scientific contents of papers, of sets of numerical or tabular data as well as, subject to the resolution of some outstanding technical difficulties, pictorial information.

The computing revolution is probably far from being completed. It gained new dimensions in the last decade with the introduction of personal computers and workstations, more and more powerful, as well as by the availability of ever more sophisticated text processing systems and the spreading of data networks, the popularity of which reflects their intrinsic usefulness in our present society.

Laser printers produce high-resolution pages of such a top quality that it is sometimes difficult to distinguish, when receiving a document, whether this is a preliminary draft, an intermediate working step or a final product coming from a publisher with the blessing of all parties involved. This is not without raising a few problems as the so-called *grey literature* (see, e.g., Smith 1991) is more and more difficult to identify. Refer also below to the paragraph on the information available from anonymous FTP (file transfer protocol) accounts.

Nobody today would however complain against the possibility of getting high-quality printout immediately, without having to go as in the past through manual typesetting and sometimes disastrous delays to obtain papers published decently in professional journals. The compuscript is particularly flexible, not only as far as its transfer is concerned, but also for the case of necessary modifications (removal of mistypings, grammatical corrections, style improvement, layout standardization, adaptation to the journal norms, and so on).

To make authors' life easier, some publishers have already produced sets of *macros* (essentially in TEX), but these are far from being identical ... and not all of them are satisfactory.

2. Broad Indications of a Survey

A *DeskTop Publishing Survey* (called hereafter DTP Survey—see Heck 1992c) has been run both by normal mail addressed to institutions and by electronic messages sent to institutions and individuals. The returned questionnaires (that could be answered either individually or collectively) concerned about 3,700 persons from more than 160 institutions in 23 countries.

Is it really feasible to draw authoritative conclusions from such a survey? Experience in public opinion polls indicates that one should not squeeze too much the data at hand. But the broad results are already quite indicative: there is obviously a majority use of TEX and associated packages (LATEX and so on) among astronomers and space scientists. The machines on which the packages run are approximately equally distributed among PCs and 'compatible' machines, DEC computers and SUN stations. This rough tendency has to be nuanced by the discussion presented in Heck (1992c) and by the comments reported from the questionnaires.

These comments concerned various aspects: policies applied in institutions or in publications, personal needs or, more frequently, general appreciations on various packages or combination packages/machines. As one can expect, there were strong supporters of a given package as well as strong adversaries. Ideally, the users want something user-friendly, WYSIWYG (*what you see is what you get*), portable, transmittable, standardized for all journals and book series, and coping with all the complications of mathematical formulae and insertions of tables, plots and figures ...

The main reasons quoted by those with authority for in-house publication series and computer managers for pushing people from their institutions to write their manuscripts in TeX are the portability of the files through networks, the extremely high ('professional') quality of the output on laser printers and, last but not least, the acceptance of manuscripts so produced by quite a few journals already.

In fact, it appears that a significant number of persons run user-friendly packages on local or private systems, and go to TeX-related ones when they have to communicate or share their files in some way. User-friendliness is definitely a key issue for respondents being dragged into using TeX by reasons of availability, possibility of easy transmission or journal policies. Therefore the DTP Survey results that are very favourable to TeX should be nuanced by such external considerations.

A frequent worry expressed is the increased fragility of the material submitted for publication, being more open to misdeeds such as tampering. Although they are not very frequent, there are enough examples of editor and referee misbehaviors to get scientists wondering how they can get their production duly protected, if we go fully to machine-readable media (see, e.g., Allen 1992 & Heck 1992b).

One can now install in a reduced space powerful equipment able to produced printed documents (including figures, graphs, drawings) of a quality rivalling a professional one. But not everybody can become a professional printer! Any professional typesetter would tell you that you do not become a typesetter just by getting a typesetting system at your fingertips (the most common heretical mistake seems to be mixing fonts with and without sérifs).

On the other hand, it is not because one has powerful DTP packages and spellcheckers at one's disposal that one dominates automatically a language and that one can write properly. A valuable style manual has already been published by the *International Astronomical Union* (IAU) (Wilkins 1989), but its current 17-line paragraph devoted to EP should definitely be updated and expanded substantially.

Aesthetics is definitely another hot issue. From the DTP Survey, the two schools among scientists are confirmed: one group for which the general appearance of a paper matters little ("what matters for us is the substance of a document") and the other group for which aesthetics is important. My point of view is that these two apparently diverging opinions are not necessarily contradictory. One can share the former view without neglecting the latter one. A well-structured text, written with both grammatical and semantic exactness, presented according to aesthetical common sense (consistency of fonts, harmonious paragraph layout, adequate insertion of illustrations, appropriate ruptures of the text flow, and so on), can only help to convey the intrinsic message of the paper to the reader, be it only at the subliminal level. This is simply one of the basic principles in communication: an elegant vector can only be in synergy with the result sought.

3. Where are We Sitting Currently?

On 1–3 October 1991, a colloquium entitled *DeskTop Publishing in Astronomy and Space Sciences* (called hereafter the DTP Colloquium—Heck 1992a) was organized at Strasbourg Observatory. This was the first open meeting of its kind where experience could be shared for the benefit of everybody and where it was certainly useful to gather experts from various sides together (essentially scientists, editors, publishers, and data centre managers) in order to make their points as to the advantages they find with given packages, the constraints they have to comply with, the requirements they would have for further developments. The reasons behind the choices that a number of publishers (some of them weighting more than 500 scientific journals) have already made, were also of importance.

A preliminary opinion poll was carried out at the beginning of 1991 essentially among scientific editors and publishers. The answer was overwhelmingly in favor of holding this meeting. It was definitely not intended to be a TEX-gathering, as there is most likely no 'best' system for astronomers and space scientists.

Some journals and publishers offer already sets of TEX macros. According to Lequeux (1991), about 15% of the papers submitted to *Astronomy & Astrophysics (A&A)* had been prepared by using the Springer-Verlag sets of macros. The proportion went rapidly up to 30% (Lequeux 1992b) and is steadily growing (60% indicated by Lequeux in September 1992 at the ALD-II Conference—Heck & Murtagh 1992). It will soon be difficult to keep track of all journals and publishers favoring TEX as typesetting system.

All this does not go however without a few worries and questions as it appears from the comments on the DTP Survey forms (Heck 1992c). Although this might simply be an unavoidable step in pioneering a new technique, people complain that the sets of macros currently available are still too unsatisfactory. As I could personally experience it, this is true especially when dealing with publications requiring figure insertion or involving huge indices. On this latter point, packages such as *Script* are still quite competitive.

At the DTP Colloquium, special sessions were focussed on the publishers' and (scientific) editors' approaches. Refer particularly to the reviews by Mitton (1992) and Lequeux (1992a) that we shall not repeat here.

Last, but not least, the points of view of data centre managers are also of fundamental importance as these places, because of their acquired expertise, will have most likely to handle ultimately the enormous amount of information arising from EP and digitization of past literature, and to integrate it in the already existing networks of databases and archives. Refer to the paper by Crézé (1992) insisting on the strong need for upgrading the efficiency of the processes feeding the information into the databases.

It is also clear that, by shifting to compuscripts, filters will have to be maintained to ensure a quality control for the specialized professional journals (exactness of the contents, originality or novelty of the contribution, repeatability of experiments, respect of the various policies, and so on). But obviously, the editing and publishing ethics will have to be somehow adapted.

Meetings of task forces interested in the development of EP and IIR in astronomy and space sciences have been taking place on both sides of the Atlantic.

On the American side, a project called STELAR (*Study of Electronic Literature for Astronomical Research*) has been undertaken as a cooperative effort between AAS, ASP, NASA and NSF (see Van Steenberg 1992) with the goal of making the refereed scientific literature accessible on-line for the astronomical community, including graphics and full-text retrieval capabilities. Plans have been made to digitize past literature. Refer also to Abt (1992) for a review of American on-line publication tests and long-range plans.

In Europe, the concerned parties met for the first time at the DTP Colloquium in October 1991 and decided to take steps to standardize electronic publishing procedures and to coordinate efforts with the American counterpart. In March 1992 in Strasbourg, scientific editors of international astronomical journals and representatives of the most important publishing companies in the fields of astronomy and space sciences gathered together with scientists and managers of data centres. TeX had become the *de facto* adopted standard for EP (see below however). Further meetings of this essentially European task force are presently being organized.

4. EP, AIR, Knowledge Bases, and Associated Tools

The ultimate aim of EP might be considered as so-called *intelligent information retrieval* (IIR) or better named *advanced information retrieval* (AIR). When we speak of IIR or AIR, we mean in fact *an added flexibility* or *additional degrees of freedom* to information retrieval, taking advantage of the fact that the material to be published appears at some stage in a machine-readable form.

From there, it can be used of course for publication in a flexible way, but advantage should also be taken of this format for other purposes such as direct archiving and storage of bibliographical information, keywords and abstract, possibly the full text, tabular and pictorial data in databases, as well as, conversely, retrieval in an intelligent way of the stored material.

What could be understood by the term *intelligent*? A loose definition of IIR could be the following: one does something more elaborate than just looking for the presence of a string of characters in a base and retrieving the data attached to it.

The concept takes its full meaning with sophisticated algorithms. IR algorithms already exist (refer, e.g., to Salton 1989, as well as to the various publications and proceedings of the ACM *Special Interest Group on Information Retrieval—SIGIR*). Refer to Kurtz (1991) for an application of multivariate data analysis to multidimensional spaces based on keywords used for indexing papers. The clustering of documents in a statistical hyperspace can be put in parallel with two aspects of the human thought: logics and association. The latter one applies to ideas, concepts, terminologies, and so on, even if the relationships involved are sometimes implicit and/or subtles. The same philosophy will be found also in the hypertext approach (see below).

The potential intelligence of a retrieval is conditioned first by the type of information that is stored, then by the way the database is structured (database engine, user interface, and so on), and finally by the tools used to search it and by the range of algorithms constructed around the database. The trend in databases is presently to shift towards the concept of *information* rather than *data*, or even more generally to speak of *knowledge bases* (Heck 1992d).

Searching for information in these bases will call for *thesauri* the construction of which is far from being obvious and immediate. They must take into account hierarchical or relational links between the terms and expressions listed (synonymities, analogies, broader or narrower concepts, and so on). They will have to be kept up-to-date, as science progresses and evolves. The *ad hoc* expertise will have to be secured. There is a risk that thesauri will always lag behind. There are already a few thesauri available presently in astronomy and space sciences, such as the INSPEC and NASA ones, not to mention those provided by various journals and by *Astronomy & Astrophysics Abstracts (A&AA)*. An important recent contribution is the *IAU Thesaurus* by Shobbrook & Shobbrook (1992).

An adequate and homogeneous indexing of documents will have to be performed in parallel. We cannot but agree with Locke (1991) when he writes: "Indexing documents properly and consistently for later retrieval is not a low-order clerical task, but a complex exercise requiring knowledge engineering skills". This aspect, that is currently grossly underestimated, will have to be carried out in collaboration with well-trained *cogniticians*, once the languages used by both parties are tuned to each other.

Refer also to Adorf & Busch (1988) and Busch (1992) for uses of 'intelligent' text retrieval systems and indications for further reading. *Hypertext* systems (Nielsen 1990) have also been investigated as means for storage, retrieval and dissemination of information. Hypertext can be characterized as a document allowing, contrary to classical texts, non-sequential and multiple 'reading' paths, selected by the consultant, which implies an interaction between this person and the infrastructure.

The navigation along these paths is regulated by the principle of idea associations (analogies), sometimes with some statistical background, but can also be piloted by situations. There are more logical capabilities in it than in a sequential reading (classical text is not always sequential or linear: footnotes and references break this flow) or in a hierarchical database.

Hypermedia systems add media facilities, such as displays, animated graphics, digitized sound, voice, video, and so on. See also Reynolds & Derose (1992).

5. Some Trends

Data, information and/or knowledge bases are intimately involved with our activities: observing (remote or traditional), data reduction, general research, publication, management of instrument time, and so on. The knowledge retrieved from them will be a combination of text, quantitative and qualitative data, n-dimensional raw and/or reduced observing material, as well as tabular and pictorial information in general.

As clearly indicated by the discussions held at the DTP Colloquium (Heck 1992a) and by the orientations taken at several meetings of task forces, the future for astronomical primary journals offering original scientific results (but this might apply to less specialized journals and other publications too), is to investigate the potentialities of electronic publishing. The future options in terms of policies, procedures, and, more down to everyday life, macros and the like, must result from a concerted effort.

Astronomers are not inventing the wheel though. Strangelove and Kovacs (1991) had already identified seven peer-reviewed electronic journals, with or without paper equivalent. The refereeing process relies heavily on a voluntary contribution by peers. As seen earlier, an ever increasing part of the papers are submitted in a machine-readable form. One could conceive a process where editors would carry out most of the current exchanges with authors and referees electronically. This is already the case for some of the existing electronic journals.

Once the 'paper' (or rather 'contribution') is approved, it would be channelled to an electronic server (or a data centre hosting it). From there, issues of the journal could be delivered automatically and free of charge (i.e., with costs diluted in the general networking funding) to subscribers as is already the case nowadays with numerous 'digests'. The availability of information would be immediate after approval by referees and editors, without printing delays.

New communication means are eliminating geographical separation and have revolutionized the way scientists transfer information to each other. Now through networks and remote logon, scientists have direct access to remote databases and archives, and can retrieve images or data they can process locally, as well as tables of data they can directly insert in a paper to be published. Conversely, once it would be accepted by a journal after appropriate screening and quality control, a compuscript could be channelled into a database from which it could be retrieved even before being published on paper, if ever. Its bibliographic information (at least its general references, the keywords, the abstract, its quotations) could be accessible immediately through the existing bibliographical databases and citation indices.

Subject to the design of specific procedures and techniques for plots, graphs, figures and pictures, even the whole paper could be retrieved. Catalogues, extensive lists of data or tabular data could also be plugged through networks into *ad hoc* bases such as SIMBAD.

Nowadays, abstracts of communications to astronomical meetings (such as the AAS ones) can already be submitted electronically. Observing proposals (such as those for the HST) follow the same way, landing also onto expert systems able to check their technical feasibility.

Alternatively advanced information retrieval tools such as `archie` (see, e.g., Emtage & Deutsch 1992), WAIS (*wide-area information system*—see, e.g., Kahle 1989 and Stein 1991), *gopher* (see, e.g., Alberti et al. 1992), or WWW (*world-wide web*—see, e.g., Berners-Lee 1992) could be used to locate the information of interest and retrieve it subsequently from such servers.

It should be noted that the journal *Acta Astronomica* maintains an anonymous FTP account from where detailed unpublished observational data can be retrieved (see, e.g., Kałuzny & Udalski 1992). Such a system is definitely less cumbersome than writing to a data centre where people would have to photocopy the material on paper or duplicate a magnetic tape (or diskette) and send it through the post. Currently, Strasbourg astronomical data centre (CDS) is already giving access through the FTP protocol to tables and catalogues (Ochsenbein et al. 1992). Actually more and more references are now given with their availability from anonymous FTP accounts and are sometimes not accessible otherwise. This is the case for a couple of references in the present paper.

At CERN, a preprint server is being currently implemented (van Herwijnen, 1992b). This initiative should be an example for large astronomical institutions or national professional societies.

6. Final Comments

It will be of fundamental importance to involve librarians in the future developments and to take their standpoints and requirements into account as they will have to participate in the modernization of the knowledge search in their realms. Surveys have already been carried out in some institutions (see, e.g., Michold 1992).

As illustrated in Heck (1992e), it is clear that we have entered a new age where librarians have a new attitude towards IR and where scientists have a new attitude towards their librarians. Compare for instance the proceedings of the ALD-II Conference (Heck & Murtagh 1992) with those of the ALD-I (Murtagh & Heck 1988) where the related matters where barely touched.

Specialists in electronic publishing outside our field claim that the electronic journal is the wave of the future and that, within less than a decade, the bulk of information will be exchanged electronically and the nature of print media will have changed drastically. However there are still a few problems to be tackled such as copyright ones, the coordination between electronic and paper publication, and, last but not least, the validation of electronic productions as legitimate media for dissemination of scholarly studies as well as for credit toward promotion and tenure in academic and other circles.

It is however important to realize that EP and DTP are not always a panacea in terms of quick availability of publication. The classical CRC technique is still sometimes much faster as we just experienced it with the proceedings of the ALD-II Conference compare to another one held significantly earlier, going through the same publisher and for which the DTP technique had been retained.

Should we rather encourage the astronomical community to allow for a variety of word processing options, rather than to promote a single standard? In order to leave some freedom to authors, an attractive solution might have to be adopted in order to allow them to use their preferred DTP package (among the major ones) and to wrap the output in an upper shell, a *markup language*, such as SGML (*Standard Generalized Markup Language*—see, e.g., van Herwijnen 1990 & 1992a) acting as a standardizing interface for the subsequent steps in the publishing process. The *electronic manuscript standard* (EMS) promoted by the *Electronic Publishing Special Interest Group (EPSIG)* should deserve all our attention too (see, e.g., van Herwijnen et al. 1992). On these matters, refer also to Wright (1992), as well as to the various issues of *EPSIG News.*

Once again, it seems obvious that the combination of desktop and electronic publishing with networking and new structuring of knowledge bases will profoundly reshape not only our ways of publishing, but also our procedures of communicating and retrieving information.

Generally evolution is accompanied by quite a few surprises and it would be dangerous—and pretentious—to play here the game of guessing what they will be. What is sure is that technological progress will play a key rôle in future orientations. But will there be forever a need of publishing which, as such, will become increasingly integrated with data reduction and analysis? There might be a time when the keyboard will not be necessary anymore and when the journals as we know them today or even publishing as such will disappear, replaced by some electronic form of dialogue with the machine and knowledge bases.

Direct computer pen writing (implying handwriting recognition) is already taking over from mouse clicking and pointing. Gesture and voice recognition will be common in the near future. The idea of direct connection between brain and computer is not new. It is used among others by *cyberpunks.* Their *cyberspace*, or knowledge space related to what is also called *virtual reality*, is where the brain navigates with a virtual body and has access to all kinds of collections and sets of data. Cyberpunks have sometimes so much puzzled government officials with their innovative—often imaginary—use of computer techniques and networking that some publishing companies have been raided by secret services (see, e.g., Kapor 1991).

Part of the last comments is still fiction, but we might expect with Jules Verne that "Tout ce qu'un homme est capable d'imaginer, d'autres hommes seront un jour capables de le réaliser" (All that a man is able to imagine, other men will be able to achieve later on).

References

Abt, H.A. 1992, in DeskTop Publishing in Astronomy and Space Sciences, A. Heck, World Scientific, Singapore, 47

Adorf, H.M., & Busch, E.K. 1988, in Astronomy from Large Databases – Scientific Objectives and Methodological Approaches, F. Murtagh & A. Heck, ESO Conf. & Workshop Proc., 28, 143

Alberti, B., Anklesaria, F., Lindner, P., McCahill, M., & Torrey, D. 1992, The Internet Gopher Protocol - A Distributed Document Search and Retrieval Protocol (FTP: /pub/gopher@boombox.micro.umn.edu)

Allen, D. 1992, Byte, August 1992, 10

Berners-Lee, T. 1992, Physics World, June 1992, 14

Busch, E.K. 1992, Byte, June 1992, 271

Crézé, M. 1992, in DeskTop Publishing in Astronomy and Space Sciences, A. Heck, World Scientific, Singapore, 129

Emtage, A., & Deutsch, P. 1992, `archie`: An Electronic Directory Service for the Internet

Heck, A. (Ed.) 1992a, DeskTop Publishing in Astronomy and Space Sciences, World Scientific, Singapore, xii + 240 pp. (ISBN 981-02-0915-0)

Heck, A. 1992b, in DeskTop Publishing in Astronomy and Space Sciences, A. Heck, World Scientific, Singapore, 3

Heck, A. 1992c, in DeskTop Publishing in Astronomy and Space Sciences, A. Heck, World Scientific, Singapore, 55

Heck, A. 1992d, in Data Analysis in Astronomy IV, V. Di Gesù, L. Scarsi, R. Buccheri, Ph. Crane, M.C. Maccarone & H.U. Zimmermann (Plenum Press, New York), 21

Heck, A. 1992e, in Astronomy from Large Databases II, A. Heck & F. Murtagh, ESO Conf. & Workshop Proc., in press

Heck, A., & Murtagh, F. (Eds.) 1992, Astronomy from Large Databases II, ESO Conf. & Workshop Proc., in press

Kahle, B. 1989, Thinking Machines Corp. Techn. Rep., 202 (FTP: /pub/wais/doc/wais-concepts.txt@quake.think.com)

Kałuzny, J., & Udalski, A. 1992, Acta Astronomica, 42, 29

Kapor, M. 1991, Scientific American, March 1991, 116

Kurtz, M.J. 1991, in On-line Astronomy Documentation and Literature, Eds. F. Giovane & C. Pilachowski, NASA Conf. Proc., in press

Lequeux, J. 1991, ESO Messenger, 63, 20

Lequeux, J. 1992a, in DeskTop Publishing in Astronomy and Space Sciences, A. Heck, World Scientific, Singapore, 93

Lequeux, J. 1992b, ESO Messenger, 67, 58

Locke, Chr. 1991, Byte, April 1991, 193

Michold, U. 1992, ESO Internal Memo, 25 pp.

Mitton, S. 1992, in DeskTop Publishing in Astronomy and Space Sciences, A. Heck, World Scientific, Singapore, 67

Murtagh, F., & Heck, A. (Eds.) 1988, Astronomy from Large Databases – Scientific Objectives and Methodological Approaches, ESO Conf. & Workshop Proc., 28, xiv + 512 pp. (ISBN 3-923524-28-5)

Nielsen, J. 1990, Hypertext and Hypermedia, Academic Press, San Francisco, xii + 264 pp. (ISBN 0-12-518410-7)

Ochsenbein, F., Florsch, J., & Halbwachs, J.L. 1992, CDS Inform. Bull., 41, 83

Reynolds, L.R., & Derose, S.J. 1992, Byte, June 1992, 263

Salton, G. 1989, Automatic Text Processing, Addison-Wesley, Reading, xiv + 530 pp. (ISBN 0-201-12227-8)

Shobbrook, R.M., & Shobbrook, R.R. 1992, The International Astronomical Union Thesaurus, 122 pp.

Smith, A.W. 1991, Science International, 45-46, 27-29

Stein, R.M. 1991, Byte, May 1991, 157

Strangelove, M., & Kovacs, D. 1991, Directory of Electronic Journals, Newsletters, and Academic Discussion Lists, Ass. Research Libraries, Washington

van Herwijnen, E. 1990, Practical SGML (Kluwer Acad. Publ., Dordrecht)

van Herwijnen, E. 1992a, in DeskTop Publishing in Astronomy and Space Sciences, A. Heck, World Scientific, Singapore, 149

van Herwijnen, E. 1992b, Are documents servers replacing journals?, preprint

van Herwijnen, E., Poppelier, N.A.F.M., & Sens, J.C. 1992, EPSIG News, March 1992, 14

Van Steenberg, M.E. 1992, in DeskTop Publishing in Astronomy and Space Sciences, A. Heck, World Scientific, Singapore, 149

Wilkins, G.A. 1989, International Astronomical Union Style Manual (Kluwer Acad. Publ., Dordrecht), 50 pp.

Wright, H. 1992, Byte, June 1992, 279

Discussion

Shaw: Your description of Advanced Information Retrieval Systems is very interesting. Will these systems work for Journals that are published in languages other than English, particularly for journals from formerly Eastern-block countries?

Heck: Do not forget French and other accentuated languages either. We are conscious of this point and it has to be solved. We investigated it already in a couple of meetings with persons involved with indexing and thesauri. Space is lacking here to discuss all aspects of this matter that is also related to multilingual thesauri and associated problems (similar forms do not cover exactly the same thing, and so on).

Schmerling: Keyword indexing requires too much time by skilled individuals. This is what drove the late Eugene Garfield to develop the citation index. The citation index can be produced rapidly by relatively unskilled people.

Heck: Thank you for reminding us of this fact. I am afraid however, that a citation index cannot quite replace a profile of a paper defined by keywords.

Hanisch: The critical distinction between DTP and EP is that one is typesetting and the other is logical mark-up. The latter is critical - authors should concentrate on the logical composition of their papers, not on the subtleties of typesetting and page layout. TeX is typesetting, LaTeX is logical mark-up. LaTeX is probably the best solution for astronomy for now, but WYSIWYG systems that support logical mark-up (e.g., SGML) are equally viable solutions to this problem.

Heck: I would not quite follow you in reducing EP to mark-up capabilities. EP implies quite a range of technical aspects (in communications, for instance), as well as policies and procedures between author, editors, referees and publishers.

As it came out of the DTP survey, LaTeX is only a partially satisfactory solution as it does not allow direct use of material produced by other DTP packages that have their strong supporters in non-negligible numbers. SGML and/or EMS seem to be the marking capabilities we have to aim at in a short term.

Astronomical Data Analysis Software and Systems II
ASP Conference Series, Vol. 52, 1993
R. J. Hanisch, R. J. V. Brissenden, and J. Barnes, eds.

Intelligent Text Retrieval in the NASA Astrophysics Data System

M. J. Kurtz, T. Karakashian, C. S. Grant, G. Eichhorn, S. S. Murray, J. M. Watson

Harvard-Smithsonian Center for Astrophysics, Cambridge, MA 02138

P. G. Ossorio, J. L. Stoner

Ellery Systems Inc., Boulder CO 80301

Abstract. In collaboration with the NASA Scientific and Technical Information System, the NASA Astrophysics Data System (ADS) is establishing a service to provide access to the literature abstracts relevant to astronomy in the NASA Scientific and Technical Aerospace Reports and the International Aerospace Abstracts (together also known as NASA RECON). The service will include several sophisticated retrieval methods, which may be combined. Included will be methods to perform relevancy ranking from natural language queries, synonym and misspelling recognition, author name translation (e.g., for multiple transliteration possibilities) and other features. The capabilities of the current release are shown, and the plans for the near future are discussed.

1. The NASA STI Abstracts Database

The use of bibliographic aids has long been crucial for astronomy, as the well worn copies of *Astronomy and Astrophysics Abstracts* and its predecessor *Astronomisches Jahresbericht* in virtually every astronomy library attest. Within the last couple of decades bibliographic information has become widely available through the use of computer databases.

Watson (1991) has reviewed the available sources of bibliographic data. Of the four sources of abstracts of scientific papers (*Astronomy and Astrophysics Abstracts*, PHYS, INSPEC, and NASA/RECON) she notes that NASA/RECON (which is one of the many names for the NASA STI abstracts database) "is the earliest and most comprehensive of databases encompassing the entire field of space-related science, including astronomy, and dates back to 1962."

Use of these abstract services has been mostly confined to librarians within the astronomical community. Typically when a scientist wants a literature search she asks her librarian, and he then logs on to one of the databases and does the search.

The goal of this project is to bring intelligent access to one of the best bibliographic services, the NASA STI collection, onto the desk of every astronomer with a set of intelligent tools to facilitate searching. The STI abstracts in astronomy are our first target: there are more than 100,000 of them, and they have nearly complete coverage of the journal, conference proceedings, and technical report literature since 1974.

2. Basic Searching and the Synonym List

Currently librarians, and others who learn how, and with access to abstract databases, can search them by asking for articles with some boolean combination of attributes, say within a date range and by a particular author. Subject matter searching by boolean combination of keywords is also popular and useful. Any bibliographic service must have these abilities to be useful.

The ADS Abstract Service will allow searching by date, author, and/or keywords. The keyword and author searches can be either by exact match (the boolean AND) or by partial match, using either inverse log frequency weighted sums or unweighted sums to rank the partial matches. An example of a partial match is a paper which is keyworded with five of a listed seven keywords.

One problem with searching for authors is that the same person may have several different names, be they sets of initials, different transliterations, or just misspellings. The ADS abstract service will permit the user to translate names into standard spellings, thus collapsing all different transliterations of a particular Russian author's name into one spelling, for example. Joyce Watson is preparing these data using a list of all names which appear in the database.

Another feature which current abstract systems have is to find articles by the boolean combination of words in the title. The abstract service will offer this capability, enhanced by the ability to use the partial match methods available for keywords and authors.

In addition to the partial match methods we are including the ability to replace words by their synonyms, where by synonym we mean words which are essentially identical for the purposes of subject matter searching. For example all these words are essentially the same for the purposes of searching, although they are (mostly) legitimately different words: abberations, aberrated, aberration, aberrational, aberrationless, aberrations. Creating this list requires many informed decisions: spectroscope and spectrograph are considered synonyms for example, while spectrophotometer is not. Joyce Watson has prepared the synonym list by inspecting the list of all words used in the database, there are about 7,000 unique synonyms used in astronomy.

3. Natural Language Searching with Factor Spaces +

The partial match methods used to search via words in the title can be extended to search for documents by matching words in a natural language query to the full text of the abstracts. In this case it makes sense to allow another weighting scheme, which accounts for the frequency of terms within documents, as well as their global distribution (Salton & Buckley 1991). The use of the synonym list makes this technique even more powerful.

Besides implementing natural language searching by partial match rankings based on word frequencies, we are also implementing natural language searching based on the classification of documents and queries into a multidimensional subject matter relevancy space called the Factor Space.

The original Factor Space work (Ossorio 1965) relied on direct human judgements of the joint relevancies of words with subject matters, for example the word redshift is highly relevant to the subject matter cosmology, but not relevant to the subject matter stellar photometry. Kurtz (1991; 1993) has developed a method to automate the creation of large Factor Spaces by an *a posteriori* statistical evaluation of a set of classified documents, essentially building a psychometric model of a librarian classifier.

The Factor Space will be created by a multivariate statistical analysis of the co-occurrence of words (synonyms) with librarian-assigned keywords in abstracts. This will be the first large scale implementation of this technique (Kurtz 1993). Jeffery (1991) has shown that a human derived Factor Space can work substantially better than one based on partial match rankings; Kurtz (1991) has shown that a functioning Factor Space can be created by a statistical analysis of the co-occurrence of words with librarian-assigned classifications (the *Astronomy and Astrophysics Abstracts* chapter headings); Deerwester et al. (1990) have shown that the retrieval system (which they call Latent Semantic Indexing) derived from a multivariate statistical evaluation of the co-occurrence of words with other words within documents of a document set can give better results than a partial match system; and Lesteven (1993) using the NASA STI astronomy abstracts has shown that a functioning Factor Space can be created from a statistical analysis of the co-occurrence of librarian assigned keywords with themselves within documents.

Each of the five search techniques (keyword, author, title, natural language query via partial match, Factor Space query) may be implemented simultaneously on the same query, and the result ranked by a weighted sum of the rankings derived from each individual query type. The weightings are assigned by the user, although as with other decisions (weighting type, synonym replacement) there are reasonable defaults.

4. Status and Schedule of the Project

Experiments with the STI data began in autumn 1991, by Kurtz and S. Lesteven; the formal building of the ADS service began in late spring 1992. Retrievals through the X/Motif user interface using only keyword, author and date were first done at the end of the summer. Currently the user interface is relatively stable and we are doing retrievals using partial match on natural language. This system will be released as the beta test in January 1993.

In early 1993 the synonym list will be implemented, followed by the Factor Space, which requires it. We will begin to keep the database up to date, including the new additions to the STI database on a regular basis. We will implement the Salton and Buckley (1991) partial match algorithm if time permits, and along with a user's guide and improvements to the user interface this system will be released by summer 1993.

Finally, here is a brief example to show the power of the system, even in its current pre-beta state. Assume one wishes to know about recent results in redshift surveys, and one knows that Margaret Geller works in this field. Make a query (not shown) asking for her recent papers. Now choose a couple of papers relevant to the request, and get their abstracts (not shown). Put the text of

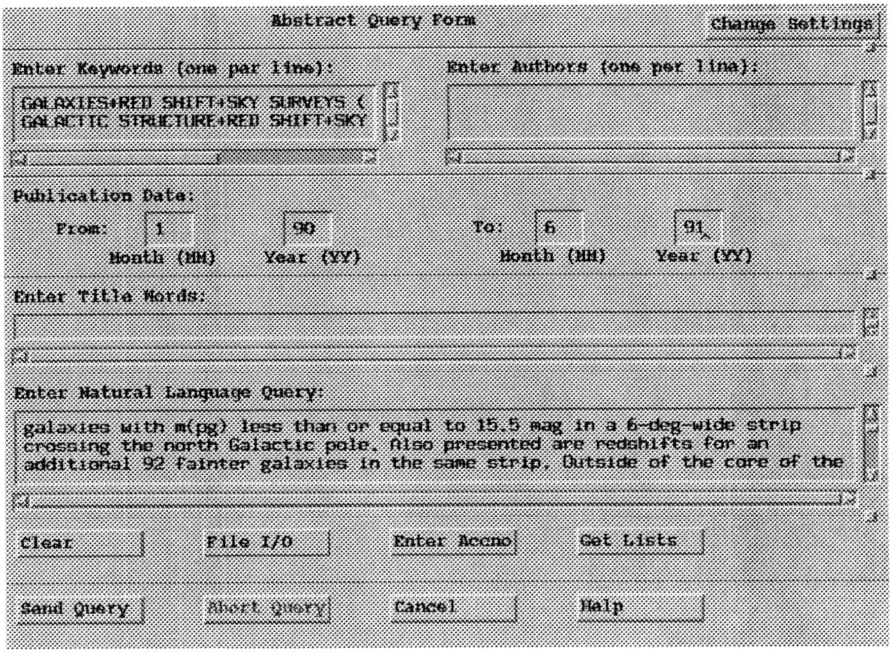

Abstract Query Form — Change Settings

Enter Keywords (one per line): GALAXIES+RED SHIFT+SKY SURVEYS (/ GALACTIC STRUCTURE+RED SHIFT+SKY

Enter Authors (one per line):

Publication Date: From: 1 Month (MM) 90 Year (YY) To: 6 Month (MM) 91 Year (YY)

Enter Title Words:

Enter Natural Language Query: galaxies with m(pg) less than or equal to 15.5 mag in a 6-deg-wide strip crossing the north Galactic pole. Also presented are redshifts for an additional 92 fainter galaxies in the same strip. Outside of the core of the

Clear | File I/O | Enter Accno | Get Lists

Send Query | Abort Query | Cancel | Help

Figure 1. A combined natural language and keyword query.

these abstracts into the natural language query box, and the keywords into the keywords box in the query window (Figure 1). Now modify the weightings so that the natural language query counts for three times the keyword query (not shown). The query returns a ranked list of 6788 documents, where the most relevant documents are near the top. Figure 2 shows the first window of abstracts returned by the query.

References

Deerwester, S., Dumais, S.T., Furnas, G.W., Landauer, T.K., & Harshman, R. 1990, Journal of the American Society for Information Science, 41, 391

Jeffery, H.J. 1991, Expert Systems with Applications, 2, 345

Kurtz, M.J. 1991, in On-Line Astronomy Documentation and Literature, eds. F. Giovane & C. Pilachowski, Washington: NASA Conference Proceeding, in press

Kurtz, M.J. 1993, in Adding Intelligence to Information Retrieval, eds. A. Heck & F. Murtagh (Dordrecht: Kluwer), to appear

Lesteven, S. 1993, this volume

Ossorio, P.G. 1965, Journal of Multivariate Behavioral Research, 2, 479

Salton, G., & Buckley, C. 1991, Science, 253, 112

Watson, J.M. 1991, in Databases and On-line Data in Astronomy, eds. D. Egret & M. Albrecht (Dordrecht: Kluwer), 199

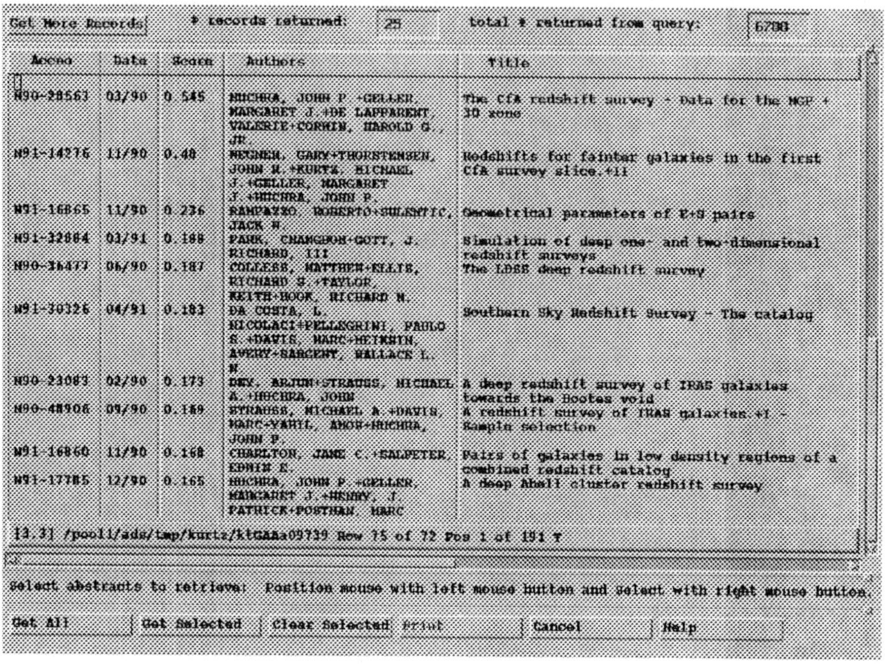

Get More Records | # records returned: 25 | total # returned from query: 6780

Accno	Date	Score	Authors	Title
N90-28563	03/90	0.545	HUCHRA, JOHN P.+GELLER, MARGARET J.+DE LAPPARENT, VALERIE+CORWIN, HAROLD G., JR.	The CfA redshift survey - Data for the NGP + 30 zone
N91-14276	11/90	0.48	WEGNER, GARY+THORSTENSEN, JOHN R.+KURTZ, MICHAEL J.+GELLER, MARGARET J.+HUCHRA, JOHN P.	Redshifts for fainter galaxies in the first CfA survey slice.+II
N91-16865	11/90	0.236	RAMPAZZO, ROBERTO+SULENTIC, JACK W.	Geometrical parameters of E+S pairs
N91-32884	03/91	0.188	PARK, CHANGBOM+GOTT, J. RICHARD, III	Simulation of deep one- and two-dimensional redshift surveys
N90-36477	06/90	0.187	COLLESS, MATTHEW+ELLIS, RICHARD S.+TAYLOR, KEITH+HOOK, RICHARD N.	The LDSS deep redshift survey
N91-30326	04/91	0.183	DA COSTA, L. NICOLACI+PELLEGRINI, PAULO S.+DAVIS, MARC+MEIKSIN, AVERY+SARGENT, WALLACE L. W	Southern Sky Redshift Survey - The catalog
N90-23083	02/90	0.173	DEY, ARJUN+STRAUSS, MICHAEL A.+HUCHRA, JOHN	A deep redshift survey of IRAS galaxies towards the Bootes void
N90-48906	09/90	0.169	STRAUSS, MICHAEL A.+DAVIS, MARC+YAHIL, AMOS+HUCHRA, JOHN P.	A redshift survey of IRAS galaxies.+I - Sample selection
N91-16860	11/90	0.168	CHARLTON, JANE C.+SALPETER, EDWIN E.	Pairs of galaxies in low density regions of a combined redshift catalog
N91-17785	12/90	0.165	HUCHRA, JOHN P.+GELLER, MARGARET J.+HENRY, J. PATRICK+POSTMAN, MARC	A deep Abell cluster redshift survey

[3,3] /pool1/ads/tmp/kurtz/ktGAAa09739 Row 15 of 72 Pos 1 of 191 T

Select abstracts to retrieve: Position mouse with left mouse button and select with right mouse button.

Get All | Get Selected | Clear Selected | Print | Cancel | Help

Figure 2. The first page of results from the query in Figure 1.

Discussion

Crabtree: How often will the database be updated?

Kurtz: Depends on future developments, eventually it will be continuously updated.

Blum: Is the technique used in other fields, is it patented, and is it effective?

Kurtz: Bell Labs—commercializing. Others use it. Other fields—20 years or more. Yes—very effective.

Unknown: What does a user have to do to start using the system? For example, does custom software have to be loaded into one's X-system?

Kurtz: If you can use ADS you can use this system—it is part of ADS, it appears as one of the options when you start up ADS.

Huenemoerder: It is easy to verify the appropriateness of items found, but how do you know what's left out? Do you—or plan to—give sample output to experts for review?

Kurtz: You can't know—it's impossible (or else you would include it). We've proposed to NSF, and if funded will perform an experiment. This is a very difficult and expensive matter.

Astronomical Data Analysis Software and Systems II
ASP Conference Series, Vol. 52, 1993
R. J. Hanisch, R. J. V. Brissenden, and J. Barnes, eds.

STELAR: An Experiment in the Electronic Distribution of Astronomical Literature

A. Warnock

Hughes STX, NASA/Goddard Space Flight Center, Code 631, Greenbelt MD 20771

M. E. Van Steenberg

NASA/Goddard Space Flight Center, Code 631, Greenbelt MD 20771

L. E. Brotzman, J. E. Gass, D. Kovalsky

Hughes STX, NASA/Goddard Space Flight Center, Code 631, Greenbelt MD 20771

F. Giovane

NASA Headquarters, Code SZE, Washington, DC 20546

Abstract. *STELAR* is a pilot project designed to study the technical and practical aspects of making the refereed scientific literature available on line. Machine readable abstractions, supplied by the NASA STI program, of articles from eight journals of interest to astronomers have been indexed and made available using WAIS as the query and retrieval mechanism for the publically-available prototype. The complete system will include the ability to retrieve scanned images of the pages of selected articles. Enhancements of the search and retrieval system, the user interface and additions to the current holdings are anticipated.

1. Introduction

Astronomical research has been transformed by the availability of wide-area networks and low-cost computing power. The American Astronomical Society has recently begun encouraging members to submit abstracts of papers for meetings by electronic mail. A number of journals now permit the electronic submission of articles for publication. The refereed literature, however, has been largely untouched by advances in information access. The result has been a tremendous growth in the volume of information published, without substantial improvement in the researchers' ability to locate and retrieve articles of interest.

It is now technically feasible to place much of the astronomical literature and documentation on-line, providing researchers with direct access to this information. More importantly, with the addition of modern text searching methods, astronomers will have the ability to quickly find articles about a particular research topic and examine them as they wish. In addition, databases of the research literature can contain products which are not currently readily available, e.g., the actual data used, and forward references to errata and to other relevant works.

While taking the first steps towards electronic publication of the research literature, it is crucial to recognize that the rigor and integrity of the scholarly journals must be maintained (Seiler 1989). The financial health of the journals must also be preserved (Boyce et al. 1992), as must their accessibility to the entire community, not just those members who happen to have network access. Issues concerning the long-term archival preservation of the literature must be addressed. Clearly, there is still much to be learned about how scientists will utilize on-line resources, and what will be its impact on the way in which research is conducted.

2. Overview

STELAR, the STudy of Electronic Literature for Astronomical Research, is a joint effort of the American Astronomical Society, the Astronomical Society of the Pacific, NASA, a number of publishers, editors, research libraries, and astronomers with additional support being provided by the American Institute of Physics, the Library of Congress, the National Science Foundation, and the University of North Carolina at Chapel Hill. It is an experiment designed to study the technical and practical issues and potential impact of placing the astronomical documentation and literature on-line.

STELAR is managed at NASA's Astrophysics Data Facility (ADF), located at the Goddard Space Flight Center in Greenbelt, MD. The initial phases of the project focus on the problems of converting existing literature for on-line access, making textual data available on-line and investigating readily available text query systems. Later phases will investigate additional text indexing and retrieval engines and the potential for applying optical character recognition to the scanned bitmaps in order to render them into machine-readable form.

A limited prototype, initially incorporating only machine-readable abstracts provided by NASA's Scientific and Technical Information (STI) program, is available for use by the general astronomical community. The completed prototype will link the abstracts to scanned bitmaps of the individual article pages.

3. STELAR Architecture

There are three components to the problem of establishing on-line documentation and literature: (1) the development of an electronic data delivery system, (2) the conversion of existing materials from printed pages to electronic files, and (3) the eventual production of new literature in a form which can be placed on-line as published.

The *STELAR* prototype system uses the highly portable and fully open, multi-disciplinary document query and delivery system known as WAIS (Wide Area Information Server) (Kahle 1989). WAIS is based on a client/server model, communicating both locally and over wide area networks like the Internet and DECnet. WAIS provides a mechanism for text-based queries of multi-media databases, and for the retrieval of relevant documents. Current WAIS implementations already include support for associating multiple data types (ASCII text and a variety of data formats) with a single document.

The fundamental data object within the *STELAR* database is an *article*, defined as a collection of items associated with a published paper. All articles contain a machine-readable (ASCII text) abstract, and additionally may include the scanned image of the first page of the published paper and the scanned images of the entire published paper. Each of these items may be retrieved individually. Queries are posed against an index built from the entire collection of text abstracts.

The notion of the article object makes the database extensible, i.e., new items can be associated with a paper without substantially modifying the query and delivery mechanism. It will eventually be possible for an article to include references to the published paper (forward referencing), errata, machine-readable versions of tables and graphs, original data and the full text of the article in a markup language (e.g., SGML or TEX).

The primary *STELAR* database incorporates machine-readable abstractions of articles in eight leading academic journals of interest to the astronomical community (ApJ, ApJS, AJ, PASP, A&A, A&AS, MNRAS and JGR). These abstracts have been supplied by NASA/STI from a database prepared for NASA's RECON system by an independent abstraction service, and are not necessarily identical with the original abstract of the published paper. The RECON system database contains abstracts from as early as the mid-1960's. No comparable source of original abstracts in machine-accessible form has been identified. The ADF will update the set of available abstracts on a regular basis.

The libraries at Goddard Space Flight Center, the Space Telescope Science Institute, NOAO/KPNO, and NRAO will work with selected astronomers to evaluate the initial prototype. Access to the bitmaps will initially be limited to test groups at the libraries to protect the copyright concerns of the societies and the journal publishers.

4. The Delivery Engine—WAIS

The WAIS system, as distributed by Thinking Machines Corp., consists of three components—a text indexer, a database server and a client program (Kahle and Medlar 1991). The text indexer builds a master index of all words occurring in a database of documents. This index is then used by the retrieval component of the server to find which documents contain the words in a query. Communication between client and server is accomplished using an information search and retrieval protocol, Z39.50-1988, developed by the National Information Standards Organization (NISO 1988).

The server runs on the computer hosting the database of documents and responds to queries. A query can either be a search of the master index or a request to retrieve a document to be sent to the client. WAIS servers currently run under Unix and VMS.

The client runs on the user's machine, and provides the user interface to the database. A free-format text query is translated into the appropriate form and sent to the server. The server processes the query, and returns the results to the client. Queries may be posed to multiple databases. WAIS clients are available for Unix (both character mode and X windows), VMS, Macintosh and MS-DOS.

The default search engine matches occurrences of words in the query with individual words in the documents, and calculates a relevance score for each document. A list of documents is then returned in ranked order. A simple extension of the search technique is the notion of "relevancy feedback." The user can select a portion of a retrieved document and submit as a query. This allows quite detailed searches without requiring the user to explicitly formulate a detailed query.

The source code in the distribution system is quite modular, and allows for replacement of individual components. Experimental servers which are capable of handling word stems and synonyms, spatial searches and boolean searches have already been implemented, as have servers incorporating the new Z39.50-1992 protocol. Other search and retrieval engines might use advanced techniques such as factor spaces or fuzzy searches. The *STELAR* project will be actively evaluating evaluating the performance of these systems.

5. Future Plans

Subject to the approval of the copyright holders of the various journals, the *STELAR* project plans to gradually make the scanned bitmaps of the article pages available to the astronomical community. NASA/STI will eventually be supplying *STELAR* with abstracts to all of the space science journals in their database.

Additional enhancements being investigated include indexing of the full text of the articles when machine-readable versions of the published articles are available (Owen 1992), making articles available in a mark-up language (TeX, SGML) (Biemesderfer 1992) or device-independent form, and the addition of errata and other forward references to the basic *STELAR* article structure.

In conjunction with this controlled study, the ADF and STI are making the abstracts and several other text databases available to the astronomical community as part of NASA's commitment to its science community. The AAS Job Register and the electronic abstracts for the Summer 1992 and Winter 1993 AAS meetings are currently available. Details may be found in Warnock (1992).

Additional information on *STELAR* and other available databases can be obtained by sending electronic mail to stelar-info@hypatia.gsfc.nasa.gov.

6. Conclusions

Interest in the electronic availability of the literature is high, particularly to the extent that it offers enhanced utility over the current paper-based literature at decreased cost. Users have already noted the ability to discover new relevant references in the course of their research with minimal time investment. The use of WAIS technology has allowed *STELAR* to concentrate on the job of assembling appropriate test materials and populating the database, rather than undertaking the task of writing a database system and user interface from scratch. The availability of WAIS clients for a wide variety of computer platforms at no cost, and the well-connected nature of the astronomical community means that *STELAR* products are already widely available.

The adoption of an existing technology is not without cost, however. Each individual user or site must obtain and install its own client software, not always a trivial task. As we add new items to the article objects in the *STELAR* database, it may become necessary to develop specialized clients. It is also not clear that the user can ever be divorced completely from the specifics of the indexing and retrieval engine. It appears that some knowledge of how the engine works is required in order for the user to adequately formulate a query that will yield the expected results.

Finally, the contents of the database itself affect the user's expectations. It is not clear that the astronomical community is comfortable with the use of a professional abstraction service in place of the original published abstracts, although such practice is common in the library profession. The availability of appropriate materials is now and will continue to be an ever-increasing subject of attention.

Acknowledgments. *STELAR* is a collaboration between NASA, the AAS, the ASP, the publishers of the astronomical journals, the Library of Congress, the University of North Carolina at Chapel Hill and librarians from GSFC, STScI and NRAO. We particularly acknowledge the valuable contributions of Jim Fullton of UNC/Chapel Hill.

References

Biemesderfer, C. 1992, AAS Newsletter 62, Special Insert, p. 7

Boyce, P.B., Pilachowski, C., & Dalterio, H. 1992, AAS Newsletter 62, Special Insert, p. 1

Kahle, B. 1989, "Wide Area Information Servers Concepts," Thinking Machines Corp. Technical Report TMC-202

Kahle, B., & Medlar, A. 1991, "An Information System for Corporate Users: Wide Area Information Servers," Thinking Machines Corp. Technical Report TMC-199, version 3

National Information Standards Organization (Z39) 1988, "Z39.50-1988: Information Retrieval Definition and Protocol Specification for Library Applications"

Owen, E. 1992, AAS Newsletter 62, Special Insert, p. 9

Seiler, L.H. 1989, Academic Computing, September 1989, p. 14

Warnock, A., Gass, J.E., Brotzman, L.E., Van Steenberg, M.E., Kovalsky, D., & Giovane, F. 1992, AAS Newsletter 62, Special Insert, p. 9

Part 2. Data Analysis Systems

Section A. Next Generation Systems and Languages

Astronomical Data Analysis Software and Systems II
ASP Conference Series, Vol. 52, 1993
R. J. Hanisch, R. J. V. Brissenden, and J. Barnes, eds.

C++, Objected-Oriented Programming, and Astronomical Data Models

A. Farris

Space Telescope Science Institute, Baltimore, MD 21218

Abstract. The fundamental features of objected-oriented programming are discussed from a C++ programming language perspective. This discussion focuses on objects, classes and their relevance to the data type system; the principle of information hiding; and the use of inheritance to implement hierarchical relationships. The basic concepts of this approach are characterized in contrast to more traditional procedure-oriented approaches. Drawing on the object-oriented approach, features of a new database model to support astronomical data analysis are presented.

1. The Structured Programming Revolution

The structured programming revolution of the 1970s was a significant turning point in the history of software development. The major goal was to find those software constructs that contributed to good program structure and to understand why they did so. The harmful effects of 'go-to' statements were recognized. Spaghetti programs, those large monolithic structures liberally sprinkled with 'go-to' statements that defied control flow analysis, were finally relegated to the dust bin of history. Focusing on concepts of program structure quite naturally lead to the concept of modularity.

What is modularity? Much of the discussion of this period was devoted to answering this question. The question can be asked in much more general contexts than software development. In fact, it is a very subtle and difficult concept to precisely define. If programmers representing the major programming languages of the period, Assembler, Cobol, Fortran, and PL/I, were to describe what they called modules, we would conclude that modularity was a very language dependent concept, having very different properties within the context of different programming languages. Pascal and C, languages that came later, added little to the discussion of modularity. Only one point was universally agreed on, that modularity was a good thing, having achieved the enviable status of a buzzword. If your code wasn't modular, it wasn't state of the art, never mind that many characterizations of modularity were very superficial.

There was an important paper published during this period. It was short, very readable, and probably ranks as one of the most frequently referenced papers on the subject of software design. "On the Criteria to be Used in Decomposing Systems into Modules", by D. L. Parnas (Comm. ACM, Dec., 1972), illustrated two methods for decomposing a simple software problem into modules. While one method was more traditional, the other embodied a principle

Parnas called 'information hiding'. The resulting modules had a well defined purpose and public interface. The internal implementation of each module was independent and isolated from its surrounding. Parnas showed that the software design resulting from employing this principle was more comprehensible and flexible. This principle has emerged as one of the most important in the field of software design.

2. C++ and Modularity

C++ (Stroustrup 1991) was developed during the 1980s, using the existing C programming language as a base. The evolution of C++ influenced the standardization process for C and the resulting ANSI C is very nearly a subset of C++. C++ offers an enhanced approach to modularity and extensive support for the concept of information hiding. It supports the object-oriented paradigm, as well as more traditional procedure-oriented approaches. The software developer is not forced, by the language, into either model of software development.

A fundamental concept of C++ is the 'class', which may be regarded as a generalization of a C 'struct'. The form of a class declaration is presented below.

```
class Angle {  // name of class
     public:   // access controls
          // data declarations ...
          // function declarations ...
     private:  // access controls
          // data declarations ...
          // function declarations ...
};

int i, j;
Angle phi, theta;
```

In contrast to data structures in C, C++ classes contain function declarations as well as data declarations. (Anything following the '//' to the end of a line is a comment.) The data and function members of the class may be designated as 'public' or 'private'. Public members may be accessed by users of the class, i. e., from environments external to the class. Private members may be accessed only by member functions defined within the scope of the class itself. The class statement above only serves to declare the class; its implementation consists also of specifying the functions declared within its scope. A newly defined class is used as an extension to the type system within C++. The name of the class, 'Angle' in the example above, serves as the name of the new type. Individual variables, or objects, as the object-oriented paradigm calls them, are instantiated from the class in the same manner as variables are declared to be instances of the built-in data types. In the example above, in the same manner that i and j are declared to be of type 'int', phi and theta are declared to be of type 'Angle'. Objects are *instantiated* from classes.

The concept of a class allows C++ to directly support a kind of modularity that is not supported in C or Fortran. A class allows for a grouping of logically related data and functions into a larger unit than a C function or Fortran subroutine. The access control labels, 'public' and 'private', support the idea of information hiding, or encapsulation, as it is called in the object-oriented paradigm. The only way to accomplish a similar functionality within C is to use the file system of the operating system, which of course, relies on concepts external to the language itself. In many large-scale C programs there is a tendency to fracture the system into a large collection of small functions. If these functions are not collected into larger units by some logical mechanism, we have a software architecture that is a polar opposite of the monolithic spaghetti programs of the pre-structured programming era, and in many respects, just as unintelligible. The concept of a class directly confronts this problem by providing linguistic support for logical groupings of data and functions.

3. A C++ Class

The code fragment below illustrates a few of the C++ statements that might be used to define the Angle class.

```
class Angle {
     public:
          Angle();
          Angle(int degs, int mins, double secs);
          ~Angle();
          int underflow() { return underflow_flag; }
          int overflow() { return overflow_flag; }
          Angle & operator = (const Angle &t);
          Angle & operator + (const Angle &t);
          Angle & operator > (const Angle &t);
     private:
          double value;
          int overflow_flag;
          int underflow_flag;
          double min_value;
          double max_value;
          void ck_range();
};
```

C++ classes may contain functions, called constructors, that initialize objects when they are created. These have the same name as the class and are automatically invoked at creation time. Destructors (functions with a tilde prefixed to the class name) are also allowed and perform housekeeping chores such as freeing any allocated storage whenever the object is destroyed. Functions in C++ are uniquely identified by their name *plus* their argument list. Thus, two distinct functions may have the same name so long as they have distinct argument lists. In the example, there are two Angle constructors, one taking no arguments that initializes the object to some default value and a second one that initializes the object to some specified value. C++ also permits in-line functions, such as 'underflow()', shown here to illustrate a technique for implementing "read-only" data. A user of the Angle class has access to 'overflow_flag' and 'underflow_flag' but cannot change them. The lines containing 'operator' are special function declarations for functions that perform the '=', '+', and '>' operations on objects of type Angle. This type of function declaration is called 'operator overloading'. Almost all the C++ operators can be overloaded in this manner, giving one the ability to define, for example, a complete set of arithmetic and comparison operations for the class. An example of how the Angle class is used is presented below.

```
Angle alpha(60,15,47.9);
Angle beta, gamma;
alpha = beta + gamma;
if (alpha.overflow())
     error("Angle overflow");
if (alpha > beta) ...
```

The Angle class adds a new type to the programming environment with a complete set of supporting operations, including construction, destruction, assignment, arithmetic, comparison, and I/O operations. Judiciously employing operator overloading allows one to implement the class so that it functions like a language extension. Furthermore, its implementation is completely encapsulated.

4. C++ and Inheritance

From biology to government, hierarchical relationships are a basic feature of the world around us. Within C++ a class, called the base class, can be used to derive a new class, called the derived class. The derived class *inherits* all the data and functionality of the base class. Right ascension and declination are kinds of angles. The C++ inheritance mechanism allows the new classes, Ra and Dec, to share all the code within the Angle class.

```
class Ra : public Angle {
     public:
          Ra(int hrs, int mins, double secs);
          char *display();
};
class Dec : public Angle {
     public:
          Dec(); // changes the range of validity
};
```

The Ra class has a constructor and display function that allow its value to be given in units of time rather than degrees. The Dec class has a constructor that changes its range of validity. Angle, Ra, and Dec form a simple two-level hierarchy. In C++, very complex hierarchies of arbitrary depth may be defined.

5. Virtual Functions and Dynamic Binding

Suppose we have a class Screen with a member function 'move_cursor'. This class is a base class for two additional classes, VT100 and VT52, each of which has its own implementation of the function 'move_cursor'. The keyword 'virtual' says, in effect, that the implementation of this function is type dependent.

```
class Screen
     virtual void move_cursor(int, int);
class VT100 : public Screeen
     void move_cursor(int, int);
class VT52 : public Screen
     void move_cursor(int, int);
...
Screen *x[2];
x[0] = new VT100;
x[1] = new VT52;
...
for (i = 0; i < N; i++)
     x[i]->move_cursor(row,col);
```

Following the class declarations, a VT100 object and a VT52 object are created using the 'new' operator, the addresses of these objects being stored in x, an array of Screen pointers. This technique enables us to treat these objects as if they were screens, which is permitted since they are, after all, kinds of screens. The 'for' loop executes the function 'move_cursor' using the screen pointers stored in x. Because 'move_cursor' is a virtual function, the function appropriate to that type is executed, the VT100 version in the case of x[0] and VT52 in the case of x[1]. Note, however, that the source code in the 'for' loop does not depend on the specific type. The decision as to which function to execute cannot be made until execution time. This property is call dynamic binding. Virtual functions actually provide another form of encapsulation. They standardize the interface to a family of functions but leave the actual execution of those functions as type dependent.

6. Procedure-Oriented versus Object-Oriented Analysis

The C++ features mentioned above are key constructs that are designed to support object-oriented programming. Classes are provided as types of objects, providing support for data abstraction. Various language features support encapsulation and information hiding, providing controlled access to class internals in a variety of contexts. Inheritance and class hierarchies allow one to model naturally occurring hierarchies and to exploit commonality that exists among classes. Virtual functions and dynamic binding allow one to standardize interfaces to type dependent member functions.

Many people involved only in small-scale software projects employ an intuitive approach to software design issues and have failed to grasp some of deeper issues encountered in more formal design methodologies. A detailed comparison of object-oriented and conventional analysis and design methodologies may be found in Fichman and Kemerer (1992). I wish to concentrate on the essential differences between these approaches. To accomplish this it is instructive to begin with a procedural approach that presents the greatest contrast.

One of the "purest" procedure-oriented design approaches was introduced by IBM in the early 1970s as part of the structured programming revolution, viz., HIPO. HIPO stood for "hierarchy plus input process output" and it came complete with forms and templates. The idea was to put the inputs on the left and the outputs on the right and then ask "What processes convert the inputs into the outputs?" This is, quite literally, *data processing*. Many more modern and sophisticated design methodologies still have this idea at their core.

To illustrate this approach, let us consider a problem familiar in astronomy: reading a FITS file and copying image data to some host specific file. The input is, of course, the FITS file. The output is the specified image file. Some of the processing steps are 1) read a FITS record, 2) determine the type of record, 3) collect the header records, 4) collect the data records, 5) convert and store keyword values, 6) determine the type of data, 7) convert and store data values, and, finally, 8) write the output record. Many existing programs are constructed, more or less, in this fashion. While one can point out many disadvantages of such an approach, it would be incorrect to conclude that it is somehow "wrong". There are a lot of problems for which this is a perfectly good model. Recipes in a cookbook, for example, consist of inputs, a sequence of hierarchically arranged processing steps, and outputs. The essence of the HIPO and, more generally, the procedure-oriented approach is to find a set of procedures.

The object-oriented approach asks a fundamentally different question at the outset. Faced with the same problem, this approach asks, "What is FITS?" In fact, this approach asks a series of increasingly specific "What is X?" questions. The object-oriented approach sees the world, or problem domain, as consisting of things that have specific characteristics, that behave in specific ways, and are related to other things. Object-oriented analysis is *the art of definition. Definition is the art of precise identification.* The job of object-oriented analysis is to discover the structure of the problem domain, i. e., to identify what things are related to the problem, their characteristics and relationships. The answer to our question above, "What is FITS?", would be something like the FITS standards document (NOST 1991). It is important to recognize that this document states, in precise detail, what FITS is. That is its sole purpose. Its purpose is *not* to tell users how to process FITS files. The fact that a reasonably intelligent person can figure out how to process FITS files by reading the standards document is not the point. The only purpose of the standards document is to precisely identify what is and is not a FITS file. The need for such precise definitions is sometimes not appreciated by software developers and almost never by users, or domain experts, who get annoyed at having to answer so many questions. In fact, precise definitions play a crucial role in procedure-oriented design approaches as well but that role is often obscured and implicit. Object-oriented analysis places the precise characterization of objects at the center of its approach. Consequently, the structure and boundary of the problem domain are clearly identified.

Unfortunately, a confusion has arisen around the object-oriented approach over what an object is, sometimes with metaphysical overtones. Furthermore, many people confuse 'object' and 'class'. This is not a new problem; Socrates had the same trouble trying to get the Athenians to understand the issue. (This point is illustrated in many of the early Platonic dialogues). The definition of the *class* human being is 'rational animal'. An *object*, or instance of the class, is Socrates. An object is a thing, an individual, an entity. One of the first attempts at a theoretical justification of this view of the world was made by Aristotle in *The Categories*. This work was based on an analysis of how we use language. His characterization of what we are calling an object is: "that which is individual and has the character of a unit is never predicable of a subject" (Categories, 1b). In modern logic, an object would be characterized as the referent of the variable bound by the existential or universal quantifiers. Having this view of an object need not, and should not, force us to embrace any form of absolute essentialism. This is an important point for the analysis process. What we identify as an object and what we designate as its characteristics will be context dependent. As the problem domain changes, so will the characterizations of our objects.

7. Object-Oriented Design

Armed with an effective object-oriented analysis of the problem, an object-oriented design defines classes that represent the entities of the problem domain. Attributes of the entities become data items within the classes; actions become member functions. Classes are related to one another by inheritance relationships, or part-whole relationships, etc. To return to our FITS problem, we would define a class that represented a FITS logical record. Then, we would define a class that represented a FITS header, which would be subdivided into primary, group, and table headers. The FITS table header would be further subdivided into ascii and binary table headers. FITS data would follow a similar inheritance pattern. Finally, we would define keyword and data field classes. Fields would be subdivided into bit, integer, real, complex, etc., representing all the FITS data types.

It should be obvious that such a class structure is almost a copy of the table of contents of the FITS standards document. This is as it should be. *A good object-oriented design is a model of the problem domain.* This is also why the object-oriented approach to software design inevitably turns to a discussion of modeling. Object-oriented programming consists of implementing the software design. The structure of the C++ source code of a properly implemented object-oriented analysis and design should be a model of the problem domain. To return to our FITS problem, we should be able to read the C++ class definition of a FITS binary table header and immediately turn to the relevant section of the FITS standards document and clearly see that it is correct. It is this basic property that is one of the keys to the claims of improved productivity attributed to the object-oriented approach.

Having a set of properly defined classes is not yet a complete solution to the problem we set out to solve. Our classes representing FITS files would be supplemented by a class representing the image data. In addition, we must also show how to use those classes to solve the problem of reading a FITS file and converting the image data. Merely having a useful set of classes corresponds to Plato's static world of forms, the world of changeless being. Plato recognized that one must also relate those forms to the dynamic world of change and becoming, if one is to gain a complete understanding. Thus the class model described above must be supplemented by a second model, sometimes called the dynamical model or functional model, depending on the methodology being used. Both kinds of models are essential to an effective system design. They are intended to capture ideas related to structure and change, fundamental categories that human beings use to model the world.

A dynamical model of our FITS problem would show a FITS logical record object being instantiated from a FITS record class, primary header and data objects being instantiated from their respective classes, and a conversion process that produces an image object that is an instantiation of the image class. In other words, the dynamical model shows how actual objects are created and behave during the execution of the program. C++ constructors and destructors are crucial linguistic concepts that form the bridge between the static and dynamical models. Together these models form a complete system design; they constitute a model of the problem domain. In comparing our object-oriented software design and the previous HIPO procedure-oriented design, one can see that a significant repackaging of the software has occurred. The global data structures and collection of procedures that operate on them have been repackaged into sets of data and functions that model the basic entities of the problem. This approach promotes comprehensibility, isolates dependencies based on logical structure, and provides for greater flexibility.

8. An Astronomical Database Model

The traditional relational database model, so long as first normal form is preserved, has two features that limit its applicability to scientific data. An item of data in a column of a table must be atomic; it must have no internal structure. In addition, an item of data must not have any direct or implied linkages to other items of data or data aggregates. These restrictions have the consequence that relational databases are unable to deal effectively with arrays of data items or to model complex relationships between collections of data. In looking at the problem of providing a more general database model for astronomical data analysis, it seems natural to base the data model on a generalization of a FITS binary table, a close cousin to a relational database table. Most astronomical data that exists in archives today is in the form of FITS tables, or some structure very close to FITS tables. Many on-line systems produce raw data in the form of FITS tables.

A FITS binary table is an ordered set of rows of data fields, organized into an ordered set of columns. A descriptive header is associated with the table and describes the table as a whole. There is also a set of keywords, each of which applies to the table as a whole. A keyword is a name plus a data value. The number of keywords is arbitrary and may be zero. Elementary data types supported by FITS binary tables include logical, bit, unsigned 8-bit integers, signed 16-bit and 32-bit integers, character strings, single and double precision floating-point numbers, and single and double precision complex numbers. A FITS binary table has an additional property that distinguishes it from a traditional relational database table, viz., any column value may be a single value or a one-dimensional array of values. Obviously, a relational database table is a subset of a FITS binary table, obtained if the number of keywords is zero and all column values are single elements.

Let us consider the following types of extensions to the basic FITS table concept. They are all related to the concept of a field and expand that concept.

First, the elementary data types of fields should be extensible. In a manner compatible with object-oriented programming, the previous list of data types should be regarded as an extensible list, which is application dependent. In astronomy, we will probably want to add such elementary data types as right ascension, declination, date, and time. Second, any field can be a multi-dimensional array of the elementary types. Third, a field can reference another table; such references to tables are to be regarded as one of the elementary data types. Finally, keywords should have the same properties as fields within columns. Of course, this latter implies that keywords or column fields may be single or multi-dimensional arrays of references to other tables, providing a mechanism for generating extremely complex relationships between tables. The concept of a set of keywords may appear to be an unnecessary and arbitrary appendage to the basic table structure. However, it has proven to be useful in providing a mechanism for assigning attributes to a table as a whole. In addition, since these values may reference other tables, it provides a mechanism for forming semantically meaningful relationships between tables.

The concepts sketched here were used to describe VLBA data being considered by the AIPS++ (Croes 1992) development group. The raw data output of the VLBA Correlator is a sequence of FITS binary tables. The main component of this sequence is the UV table. Its rows contain data items familiar to interferometry, viz., U, V, W, date, time, baseline, array, source number, frequency id number, data integration time, weight, and finally the interferometer data. This interferometer data field is actually a $2 \times 4 \times 128 \times 4$ four-dimensional matrix. However, while this table is the most significant component, it requires several other tables for proper interpretation and analysis, e.g., a flagging table, a calibration table, and an antenna table. Additional tables, which are constant over relatively long periods of time, several days for example, are also required, viz., an array geometry table, a frequency table, and a source table.

Using the concepts embodied in the extended tables, the most natural approach to this VLBA data is to view it as a table of tables. The VLBA data tables are separated into two groups, time-invariant tables and time-variant tables. A collection of VLBA data can be viewed as consisting of a list of keyword fields that reference time-invariant tables and include the array geometry table, frequency table, source table, and gain curve table. The column fields of the collection of VLBA data each reference a time-variant table. The rows of this table are all the tables that are relevant to that period of time, which is of the order of minutes, including the UV data table itself mentioned above, antenna characteristics table, calibration table, flagging table, bandpass table, baseline correction table, phase-cal table, interferometer model table, CALC table, weather table, opacity table, ephemeris table, and sampler statistics table. The rows of this table of tables are ordered by increasing time.

It should be noted that this VLBA data table consists of a collection of references to other tables. As such, it is a high-level structure that expresses relationships between groups of tables. The strength of this approach is that it provides a way to express such relationships. This way of describing VLBA data, especially the division of tables into time-variant and time-invariant, is very natural and intuitive. Within the framework of existing FITS or relational data structures, there is simply no way to express such interrelationships between tables. It is also fairly easy to change these relationships. All that is required is to restructure a set of references to other tables, not the actual tables that are referenced.

Within this scheme there is a natural way to form the concept of classes of tables. All tables are completely defined within the context of the database manager via a table descriptor. The table descriptor may leave unspecified such information as the number of elements in an array or the actual values in keyword attributes, specifying only their data types. When an actual table is constructed, it is instantiated from a table descriptor, and any unspecified values of attributes are specified. Such table descriptors become descriptions of classes of tables, viz., all those tables that have been created using that particular table description. This concept of a table description can also be used to support inheritance relationships. There are several advantages to such a concept. It can be used to refine and restrict search operations; one can search tables of a particular type rather than all tables. It can also be used to simplify and generalize application programs. Through a suitable scheme of C++ operator overloading, operators can be defined in such a manner that application programs are not dependent on the sizes of arrays. This means that application programs need not be dependent on a fixed number of frequency channels or time samples, a problem that has plagued data analysis programs in the past. Finally, this concept of a table type can be used as an integrity check. If an attribute value references a table, the table descriptor specifies a table type. When an instantiated table specifies a table reference for that value, it must conform to that table type. This mechanism can be used to preserve data integrity, in the same manner as specifying an elementary data type.

Acknowledgments. I would like to acknowledge many helpful discussions with Ron Allen on the topics presented in this paper. I would also like to thank the members of the AIPS++ project, especially, Ger van Diepen, Brian Glendenning, and Dave Shone, for many helpful discussions on the data management issues.

References

Croes, G.A. 1993, this volume

Fichman, R., & Kemerer, C. 1992, "Object-Oriented and Conventional Analysis and Design Methodologies, Comparison and Critique", IEEE Computer, October issue

NOST, November 6, 1991, "Implementation of the Flexible Image Transport System (FITS)", Draft Standard, NOST 100-0.3b, NASA/OSSA Office of Standards and Technology, Code 933, NASA Goddard Space Flight Center, Greenbelt, Maryland

Stroustrup, B. 1991, The C++ Programming Language, Second Edition (Addison-Wesley)

Discussion

Kurtz: Would you please discuss your database project more?

Farris: This topic has been included in the paper. A data management system employing the ideas presented here is under development by the AIPS++ project.

Adorf: I appreciate everything you said about object-oriented and functional approaches. One thing which is lacking from the C++ language is dynamism. C++ is a static language; it is not interactive. You have to compile and link your programs. So, it does not support rapid prototyping or quick reactions to up-coming data analysis problems. Would you please comment on recent developments in the area of dynamic languages?

Farris: Indeed, in this sense, C++ is a compiled language and not an interactive one. Furthermore, C++ is a statically typed language; the compiler must have complete information about types. One can create new objects at execution time but not new types. In its evolution, C++ has remained focused on being a statically typed, compiled language and has maintained a commitment to provide reasonable run-time efficiency. (C++ has been used for high-performance computing and real-time applications.) For this reason, it has avoided a more dynamic and less efficient run-time environment.

Smalltalk, an interactive object-oriented language, has played a major role in the development of the object-oriented paradigm and is widely used as a prototyping tool. In fact, some institutions use Smalltalk and C++ together. One organization I have heard about puts their new programmers who are not familiar with the object-oriented approach to work on developing prototypes on Smalltalk and teaches them C++ later. The rationale behind this approach is that Smalltalk forces you to use the object-oriented approach.

Another development to watch for, in the area of dynamic environments, is interactive query languages in conjunction with object-oriented database management systems. These are in their infancy now but a lot of research is being done in this area.

Astronomical Data Analysis Software and Systems II
ASP Conference Series, Vol. 52, 1993
R. J. Hanisch, R. J. V. Brissenden, and J. Barnes, eds.

On AIPS++, A New Astronomical Information Processing System

G. A. Croes

National Radio Astronomy Observatory[1]*, 520 Edgemont Road, Charlottesville, VA 22903-2475*

Abstract. The AIPS system that has served the needs of the radio astronomical community remarkably well during the last 15 years, is showing signs of age, and is being replaced by a more modern system, AIPS++. As the name implies AIPS++ will be developed in an object-oriented fashion, and use C++ as its main programming language. The work is being done by a consortium of seven organizations, with coordinated activities worldwide.

After a review of the history of the project to this date, from management, astronomical and technical viewpoints and the current state of the project, the paper concentrates on the tradeoffs implied by the choice of implementation style, and the lessons we have learned, good and bad.

1. Introduction, Pre-History

AIPS has dominated the scene of processing radio interferometer data for the last decade. Although intended initially for the processing of VLA data only, it was subsequently extended to process data for the VLBA, the Australian Telescope (AT) and the British MERLIN. It has grown into a fairly large system, about 600,000 lines of primarily FORTRAN77 code. Its great success speaks well for the genius of the people who created it under the leadership of Eric Greisen and Bill Cotton.

Despite its excellent track record, AIPS began to show the signs of old age during the last few years. It was designed originally as a FORTRAN-based system on a small computer and, to this day, betrays these humble beginnings. The most frequently heard complaints from its users were:

- AIPS is pretty buggy: when a new path through the code is taken, the program often fails;
- It takes quite a while to learn how to extend AIPS, and even when mastered, the job remains difficult and error-prone;
- AIPS is difficult to maintain: once an error has been signaled, it is far from easy to correct it;

[1]NRAO is operated by Associated Universities, Inc., under cooperative agreement with the National Science Foundation.

- AIPS is somewhat old-fashioned in its user interface and does not exploit modern capabilities fully.

The main causes of the problems were, of course, inherent in the kind of machines available in the late 70's, when the AIPS architecture was laid down, and in the poor programming language used: FORTRAN. The latter, although the language of choice for scientific programming, is notoriously poor for building large systems. In particular the need to have large and complex collections of global data (COMMON blocks) has proven to be fatal.

About five years ago, a coincidental disturbance in NRAO brought the above issues into focus, and the NRAO Director, Paul Vanden Bout, convened a committee to advise him on the course to take with regard to this software. The committee, chaired by Tim Cornwell, consisted of both NRAO staff members and outsiders. It recommended, in short, that:

- All development of data processing software in NRAO should be coordinated by a new Assistant Director for Computing;
- AIPS should be re-designed and re-implemented following certain general guidelines;
- An equal amount of attention should be devoted to single-dish software.

The recommendations were accepted, Paul Vanden Bout implemented the first recommendation at the end of 1990, and the quest for a better AIPS began.

A small team set out to define the global parameters for the project, and concluded quickly that the only way to avoid the problems that AIPS had, was to use an object-oriented approach for the new AIPS. In order to understand what happened next, it is necessary to have a general appreciation of what that approach entails. This will be the subject of the next section. We will resume our account of what happened in our project later in the paper.

2. The Object-Oriented Approach To Programming

2.1. Introduction

The fashionable phrase "object-oriented" refers to a specific methodology for designing computer programs. It utilizes an explicit model of the computational problem to be solved, defined in terms of a hierarchy of self-contained sub-units: objects. The approach uses two fundamental mechanisms: encapsulation and inheritance which I will try to explain. An important implementation consideration for large programs is polymorphism. Finally, I will discuss design ideas.

In order to implement an object-oriented design, one would prefer to use a programming language that provides direct support for the basic concepts and generally reject one that is incompatible with the approach. You will find our reasons for choosing C++ as our main implementation language for AIPS++ at the end of this section.

2.2. Encapsulation

Encapsulation refers to the principle that a computational unit must be self-contained. It provides a well defined functionality to its environment through a number of precisely defined interfaces. The way the functionality comes about, i.e., the internal implementation of the unit, is entirely opaque to its environment.

On closer examination, the computational unit is defined by a finite set of numbers and some functions that operate on these numbers. A concrete instance of such a unit has values defined for each number and is called an object. The symbolic form of the unit is called a class. It has variables defined for the numbers and specific functions in terms of the variables. A class defines therefore the set of all possible objects of a certain type and the set of all functions that can be applied to an object. A simple example is the class of, say, integers. The functions defined are addition, multiplication, etc., and objects are 5, 10, 2122.

The two main reasons for using encapsulation, and hence classes and objects, are:

1. The implementation of a class can be changed without affecting its environment. This allows one to
   (a) Construct quick "stubs" (rough outlines) for prototyping, filling them out with more detailed coding later;
   (b) Optimize a class for a particular computer;
   (c) Change the algorithms used for the implementation of a class.
2. The integrity and maintainability of large programs improves greatly, because
   (a) All global data have a scope limited to a finite, enumerated set of functions. The existence of unscoped global data is a limiting factor for the size of programs that can be developed in practice;
   (b) Individual classes can be tested much more thoroughly than routines referring to unscoped global data. This leads to significantly more reliable systems for a certain size.

A note may be appropriate here with regard to the naming of classes. In order to remember the kind of functionality a class provides, one tends to give it a name of a real-life object, such as "manager," "director," "slave," etc. The functionality provided by these classes must still be defined in a dictionary just as if they had been given names consisting of random sequences of characters. This activity of having to define "manager," "person," "user," etc., has given rise to mis-understandings by those related only peripherally to the design effort.

2.3. Inheritance and Polymorphism

Inheritance exploits the similarities that exist between various classes and extracts their common aspects (data and functions). For instance, if one would have cars of various makes, it would be efficient to encapsulate the common capabilities in a class "car" and let specific sub-classes, such as "Ford" and "Volvo," inherit them from "car." One can reduce the amount of code that needs to be written and tested considerably by using inheritance.

Polymorphism uses inheritance to gain a crucial improvement in extensibility of a given system. A "parent class" can contain a virtual function, i.e., a function that can (and in some cases must) be re-defined by its children. For instance, a class "figure" may have a virtual function "draw" which is defined by all its children "square," "contours," "graph," etc., in a very different manner. When the environment specifies "draw figure," the system determines at run-time which of the various draw routines to call, depending on the type of the figure. The implication of this in that one can define a new type of figure, e.g., "histogram," compile it and add it to the system without changing a single line of code in the existing system.

The use of inheritance is closely related to the use of classification in science. There is one subtle difference. In science one always starts out with a large collection of objects which one splits into classes (encapsulation), which are then hung into a hierarchy of more and more abstract classes (inheritance). In the design of computer programs, the design of classes and of the inheritance hierarchies occurs simultaneously. This is one of the reasons why an object-oriented design must always go through a number of iterations.

2.4. Design Tools

Although the general object-oriented design methodology is now well understood, its application to a particular computational problem is still far from easy. It requires the use of tools ranging from simple cards to elaborate graphical support systems. The main purpose of these tools is to enable quick (and sometimes drastic) changes in a design and to communicate that design to others—potential users and implementors alike.

Various researchers have been advocating their particular set of concepts, symbols and graphics. We have examined and tried the approaches advocated by Coates and Yourdon (1991), by Booch (1991), and OMT proposed by Rumbaugh et al. (1991). We found the first one to simplistic. We settled on OMT, as the reference made the impression (confirmed later) that OMT had been used in a great variety of real applications, where it had worked well.

There is a small learning curve using the OMT. It depicts a rich variety of associations that can exist between objects and classes for which it uses a correspondingly rich variety of symbols. Communication with a potential user of the system, necessary to verify the correct interpretation of program requirements, can only be successful if that user is willing to learn how to interpret the graphical depiction of a design.

2.5. Why C++

Considering what we have noted above, we had to make an early decision on which implementation language to use for AIPS++. Although the decision to use an object-oriented design methodology was the more important one, the choice of the main implementation language does have an impact on the approach one takes.

Many traditional languages now support encapsulation, i.e., a number of "global" data to which only a few, specified routines have access.: FORTRAN 90

(Brainerd et al. 1990), PASCAL (i.e., MODULA II and MODULA III), LISP (i.e., CLISP), C (i.e., C++ (Stroustrup 1991)), and ADA. Even good old FORTRAN 66 or 77 could do this if one were willing to use a pre-processor (Croes 1988). In addition, there are brand new languages that support this methodology, such as SMALLTALK and EIFEL (Meyer 1988). FORTRAN still requires fairly heroic measures to instantiate multiple instances of objects, but it can be done.

On a mixture of practicality and principle we ruled out all but C++, EIFEL, and FORTRAN. The latter, of course, does not provide support for inheritance, and polymorphism is simply impossible. Even so, we decided to try how far we could come with it as it is the programming language of AIPS. We were inclined to go with C++ as it provides full support for the implementation of all aspects of an object-oriented design and it comes with broad industry support as reflected in the availability of debuggers, browsers, and other CASE tools.

Early in 1992 Bill Cotton did an investigation of FORTRAN (Cotton, Jr. 1992). It confirmed our earlier suspicions. Although FORTRAN was finally not accepted as the language of choice for AIPS++, the investigation turned out to be quite useful. It gave rise to a significant improvement in the programmability of AIPS and has helped to create better VLBA modules in the old package faster.

Chris Flatters and Brian Glendenning investigated EIFEL in parallel with the FORTRAN effort. It proved that EIFEL is a very powerful, elegant language, suitable for our purposes. We were a bit afraid, though, of the rather narrow support base for EIFEL and the apparent lack of CASE tools. This investigation showed that EIFEL provides an excellent training ground for persons who wish to become familiar with object-oriented programming.

3. A Brief History of the AIPS++ Project

Now armed with full 20/20 hindsight we can examine what actually happened in the AIPS++ project, in particular some of the mistakes we made.

3.1. The Early Months

Considering the fact that we are trying to develop a system that covered a wide variety of instruments (we are in the final stages of building the Very Large Base Line Array, are starting on the Green Bank Telescope and hopefully will be building a Millimeter Array later in the decade), it was pretty natural to consider the possibility of similar organizations joining us in this effort. I was in Australia in November of 1990 on other business, and took the opportunity to discuss this with the local staff and management. A rather positive reaction led me to further visits to the Netherlands and the UK and getting encouraging reactions there as well, we called a meeting in June of 1991 to discuss the action we should take. The two day meeting covered a wide variety of subjects, and two crucial results emerged:

- The effort would be a joint effort, and
- It would be based on an object-oriented approach with C++.

Founding members of the cooperative venture at the meeting were: the Dutch organization (NFRA), the Australian Telescope, Jodrell Bank, the BIMA organization, and the NRAO. A few months later the Canadian Herzberg Institute and the TATA institute of India joined as well. An agreement was drawn up and signed by all parties. It specified that

- All parties must contribute a minimum of two Man Year Equivalents (MYE) over a period of two years (NRAO to provide ten MYE) and would station at least one man for half a year in Charlottesville, Virginia, the NRAO headquarters;
- There would be a steering committee consisting of members of the participating organizations to oversee developments;
- All participants would pay their own way.

So far, so good. We started making arrangement for housing a dozen or so staff and providing them with work stations, desks, etc., and asked the staff that was to come to Charlottesville to get some training in C++. This was our first mistake. We should have asked for a training in object-oriented design, this being the more difficult subject to learn and one that was most appropriate for the forthcoming effort.

In September, a group of half a dozen scientists led by Robert M. Hjellming started to write the specifications for AIPS++. They produced this document in early December. It was sent around to the other participants with a request to produce their own requirements. Most of them complied, specifying alternatives and additions to the document produced in Socorro by the end of 1991. Robert Hjellming made a determined effort to combine these documents into a single requirements specification, but had to struggle with various semantic and even conceptual inconsistencies. He finally, by the end of February, came up with a compromise document that has served as since then a definition of the scope of the work that needed to be done.

3.2. The Middle Months

In early January the first taskforce descended on Charlottesville. It consisted of Mark Calabretta and Bob Sault of the Australian Telescope, Dave Shone of Jodrell Bank, Friso Onlon of The Netherlands, Sanjay Bhatnagar of the TATA institute, Mark Stupar and Peter Teuben of BIMA, Lloyd Higgs of DRAO in Canada, and Brian Glendenning, Robert Hjellming, Mark Holdaway, Chris Flatters, and Bob Payne of the NRAO. After a first orientation on the user requirements and a course in advanced C++ (!), half the group took off for Green Bank, West Virginia, where they were joined by Tim Cornwell (NRAO), Roger Noble (UK), Johan Hamaker (Netherlands) and Rick Fisher (NRAO). They were given two weeks to come up with a first analysis and design for the calibration and imaging parts of AIPS++.

One glaring error came to light during this session. We realized early on that we needed a graphical tool to communicate with one another on the analysis, and the only tool we could find on short notice was a simple one for the Coad/Yourdon method, which was therefore adopted by default. This was clearly the wrong way around to select a design methodology and cost us a few months of progress as we found the Coad/Yourdon method not the most suitable for our problem.

The group, chaired by Lloyd Higgs, produced a report which was the basis of a subsequent prototype. In the mean time we had discovered the OMT method and a suitable graphical support tool for it. In the last three months of the joint effort in Charlottesville virtually the entire group was busy with analysis and design, until, when everybody went home again in July we had a reasonable basis to start parcelling out work.

The more formal organization of the AIPS++ group had taken form, and Gareth Hunt had joined the group in April as the Project Manager. Another important development was the start of a Project Book in which all major decisions and considerations for the AIPS++ project were recorded. This has now grown, under the editorship of Brian Glendenning and Robert Hjellming to a volume of hundreds of pages.

3.3. The Last Six Months

In the beginning of the next six months, most people went on holidays and caught up on work that had accumulated in their absence. The AIPS++ group was visited by Walter Jaffe of Leiden University (user interfaces), Russell Redman of HIA in Ottawa (single dish) and Tony Willis of DRAO (gridding), each for several months.

In September work started up again in earnest, the lines of communications were established, and detailed design and coding started. Later in this paper I will show the progress to date, that is as of December, 1992. The only thing I can say here is that the remote cooperation would never have worked without those first few hectic months of daily close contacts, and it may even be desirable to repeat the experience at some later date, albeit on a smaller scale.

4. Managing When the Sun Never Goes Under

The fact that the AIPS++ development involves the cooperation of many organizations spread over the world has important advantages, in particular the validation of the computational model in a wide variety of environments and the contributions from experts with very different backgrounds. It also has serious disadvantages. The most important one of these is the need to provide for extensive communications.

4.1. The AIPS++ Organization

The first consequence of this situation is that the Consortium needs a well-defined organization with clear roles for the staff involved in it. The result was the definition of a coordinating Center, located incidentally in Charlottesville, Virginia, and led by three officers. One of these is the Project Manager (Gareth Hunt), whose main duty is to direct and organize an adequate flow of information. The other two are the Project Astronomer (Robert Hjellming), who decides on what needs to be done, and the Project Computing Scientist (Brian Glendenning) who decides on how things are done. In order to be effective, each participating organization requires the equivalent of the Project Manager locally to act as a focal point for communications. The organization of the center is done, that at the other sites is still coming into place.

There is a steering committee, consisting of one representative of each of the participating organizations, that provides an overview function. One of its members provides for frequent contacts with and supervision of the officers of the Center.

4.2. The Version Control System

Our most important means of communication is the version control system developed by Mark Calabretta of the Australian Telescope, discussed later. It keeps everybody in the system, both at the Consortium sites and at other selected locations informed on all developments on a daily basis.

4.3. Contracts

In order for all parties involved to be clearly aware of what is expected of them, we have set up a system of contracts. These specify as unambiguously as possible the area that the participant should work on. The work is specified in three stages: a design in terms of OMT diagrams, a programmers' interface in terms of header files, and a tested implementation, including the test programs used. The end product of each stage is subject to a peer review. As a consequence, the contract for each participant also specifies the reviews that it has to carry out.

Each activity in a contract has a deadline attached to it. This allows the project manager to schedule progress and to draw up a TimeLine. Most of the contracts are now in place, and progress in the project is reasonably under control.

4.4. E-mail

A project like AIPS++ would be quite difficult to do without e-mail. We use it to discuss anything from user requirements to code reviews, to disseminate information to limited groups of people through exploders and discussion groups and for person to person communications. E-mail does have one serious limitation. Its narrow bandwidth does not support coherence in a group that is so far flung, and in particular at sites where there are only one or two persons engaged in the project a sense of aloneness is noticeable.

4.5. Personal Contacts

In order to overcome the e-mail limitations we have resorted to fairly regular personal contacts. We are using telephones, but you can imagine the strange times at which we have to have our conversations if we want India, the UK, Australia and the USA talking together on a conference call! A more effective, but also more expensive, way is to hold work meetings. We recently had one to reconcile the single dish and interferometer calibration models, in which Dave Shone of the UK and Russell Redman of Canada participated, and we will have a larger scale working meeting probably in April.

5. Where Does the AIPS++ Project Stand Now

The version control system, written by Mark Callabretta of the Australian Telescope, has been working now for over half a year and is the lifeline on which all of AIPS++ depends. It uses RCS, GNUmake, an extensive set of scripts and a C++ compiler to manage all of the AIPS++ source, libraries, executables and documentation. It works both on Sun and IBM RS/6000 machines. Its function is to provide the complete AIPS++, updated daily, at all Consortium sites. It does this by "inhaling" all new material in order to update the master copy in Charlottesville, and "exhaling" the resulting new version back to the sites. This material will also be available shortly by anonymous ftp.

Tim Cornwell has provided a mathematical foundation for the analysis and design effort. Robert Hjellming, Dave Shone and many others have used this to establish a model that covers both single dish and interferometric data. The breakdown of the model into an hierarchy of classes down to a fairly detailed level is proceeding apace. A number of AIPS++ memos have been issued to cover progress in this area.

Brian Glendenning has written and tested the classes that cover vector, matrix and cube mathematics. They still need to be fine tuned for efficiency and extended with linear algebra methods.

The definition of the extensions to a regular RDBMS required for handling the complex data structures that represent a normal set of interferometer raw and derived data was developed by Allen Farris of the StSci. A first implementation of a table class that handles these structures was done by Ger van Diepen of the NFRA. This is now part of the AIPS++ source database. An investigation into efficient methods to store and retrieve multi-dimensional data structures to and from disk using grid files or related methods is being done by John Karpovitch of the Computer Science Department of the University of Virginia. Finally, Allen Farris is currently writing a set of utilities to read and write FITS files into/from the above table objects. All of these either are done or will be done by about the end of 1992. They will allow us to construct a prototype during early January that can be tested with real, observed data.

Classes that cover the transformation of a full range of astronomical coordinate systems have been defined by Mark Calabretta. He expects to have them finished by the end of 1992.

We expect InterViews (Linton et al. 1992) to be the basis for the development of user interfaces, graphics and visualization in AIPS++. A team at the University of Illinois intends to re-develop their MXV system (BIMA/NCSA 1992) in the AIPS++ context on this basis and have a working system by the middle of 1993. Paul Shannon in Charlottesville has just started on the development of a hypertext help facility using Texinfo, a GNU hypertext, and InterViews. He expects a working system by the end of February.

Tony Willis of DRAO has finished a gridding class and a FFT utility.

Darrell Schiebel of Charlottesville has produced C++ coding and documentation standards. He has just finished template and exception handling facilities and is currently writing a documentation extraction facility that will generate hypertext documentation for programmers from comments embedded in the source.

As to the somewhat longer future: we expect to begin constructing major applications by the middle of 1993 and to have the first sub systems in the hand of our users by the end of 1993.

6. Conclusions

This paper can hardly be more than a snapshot of an ongoing development. Generally speaking, progress is roughly in line with what we predicted 18 months ago, although individual events have gone quite different than expected. Also, our views of the future, as listed at the end of the previous section, still appear to be realistic. Note that the AIPS++ project has two distinct deliverables: class libraries by means of which new applications can be written quickly, and pre-cooked applications which together allow routine astronomical data processing.

Although we are happy with what has transpired so far, due to the enthusiastic collaboration of a large number of very talented people, AIPS++ has not made it over the hump yet. The crunch will come with the tuning of the basic data base and mathematical routines to achieve an efficiency comparable to that of FORTRAN. Results reported in the literature are encouraging. Thirty five years ago we faced the same dilemma of programmability versus efficiency with the switch from assembler to FORTRAN, and you all know what the outcome was. In the long term we know on whose side the angels are, but in the short term we still have a mountain of work ahead of us.

References

BIMA/NCSA 1992, The Miriad X Vizualizer User Manual, Version 2.1, Sept.

Booch, G. 1991, Object-Oriented Design With Applications (Grenjam/Cummings Publishing Company Inc.)

Brainerd, W.S., Goldberg, C.H., & Adams, J.C. 1990, Programmer's Guide to FORTRAN90 (McGraw Hill Book Company)

Coad, P., & Yourdon, E. 1991, Object-Oriented Analysis, Second Edition (Yourdon Press)

Cotton, Jr., W.R. 1992, "Object-Oriented Programming in AIPS FORTRAN", AIPS Memo 78, available from NRAO

Croes, G.A. 1988, Informal Contribution to the Initial AIPS++ Project, Based on the FORCE Pre-Processor, DRAO, Penticton, Canada

Linton, M.A., et al. 1992, InterViews Reference Manual, Version 3.1-Beta, Stanford University

Meyer, B. 1988, Object-Oriented Software Construction (Prentice Hall) – The first few chapters of this book provide an excellent introduction to object-oriented programming.

Rumbaugh, J., et al. 1991, Object-Oriented Modeling and Design (Prentice Hall)

Stroustrup, B. 1991, The C++ Programming Language, Second Edition (Addiscon Wesley Publishing Company)

Discussion

Berczuk: How did you come to choose the Rumbaugh methodology for the AIPS++ project (as opposed to any other one)?

Croes: They were impressed with the fact that GE had built a system using it.

Shaw: To what extent do you intend AIPS++ to be ported to various machine architectures?

Croes: We intend AIPS++ to be as portable as AIPS, i.e., to any machine of interest to our users. We are, in the first round, aiming at POSIX compliant systems and developing on SUN SPARC stations and IBM RISC machines. We are carefully isolating all machine dependent aspects, including architectural considerations, such as those for massively parallel machines.

J. Williams: You stated you were using the Rumbaugh method. Is there a particular case tool to support this methodology?

Croes: There exist at least two CASE tools that support the Rumbaugh method. One is the tool developed by the Rumbaugh group at GE. The other, which we are using, is Objectmaker, which supports a wide variety of design methodologies in addition to the Rumbaugh method.

Adorf: I am impressed with your so far successful effort to organize a world-wide collaboration for a distributed software development project. I may have missed the point, but how exactly did you organize remote, electronic communications and what was your experience? Did it work well, or were you still hindered by technical (network) problems?

Croes: We are using Internet and a homemade system based on RCS, gmake and a number of scripts. It works very well indeed, even in cases where there were severe disruptions in the network (e.g., with India). The whole system is, of course, freely available to any interested party.

Astronomical Data Analysis Software and Systems II
ASP Conference Series, Vol. 52, 1993
R. J. Hanisch, R. J. V. Brissenden, and J. Barnes, eds.

Programmability in AIPS++

R. M. Hjellming

National Radio Astronomy Observatory[1], *Socorro, NM 87801-0379*

Abstract. AIPS++ is a software system being developed for processing of data from radio and other telescopes. Since it is being implemented in C++ using object-oriented techniques, the issue of programmability has more than the normal number of levels of application. In this paper we discuss the planned programmability in AIPS++ from the point of view of the astronomer "user", the programmer coding "outside" AIPS++, and the programmer coding inside AIPS++ with C++. We emphasize that in the latter area there is a tremendous difference between programming with extensive libraries of C++ classes and programming where the design of classes is paramount; and that even more important than "another" system with "another" acronym is the development of classes for astronomical purposes inside and outside AIPS++.

1. Introduction

AIPS++ is an Astronomical Information Processing System being implemented in C++ (Croes 1993) using object-oriented techniques (Farris 1993), which is intended to replace the functionality of AIPS (Astronomical Image Processing System) for radio astronomical data reduction, imaging, image analysis, and image display. Largely because most astronomers are only familiar with programming in FORTRAN, but partly because user programmability at the command language level has become more important to astronomers, amongst the principal questions asked about AIPS++ is "how easy will it be to program". In this paper I wish to focus on various aspects of this programmability.

Implementation of AIPS++ in C++ means that extensive libraries of classes are being designed and implemented. This fact, and the fact that learning how to design classes (think in terms of objects) is new to most people, adds new dimensions to the programmability issue. The result is that programmability involves at least four levels of expertise:

- astronomer user "programming", both interactively with a command language, and with scripts;
- programming outside AIPS++ with access to AIPS++ data files;

[1]The National Radio Astronomy Observatory is operated by Associated Universities, Inc., under a cooperative agreement with the National Science Foundation.

- astronomer programming inside AIPS++ using C++ classes; and
- programming in AIPS++, with C++, at a lower level with class design.

In this paper we will briefly discuss all four levels, but we mainly wish to point out the considerable difference between the third and fourth levels. Programming with extensive libraries of classes will be easier and more powerful than any previous form of programming that astronomers are familiar with.

2. Astronomer User "Programming"

For most users of any data reduction system the "friendliness" of the user interface, and the match between what the user wants to do and what the system is designed to do, are the most important elements of the system. AIPS++ is planned to allow multiple, "plug-compatible" user interfaces to control the same processing tasks. The most basic user interface is planned to be an IDL-like, interactive (command line), and script programming environment. It is planned to allow the user to control the scope of "packages" in each AIPS++ session, in addition to selection amongst available user interfaces. Graphical displays will be integrated into all packages and basic X-windows with multiple mixing and matching of text, plot, and image display will be the most commonly used display setup. Serious consideration is being given to a graphical, data-flow user interface with programmability with, and inside, "icons" (Khoros-, AVS-like).

It is hoped that the IDL-like programmability with the control language will satisfy the programmability needs of a large fraction of astronomers.

3. Programming from Outside AIPS++

Because AIPS++ is aimed at a certain level of POSIX-compliant systems, we plan the system architecture to allow UNIX-level execution of AIPS++ "tasks". In addition, we recognize that astronomers use a range of data processing systems and commercial packages, so data input and output in simple table, FITS, etc., formats will be emphasized. This will allow as high a degree of compatibility as possible with other commercial and non-commercial data analysis systems

Most astronomers still program in FORTRAN, so we are planning a set of FORTRAN I/O subroutines to access telescope and image data from external FORTRAN programs. However it will be the sole responsibility of the person programming, or using, programs outside AIPS++ to NOT corrupt telescope, image, and other data structures inside AIPS++.

4. Programming Inside AIPS++ With C++

The most important aspect of programming inside AIPS++, that will have to be seen to believed, is that most astronomers should be able to program new things using basic and application class libraries, requiring only a simple level of knowledge of C++ programming. Class libraries for telescope and image association data handling, mathematical transformations, table handling, and high level graphics, will provide a more powerful set of programming "tools" than one has ever had. This is largely because objects couple data and operations on that data so that much of the normal drudgery of book-keeping and programming control structures is minimized. At the level of programming using class libraries you need to know the functionality of the classes you use, but you do NOT need to know how to develop object-oriented software

Class design and low level C++ programming will always require a high level of expertise at object-oriented programming in general, and C++ in particular. However, few astronomers will face this need once a rich enough set of C++ class libraries are available for use inside, and outside, AIPS++.

5. Mathematical Classes

Most data processing can be decomposed into: organizing data into associations, files, and arrays or tables; operations on the associations and files; and mathematical operations on *ARRAYS* (scalars, vectors, matrices, cubes, ...) and *TABLES* (data structure with columns of the same data type).

For this reason extensive effort is being devoted to such mathematical data processing components (classes). This will be based on two types of basic classes: array classes which implement basic operations on arrays (multiplication, inversion, determinant, ...); and table classes which allow complicated storage of data arrays (including strings and other specially defined data types) as a function of data coordinates. Table classes will allow: rows with pre defined fields for each column; linear operations for arithmetic between row elements and between rows of different tables (calibration application, interpolation, ...); sorting; coordinate transformations; data transformations; and display operations on columns and column sub-sets.

Beyond this, higher order math classes will then inherit the properties of array and table classes to do a large fraction of the real processing work. Obvious examples, some of which have already been prototyped as part of AIPS++ development, are:

- specialized matrix classes (banded, general, Hermitian, Hermitian banded, skew symmetric, symmetric, tridiagonal, lower triangular, upper triangular);
- GridTool class for gridding and de-gridding between tables and n-D arrays;
- FFTTool classes;
- linear algebra classes (including LU factorization);
- statistics classes (linear least squares fitting with error analysis; histogram classes; standard distribution generation with noise)
- decomposition (SVD, Cholesky, QR, LU);
- bilinear interpolation in matrices and cubes;
- spline fitting and interpolation;
- Gaussian (and other) functional component fitting;

- root computation;
- non-linear fitting and minimization (iterative substitution, conjugate gradient, steepest descent, ...); and
- polynomial classes for various standard and orthonormal polynomials.

An example, to illustrate the sort of programming one can do with a powerful set of classes, is the following which uses class libraries to do a fairly sophisticated level of data fitting with error analysis.

```
/* Programming with C++ Classes - a non-trivial example:
   Polynomial Least Squares Fitting with Error Analysis
   Solving for X in AX = y where y is a vector of data as a
   function of an independent variable vector x.
*/
#include <fstream.h>
#include <dgenfct.h>          // Basic I/O and matrix libraries
#include <dgenmat.h>
#include <rstream.h>
#include <pstream.h>
#include <polylsq.h>          // Class doing all the work

int main {

  PolyLeastSquareFit x;     // Invoke constructor for object x

  x.calcfit();              // Solve for fit to polynomial

  x.erroranal();            // Do error analysis for fit
}
```

This is the sort of coding an astronomer would have to do to use these classes. All that he needs to know is the public information in the following header file for the PolyLeastSquareFit class.

```
Class PolyLeastSquareFit{
  Public:
    PolyLeastSquareFit();       // Constructor.  Asks for order of
                                //      polynomial and data input
    void fileInput();           // File input of (x,y) data table
    void kbdInput();            // Type in (x,y) data table
    void calcFit();             // Find Atranspose, G = Atranspose*A,
                                //  Ginv, Xsoln = Ginv(Ainv*y), ysoln
                                //  = A*Xsoln, using matrix classes
    void errorAnal();           // Compute rms, sigma vectors and
                                // correlation coefficient matrix
  Private:                      // hidden data and member functions
    int n;                      // number of data points
    int order;                  // order of polynomial fit
    DoubleGenMat *y;            // pointer array of y data
    DoubleGenMat *a;            // pointer to "A" matrix
    DoubleGenMat *calcValues;   // pointer to calculated values
    DoubleGenMat *X;            // matrix of polynomial coeff.
```

```
    double errorSqd();          // member function for finding
                                //         sum of errors squared
    double rms();               // find rms
    double std();               // find standard deviation
    double corrCoeff();         // find correlation coefficients
}; /* Followed by code for constructor and all member functions */
```

Note that in the private declaration portion of the class the sort of details one usually must cope with in programming are hidden because they are things the astronomer does not need to know or deal with.

In order to emphasize the programmability aspects of AIPS++ in terms that can be commonly understood, we have not discussed the applications classes being developed for manipulating, calibrating, imaging, etc., radio (and other) astronomical data. These can be viewed as the next layer of classes on top of both the mathematical and table (data base) classes and higher level math classes. In the end, all these things, hidden inside what will be a hopefully friendly user interface, will be the program entity called AIPS++.

6. Conclusions

Programming with class libraries developed for AIPS++ should be both powerful and easy. All classes developed for AIPS++ will be copyrighted with a Gnu-like copyright, but will be available in the public domain via anonymous ftp. For most astronomers the planned IDL-like programmability of the command language will be all that is needed. Import and export between AIPS++ and other commercial and non-commercial packages will be highly supported. However, of all these things it is the development of class libraries for further development inside and outside AIPS++ that will constitute what has been called the "freedom layer" of software development, and which will do the most the change the way software of this type is developed in the future.

References

Croes, G.A. 1993, this volume

Farris, A. 1993, this volume

Discussion

Adorf: My question relates to the intelligibility and verifiability of algorithms. Do you intend to publish your algorithms?

Hjellming: We plan to provide mathematical and algorithmic descriptions of AIPS++ classes, tasks, etc., as part of the documentation of the system.

Hsieh: Much of the usefulness and power of C++ depends on the class definitions. Do you plan to make your AIPS++ classes available to the community before the release of AIPS++?

Hjellming: Yes. All AIPS++ classes will be available via anonymous ftp.

Astronomical Data Analysis Software and Systems II
ASP Conference Series, Vol. 52, 1993
R. J. Hanisch, R. J. V. Brissenden, and J. Barnes, eds.

IRAF in the Nineties

Doug Tody

National Optical Astronomy Observatories[1], *Tucson, AZ 85726*

Abstract. The IRAF system (Image Reduction and Analysis Facility) has been under development since 1981 and in use since 1984. Currently, in 1992, IRAF is a mature system with hundreds of applications which is in wide use within the astronomical community. After a brief look at the current state of IRAF, this paper focuses on how the IRAF system is expected to develop during the coming decade. Certain key new technologies or trends which any new data analysis system will need to deal with to be viable in the nineties and beyond are discussed. An overview of the planned enhancements to the IRAF system software is presented, including work in the areas of image data structures, database facilities, networking and distributed applications, display interfaces, and user interfaces.

1. Introduction

The IRAF data reduction and analysis system has been around since 1981. Today IRAF is a mature system with hundreds of applications which is supported on all major platforms. Many institutions, projects, and individuals around the U.S. and the world have developed software for IRAF. Some of these packages are comparable in size to the IRAF core system itself.

At the present time there are half a dozen large groups developing software for IRAF, plus many individuals or small groups. Coordination of the work being done by the large groups is the responsibility of the IRAF TWG (Technical Working Group), an interagency group which oversees IRAF software development. Scientific review of IRAF development is provided by an IRAF User's Committee, which oversees IRAF as a whole and which reports to NOAO, by additional User's Committees reporting on the various projects developing large layered packages for IRAF, and by the staff and management of the institutions funding IRAF development. IRAF is used primarily by the ground based astronomy (NSF) and NASA space astrophysics communities.

A list of the IRAF layered packages currently installed at NOAO/Tucson is shown in Figure 1, to illustrate the variety of packages available. This list is not all-inclusive, i.e., there are additional layered packages available for IRAF other than those shown here, which were the ones which happened to be installed at

[1]NOAO is operated by AURA, Inc. under cooperative agreement with the National Science Foundation.

NOAO when this figure was prepared. The standard IRAF distribution itself, consisting of the core IRAF and NOAO package trees, contains about 50 additional packages, or several hundred tasks, totaling approximately 1.3 million source lines. The core IRAF system includes the IRAF system software (host system interface, run time and programming environments, command language and other user interfaces, and core applications) and is required to compile and run any layered software.

adccdrom	tools for accessing ADC CD-ROM
ccaccq	IRAF CCD data acquisition
color	prototype RGB rendering tasks
ctio	CTIO local tasks
demos	IRAF demos
ftools	FITS tools package
grasp	GONG data processing (helioseismology)
ice	IRAF CCD data acquisition
iue	tools for importing IUE spectral data into IRAF
mem0	maximum entropy image restoration
nlocal	NOAO/Tucson local tasks
nso	Solar astronomy
spptools	SPP programming utilities
steward	Steward observatory local tasks
stsdas	STScI (HST) data processing
tables	STScI table tools package
vol	volume rendering
xray	SAO x-ray data analysis package

Figure 1. IRAF layered packages installed at NOAO (Dec. 1992)

As of late 1992 the current release of IRAF, which is still in distribution, is version 2.10. As of December 1992 there were a total of 1068 logged distributions of the previous version of IRAF, V2.9, of which 196 distributions were tape distributions mailed to the user at cost, and 872 distributions were downloaded via anonymous ftp from the IRAF network archive on iraf.noao.edu. An unknown number of additional distributions were downloaded via DECNET network transfer (we don't know how many, as we log only ftp file transfers). These statistics count only distributions leaving NOAO; since the system is freely available, we have no way of recording redistribution of the system at remote sites or within large institutions. IRAF site support traffic totals over 5000 e-mail messages or phone calls per year, counting both incoming and outgoing messages. Based on the number of distributions and the site support traffic we estimate there are currently several thousand active users of IRAF.

The remainder of this paper focuses on where IRAF is headed over the remainder of this decade. We look first at some key new technologies that we feel IRAF (or any modern astronomical data analysis system) must use effectively to be competitive by the end of the decade. Some pitfalls that we feel system developers would be wise to consider are also discussed. Finally, we summarize the work planned for the next few years to enhance the IRAF system software to meet these new challenges.

2. Key New Technologies for the Next Decade

IRAF is a long term project. Most of the software or hardware technologies upon which IRAF is based typically have a lifetime of only 5 or 10 years—considerably less than the expected lifetime of the IRAF software. To remain up to date it is necessary to make use of new technology, but one must do so carefully to avoid tying the software irrevocably to a technology which will one day become obsolete. It is not easy to change a large system once a direction has been chosen, so one must take the long term view, always planning 5 to 10 years in the future, trying to visualize what future computer systems will be like and what we want our software to look like on those systems.

2.1. Key New Technologies

The following are some key new technologies and technological issues or trends which we feel any modern data analysis system should be concerned with.

User Interfaces As computer systems become more powerful, software systems are becoming larger and more complex. People do a lot more with computers now than they did a few years ago. Functionality and efficiency, while still important, are no longer the overriding concerns they once were. The issues of managing complexity, and ease of use, are increasingly important concerns. The challenge of user interface design is to make complex systems comprehensible, intuitive, and easier to learn and use. Sophisticated user interfaces will make our software more pleasant to use, and allow more complex and sophisticated applications to be written. The days are past when the user interface can be taken for granted when designing new software.

High Level Languages Our common everyday computers are becoming so powerful that most of the compute cycles now go to waste. At the level of compiled code, our computer languages and software systems are becoming increasingly complex, to the point where it may take an expert with years of training to deal with them. It may be that the time is rapidly passing when casual users will do very much programming with general purpose compiled languages like Fortran, C, C++, and so on. The trend is towards higher level interpreted languages which are tailored for a particular type of application. Whether these languages are syntax driven, visual, or whatever does not really matter; in general the optimum type of language depends upon how the language will be used, and languages should be customized for a particular application. In the future, users will still develop custom applications, but they will increasingly do so using sophisticated, high level, application specific custom languages which are embedded in feature-rich data processing environments.

Networking and Distributed Objects and Data Our computers are getting powerful enough that for many applications, further gains in compute power won't make a whole lot of difference. A powerful computer and sophisticated software aren't worth much unless one has data or other raw information of some type to process, analyze, or query. Fortunately a new way has been found to expand the capability of a computer system: the growth of global networks is opening up a whole new dimension on what we can do with computers. It is already the case that one can do wondrous things with even the simplest hand held computer—so long as it is connected to the global networks. The networks give us access to an inconceivable amount of data or information of various types. Not only do the networks provide access to vast amounts of raw information, they make it possible to export arbitrary *services* via the network. Rather than export data or software, one can now export services which remote clients can access at runtime to do any number of interesting things. We are only just starting to learn how to make use of the global networks, but it is already clear that the networks will change forever how we do computing, and how we use computers.

Object Oriented Software Structure and Methodology Every few years something new comes along (e.g., AI, CASE, HyperCard, etc.) which proponents claim will revolutionize how we do computing. Most of these "silver bullets", useful though they may be in some applications, are oversold and after a time something else gets all the press. The latest such hot item is object oriented programming (OOP). The journals are full of talk about object oriented languages, databases, programming tools, and so on. This time it is not just another over-hyped product though. The object oriented approach to software development and software design is probably the most important development in software engineering since structured programming in the 80's. In fact it is a natural outgrowth of the best software practices of the 80's.

What is important about object orientation is not a particular language, commercial product, or other tool, but the concepts and methodology underlying the object oriented approach to software development and systems design. In particular, the object oriented approach places a special emphasis on the *conceptual modeling* of the objects (classes) comprising a software system. This emphasis on conceptual modeling, and the encapsulation or information hiding that is a natural part of the object oriented approach, are fundamentally important in dealing with the complexity of large modern software systems. Other elements of the object oriented approach such as subclassing and inheritance are probably fundamental to a true object oriented software structure, but this is a fairly specific software structure, and not necessarily the best one for all applications. Like any technology, OOP will be better for some things than for others, and we are still learning the limitations of this new technology and where it can be used most effectively.

Database Technology There is nothing very new about database technology. To date though, astronomy has done little with database technology, beyond its obvious use for indexing data archives. We think that there is much that could be done by combining, e.g., database technology with a graphical user interface to perform sophisticated queries of the catalogs produced by astronomical analysis programs such as are produced by image classification, source detection, and stellar or galaxy photometry programs. Furthermore our data sets are becoming larger and increasingly more complex, as is the way we access data, especially when we take network access to remote databases into account. Database technology will eventually have to be brought into play to effectively manage this increased complexity. Conventional relational database technology will continue to be important for large data archives and for some types of catalogs produced by analysis programs, but object oriented database techniques will be better suited to the complex data objects dealt with by our online analysis systems.

2.2. Concerns

In the process of employing all this new technology there are a number of things to watch out for.

The Coming OS Wars UNIX is king right now, but will this be the case ten years from now? Ten years ago the dominant system in astronomy was the VAX running VMS. Today it is the UNIX workstation. UNIX is a very good system and it (or more properly its descendents) might still be the dominant system ten years from now, but this is by no means certain. There is a very real possibility that the dominant system for astronomers ten years from now could be the PC. Not the PC we have now, but the PC we will have then, when a PC is more powerful than the workstation of today, cheaper, portable, fully connected to the networks, and capable of running "shrink wrapped" personal applications in addition to specialized astronomical software. The engineering workstations and servers of today will still be around, and they will be more powerful than ever, but an increasing share of scientific computing is likely to be done on mass market PC systems.

If the PC does so well then we cannot be certain that UNIX will still be the predominant operating system ten years from now. We might instead be using an operating system which was designed for the mass market, such as Windows/NT, or conceivably even some future version of MacOS (most likely a mixture of all these). UNIX may be more powerful, more elegant, more technically superior, and less proprietary, but those criteria will not necessarily prove as compelling to the mass markets as they have to the academic and engineering markets. Even within the UNIX community there is still considerable variation in what we call UNIX, and despite efforts like POSIX there is no real evidence that this situation will ever change. On the contrary, UNIX systems are becoming increasingly complex and small differences are correspondingly magnified.

Window Systems In the past the main concern when porting software to a new platform was the operating system. Operating system differences are still a concern, but not as big a concern as in times past. A possibly more significant problem, and one which is perhaps being overlooked by many folks now writing software, is the window system, or in the case of X, the window system toolkit. Modern window systems are comparable in complexity to operating systems but the technology is much newer, and is still evolving rapidly. It is likely that any window system specific software written today will have to be thrown out and rewritten a few years from now. Most window system specific software today would have to be largely rewritten to "port" the application to a different window system on a different platform. Despite these problems, most window system applications written today are monolithic applications with the application specific functions and user interface code tightly interwoven. Window systems may prove to be the "assembly language" of the 90's.

Computer Languages In the past ten years the major players in the general computer languages arena have changed, but the game has not. We have seen

Fortran 77, K&R C, ANSI C, Fortran 90, and lately C++, with others such as Ada, Pascal, and Objective C on the sidelines. Computer languages are constantly evolving. Even given language standards, implementations of a language by vendors on different hardware vary considerably. This is unlikely to ever change.

The evolution of computer languages is not a problem so long as one is content to write disposable software. If the projected lifetime of a body of software is ten years or longer, and the body of software is large enough that rewriting it may not be practical, the evolution of languages is a serious problem which may eventually cause the software to become obsolete, along with the technology it has been tied to. Even in the short run, the variation in language implementations on different platforms can be a serious support headache if the body of code is sufficiently large.

3. Major IRAF System Software Enhancements

In this section we describe the work being done to enhance the IRAF system software. This is a long term effort extending over a period of years. Some of the work discussed has already been completed, but much of it is either in progress or still to be done.

The work presented here attempts to exploit the new technologies discussed in the last section, while avoiding the pitfalls that can come from tying software too closely to a particular technology. Existing technology is only useful up to a point; much of the work discussed in this section is an outgrowth of the work already done on the IRAF system, and reflects problems some of which appear to be unique (at some level) to astronomical data analysis.

The reader is assumed to already have some familiarity with the current IRAF system software. The software described here is very extensive and it is impossible in a short review article like this to go into very much detail, or explain all the terminology used.

3.1. Image Structures

The term "image structures" refers to the representation of the primary data type in IRAF, the *image*. In IRAF an image is not a simple picture, but an arbitrarily complex data object. An image consists of an N-dimensional logical data raster (sampled data array) plus various bits of associated information, some of which may be quite complex objects in their own right. The logical data raster need not be physically stored as a sampled data array, for example in the case of event data the data is stored as an event list and sampled only when the image is accessed. The information often associated with a data raster includes history information, any attributes computed by analysis of the image data, world coordinate systems, pixel or region masks, uncertainty or "noise" information, and so on.

A common example of an image is a raw data image, i.e., an astronomical observation. Examples of raw data images are a 1D spectrum, a 2D CCD data frame, or a 3D Fabry-Perot or radio spectral image cube. Images of dimension higher than 3 are rare in astronomy. Typical astronomical data sets can be quite large, e.g., several gigabytes, perhaps consisting of thousands of small spectra, or several hundred large 2D images. Individual images of 32 megabytes or larger are occasionally seen.

high level image class	
`IMIO`	image i/o
image header access	
`IMIO`	header access
`DFIO`	datafile manager (*)
`FMIO`	file manager
image kernels (internal to IMIO)	
`IKI`	image kernel interface
`OIF`	old (original) image format
`STF`	HST image format
`PLF`	pixel list image format
`QPF`	QPOE (event list) image format
`FTF`	FITS image format
new format	new DFIO based image format (*)
others	HDS(?), "PC" image formats (*)
auxiliary classes	
`MWCS`	world coordinate systems
`PMIO, PLIO, MIO`	pixel masks or lists
`QPOE`	event list data files
`NFIO`	noise function package (*)

Figure 2. New Image Structures

When IRAF was first released some years ago the only image format consisted of a pair of files per image, one for the header and one for the pixel matrix, with the header file consisting of a fixed binary structure plus a variable number of FITS cards (not a terribly flexible or efficient structure). Over time several alternative image formats have been added, as well as support for some of the auxiliary data objects associated with images. It has become increasingly difficult to store all this information in the simple data structures provided by the older IRAF image formats. The purpose of the new image structures project is to provide a general, well integrated hierarchy of image object classes for flexibly and efficiently representing a wide variety of image data.

The major components of the new image structures are summarized in Figure 2. The new image structures project is further along than most of the other new software discussed in this paper; everything listed has been implemented except for the items marked with an asterisk. This is a *big* project; some of the subsystems listed here, e.g., MWCS, QPOE, etc., are major projects in their own right, and in total the new image structures code will likely exceed 100K lines, not counting the lower level IRAF classes or other library code.

A key feature of the implementation of the image interface in IRAF is the *image kernel.* An image kernel is the only part of IRAF that knows anything about how an image is stored externally, i.e., the physical image format. The image kernel implements a mapping between the physical image format and the logical view of an image implemented in the runtime image descriptor used to access an active image object. The image kernel can provide a standard interface to a wide variety of image types, including odd things like photon event lists, image masks, or image display server frame buffers, in addition to various standard image raster disk file formats. In principle IRAF can be integrated with any external image processing system by implementing an IKI image kernel for the image format defined by the external system. The best example of this is FITS. The FITS image kernel also allows archival data, e.g., on CD-ROM or on a remote network server, to be directly accessed by IRAF programs.

The main work remaining to be done to finish the new image structures project is to implement a new standard IRAF online image format based on the general datafile manager (DFIO, discussed in the next section). This will replace the existing OIF image format. The new format will make it possible to simply and efficiently group complex objects such as world coordinate systems, pixel masks, and compressed pixel uncertainty arrays with pixel arrays to form the objects we call images.

3.2. Database Facilities

Managing complex data structures such as the image structures in a flexible and efficient manner, while providing features such as data independence, machine independence, a data recovery capability, transparent storage of arbitrarily large data elements, capabilities for storing complex objects (not just simple tables), indexing for efficient lookup, and a good integration with the higher level IRAF software, is a complex and demanding problem. The low level interface planned to provide this capability for IRAF is DFIO, the data file manager. DFIO is layered upon FMIO, the file manager, which is in turn layered upon IRAF binary file i/o (FIO). DFIO is a medium level interface designed for embedded applications, e.g., it will be used internally within IMIO to store image data. For the most part IRAF applications will not use DFIO directly, rather they will deal with data at a higher level, e.g., via the image class.

One of the types of data DFIO will be capable of storing is the table, as in a relational database. Hence, in addition to its use as an embedded interface within IRAF system software to store complex data objects, DFIO will provide a traditional relational database capability for applications such as catalog access. Since IRAF already provides a builtin networking capability, DFIO will automatically be usable in client-server applications to provide a distributed database facility. To be able to access external, non-IRAF databases, a database server architecture will be used (similar to the use of image kernels by IMIO).

3.3. Networking and Distributed Applications

Networking is an integral part of IRAF and IRAF has always been able to support distributed applications. In IRAF all access to external resources is via the IRAF kernel. The IRAF kernel has a builtin remote procedure call facility allowing kernel procedures to execute either locally or remotely. (This includes *all* kernel procedures that access a named external resource, be it a file, directory, tape drive, image display, process, or whatever). In a local reference the procedure is executed directly; when the resource being accessed resides on a remote node a custom RPC protocol layered upon the IRAF networking driver is used to remotely execute the kernel procedure. The only system dependent part of all this is the networking driver, which can use any standard message or stream oriented transport layer, e.g., TCP/IP, DECNET, and so on. Since the IRAF kernel provides a standard host interface, the routing or leaf nodes in a distributed IRAF process tree can execute on host machines running any operating system to which IRAF has been ported. It is even possible to transparently route RPC calls between different networks, e.g., the Internet and SPAN.

There are still some significant enhancements planned for the IRAF networking system but these are for the most part comparatively minor evolutionary enhancements. One of the most interesting enhancements being considered is some sort of interface to non-IRAF servers, e.g., ftp or WAIS servers. This would not provide the full capability of the IRAF kernel, but might work for simple directory and file access, and would allow any IRAF application to transparently access arbitrary servers on the network whether or not they provide an IRAF kernel server.

3.4. User Interfaces

In general, the IRAF system circa 1992 is very strong in terms of the functionality provided, but is weak in the area of user interfaces. This directly reflects the priorities for IRAF development in the late 1980's, which emphasized getting numbers out of the data. This meant new applications, and due to the common environment, enhanced system support for these applications (e.g., the new image structures). In the early 1990's, with a wealth of software now in the system and more people than ever using IRAF, the emphasis has shifted towards ease of use and improved user interaction and data display.

Enhancements to the IRAF user interfaces are planned in many areas. Two of the most exciting are a general GUI (graphics user interface) capability, available to any IRAF application and capable of making full use of the advanced capabilities of modern window systems, and less obviously, something called minilanguage support.

Modern window systems are remarkably complex software systems, and the field is still evolving rapidly, with many quite different window systems and window system toolkits being developed or in use. Using window user interfaces effectively in scientific applications is challenging, as if one is not careful and the wrong approach is taken, programmers may spend all their time struggling with complex window system software and not get any science software written. Due to the complexity of the field, learning a particular toolkit or user interface builder is time consuming and a considerable investment in time is required to make use of any particular tool. A wrong decision could result in a great deal of wasted time, particularly for a large project where many people may be developing software.

After considerable time spent studying window systems and graphics user interfaces we think we have found a solution to this problem. It is called the *widget server*. In the widget server architecture the application and the user interface are in two separate processes. The application is a type of minilanguage with a simple parsed command line interface. The user interface resides in the widget server process. When an application starts up it downloads a text file to the widget server containing the user interface to be executed. This defines all the widgets forming the user interface as well as the code to be executed (interpreted) while the user interface executes. During execution, the

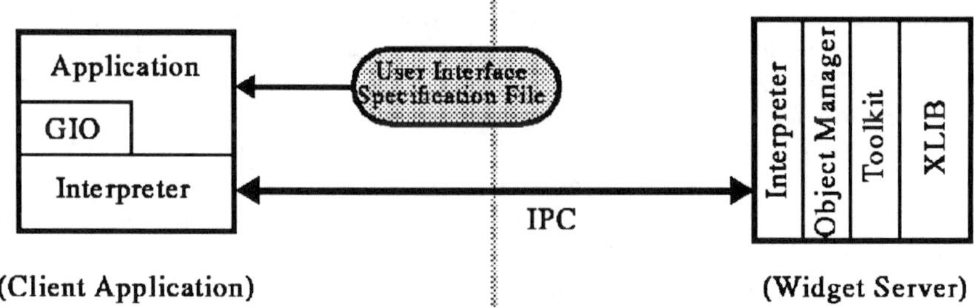

Figure 3. Widget Server Architecture

user interface (widget server) and client application exchange commands and data via interprocess communication.

This architecture has many advantages, e.g., a complete separation of the user interface and functional code, and a high level interpreted interface to the window system for the programmer, making it easy to develop GUIs (and easy for the user to customize the user interface). Since only the widget server knows about a particular window system or toolkit, the widget server also provides window system and toolkit independence, allowing a new window system or toolkit to be supported merely by implementing a new version of the widget server. The widget set provided by the widget server will include the standard toolkit text, button, scollbar, list, geometry, etc., widgets, plus some custom widgets (such as graphics and image display widgets) tailored to IRAF applications.

As powerful as the graphics user interface can be, it is not the only way to do a user interface, nor is it necessarily the best type of user interface for all interactive applications. A quite different type of applications user interface which is at least as powerful, and also well suited to complex applications, is the context-based minilanguage. A minilanguage is a single program (IRAF task) with a syntax driven, command line user interface. The program maintains an internal state and successive input statements modify this state. The syntax, command or function set, and internal data structures are customized for each application. The individual functions are usually simple, but arbitrarily complex operations can be performed by stringing together sequences of commands or expressions. By designing an appropriate syntax very powerful applications-specific languages can be devised.

A good language will be extensible, allowing users to define new procedures, link to external compiled routines, or interface external IRAF or host tasks so that they appear as functions in the minilanguage. It will even be possible to combine a graphics user interface with a minilanguage. For example, the combination of the widget server with a minilanguage will provide both a fully featured GUI capability and a powerful interpreted computing engine, both programmable by the user without need to resort to low level compiled languages. When all this is layered upon the IRAF environment, providing well integrated access to powerful facilities such as the IRAF image structures and a wealth of existing external tasks, the result will be high level applications of unprecedented power, flexibility, and sophistication.

Acknowledgments. Without the contributions of many people over the years, the IRAF system we have now would not exist. The author particularly wishes to thank Frank Valdes and Lindsey Davis of the NOAO IRAF group, who wrote much of the IRAF software. STScI and SAO have made major contributions to the system over the years and the IRAF project would not be the same without their involvement. A grant from the NASA astrophysics data program has made all the difference as IRAF use continued to grow while the NOAO budget continued to shrink. Finally, we wish to thank the NOAO directors and scientific staff for their continuing support and enthusiastic use of IRAF, and for their help in making IRAF a better system.

References

Hanisch, R.J. 1991, "STSDAS: The Space Telescope Science Data Analysis System" in Data Analysis in Astronomy IV, eds. V. Di Gesù, L. Scarsi, R. Buccheri, P. Crane, M.C. Maccarrone, & H.U. Zimmermann (Plenum Press, New York), 97

Olson, E.C., & Christian, C.A. 1992, "The EUVE Guest Observer Analysis Software" in Astronomical Data Analysis Software and Systems I, A.S.P. Conf. Ser., Vol. 25, eds. D.M. Worrall, C. Biemesderfer & J. Barnes, 110

Tody, D. 1986, "The IRAF Data Reduction and Analysis System" in Proc. SPIE Instrumentation in Astronomy VI, ed. D.L. Crawford, 627, 733

Worrall, D.M., Conroy, M., DePonte, J., Harnden, F.R., Mandel, E., Murray, S.S., Trinchieri, G., VanHilst, M., & Wilkes, B.J. 1992, "PROS: Data Analysis for ROSAT" in Data Analysis in Astronomy IV, eds. V. Di Gesu et al. (Plenum Press, New York), 145

See also the many technical papers describing the IRAF software, available in `iraf.noao.edu:iraf/docs`.

Discussion

Kibrick: One of your slides indicated the possibility of supporting HDS format under the Image Kernel interface. Is this the Starlink/Figaro HDS format?

Tody: Yes. This would be feasible, although there are no immediate plans to do this. HDS is very complex, and IRAF cannot support the full complexity of information which HDS can represent. Nonetheless, most of the data from an HDS file could be mapped into IRAF.

Astronomical Data Analysis Software and Systems II
ASP Conference Series, Vol. 52, 1993
R. J. Hanisch, R. J. V. Brissenden, and J. Barnes, eds.

Scientific Computing in the 1990s—an Astronomical Perspective

Hans-Martin Adorf

Space Telescope–European Coordinating Facility, European Southern Observatory, Karl-Schwarzschild-Str. 2, D-8046 Garching b. München, FRG. (Internet: adorf@eso.org—SPAN: ESO::ADORF)

Abstract. The compute performance, storage capability, and degree of networking of modern computer hardware have enormously progressed in the past decade. These hardware advances are not paralleled by an equivalent increase in software productivity and usability. Bridging this gap, i.e., effectively harnessing the new computer technologies, is required to analyze the increasingly complex scientific data sets, including those from astronomical observatories. While helping to gain full value from investment in astronomical facilities and personnel, a breakthrough in astronomical software could advance the art of computing on a broad front. A *discussion forum* is proposed to determine the characteristics of the software shortfall, to delineate technical and programmatic opportunities for realizing the full power of computer hardware, and to clarify the benefits that might accrue from a software initiative in astronomy.

> One Mbyte of RAM now costs less
> than two lines of source code.
> — *N. Plant, 1992*

1. Introduction

It is common-place to say that the computing world has enormously changed from the early 1980s to the early 1990s, as we see, e.g., vastly *increased capabilities of computer hardware* (compute power, external and internal storage, mice and trackballs, high-resolution black-and-white or colour output devices, local and wide-area networking, etc.), and also different *user patterns* (e.g., batch-processing before versus multi-threaded interactive processing now, mobility, etc.).

On the other hand, the methods we are employing for astronomical software construction have not changed that dramatically (cf. Albrecht, Adorf & Richmond 1986). We still use *simple static programming languages* (not collection-oriented, without vector operators, mostly without an object-system, without event-handlers, but some with explicit memory management capabilities!), fairly *primitive software development environments* (e.g., without language-sensitive editors), and *run-time systems* with seams (i.e., with a wide gap between the interactive high-level "command" language and the efficient, but low-level implementation languages), to name a few areas where amazingly little has changed over the years, despite obvious, major shortcomings.

In the 1990s ground-based optical astronomy is expected to produce data frames 10 to 100 times larger than those regularly seen today and at rates which are again factors higher than current ones.

One example is an image dissecting spectrograph, proposed for one of the later ESO-VLT units, which is supposed to cover 4x4 arcsec^2 on the sky at a spatial resolution of 1/10 arcsec and with 10,000 pixels in the spectral dimension (Wampler 1992, *pers. comm.*). At a sampling rate of 2 pixels per resolution element and 16 bits = 2 byte depth per pixel, a single frame will have a data volume of 1,600 x 10,000 x 4 x 2 byte = 128 Megabyte. Such data volumes, each equivalent to one sixth of a digitized Schmidt plate (assuming 20 kilopixels squared, with two bytes per pixel, i.e., 800 MB), can only be analyzed with fast and automatic, robust and reliable algorithms.

Also, at least one 2m-class sky patrol telescope is currently being envisaged (West 1992, *pers. comm.*) whose focal plane is paved with CCDs. The computationally demanding requirement is being discussed to analyze the incoming very large data volume virtually in real-time in order to allow rapid follow-up observations with a large telescope. Even when one takes into account the current approximately annual doubling of the floating point (FP) performance of scalar machines, it is difficult to imagine how parallelism can be avoided (cf. Adorf 1993).

2. The Idea

Given the envisaged increase in user demands, and regarding the rapid turnover in one area of scientific computing (hardware) and the relative inertia in the other (software), there is a gradually perceived need to discuss (or re-discuss) the fundamental assumptions that have implicitly or sometimes even explicitly guided the design and development of astronomical software such as the large data analysis systems we see in place today. The discussion should address the question how we can make best use of all the improved hardware capabilities and our limited resources in order to better serve the user demands of the near future.

It is the purpose of this contribution to lay out the topics, goals and rules, as well as to kick off the envisaged discussion.

3. The Topics

In principle, all topics that are of interest to scientific/astronomical computing in general, are of interest to the discussion forum, too. Here is a fairly randomly selected set of questions[1] that I consider interesting: (1) Should we continue to provide general and therefore necessarily *large and complex software modules*, which are subsequently specialized by the user to his/her particular problem via

[1]The original poster contained an additional series of 25 "food for thought" questions, omitted here for space reasons, which can be obtained upon request.

assigning values to several control parameters? (2) Does it make sense to *conceal the building blocks* of larger modules from the scientific user considered to be an expert in his/her field? (3) Should we start assembling *astronomical subroutine libraries* containing accepted, well-documented (i.e., published), well-tested algorithms encoded in a re-usable form? (4) Should we place more emphasis (and spend more resources) on issues such as *cross-architectural portability*, thus facilitating the usage of vector or massively parallel computers, and less on, e.g., user-interfaces? (5) Should we support better than now a *synthetic, "bottom-up" approach* where the data analyst is given a set of fundamental, scientifically well-described (i.e., published), small, reliable modules from which to assemble the algorithm (or sequence of algorithms) needed for the particular problem at hand?[2]

An important decision yet to be made, presumably in the initial phase of the formal discussion, concerns the balance between breadth and depth of the discussion. It is supposed to start out quite general and, after an identification of the most important topics, to be limited in scope.

4. The Goals

The goal of the envisaged discussion forum is to generate, if possible, a *consensus about where astronomical software development should be heading* in the coming years. If the broad goal stated above is not achievable (for instance, because the requirements for a real-time telescope control program turn out to be so vastly different from those for an off-line data analysis program, that no consensus is feasible), than the discussion might be restricted to topics that are of interest to data analysis in a broader sense, i.e., high-performance (vector and parallel) computing, cross-platform and cross-architecture portability, programming languages and environments, algorithms and subroutine libraries, etc.

5. The Means

With wide-area computer networks in place, it is timely to consider a novel form of the discussion forum in question. I propose to lead it remotely using electronic mail. The discussion is supposed to be limited in access, moderated and to extend over a finite time interval, with several interaction points and several "activity" periods in between.

The way the discussion is envisaged to progress can be likened to a remote chess tournament, except that such a tournament is a bi-lateral game, ours is a multi-lateral affair. In setting up the rules (see below) I have been inspired to some extent by the example set by Guy Steele when he standardized the Common Lisp language a few years ago (Steele 1990).

In another respect I am influenced by the way a standard scientific panel discussion proceeds, except that I propose to take advantage of the special conditions under which a remote, electronic discussion takes place: firstly, written

---

[2]Under such circumstances, e.g., the user-interface question might largely disappear.

instead of verbal communication; secondly, delays between sending and receiving contributions; and, thirdly, being located in our home offices. Therefore: A *print-out* of broadcast material for serious contemplation is encouraged.[3] Every discussant is asked to *discuss his/her viewpoints* with his/her colleagues at home and/or abroad before sending in a contribution.[4] The *usage of a computer* to perform a little calculation, to generate a drawing, to carry out a small simulation, etc., is encouraged. *Consultation of the library* and the inclusion of references is strongly encouraged; a bibliography of all references is required in this case.[5]

Current plans foresee four categories of participants: the moderator, a small number of panel members, a larger audience and some external consultants.

The *moderator* initiates, re-convenes and closes the discussion, handles incoming and outgoing e-mail, and is responsible for the final report and joint paper.

Panel members, astronomers or software people experienced in astronomy, interact directly with the moderator and vice versa. Panel members are the only ones, whose contributions will be circulated to all others.

The *audience* (they actually don't listen, but read) consists of people who are generally interested in the area, but do not belong to the panel. They will receive the moderated contributions of the panel members (via the moderator), but will usually not contribute directly to the discussion. Members of the audience can of course always interact personally with one of the panel members, i.e., to provide a "hallway" comment and thereby indirectly influence the debate.

The *consultants*, usually non-astronomer experts in a certain area of interest, do not participate in the general discussion, but, as the name suggests, are available in the background for answering questions or clarifying debated issues.

6. The Rules

In order to give the whole enterprise a recognizable and manageable form and to guarantee a tangible outcome, I suggest to proceed according to the following ten rules[6], which comprise deterministic and stochastic elements. (After all we want to learn something, don't we? So expect some surprises.)

R1. The moderator will select the *participants* of the electronic discussion group which is supposed to be rather small and moderate the discussion.

R2. Four to six *topics* will be discussed in sequence.

R3. The moderator prepares a *list of topics* to discuss, but the final decision which topic to discuss next will depend on the outcome of the previous discussion and the suggestions of the participants.

---

[3]However, the usual rules for unpublished material apply.

[4]A major thought of a colleague deserves an explicit acknowledgment, of course.

[5]A participant, whose referenced material is not readily accessible to all other participants, should be prepared to deliver a copy to the moderator (either per e-mail or per FAX) so that an effort could be made to re-distribute it in a suitable form.

[6]These rules are not cast in iron, and minor changes can be negotiated before the discussion formally starts.

R4. Every topic will be dealt with in a *fixed period* of a few weeks.

R5. The moderator will start a *new topic*, e.g., by throwing in a number of (sometimes perhaps controversial) statements defining a position and presumably also some questions.

R6. There will be presumably *two rounds per topic*, so every panel member gets a chance to react to the statements of his/her colleague(s). A deadline for contributions will be announced for every round.

R7. All *contributions* should be e-mailed in parallel to the moderator. They can be either in plain ASCII or in LaTeX-format, particularly when formulae are included. Contributions may be structured with headers to facilitate reception. The inclusion of figures or even images (in PostScript format) is strongly encouraged. Off-the-record comments are allowed and will not be distributed.

R8. The moderator will put the received contributions verbatim into a linear sequence and *distribute* the resulting file in a common format (presumably LaTeX) to all members of the discussion group. The moderator will attempt to summarize each topic, whether controversial or not.

R9. As the immediate result of the joint efforts there will be a *report* closely documenting the discussion as it proceeded; it should be similar to the write-up of a round-table discussion, but hopefully more profound. The report may be used by all participants for in-house purposes.

R10. The report will form the basis of a *joint paper* which the moderator will edit. Every panel member will have a chance to (moderately) edit/shorten his earlier contributions in order to minimize possible misunderstandings, to emphasize or de-emphasize some points in light of the discussion, and to make the paper as interesting, crisp, and solid as possible. (Participants with only marginally contributions will drop from the list of authors to the acknowledgments.) The decisions how closely the paper's form should resemble the discussion and where to publish it will be made after the report is ready.

7. Conclusion

The whole enterprise proposed here will be an experiment of course. Apart from learning something about the topic(s) dealt with, I hope that we shall also all learn something about a novel form of ecological, global scientific exchange.

Acknowledgments. I wish to particularly thank Bob Brown, STScI, for his suggestions concerning a draft of this contribution.

References

Adorf, H.-M. 1993, this volume

Albrecht, R., Adorf, H.-M., & Richmond, A. 1986, in Proc. SPIE Instrumentation in Astronomy IV, ed. D.L. Crawford, 225

Plant, N. 1992, Physics World 5, No. 9, 43

Steele, G.L. 1990, in Common Lisp – The Language, 2nd Edition, Digital Press

Astronomical Data Analysis Software and Systems II
ASP Conference Series, Vol. 52, 1993
R. J. Hanisch, R. J. V. Brissenden, and J. Barnes, eds.

Neural Networks: Letting Your Software Think for Itself

D. Bazell

General Sciences Corporation, 6100 Chevy Chase Drive, Laurel, MD 20707

I. Bankman

Johns Hopkins Applied Physics Laboratory, Laurel, MD

Abstract. Certain problems in astronomical data analysis do not lend themselves to algorithmic solutions. Pattern recognition and classification problems are two examples. The use of artificial neural networks in the solution of such problems provides an alternative approach. We are developing Neural Network Prototyping Package within the IRAF environment to provide the astronomical community with a powerful tool for solving problems of this sort.

1. Introduction

Neural networks are rapidly becoming an important method of computing and data analysis that can be used to complement the traditional numerical and algorithmic techniques. They are modeled after the networks of neurons in the brain where a large number of relatively simple units communicate with each other by means of connections of different strengths to form a computing machine capable of a wide variety of tasks. While the solutions to many problems can be defined in terms of an algorithm—flat fielding an image or calculating a power spectrum—there is a multitude of problems that have not yielded to an algorithmic approach. These problems often lend themselves to being solved by using a neural network. For example, neural nets are capable of classifying patterns, recalling complete patterns from partial cues, solving constrained optimization problems, adaptive filtering of data and various image processing tasks. These topics are all of interest in astronomy and astrophysics and we will discuss several of them in more detail below.

Neural Networks are fundamentally different from traditional algorithmic methods. A given network paradigm can be coded once and used repeatedly for widely different applications. A neural network paradigm is characterized by the network architecture, the transformation properties of the individual nodes (the activation function) and rules by which the network is trained or learns. This last property, the ability to learn, is one of the basic points that sets neural networks apart from other types of computation. When we speak of a network learning a task we mean that the values of the weights that connect nodes are changed in response to examples presented to the network in order to improve its performance in a given task, e.g., distinguishing various objects based on their shape.

The examples used to train the network are called the "training" set. Once the network has learned to produce certain output values the training phase has ended and the weights are fixed. The "test" data are then used to verify the network behavior. In the test phase one can evaluate how well the network was trained by observing the number of correct and incorrect results. If training was adequate then new data can be applied to the trained net to produce the desired output.

The individual nodes in a network typically sum all of their inputs and then pass the sum through some nonlinear transfer function such as a sigmoidal curve. Other transfer functions can be used such as a linear ramp with a threshold (threshold logic unit) and a step function (hard limiter) depending on the type of network desired. Product nodes are also used wherein the inputs to a node are multiplied together and then passed through a transfer function.

Different neural network architectures are useful for different tasks. One commonly used network is the *backpropagation* type (Rummelhart, Hinton & Williams 1986) where there are several layers of nodes, commonly three but possibly more, where the nodes in one layer are connected by weights to the nodes in the next layer. Using a sigmoidal transfer function one can derive an equation for updating the weights in such a way as to minimize some objective function, commonly taken to be a function of the difference between the actual output of the network and the desired output. In this way the backpropagation network "learns" to produce a given output for a given input.

Also of particular interest to this project is *Adaptive Resonance Theory* (ART) (Carpenter & Grossberg 1988). For this network, patterns are presented sequentially to the input nodes. The first pattern forms the example for the first cluster. The next pattern is compared to the first using a certain algorithm. If the "distance" between patterns is less than some threshold, the second pattern is classified with the first. Otherwise it forms the example for a new cluster. The number of clusters can grow and will depend on the threshold value and the distance metric used.

Another basic property of neural networks is their ability to generalize from examples they have seen. If it were not for this generalization property, a neural net would be little more than a look-up table. Generalization allows a neural net to look at a new example, say an image of some object, and to determine how to classify the new example based on its previous experience, i.e., based on previous examples it has seen.

From the properties of neural networks just discussed one can get an idea of their utility for astronomical data analysis. However, the astronomical community is just becoming familiar with neural network concepts for a limited number of applications. We feel that the integration of neural network architectures into familiar data reduction environments will provide the astronomical community with a new tool that they will soon find indispensable.

2. System Development

There are numerous neural network architectures for which we already have Fortran and/or C code available. These include the standard backpropagation networks, Bidirectional Associative Memory (BAM), the Kohonen self-organizing map and Adaptive Resonance Theory (ART) networks. The prototype system will include tools to develop feedforward/backpropagation networks and ART networks. We will make use of previously existing code when possible and use the information gained from our exploration of other neural network environments to develop the user interface and development tools for the system. We plan to develop a system that is general enough to allow additional network architectures to be added at a later date.

Among the tools we plan to make available to the user of our system are the following:

- Provide a simple tutorial for developing each type of network to accomplish a certain task example. Training and test data will be provided for the sample problems.

- Specify network parameters such as the number of input and output nodes, the number of hidden layers and nodes in the hidden layers for backpropagation, initial values for weights, and learning rate and termination criteria.

- Normalize the input data in various ways. Neural networks generally require normalized input values. The type of normalization depends upon the network. For backpropagation networks inputs are often linearly scaled between zero and one while self-organizing networks require the length of the input vector to be normalized to one.

- Train a network given network specifications and training data (relevant only to the case of supervised learning). This task will present the test data to the network and begin iterating. It will be possible to access training parameters from this task. Weights connecting nodes will be altered during the training phase.

- Test the trained network, given test data. This is similar to the previous task but weights are fixed. Previously determined weights can be used if a network has already been trained.

- Analyze the current state of the network while it is being trained. This will include, for example, the ability to graphically examine the values of weights and activations every n_{iter} iterations using graphs or Hinton Diagrams and to plot differences between expected and actual output for a given node as a function of iteration number or as RMS error (sum over all output nodes).

- Provide general online help and help that is specific to the current task being performed. This will include guidelines on choosing a network paradigm, preprocessing data, setting initial parameters values and analyzing the state of a network.

3. Scientific Applications

We intend to use several specific applications of astrophysical interest as test cases for our neural network environment. We chose these applications partly for their general interest to the astronomical community and partly based on our assessment of the prospects for successfully applying a neural network approach to the problem. Similar applications have been successfully developed in other areas such as interpretation of medical images for diagnosis, adaptive filtering in communications using backpropagation networks with a single hidden layer, and inverse filtering. Our purpose is to demonstrate the utility of the neural network approach to astrophysical problems and to provide an idea of the types of problems that can be solved with neural networks.

Spectral Classification Automatic classification of astronomical spectra is a topic that has generated much interest and controversy over the years. With the advent of space based telescopes like IUE, IRAS, COBE and HST the volume of data is growing beyond our ability to analyze and classify it by traditional means. An automated scheme is needed that will allow astronomers to process large quantities of data while focusing in on the interesting science.

From the neural network point of view there are two distinct approaches to automated classification that can be followed: supervised and unsupervised learning. The first approach involves the development of a network for supervised learning in which a set of pre-classified spectra are fed into the network to train it. This can be accomplished using a feedforward network with one input node for each wavelength bin. It will be necessary to experiment with the number of nodes in the hidden layer and the number of interconnections to find an acceptable learning rate and an adequately robust classification ability. The network will be trained to output a signal for each spectrum in the pre-classified set. For example it could produce a "one" at a certain output node while leaving the other output nodes zero. This would indicate that the input spectrum belongs to a certain class. Once the network has been adequately trained, it will be given test spectra, i.e., an independent set of spectra of known class, and the output will be monitored to determine the fidelity of the classifier.

A second approach to this problem involves unsupervised learning. In this case the network determines how to classify the spectra by determining, based on some internal metric, how different the current spectrum is from the previously encountered examples. The network will then group the current spectrum with others in the appropriate class, or create a new class if the differences are larger than allowed by some "vigilance" parameter. This type of classification can most easily be done using a network based on Adaptive Resonance Theory.

Object Detection Neural networks tend to be especially good at pattern recognition problems. By pattern recognition we mean distinguishing between objects based on certain characteristics which might include morphology or spectral properties. An additional project that will serve as a test case for our package is the development of a neural network to detect and differentiate objects in images. Examples of objects of interest include stars, galaxies, nebulae (many separate classes exist) and cosmic rays. The procedure will be to feed the network an image and have it output the coordinates and classification of each object in the image. This will be useful for obtaining PSFs (from stars only), collecting statistics on different objects (stars, galaxies and other extended objects) and detection of cosmic ray hits as a first step in their removal.

As an initial attempt at object detection we plan to use a backpropagation network. We will create a training data set by examining images containing known objects and extracting subimages to isolate particular examples of these objects. Testing the network will involve two steps. First a test data set will be created containing only images of isolated objects. With these images as input we will determine how well the the network can detect and classify different types of objects. Second, test images containing a variety of objects will be used as input to the network. With this test data set we will determine how robust the network is when confronted with more crowded fields, perhaps containing overlapping objects.

Deconvolution of Images The application of neural networks to the deconvolution of images has been examined by a few groups for the rather restricted application of removing motion blurring from images using a Hopfield network (Zhou et al. 1988).

An additional project of interest is the use of a backpropagation network to deconvolve the point spread function from a given image. The network would be trained using sample PSFs with the output being a strongly peaked function such as a Gaussian, positioned at the central pixel of the PSF. Once the network has been trained to carry out this transformation, test cases involving extended sources will be examined.

The thrust of this approach is to have the neural network act as an inverse filter. Because of the large size of many images, e.g., 512 $\times$ 512, a practical approach is to design the network with an input layer that receives a subsection of the image, say a 21 $\times$ 21 kernel. The input layer will then have 441 nodes rather than 262144 nodes, a significant reduction in complexity. The network would then process this subsection and produce an output value for the central pixel in the kernel. When the kernel is moved one pixel at a time, all pixels in the image will be processed, but the size of the network and training time can be cut significantly.

Acknowledgments. The authors would like to thank Dr. Robert Hanisch for encouragement and support during the early stages of this project.

References

Carpenter, G.A., & Grossberg, S. 1988, Computer, March, 77

Rummelhart, D.E., Hinton, G.E., & Williams, R.J. 1986, Nature, 323, 533

Zhou, Y., Chellappa, R., Vaid, A., & Jenkins, B.K. 1988, IEEE Trans. ASSP, 36, 1141

Astronomical Data Analysis Software and Systems II
ASP Conference Series, Vol. 52, 1993
R. J. Hanisch, R. J. V. Brissenden, and J. Barnes, eds.

Khoros Software Specification Format and Interoperability

A. H. Rots

Universities Space Research Association/Laboratory for High Energy Astrophysics, Code 668, NASA Goddard Space Flight Center, Greenbelt, MD 20771

Abstract. Khoros defines formats for User Interface Specification (UIS) and Program Specification (PS) files. From such files, its code generator, Ghostwriter, creates source files and documentation. The great strength of the system is that the code fragments that make up part of the PS file are purely generic. All Khoros-related code is created by the code generator; this includes all user interface code. One could imagine using these specification files as the basis for other software systems, thus defining a universal application code format.

1. Introduction

There is a clear trend discernible among software designers to loosen the bonds that tie particular applications to particular software packages or environments and move toward more "open" systems. There are two ways to achieve this:

- Interoperability at the executable level: making sure that executables from different packages can be accessed from each other's user interfaces.
- Code sharing: writing "generic" applications that are integrated into particular software packages through the use of code generators, taking a considerable burden off the application programmer's shoulders.

I will show in this paper that the first approach can be fairly trivially implemented for Khoros and, in fact, has been applied. For the second approach, I will argue, the Khoros specification (meta) formats are sufficiently generic and complete that they may serve as a universal application specification format for any of the software packages and environments currently in use in the astronomical community.

2. Overview of Khoros

Khoros is an extensive software package, available for free by anonymous ftp from the University of New Mexico's Electrical Engineering and Computer Engineering department. The Khoros group is led by Dr. John Rasure. The system is described in the papers by Rasure and Young (1992), Argiro and Rasure (1992), Young and Rasure (1992), and references therein. Khoros operates on virtually all Unix platforms with X-windows Version 11, Revisions 4 or 5.

Khoros provides an integrated environment for data processing and software development. I shall focus in this paper on the "command and control" interface, i.e., that part of the user interface that controls the execution and execution order of modules, and the transfer of input parameters to these modules, but not interactive execution within a single module. I will first provide a brief overview of Khoros's architecture and services. Its salient elements are:

- A User Interface Management System (UIMS), consisting of a meta-language to describe user input parameters, and a code generator to transform the meta-language instructions into viable code.
- A code generator that turns generic code, as written by the application programmer, into Khoros-specific source and header files.
- A documentation extractor that assembles various bits and pieces of documentation information into a man-page.
- An imake facility that automatically generates appropriate Makefiles.
- A library of I/O functions that provide transparent data transport services.
- A library of X-window tools.
- A visual programming user interface (Cantata).
- A fair number of application modules in the areas of signal processing, image processing, format conversions, and data display.

From the user's point of view, Khoros modules may be activated in any of the following modes:

- Through Cantata, the visual programming tool.
- Using the Graphical User Interface.
- From the command line, providing a full specification.
- From the command line, asking to be prompted for input parameters.
- From the command line, directing input to an ASCII file containing input parameter values.

Cantata provides the user with a workspace in which individual applications or modules may be positioned upon selection from various application menus. The modules are represented as "glyphs" which have at least three action buttons (destroy, start/stop execution, GUI) and which may be connected by data streams. Execution control glyphs are provided and Cantata takes care of automatic execution scheduling. Execution across heterogeneous networks is supported transparently. The user may at any time specify (and change) for each module the network node it is to run on, and for each data stream the data transport mechanism to be employed. Analysis applications may thus be built from building blocks consisting of the individual application executables in a Legoblock-like manner, using a visual interface. The "workspaces" so constructed may be saved and restored, and are very easily modified and customized. The approach is comparable to what is used in systems like apE, AVS, and Explorer. Cantata actually achieves execution for each module by constructing a Unix command line and sending that off to the operating system.

Khoros's data transport library provides transparent access to different physical transport mechanisms with run-time binding. This means that it looks to the application programmer as if (s)he is performing I/O to and from a disk file, while in reality the user may choose immediately prior to execution from a menu of physical transport mechanisms. Currently disk files and shared memory are supported. Khoros 2.0 which is slated for public release in the summer of 1993 will provide support for pipes, sockets, streams, etc., as well. Khoros 2.0 will also contain data services support in a more object-oriented sense, by providing transparent format conversions between what the application expects to see and what the input "file" is offering. It should be emphasized, though, that the use of these utility libraries is optional; one can use anybody's I/O library in Khoros application code.

Although C is the main language, Khoros supports Fortran code as well. Khoros 2.0 will provide more explicit C++ support.

3. Khoros Specification Formats

Each Khoros application module is defined by the contents of two ASCII files: a User Interface Specification (UIS) file and a Program Specification (PS) file. These two together contain all the generic non-Khoros information that the Khoros code generator needs to create proper source and header files. The system is based on the philosophy that the code generator should know all Khoros-related information, so all it needs from the application programmer is the "generic" part of the application.

3.1. User Interface Specification (UIS) File

The two most important elements in this file, for the purpose of this paper, are the lines that define input parameters and the lines that define actions.

Each input parameter is defined by a single line in the UIS file. This line contains fields that specify the parameter's type (integer, float, string, file name, etc.), options (such as whether the parameter is optional), upper and lower bounds, default value, name, explanation (documentation), label, and widget geometry (for the GUI). This type of information is very similar to what one finds in the input or parameter files of other software packages.

Action buttons include especially an "Execute" and a "Help" button. The latter brings up a window containing the text of the automatically generated manpage. The former contains the basic part of the command string that is sent to the operating system to initiate execution. It should be pointed out that, although this string will usually just consist of the name of the module, one has the option to put more information in it.

UIS files may be prepared through a graphical interface, by selecting and positioning various widgets, or by editing the ASCII text of the file with an updated display of the GUI. Currently, only Athena widgets can be used, but Khoros 2.0 will provide support for Motif and OLIT.

3.2. Program Specification (PS) File

The PS file consists of a number of ASCII text fragments, each identified by a key. Some of these are documentation fragments, others are pieces of code. The architecture of a Khoros module consists of two major elements: a "library" function which contains the actual algorithm or application and a "main" which is largely written by the code generator to provide a Khoros-compatible wrapper around the generic code. Obviously, the library function has to be present in its entirety in the PS file, but for the main only small fragments are required, such extra includes, the call to the library function, and taking care of any interdependencies between input parameters; the automatically generated code will take care of getting the parameter values and checking their bounds, but cannot resolve interdependencies.

Khoros provides a menu-based software development tool to edit the contents of the various text fragments. However, one may also edit the PS file directly with one's favorite editor, or for that matter, even the source files generated by Ghostwriter.

4. Ghostwriter: the Code Generator

Ghostwriter, Khoros's code generator, takes in the two specification files, as well as a modest configuration file, and performs the following tasks.

- Collect all documentation fragments and prepare the manpage. This manpage is available from the command line through a utility similar to "man," and in real-time from the "Help" button.
- Write a function that takes care of all input parameters: check their presence, obtain and check their values, and put them into a data structure.
- Generate all header and source files required to build the module. This includes writing the "main" wrapper.
- Generate a Makefile. Most of the necessary information is gleaned from various Khoros system configuration files.

5. Interoperability

As we are inexorably moving toward more open systems, there will be increasing pressure on us to answer the question how to use some useful application, available in one software package, in the context of another package. Not *whether*, but *how*. As a definition of "software package" I will use "user interface environment," since that is the way users tend to think of it. There are, in essence, two ways to achieve interoperability: by making executables available to different user interfaces; and through code sharing.

5.1. Sharing of Executables

One can achieve interoperability by allowing executables to be activated from different user interface environments. This largely amounts to translating the command line syntax.

Scripts Peter Teuben (BIMA, Univ. of Maryland) has provided a Khoros user interface to Miriad and AIPS++ executables by inserting a script name "under" the Khoros "Execute" button. The script translates the syntax of the Khoros command line to that of the other system. At the same time, he constructed a simple translator that translates Miriad or AIPS++ input files to Khoros UIS files.

Dummy Programs I have, for a proof-of-concept exercise, written some simple programs that operate from the IRAF CLI and construct a Khoros command line and send it to the operating system. This enables me to execute Khoros applications from an IRAF environment. The generation of code and parameter files can easily be automated.

5.2. Code Sharing

A more powerful technique is to actually be able to share the core *generic* application code. I submit that the Khoros UIS and PS files contain all information necessary to build an application in almost any environment. Consequently, it should be possible, if not trivial, to produce alternative "Ghostwriters" for, say, IRAF and AIPS++. I have concluded that the job for IRAF is fairly minor.

6. Conclusion

I argued in the previous section for adopting the Khoros (UIS and PS) specification formats in all software packages, and for each package's having its own code generator for creating proper code from these files containing all generic information necessary to construct an application. As a result, the community would end up with a universal code format for applications which can be inserted in any software environment the user is interested in. Since the UIS and PS files only contain generic code, it would actually provide an important service to each of the programming environments as well by relieving all application programmers (and in particular user-programmers) from the burden of coding the package-dependent parts over and over for each module. Experience has shown that programmer productivity increases significantly in the Khoros environment, mainly for this reason.

The idea is not new, of course, and there are many ways in which this could be achieved. However, I feel there is no reason to reinvent the wheel in the astronomical community for something that is available off-the-shelf for free outside the community. In addition, there is the added advantage that the powerful Khoros UIMS, as well as the applications and utilities that come with it, are immediately available for use.

If we can settle the user interface aspects of interoperability, the next issue will be data exchange. I suggest we take a hard look at the data services that will be available in Khoros 2.0, since they may well provide a transparent data exchange functionality that will include, but at the same time go far beyond, simple FITS format standardization, bringing the rest of the world within our reach.

References

Argiro, & Rasure 1992, An X Windows Based Application Programming System, Xhibition, June 15, 1992, 151

Young, & Rasure 1992, An Open Environment for Heterogeneous Distributed Computing, Xhibition, June 15, 1992, 159

Rasure, & Young 1992, SPIE/IS&T Symposium on Electronic Imaging, SPIE Proceedings Vol. 1659

Astronomical Data Analysis Software and Systems II
ASP Conference Series, Vol. 52, 1993
R. J. Hanisch, R. J. V. Brissenden, and J. Barnes, eds.

An Object-Oriented Data Reduction System in Fortran

Jeremy Bailey

Anglo-Australian Observatory, PO Box 296, Epping, NSW 2121, Australia

Abstract. A data reduction system for the AAO two-degree field project is being developed using an object-oriented approach. Rather than use an object-oriented language (such as C++) the system is written in Fortran and makes extensive use of existing subroutine libraries provided by the UK Starlink project.

1. Introduction

Object-oriented programming (OOP) techniques have a number of well established advantages in allowing for easier reuse of existing code, and greater reliability. Generally we associate object-oriented programming with specialized object-oriented languages such as C++. Since most existing astronomical data reduction code is written in non OOP languages (mostly Fortran) it might be thought that object-oriented techniques are not relevant unless we are prepared to go to the lengths of totally rewriting our system in a new language (as the AIPS++ project is doing).

The data reduction system being developed for the AAO two-degree field project (a 400 object fibre system feeding 2 CCD spectrographs) shows how an object-oriented approach can be used in a non-OOP language (in this case Fortran).

2. The System

To produce an object-oriented system it is necessary to have a way of representing the data associated with an object, and a way of coding the methods of the object. Fortran itself has very limited facilities for data structuring so we have used the Starlink NDF package to represent the data of the objects. NDF, which is itself built above the Hierarchical Data System (HDS) can represent N-dimensional data arrays with associated axis, quality, and variance information so can easily handle a wide range of astronomical datasets. Most importantly the data structures are extensible allowing the addition of arbitrary extensions to represent the data of more specialist classes.

Using the NDF/HDS method of representing the object's data has a number of advantages over the use of C++ objects.

- Since the objects are stored in HDS files they are automatically *persistent objects* (i.e., they continue to exist outside the context of any one program).

C++ objects normally exist only within an individual program.

- The objects can be moved even between machines with different architectures, since HDS automatically handles differences in byte order, floating-point format, etc.

- The data format is compatible with existing NDF based data reduction software such as Figaro and ADAM.

The methods of the objects are provided using Fortran subroutines with a standard calling sequence. Each class has a subroutine of the following form, this example being for the NDF class which is the base class for all NDF objects.

```
      CALL CLA_NDF(OBJECT,METHOD,ARGS,STATUS)
```

Where OBJECT is the NDF identifier of the object for which the method is to be invoked, METHOD is a character string specifying the name of the method to be invoked, ARGS is a structure containing the arguments of the method and STATUS is an integer status argument.

It is crucial for a number of reasons that all class subroutines have the same sequence of call parameters, so the packaging of all the arguments into a single structure is essential. This is achieved by using the SDS memory based hierarchical data system. Arguments are loaded into the argument structure by a series of ARG subroutine calls preceding the call to a class routine

A simple example of a program using the class library is as follows:

```
      CHARACTER*40 FILE,DEVICE
      INTEGER OBJECT,ARGS,STATUS

*  Get File name and device name
      PRINT *,'Enter name of file to display'
      READ(*,'(A)') FILE
      PRINT *,'Enter name of display device'
      READ(*,'(A)') DEVICE

*  Start HDS
      STATUS = 0
      CALL HDS_START(STATUS)

*  Get object from file
      CALL ARG_NEW(ARGS,STATUS)
      CALL ARG_PUTOC(ARGS,'FILENAME',FILE,STATUS)
      CALL CLA_NDF(OBJECT,'READ',ARGS,STATUS)

*  Display the object
      CALL ARG_PUTOC(ARGS,'DEVICE',DEVICE,STATUS)
      CALL ARG_PUTOL(ARGS,'INTERACTIVE',.TRUE.,STATUS)
      CALL CLA_2D(OBJECT,'PLOT',ARGS,STATUS)

      END
```

This program is a plotting program for two dimensional images. It makes use of two subroutines from the class library. CLA_NDF is called with the READ method, which opens a file for read access and returns its NDF identifier in the variable OBJECT. Then CLA_2D is called specifying the PLOT method of the 2D class to plot the data. In each case the necessary arguments are loaded into the ARGS structure first with ARG_ calls.

3. Inheritance

Inheritance is built into the class library by structuring class routines as follows.

```
IF (METHOD .EQ. 'METHOD1') THEN
    CALL method1_routine(OBJECT,ARGS,STATUS)
ELSE IF (METHOD .EQ. 'METHOD2') THEN
    CALL method2_routine(OBJECT,ARGS,STATUS)
.
.
ELSE
    CALL CLA_base(OBJECT,METHOD,ARGS,STATUS)
ENDIF
```

All methods which are recognized are handled by routines which are specific to the class. Any others will simply be passed to the base class by calling its class routine.

4. Polymorphism

In addition to the individual routines for each class there is a generic routine CLA_GENER. This routine determines the class of the object and then calls the appropriate class routine. Objects generated by the class library will normally be labeled with their class but NDF objects created by other software may not be so labeled. In the latter case CLA_GENER attempts to determine the class by examining the structure of the object (e.g., number of dimensions, presence of extensions, etc.).

The example program above could therefore be made more general by calling CLA_GENER in place of CLA_2D. PLOT is a polymorphic method, present in a number of classes, and therefore the plotting program would now be able to plot data of various different types. Moreover, the program would automatically gain new functionality as new classes are added to the class library.

5. GROUP Objects

One important class that has been implemented in the current system is the GROUP class which can represent a group of NDF datasets as a single object. The GROUP object file can either contain the data for all the objects in the group, or more usually, can simply contain pointers to individual files which contain the objects in the group. Group objects simplify many common operations in data reduction. For example, to flat field a number of CCD images, the images can be combined into a group and the flat fielding operation can then be performed by using the ARITHMETIC method on the group to divide by a flat field frame. This will have the effect of dividing each of the images in the group by the flat field frame.

The GROUP class also has a method to handle combination of a set of images by median combination or a variety of other algorithms to form a single image.

6. User Interfaces

The class library itself is simply a library of subroutines that can be used in any Fortran program and does not depend on any particular software environment. It has been used effectively to develop applications that run in the ADAM environment using the ADAM parameter system and command language. However, for the 2dF project we are developing a graphical user interface based on OSF/Motif.

A graphical user interface maps well onto an object oriented system, since the concept of selecting an object (by clicking on it) and then choosing what to do with it (e.g., from a menu) can easily be translated into invoking a specified method for an object.

A preliminary version of the user interface has now been implemented which presents the user with a scrolling list of objects (i.e., files) and ways of navigating through the directory structure to select the required working directory. Double clicking on an object invokes the PLOT method for the object putting up an interactive display window containing a suitable display of the object's data.

Arithmetic operations are provided through a calculator widget which contains buttons for $+,-,\times$ and $/$. An arithmetic operation is carried out by clicking on the first object, then on the operation button and then on the second object. Clicking on the $=$ button brings up a dialog box in which the resulting file name can be entered. Other operations are selected from a Commands menu.

A powerful feature of the system is the incorporation of group objects into the user interface. A multiple selection can be made from the list of objects, and this results in a temporary group object being created which contains the selected files. Any selected operation will then apply to the group, which in most cases means that it will apply to all the objects in the group.

Discussion

Kibrick: How much of the StarLink software environment is required to support the use of the Fortran callable class libraries?

Bailey: Primarily HDS and NDF. Probably about a total of 7–8 libraries.

Silberberg: If you were to start this project again, would you go to all the effort to make Fortran object-oriented or would you use a language like C++?

Bailey: Much more work would be involved in doing the project in C++ since we would have to convert many of the subroutine libraries that the system uses. Making Fortran object-oriented was not a lot of effort given the existence of HDS/NDF to represent the objects.

Astronomical Data Analysis Software and Systems II
ASP Conference Series, Vol. 52, 1993
R. J. Hanisch, R. J. V. Brissenden, and J. Barnes, eds.

The Keck Keyword Layer

A. R. Conrad and W. F. Lupton

W. M. Keck Observatory, P. O. Box 220, Kamuela, HI 96743

Abstract. Each Keck instrument presents a consistent software view to the applications programmer. A set of "plug compatible" applications has been developed that can be used with any instrument. Because image capture software uses the same keyword library to collect data for the image header, a given observation can be "replayed" by extracting keyword-value pairs from the image header and passing them back to the control system.

1. Introduction

In the first section we give the advantages of the keyword approach. These include: replay, a consistent `API`, and "plug compatible" user interface tools. A brief history of the events which lead up to the keyword philosophy is given in the following section. A few examples are given in section 4. The final three sections give the current status, known problems, and future plans for the keyword layer.

2. Advantages

Each Keck instrument presents a consistent software view to an applications programmer (see Figure 1). When a programmer sits down to write an application on top of one of the Keck keyword libraries, he or she is faced with the same half dozen functions, regardless of which subsystem the keyword library controls. The keyword layer spares the applications programmer from having to learn the complexities of the underlying task structure.

Image capture software uses the same function library to collect data for the image header. Because the image capture software and the instrument control software are built on top of the same keyword layer, a given observation can be "replayed" by extracting keyword-value pairs from the image header and passing them back to the control system.

The small set of functions used to access a keyword library are bound dynamically via shared libraries. As a result, generic applications can be shared by all keyword based instruments (see Figure 2). The tools shown in Figure 2 are simple "keyword free" tools that will plug into *any* keyword library. More specific tools, that are "keyword sensitive," only work with particular keyword libraries. For example, the exposure control application *xpose* can only be plugged into one of the two optical spectrographs.

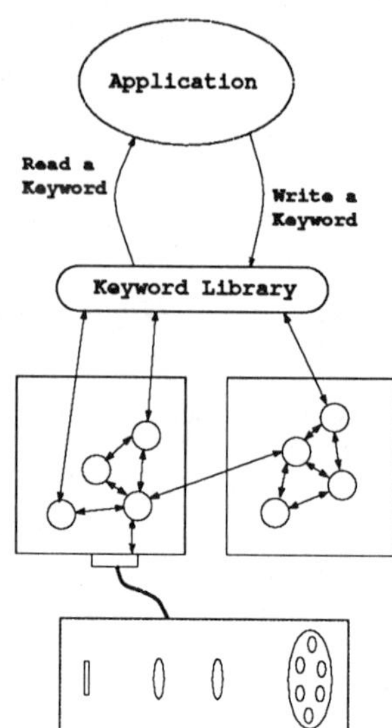

Figure 1. A consistent API is presented to application writers

3. History

In early 1990, the software design for the initial five Keck instruments was approximately half complete. Those involved were beginning to worry that the five design efforts were going in different directions. Although we were willing to tolerate some disparity, we all agreed that it was essential to standardize two areas: the `keywords` for image headers and the `command language` for controlling the instruments. A "czar" was appointed to mandate keywords and commands. After polling the five instrument teams, the approach being taken by the team at Berkeley Space Sciences Laboratory stood out as unique. Their approach was to provide a single command, *go*, with a single argument: a file containing keyword/value pairs. The primary motivation was the ability to replay an observation from a FITS header. With the addition of a few more commands (*show* and *modify*), the idea was sold to four of the five instrument teams.

Later in the year the design was extended as the basis for an API. The *show* and *modify* commands became thin wrappers around standard subroutines for reading and writing a keyword.

To accommodate X window applications, the keyword API was extended to handle asynchronous events in the event-loop style of X toolkit applications. This occurred in early 1992.

Meanwhile, CCX, an event-loop mechanism for cooperating control system processes, was being developed at Keck. During spring and summer of 1992, CCX was generalized to encompass the event-loop needs of keywords and messages. The resulting system has been dubbed the *Keck Task Library* (KTL, Lupton & Conrad 1993).

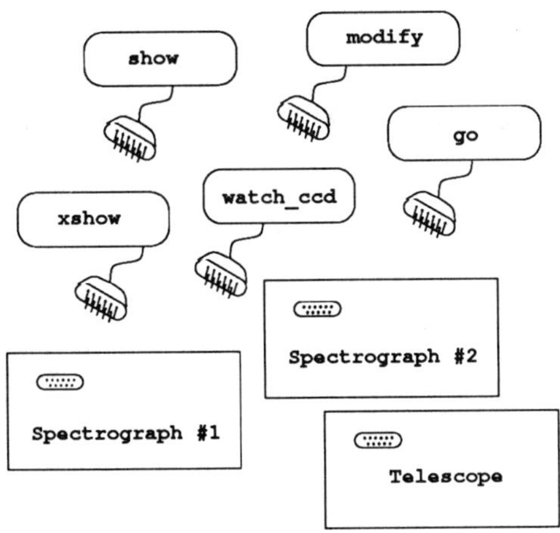

Figure 2. Plug compatible applications

4. Examples

Following is a condensed excerpt of the C source for the *show* command:

```
ktl_open( service, "keyword", 0, &khand );

for ( i = 1; i < argc; i++ ) {
    ktl_read( khand, KTL_WAIT, argv[i], NULL, &data, NULL );
    ktl_ioctl( khand, KTL_FORMAT | KTL_BININ | KTL_ASCOUT,
               argv[i], &data, &form );
    printf( "%30s = %s\n", argv[i], form );
}

ktl_close( khand );
```

The *service* is the name of a subsystem, such as a spectrograph. It is passed to *ktl_open* which invokes *dlopen* to open the shared library and load the function addresses for *ktl_read* and *ktl_ioctl* into the opaque handle *khand*. Then for each keyword, we read its value, convert it to a printable format, and print it.

Now consider a more sophisticated keyword application. Suppose we are writing an X application that:

1. continuously displays total exposure time and elapsed exposure time.

2. provides a button to open the shutter.

Figure 3 shows the KTL structures created to support this task. The elements on the *interest* list are used to associate a callback to be invoked whenever either of the time keywords change value. The interest list will persist throughout the lifetime of the program. The elements on the *subcontext* list will exist only during the time between the the shutter button push and the notification that the shutter move is complete.

KTL supports a *message* style in addition to the keyword style. The same structures, shown in Figure 3, would be created by a *message* application that expressed interest in two *message* types and was at the same time sending a command *message* (KTL, Lupton & Conrad 1993).

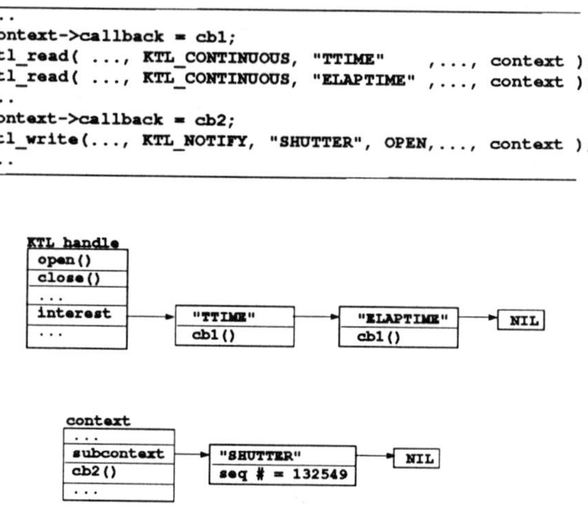

Figure 3. KTL structures for keyword callbacks

5. Current Status

Currently we have five operational keyword libraries.

- `musicFiord`. Used for the High Resolution Spectrograph, the Low Resolution Imaging Spectrograph, and Telescope Control.
- `rpcFiord`. Used for the Long Wave Camera and the Long Wave Spectrometer.

Three of these were built on top of `musicFiord`, a keyword library support package based on the Lick MUSIC system (Kibrick et al. 1993, Stover 1989). The other two were built on top of `musicFiord`, a keyword library support package based on Sun's *Remote Procedure Call* mechanism. Table 1 summarizes the differences between the two packages.

	`musicFiord`	`rpcFiord`
Message System	music	RPC
Keyword Knowledge	Client-side	Server-side
Language	C	C++
Process Weight	Heavy	Light

Table 1. Keyword support packages

6. Problems

6.1. Unfriendly User Interface

Often the keyword layer is misinterpreted as a user interface. Potential users imagine that they will have to learn hundreds of obscure keyword names to perform simple operations with an instrument.

The keyword layer is a programmer's interface. Because the keyword layer simplifies writing X applications, instruments which provide a keyword library are more likely to present user interface tools that are friendly. Ideally, first time users will not know that the keyword layer exists.

On the other hand, experienced observers can make use of the keyword layer to write simple scripts; for example, a script that rings a bell when the dewar temperature drops below a given temperature.

6.2. No Aggregate Structures

The lack of aggregate structures is a common complaint. This restriction stems from the initial motivation for the keyword layer: the ability to "play back" a FITS header. Thus far the lack of aggregate structures has not been a serious impediment. In fact, by forcing the simplicity of a flat data dictionary, we have avoided the proliferation of structures within structures that often occur in what should be a simple API.

6.3. Event Loops are Passé

There are those that argue that the event loop style of an X toolkit application will give way to blocking I/O with lightweight processes. This may be true, but not in the time frame of this project.

7. Future Plans

The bulk of the work required to provide keyword libraries is done. The task of building X applications and replay tools on top of this foundation lies ahead.

References

Lupton, W.F., & Conrad, A.R. 1993, this volume

Kibrick, R.I., Stover, R.J., & Conrad, A.R. 1993, this volume

Stover, R.J. 1989, UCO/Lick Technical Report 54, MUSIC - a Multi-User System for Instrument Control

Astronomical Data Analysis Software and Systems II
ASP Conference Series, Vol. 52, 1993
R. J. Hanisch, R. J. V. Brissenden, and J. Barnes, eds.

A New Programming Metaphor For Image Processing Procedures

O. M. Smirnov

Institute of Astronomy of the Russian Academy of Sciences, 48 Pyatnitskaya st., Moscow 109017 Russia

N. E. Piskunov

Observatory and Astrophysics Laboratory, University of Helsinki, Tähtitorninmäki, SF-00130 Helsinki, Finland

1. Introduction

A mature image processing system is usually expected to provide two levels of programmability. On the lower level, there is an Application Program Interface (API) which must let users write their own programs, usually in C and/or FORTRAN, and integrate these programs into the system. On a higher level, the system should provide means for gathering separate programs together into procedures (scripts). Traditionally, this is some sort of interpreted command or macro language.

Unfortunately, all command languages seem to be downright retarded when faced with some recent developments in hardware and, more important, users' expectations. These days, people want their software to provide two or more of the following: a Graphical User Interface (GUI), a gentle learning curve, rapid prototyping, real-time data acquisition and analysis, distributed processing and/or user interface, support for concurrent systems, etc. A command language has difficulties with all of these. In the old days, when rodents were rodents and not pointing devices, and images were columns of green digits on a terminal screen, it seemed a good enough solution. Not now, though.

When an idea enters such hard times, it's usually best to classify it as fundamentally incorrect and think of a better one, instead of adding all sorts of extensions to handle new complexities until the whole thing collapses under its own weight. But thanks to "upward compatibility", that scourge of innovation, such an approach is just about impossible, unless a new system is started from scratch. We got this opportunity with PCIPS.

PCIPS (Smirnov et al. 1992; Smirnov & Piskunov 1993) was initially developed for all sorts of image processing on modest hardware—a PC AT running MS-DOS—and is now being overhauled for a multitasking environment (Windows). The result of this will be a new system, which we'll call IPS+ for now, with better (we believe) programming capabilities than can be provided by a command language. What follows is a description of these capabilities.

2. The Basic Idea

Let's try to understand what is fundamentally wrong with the idea of a command language (script). Practical issues aside, a scheme with scripts looks very convolved. In it, image processing procedures—members of a very wide class endowed with rich properties—are represented by commands of some formal grammar, a very narrow and rigid class. Also, the representation is highly unnatural to the mind. Its only virtue is that it mimics the computer's way of thinking, but then machine language is even stronger on that point, and we don't use it much.

In short, scripts are not a natural programming metaphor. In this sense, C and FORTRAN are not natural metaphors too. But they are used to represent such general procedures that current software technology just can't offer much of an alternative. With image processing procedures, it can.

The alternative metaphor is a flowchart of sorts: a bunch of modules (*applications*) that perform separate processing tasks, connected by *pipes* that carry data between them. In IPS+, we call it a *factory*. This metaphor does not cramp flexibility the way scripts do, and it is intuitive to the mind. If you look back to our list of requisites which scripts have difficulties with, you'll find that factories can handle them easily—if that's not obvious now, it should be by the end of this paper.

IPS+ will not be the first image processing system to implement such a concept. Credit for that probably belongs to *Khoros* (Rasure & Williams 1991), where the idea was given different treatment and took the form of a visual programming language.

3. The Hurdles

To implement the factory programming metaphor in a productive way, some hurdles had to be cleared. First, there's the requirement for different perspectives: when part of a factory, an application is best thought of as a *static* box with inputs and outputs, but when it's running on its own, it's a *dynamic*, interactive process with a GUI. Second, there's the compatibility problem: when on the receiving end of a pipe, an application must be compatible with all kinds of input (i.e., images of different type or size). Otherwise, the user will be limited in the kind of connections that he can create. Also, these problems should be solved transparently to the application programmer—new applications should be easy to create, without regard to GUIs and pipelining. And, of course, the user interface deserves special attention because it must allow the user to assemble factories easily and efficiently.

With these considerations, the API becomes a very important component of the system. IPS+ will inherit PCIPS' API—AIS (Application Interface and Services), which was designed with factories very much in mind. Under it, applications never interact directly with the user or with disk storage. Instead, they call AIS to accomplish these tasks. This is covered in detail in a paper on PCIPS (Smirnov & Piskunov 1993).

To receive input and produce results, IPS+ applications dynamically generate *requests*. The normal behaviour of IPS+ is to pack these requests into a

GUI, present them to the user, get input, and report results back to the application. To build a factory, the user connects requests from different applications by pipes. Connected requests are treated differently—instead of querying the user, IPS+ serves them by obtaining input from, or sending output to the pipe.

This concept is somewhat similar to the way that Unix programs can be combined by redirecting their input/output streams. When constructing a factory, a user is essentially redirecting requests. Factories are an extension of this idea: applications produce multiple requests dynamically instead of having just two standard streams.

Compatibility is ensured via type/size conversion. When requesting images, the application specifies the type/size that it is prepared to handle, and IPS+ assumes responsibility for converting the input images to this format. The specification can range from the simple ("I can handle any format") or the rigid ("256×256 16-bit integer"), to the flexible and complex ("two 16-bit integer images of the same size, and an image of any type, but of the same size as the first two"—this is much easier to program than to say).

4. An Example

Factory construction is best illustrated by an example. Consider a simple and purely hypothetical data reduction setup. One application acquires frames from a CCD (we'll call it ACQ from now on), another finds and removes cosmic ray hits (CR), a third does normalization by flat field (FF), and a fourth finds and extracts objects from the frames (OBJ). These applications generate a number of input and output requests: ACQ accepts acquisition parameters, while producing a flat field and a data frame, CR takes the data frame and a detection criterion to produce a cleaned data frame, FF takes the cleaned data frame and the flat field to produce a normalized frame, and OBJ takes the normalized frame and produces objects. Some of these requests are grouped into *lists* (also called *dialog boxes*), which means that they are generated and served at the same time. Also, the OBJ application displays some images while running, i.e., it also produces a *viewport.*

To build a factory, the user simply specifies that the result of an output request is to be sent to another input request. From his point of view, this process is easy and intuitive ("plug and play"), and is certainly less demanding than having to write a script—it can be done in seconds with the mouse.

The factory's user interface consists of the unconnected ("loose") requests, plus the viewport. It is concurrent by nature. ACQ runs continuously, and the other applications are only active while there's something for them to do. When frames become available, ACQ dumps them in the pipes, and goes back to controlling the CCD. Meanwhile, the frames are processed by CR, FF and OBJ. If ACQ produces frames faster than they are processed, IPS+ will pile them up in their respective pipes until a break or pause in the acquisition allows the rest of the factory to catch up.

In effect, the user can create a concurrent procedure for real-time data acquisition by simply attaching four pipes and specifying that ACQ is to run continuously. This is done faster than you can repeat "rapid prototyping" five times.

5. Advanced Programming Tools and Constructs

For sophisticated data flow control within factories, IPS+ will include a set of pseudo-applications called *pidgets* (**pi**pe **gadgets**). Pidgets distribute objects from one or more "in" pipes into one or more "out" pipes, while accepting and/or generating some sort of control data. They provide the functional (though not conceptual) equivalent of a command language's looping, branching, condition testing and procedure call constructs, as well as additional capabilities unique to factory programming. Actually, being a totally different metaphor, a well-programmed factory usually relies on a completely different set of programming paradigms, so natural parallels with command languages are impossible to establish. IPS+ will support a Pidget Interface to facilitate the creation of custom pidgets.

An important tool is the *history analyzer*, providing factory CASE capabilities. By scanning the log of an interactive session and analyzing the actual data flow within it, it will be able to come up with ready prototypes of factories that were implicitly constructed by the user during processing. This is made possible because a factory is a natural representation of the data flow.

6. Other Advantages

If we ignore its internal structure and look at it from the outside, a factory resembles an ordinary application a lot—it also produces requests and/or viewports. What's important is that it looks that way from the system's point of view too. Consequently, any factory can be used as a component of other factories or on its own. Users don't even have to know whether a particular application is the "real" C-or-FORTRAN variety, or a factory that utilizes several separate applications.

Other goodies are self-tuning and true pipelining. If self-tuning is on for a particular application, IPS+ will allocate it more CPU time when its pipes pile up data, resulting in a more balanced factory. True pipelining is possible when two interconnected applications access images in a controlled way, i.e., by "giving" IPS+ a tool (a call-back function) for handling the whole data sequentially in one pass. Under these conditions, the second application does not have to wait for the first one to finish processing a whole image, because it is able to receive data in parts.

IPS+ also provides means for interprocess communication, via the implementation of dynamic requests. A dynamic request is not retracted when serviced, like a normal one is. Instead, it remains active and can receive more input at any time, interrupting the application and giving control to a specific function to handle the input. In this way, it's somewhat similar to a hardware interrupt. Dynamic requests require more effort from the application programmer, but when connected by pipes, they allow applications to maintain "conversations", instead of the "letter exchange" of normal requests.

A very effective configuration for parallel processing is a network. IPS+ will support both distributed processing and distributed user interfaces. Several sessions of the system running on different network nodes will be able to establish communication, after which they can be configured (from one machine!) to execute different parts of a factory. Some pipes or requests will then send data over the network. The whole user interface of a factory can either be concentrated on one master computer, or, with very complex factories, spread over several network nodes, with several people at once cooperating in one procedure.

7. Implementation

It would be safe to say that the kernel of factory programming is easier to implement than a command language. Because a natural programming metaphor preserves many properties of the object being represented (an image processing procedure, in this case), it sort of takes on a life of its own, and a lot of problems simply fail to arise. However, factory capabilities require a very sophisticated user interface, and that complicates the project immensely. Some relief is provided by C++ and an object-oriented architecture, though.

We plan to complete the first version of IPS+ in 1993. We're ready to answer all questions about the system. Our e-mail addresses are `oms@airas.msk.su` (Oleg Smirnov) and `piscounov@cc.helsinki.fi` (Nikolai Piskunov).

Acknowledgments. Financial support from the Smithsonian Astrophysical Observatory, which made this paper possible, is gratefully acknowledged.

References

Rasure, J., & Williams, C. 1991, J. of Visual Languages and Computing, 2, 1

Smirnov, O.M., & Piskunov, N.E. 1993, this volume

Smirnov, O.M., Piskunov, N.E., Afansyev, V.P., & Morozov, A.I. 1992, in Astronomical Data Analysis Software and Systems I, A.S.P. Conf. Ser., Vol. 25, eds. D.M. Worrall, C. Biemesderfer & J. Barnes, 344

Astronomical Data Analysis Software and Systems II
ASP Conference Series, Vol. 52, 1993
R. J. Hanisch, R. J. V. Brissenden, and J. Barnes, eds.

SPPTOOLS: Programming Tools for the IRAF SPP Language

Michael J. Fitzpatrick

IRAF Group, NOAO[1], *PO Box 26732, Tucson, AZ 85726*

Abstract. An IRAF package to assist in SPP code development and debugging is described. SPP is the machine-independent programming language used by virtually all IRAF tasks. Tools have been written to aide both novice and advanced SPP programmers with development and debugging by providing tasks to check the code for the number and type of arguments in all calls to IRAF VOS library procedures, list the calling sequences of IRAF tasks, create a database of identifiers for quick access, check for memory which is not freed, and a source code formatter. Debugging is simplified since the programmer is able to get a better understanding of the structure of his/her code, and IRAF library procedure calls (probably the most common source of errors) are automatically checked for correctness.

1. Introduction

SPPTOOLS is an IRAF[2] package designed to aide in developing and debugging SPP tasks and packages. It is hoped that both novice and experienced programmers will benefit, and that productivity can be increased by automating some of the more mundane aspects of task development. This package is still experimental and not all of the tools are fully developed. In some cases there is still some "post-processing" to be done by the user to make sense of the output. For example, the **spplint** task lists program errors with line numbers that do not refer to the SPP code being checked, but rather the numbers refer to the translated **C** code that is produced. Other tasks are more complete and make no additional demands on the programmer.

The **SPPTOOLS** package tasks can perform the following functions:

- Automatic VOS (Virtual Operating System) interface checking.
- Reformat source code to follow common IRAF standards and conventions.
- Print an indented list of calling sequences within tasks.

[1]National Optical Astronomy Observatories, operated by the Association of Universities for Research in Astronomy, Inc. (AURA) under cooperative agreement with the National Science Foundation

[2]Image Reduction and Analysis Facility, distributed by the National Optical Astronomy Observatories

- Quick lookup of identifiers in sources.
- Create an external package from a standard template.
- Rename an external package.
- Check for missing memory procedure calls or erroneous usage.

IRAF Site Managers may also find this package useful since it allows them to quickly scan unfamiliar code if debugging a problem, and creating external packages for locally written software is also simplified. It is also hoped that, along with documentation of the SPP language, users will find it easier to program in SPP and choose to do software development in the IRAF environment.

Since some of the utilities in the package are derived from (or are dependent on) Unix utilities, there may be problems running on non-Unix systems such as VMS (the **spplint** task is actually implemented as a C-shell script). The package is still experimental, the more successful tasks will be rewritten to run on all supported IRAF platforms.

2. Task Overviews

SPPLINT - This task checks the source code for possible VOS interface violations, as well as internal consistency checks for argument type and number, and unused variables within procedures. Function prototypes may be created for the source code which can then be compared to a prototype file for the IRAF VOS. Fortran sources may be checked with SPP sources, and processed **C** and Fortran code can be produced. This task is analogous to the Unix *lint* utility for **C** program sources.

SPPCALLS - Prints an indented list of calling sequences that allows the programmer to see the "flow" of a program. When examining unfamiliar code, or even large amounts of familiar code, it is often useful to have an idea of the calling sequences to get a better understanding of the code's structure. Debugging is simplified because the programmer is able to more quickly isolate the problem procedure by tracing the calling sequence that's printed rather than tracing the code itself. Recursive constructs, which are not permitted in SPP, are also trapped. This task is analogous to the Unix *cflow* utility for **C** program sources.

PKGCREATE - One of the major strengths of IRAF is it's programming environment, users can write their own reduction or analysis software for distribution that can then be exported to outside sites as an external package, or kept locally for personal use. It's been difficult however for novice users to create a properly configured external package (due perhaps to a lack of understanding of the basic package structure). The **pkgcreate** task uses a standard package template and automatically renames or edits the files that properly define the package, and will also copy source code to the new package in a first attempt at completing the final package. The work required to complete the package is minimal.

PKGRENAME - This task is a simple utility for renaming external packages (e.g., to change a revision level if that was part of the package name, or to avoid duplicate packages). The package files are edited and renamed as required, but any outside reference to the package (such as it's definition in hlib$extern.pkg or in a parent package) are unchanged.

SPPFMT - The importance of coding style is often overlooked when developing new software. It's not uncommon that during the lifetime of a piece of software several programmers will be responsible for it, having a standard coding style can greatly reduce the amount of time spent maintaining the code. To this end, the **sppfmt** task can be used to reformat code to the style proposed by the "IRAF Standards and Conventions" document by Downey et al. Indention levels and spacing within lines are adjusted as needed, comments are put in a common location, and lines will be split or joined when necessary. Backup files of the original source code may be created for safety. This task is analogous to the Unix *indent* utility for **C** program sources.

CHCOUNT - Since the code is modified by the **sppfmt** task, it's very important that no new bugs are introduced. The **chcount** task simply compares the number of characters in the original and modified source to assure the user that nothing was "lost in the translation." Suspicious files are flagged for the user to check at a later time.

MKID - Scan the source code and create an identifier database. SPP procedures, local variables, macro definitions and the files in which they occur are included in the database.

QID - Query the ID database, producing a list of files that contain a named identifier (possibly specified as a substring), or a context in which that identifier is used (similar to the using the IRAF command "match *id* *.x"). Using the ID database is faster than scanning the code by hand (or with other host level or IRAF utilities).

IID - Interactively query the ID database.

FID - Query the ID database for specific files.

MEMCHECK - The **memcheck** task (which is still in development) can be used to find possible memory errors. The task outputs a table of pointer usage containing the procedure in which the pointer was allocated/freed and it's type.

3. Future Work

Although well over a million lines of SPP code have been written, SPP itself is not as fully supported by tools as are other languages such as **C**. While the language itself is rich with routines in the VOS that make it fully functional, and utility libraries such as **gtools** that make it useful, common programming necessities such as a source-level debugger are not yet available. SPP, as the acronym implies, is a pre-processor language, the code which is actually compiled can be virtually anything (several recent IRAF ports have used processed C code rather than the usual Fortran). Programmers actually debug this processed language and then have to mentally translate back to the line of SPP code causing the problem.

The SPP language is presently not fully implemented (many people notice a lack of "real" pointers and structures), doing so with a complete compiler would greatly simplify the writing of a source-level debugger. Usage errors, such as those trapped by **spplint** currently, could be caught at compile time resulting in more error-free code, and an execution profiler to help in optimizing programs could eliminate coding inefficiencies. All of this would result in a truly portable programming language that is no more difficult to use than any other language in use today.

4. Availability

The **SPPTOOLS** package is available via anonymous ftp from iraf.noao.edu (IP address 140.252.1.1) in the "iraf.old" subdirectory. Retrieve the files (as binary) "readme.spptools" and "spptools.tar.Z". An example session would look something like:

```
% ftp -i iraf.noao.edu
..... login as 'anonymous'
..... use your e-mail address as the password
ftp> binary
ftp> cd iraf.old
ftp> mget readme.spptools spptools.tar.Z
ftp> quit
%
```

Please send any comments, suggestions, or bug reports to the author at either `fitz@noao.edu` or `5355::fitz`.

Acknowledgments. Some of the tasks in the SPPTOOLS package are based on public domain programs originally written as C programming tools. I am indebted to the original authors for making the source code available. This work was supported in part under the "SAO Software Environments" grant, NRA 89-OSSA-8.

References

Downey, E., et al. 1983, IRAF Standards and Conventions, KPNO document

Tody, D. 1983, A Reference Manual for the IRAF Subset Preprocessor Language, KPNO document

Part 2. Data Analysis Systems

Section B. Software Systems

Astronomical Data Analysis Software and Systems II
ASP Conference Series, Vol. 52, 1993
R. J. Hanisch, R. J. V. Brissenden, and J. Barnes, eds.

The Evolution of the Figaro Data Reduction System

K. Shortridge

Anglo-Australian Observatory, P.O.Box 296, Epping, NSW 2121, Australia

Abstract. This paper describes some of the history of the Figaro data reduction system. Figaro uses hierarchical data structures to provide flexibility in its data file formats, and these have changed over the years requiring the provision of more abstract data access routines. Figaro applications are now much easier to write, and are able to provide easy access to such things as data quality and error information. Applications are now able to be invoked in a variety of ways and on an increasing variety of machines.

1. Introduction

Figaro is a general-purpose data reduction system that originated at Caltech in late 1982 (see Cohen 1987) and has changed considerably since then, now being used at a number of sites throughout the world. It has always been closely connected with the Starlink project in the UK, taking ideas from Starlink and also feeding some back. It was adopted as the standard Starlink spectroscopy package some years ago, and many of the changes made to it over the years have been influenced by the need to adapt to the evolving set of Starlink standards, particularly in the matter of file formats. The first version of Figaro was a VAX/VMS system, making full use of VMS facilities such as file mapping. In recent years, however, it has moved towards being a portable system. A SUN version has been produced at Caltech for use at the Keck telescope, and more recently effort in the UK and in Australia has gone into extending this work to provide a system capable of running on a wider range of machines.

2. A Brief Overview of Figaro

Figaro emerged at Caltech in response to a need for a uniform set of data reduction programs that would be flexible enough to be able to handle a wide range of types of data. Instruments were producing multi-dimensional data arrays (at AAO the FIGS infra-red instrument would later produce four-dimensional data hypercubes for Figaro to handle); axis information had to be encoded in as general a way as possible, to deal with both the linear and non-linear axis values encountered at different stages of data reduction; assorted ancillary data arrays would be needed. At the time, the Starlink project had produced an initial design study for a subroutine library that would be used to access data elements held in a self-defining, hierarchical, data structure. Such a structure contains not only the data, but also a detailed description of the structure itself.

A program looks for data in a file by name (knowing, for example, that the array '.AXIS[1].DATA_ARRAY' contains the X-axis information) rather than by assuming anything about the byte-by-byte layout of the file. The adoption right from the start by Figaro of such a hierarchical, extensible, data format has given it great flexibility. A programmer who needs, for example, to add a list of echelle order numbers to a data file can simply add such an array to the file structure—naming it sensibly, one hopes. Items in a data structure can be manipulated by calls to the data system routines, being renamed, created, deleted, resized, in the course of program execution.

Another important aspect of the Figaro design taken from Starlink was that programs were not allowed to communicate directly with the user. Once a program has been started, all the details of its operation, such as the names of the files it is to process, are treated as program parameters and the program obtains their values through calls to parameter system routines. The parameter system will prompt the user as necessary, and the Figaro parameter system has always been regarded as having a particularly friendly 'feel'.

Interestingly, when Figaro was designed, it was not clear how Figaro programs were to be run. Differing preferences were expressed for, amongst others, combining them into monolithic programs, or for running them in sub-processes spawned by an—at the time—undefined command language, and even for linking them into a FORTH system and treating them as FORTH words (FORTH being particularly popular at Caltech at the time). To leave these options open, all Figaro applications were written as simple subroutines that could be invoked in any of these ways. They were tested by wrapping them up in an automatically generated main program which was run from the operating system command line and which then called the application. This mode of operation proved astonishingly popular and most Figaro programs are still run in this way, even though alternative methods are now available. Wanting to defer decisions about the overall environment was another reason for abstracting user interaction into the set of parameter system routines. The Figaro Programmer's Guide (Shortridge, 1990) has a history section that discusses this in more detail.

The data access routines used in the first Figaro programs came from the DTA package, a design based on the original Starlink concepts for what would become HDS (Starlink's Hierarchical Data System). The data structure conventions used were Figaro's own. It is these conventions that determine, for example, what the main data array in the structure is called. Systems such as HDS and DTA make it possible to handle hierarchical data structures in disk files, but they do not define how they should be used.

Line graphics used Tim Pearson's PGPLOT package, and image display used Mike Lesser's TVPCKG. These completed the original Figaro system.

3. Evolution of the Ways Data are Handled

In 1984, I moved to the Anglo-Australian Observatory, AAO, and Figaro was introduced both there and at Starlink. The DTA library was recoded to become just a thin layer on top of the now-available Starlink HDS. New applications were written by astronomers in Australia and the UK. It was always relatively easy to write a new Figaro program, and it became easier, perversely, as a result of the need to support increasingly complex data structures.

Starlink defined, after considerable open discussion, the NDF conventions (Currie et al. 1989) for the representation of data in hierarchical structures. Figaro was modified to support both these and its original data conventions. The following shows the output of a Figaro program ('EXAM') listing the structure of an NDF data file:

```
HORSE
  .TITLE          Char    Horsehead nebula
  .BAD_PIXEL      Logical 0
  .DATA_ARRAY     Array
    .VARIANT[16]  Char    SCALED
    .SCALE        Double  3.906E-3
    .ZERO         Double  127.6
    .DATA[256,256]  Short -4497 899 -1670 -5525
                              .... 4754 4497 1670 2955
  .AXIS[1]        Axis
    .DATA_ARRAY[256]  Float 1 2 3 4 5 6 7 8 9 10 11
                                      .... 251 252 253
  .MORE           Ext
    .FITS[80,3]   Char    INSTRUME= 'RCA CCD '     /
  .VARIANCE[256,256]  Float 110 131 121 106.0 87
                                  .... 146 145 134 139
```

This file also demonstrates another feature added to Figaro when it moved to version 3.0. It contains error data (the 'VARIANCE' array), and support for both such error information and data quality information was added to all the main Figaro routines at the time. (Obviously, it was always possible to add such an array to a structure, but most applications prior to version 3.0 would simply ignore it.)

The main data array in this NDF format example is called 'DATA_ARRAY', but this is not a simple floating point array. It is a scaled array, containing the data as two byte integers with scale and zero values. A number of different such ways of holding data are defined by the NDF conventions. Originally, the high level code of a Figaro program would assume the data was a floating point array and would access it directly using the conventional main data array name. To support structures as complex as those shown (not to mention the need to support the old Figaro structure conventions as well), it was necessary to introduce a new layer of data access routines, and this became the DSA (Data Structure Access) layer. A call to a routine such as DSA_MAP_DATA will return a pointer to a copy of the main data array in the format specified in the call (a single precision floating point array, for example), no matter how the data is actually held in the file. For the scaled array in the example, DSA will apply the scale and zero values to produce a floating point array.

Given these routines, a complete Figaro program can be remarkably simple. For example, here is a program (with only the comments removed for reasons of space) that adds a constant to an image:

```
      SUBROUTINE ADDCONST
C
      INCLUDE 'DYNAMIC_MEMORY'
      REAL FMAX,FMIN
      PARAMETER (FMAX=1.7E38,FMIN=-1.7E38)
      INTEGER ADDRESS,DIMS(10),DPTR,NDIM,NELM,SLOT,STATUS
      REAL VALUE
      INTEGER DYN_ELEMENT
C
      STATUS = 0
      CALL DSA_OPEN (STATUS)
      CALL DSA_INPUT ('IN','INPUT',STATUS)
      IF (STATUS.NE.0) GO TO 500
      CALL PAR_RDVAL ('CONSTANT',FMIN,FMAX,0.0,' ',VALUE)
      CALL DSA_OUTPUT ('OUT','OUTPUT','IN',0,0,STATUS)
      CALL DSA_DATA_SIZE ('IN',10,NDIM,DIMS,NELM,STATUS)
      CALL DSA_MAP_DATA ('OUT','UPDATE','FLOAT',
     :                                  ADDRESS,SLOT,STATUS)
      DPTR = DYN_ELEMENT(ADDRESS)
      IF (STATUS.NE.0) GO TO 500
      CALL ADDIT (DYNAMIC_MEM(DPTR),NELM,VALUE)
  500 CONTINUE
      CALL DSA_CLOSE (STATUS)
      END
C
      SUBROUTINE ADDIT (ARRAY,NELM,VALUE)
      IMPLICIT NONE
      INTEGER NELM
      REAL ARRAY(NELM),VALUE
      INTEGER I
      DO I=1,NELM
         ARRAY(I) = ARRAY(I) + VALUE
      END DO
      END
```

This example, together with all the missing comments and the 'connection file' used to describe the parameters to the system, can be found in the Figaro Programmer's Guide. The code that actually does the adding of the constant is trivial, of course, being only one line in the main routine (anything can be done in one line, of course, if that line is a subroutine call), but the rest of the code handles: getting the name of the file from the parameter system; opening that file; getting the value of the constant from the parameter system; getting the name of the output file; opening it if it differs from the input file; locating, converting and mapping the main data array (and handling any pixels flagged as containing bad quality data); finally closing down the files involved, checking whether any error information or data range information in the file is still valid. DSA is doing a lot of work behind the scenes here, mainly on the ancillary data held in the Figaro data structures.

A standard Figaro application will do more, of course, but astronomers wanting to apply some private algorithm to their Figaro format data can write programs such as this with some ease, and many have done so.

4. Evolving to Other Environments, Other Platforms

Figaro applications continue to be run directly from the command line, but other ways of running them are now available. It is possible to link all the main Figaro applications into a monolithic program that can be run by Starlink's ADAM system. Procedures written in the ADAM command language, ICL, can then trigger the execution of Figaro applications from that monolith. Since ADAM was designed from the start to be a data acquisition system, this provides a means of incorporating Figaro applications into a data taking environment. This required the writing of a different main program to call the Figaro application subroutines, and a new version of the Figaro parameter system that made use of the standard ADAM parameter system was needed, but that was all.

Another interesting way of running Figaro applications is provided by the 'Callable Figaro' system. This allows a program to be linked with the Figaro applications in such a way that it can make a call such as:

```
CALL FIGARO ('IADD IMAGE1, IMAGE2, RESULT')
```

The character argument passed to the FIGARO routine can be any Figaro command. This allows Fortran programmers to construct very complex sequences of Figaro operations in a way that many find very comfortable. Callable Figaro is not widely used, but is has proven very popular with those that do use it.

At the same time, Figaro has not remained tied to VMS, although it has been slow to make the full changeover to UNIX. At Caltech, Sam Southard has released versions ported to the SUN and to the Convex. The whole Figaro code is being revised in Australia at the moment as a collaboration between AAO, the University of New South Wales, and Mount Stromlo Observatory (ANU). This revision builds on Sam's work to provide a more widely portable Figaro system. Starlink is also involved, and in Edinburgh Horst Meyerdierks has already been able to release a cut-down version of Figaro that runs on SUNs and DECStations under Starlink's portable ADAM system.

Figaro remains a popular data reduction system, probably because of the combination it provides of easy user-programmability and the flexibility of its data structures. Whether its popularity continues will depend on how successfully it manages to make the transition to being a fully portable system.

References

Cohen, J.L. 1987, 'The FIGARO Package for Astronomical Data Analysis' in Instrumentation for Ground-Based Optical Astronomy, 448

Currie, M.J., Wallace, P.T., & Warren-Smith, R.F. 1989, Starlink Standard Data Structures, Starlink General Paper 38

Shortridge, K. 1990, Figaro Programmer's Guide, Anglo-Australian Observatory

Astronomical Data Analysis Software and Systems II
ASP Conference Series, Vol. 52, 1993
R. J. Hanisch, R. J. V. Brissenden, and J. Barnes, eds.

Multi-frequency Data Analysis Software on STARLINK

P. M. Allan

Rutherford Appleton Laboratory, Chilton, Didcot, Oxon OX11 0QX

Abstract. This paper gives an overview of many of the data reduction and analysis packages available from STARLINK. It then goes on to give an example of how they can be put together to form a complete data processing pipeline from the telescope to the published result. Any or all of the software described in this paper can be obtained from the address given at the end of this paper.

1. Introduction

The STARLINK project is unusual in the world of astronomical software in that we provide hardware and manpower to STARLINK sites, as well as software. The implication for software developers is that users have day-to-day contact with their STARLINK site manager who is the first port of call when reporting a bug. This leads to a closer connection between the users and the developers of the software than is the case with other projects.

The hardware bought by STARLINK is currently a mixture of VAXs, Suns and DECstations, although we plan to phase out the VAXs over the next couple of years.

The STARLINK project was set up in the late 1970's to provide facilities for optical image processing. However, over the last decade, our role has expanded such that we now provide packages to handle data over the entire electromagnetic spectrum.

2. What *is* STARLINK Software?

There are several possible definitions of what constitutes 'STARLINK software'. The STARLINK software collection is well defined, but includes all sorts of utilities, such as LaTeX, that we distribute, but have no hand in writing. It this paper, I shall use the term 'STARLINK software' to refer to the programs used for astronomical data processing and the subroutine libraries that are provided to allow users to write their own STARLINK programs. It is a tenet of the STARLINK philosophy that users will want to write their own software on occasions, and that they should submit their best programs to STARLINK for inclusion in the software collection.

3. Major STARLINK Packages

There follows a list of the major STARLINK data reduction and analysis packages. The list is broken down into wavelength regions for convenience, although many of the packages have a much wider range of applicability than this crude categorization implies. Also the one line descriptions necessarily do not do justice to the software.

3.1. Radio

AIPS The de facto radio package from NRAO

3.2. Millimeter

JCMTDR A Figaro-like package for handling JCMT data

SPECX A mm spectral reduction and analysis package

3.3. Infrared

CGS4 Spectral data reduction package

IRAS Image reconstruction from raw IRAS data

IRCAM Image processing for IRCAM data

3.4. Optical

APIG Fitting Interstellar Absorption Line Profiles

CCDPACK Bulk CCD data reduction

DAOPHOT Peter Stetson's photometry package

FIGARO A spectroscopy package that does much more besides

KAPPA Kernel Applications Package—STARLINK's cornerstone image processing and visualization package

IRAF The NOAO package

PISA Position, Intensity and Shape Analysis

PHOTOM Aperture photometry

SPECDRE Spectroscopy facilities not found in Figaro

STARMAN Alan Penny's photometry package

TSP Time series and polarimetry

3.5. Ultraviolet

DIPSO A spectrum analysis program

IUEDR IUE data reduction

IUEDEARCH IUE de-archiving program

USSP The uniform low dispersion archive software for IUE data

3.6. X-ray

ASTERIX A general purpose X-ray data reduction program

HXIS Hard X-ray imaging spectrometer data reduction

There are many other utilities in the STARLINK software collection. These cover statistical analysis, interactive graphics, text processing, programming tools, etc. If you are interested in any of these, contact the STARLINK librarian for more details.

It is immediately apparent from the above list that STARLINK software is gathered from a wide range of sources. For some packages such as AIPS, STARLINK merely acts as the UK distributor and as the first port of call for problems. DIPSO is an example of a package that was written by a UK astronomer and was submitted to STARLINK for inclusion in the software collection. The responsibility for maintaining the package resides with the original author, although STARLINK provides assistance from time to time. The Figaro package was written by Keith Shortridge using some of the STARLINK subroutine libraries, but there have been several add-on packages written by others and STARLINK has an applications programmer who spends a lot of time working on Figaro. Finally there are packages such as CCDPACK, which was written by a STARLINK contract applications programmer to fully comply with the STARLINK programming standard and the code is vetted by the head of applications at RAL before it is released.

4. An Example of Multi-frequency Data Analysis

Consider a hypothetical, but typical, scenario for analysing some data gathered from several telescopes. I have some observations of my favourite active galaxies from MERLIN, from the CCD camera on the WHT and from ROSAT. I wish to produce images of the galaxies in each of the wavelengths and to overlay the pictures. Since I know that the X-rays come from a point source, I will chose to display the optical data as a grey scale, the radio data as a contour map drawn in red and the X-ray data as a false colour image.

The radio data is processed on a Sparcstation 2 using AIPS. Once I am happy with the final result, I will write the data in FITS format for later processing.

The optical data is read in with the STARLINK KAPPA package and is processed with the CCDPACK system on a Sparcstation. This automatically does the bias and flat field corrections for all of my data frames. It also generates and processes the variance of the data, which will become important later on.

The ROSAT data is processed using the ASTERIX package. For this task I will use a VAX as ASTERIX is not yet available on Unix[1].

I now have the radio data in FITS format on an Exabyte, and the optical and X-ray data in STARLINK HDS files. I re-sample the X-ray data on my DECstation using KAPPA so that the optical and X-ray images have the same

---

[1]A port is expected by the date of publication.

scale and centre and display them on the screen of my workstation using KAPPA. The fact that the HDS files containing the optical data were created on a Sun and the X-ray data were created on a VAX is irrelevant as the Sun and VAX NFS serve their disks to my DECstation and HDS will automatically perform any data format conversions that are necessary. Next I read in the radio data using KAPPA, resample it and overlay the contour map on top of the optical and radio data. To finish the initial processing, I use KAPPA to get a hard copy of the display window and send the picture to reprographics to get some slides made up for next week's seminar.

On examining the data more closely, I discover that there appears to be enhanced optical emission from a region of a galaxy where there is particularly strong radio emission. However, the optical emission is weak. Can I believe it? By looking at the strength of the optical emission and comparing it with the variance of the data, I can obtain a value for the statistical significance of the supposed emission. Astronomers are a cautious bunch and will no doubt not believe me and will ask about systematic errors, but at least I know exactly how many more hours of observing I will need to confirm the detection.

5. Where to Get It

STARLINK software can be obtained from the STARLINK software librarian who has an address of **star@star.rl.ac.uk** (IP address = 130.246.32.1). The entire collection requires over 500 MB of storage space and is usually distributed by tape. However, individual packages can be made available across the network. There is also an anonymous ftp account on **starlink-ftp.rl.ac.uk** that you can use to get hold of many STARLINK packages.

Discussion

Worrall: Do you incorporate much software written by individual astronomers in the general community? If no, what level of support do you provide in verification and documentation, and what has been your experience?

P. Allan: (A partial answer to this question was added to the main body of this talk after it was given. This is a more complete answer).

It is a fundamental tenet of the STARLINK philosophy that astronomers will sometimes need to write their own software, and we encourage them to submit this to STARLINK for incorporation into the software collection. In the early days of STARLINK, most of our software was obtained in this manner. This is less true today, but we still regard this as an important source of software. If an astronomer submits software to STARLINK, then we expect them to provide adequate documentation as part of the submission. We do not have sufficient manpower to give submitted packages major verification tests, but we do try to ensure that they work satisfactorily before we distribute them. Our experience with this form of software is generally quite favourable and some of our most used packages are ones submitted by users.

Hanisch: How do you handle the overlays of optical, radio, and X-ray data? Do you have an automated and consistent method for handling the world coordinate systems associated with data sets from different instruments and observatories?

P. Allan: We do have a system for handling world coordinates, but this is not used by all applications programs. This means that the ease with which you can overlay different data sets varies with the software package that was used to reduce the data. The fundamental problem with completely automating the process is indeed the fact that different observatories provide astrometric information in FITS header records in different ways and it is sometimes the case that there is insufficient information to do the job properly.

Astronomical Data Analysis Software and Systems II
ASP Conference Series, Vol. 52, 1993
R. J. Hanisch, R. J. V. Brissenden, and J. Barnes, eds.

The STARLINK Software Collection

R. F. Warren-Smith and P. T. Wallace

Rutherford Appleton Laboratories, Chilton, Didcot, Oxon., UK.

Abstract. The UK's STARLINK project develops and distributes a collection of software applicable to a wide range of problems in Astronomy; it covers most wave-bands, caters for a variety of instrumentation and ranges from programming tools and libraries through to large packages of applications. This paper briefly describes the STARLINK project as a whole and highlights several active areas of software development and current interest. Details are also given of how to obtain copies of STARLINK software.

1. Introduction

STARLINK is:

- a network of computers used by UK astronomers;
- a collection of software for the calibration and analysis of astronomical data;
- a team of people giving hardware, software and administrative support.

STARLINK is funded by the UK's Science and Engineering Research Council to provide UK astronomers with computers and software for interactive data analysis. At present there are 23 STARLINK sites, serving about 1500 users. The computers, about 200 in number, are interconnected through the JANET network which is used to exchange messages, software and data between sites, and to provide international access. STARLINK also maintains a collection of astronomical software amounting to about 500 Mbytes in size. All STARLINK sites are configured in the same way and run identical software, so that astronomers can move from site to site and find a familiar computing environment.

The multi-user VAX/VMS computer architectures which were the basis of STARLINK throughout the 1980s are rapidly being superseded by networks of UNIX-based workstations. The transition will be complete in 2–3 years and any inconvenience to users is being minimized by releasing UNIX versions of the existing VMS software.

STARLINK's management team is located at the Rutherford Appleton Laboratory near Oxford. This group develops and distributes software and documentation, arranges hardware maintenance and plans and purchases new equipment. Contract staff are also provided at other STARLINK sites to develop applications software, look after the equipment, install new software, and to give on-the-spot expert help to users.

Outside the UK there are well over 100 sites that have some connection with STARLINK, ranging from installations at major UK telescopes, which are STARLINK-compatible and managed like STARLINK sites, to institutions which run STARLINK software.

STARLINK's software collection contains a number of major packages covering a wide range of astronomical data reduction and analysis techniques. At the core of most of these packages is a common 'software environment' known as ADAM, which was originally designed for on-line telescope control and data acquisition at the UK's overseas observatories. ADAM is now paying dividends in the area of data analysis by simplifying programming and support, by ensuring portability of software *and data* between different computer systems, and by allowing astronomers to intermix packages during data analysis.

A few active areas of software development and current interest are briefly described below.

2. Maximum Entropy Reconstruction of IRAS Data (MEMCRDD)

STARLINK has distributed a suite of software for analysis of data from the Infra Red-Astronomical Satellite (IRAS) for some time, but a new program MEMCRDD has recently been added to create images from time ordered IRAS detector data using a maximum entropy technique.

MEMCRDD is based on the Gull and Skilling MEMSYS package and can improve the resolution of IRAS images by a factor of 2 to 3. It also provides estimates of the uncertainties in integrated fluxes derived from the improved images and is particularly suited to resolving merged sources.

3. Interactive Data Plotting (PONGO/KAPPA)

Development is being carried out on a new PGPLOT-based interactive data plotting package called PONGO, designed to complement the image-display and graphical applications already present in other STARLINK packages.

PONGO has all the features you might expect for drawing and annotating various styles of graph, but it also has a number of specifically astronomical features built into it. These include the ability to handle astronomical coordinates and to plot positional data using various map projections.

PONGO is also a tool for annotating graphics produced by other applications, a task which it performs with the help of the Applications Graphics Interface (AGI), a database of graphical context information. To give a simple example, a series of images displayed using the Kernel Applications Package KAPPA might be arranged into a grid by first dividing the display surface into separate plotting regions known as "pictures". Communicating via AGI, which holds information about these pictures and their coordinate systems, PONGO could then be used to draw coordinate grids or other annotation on to the displayed images. It can also be used to produce complete plots of its own within selected pictures, leading to a wide range of layout possibilities.

4. Mosaic Generation from Optical/IR CCD frames (CCDPACK)

Development is also underway to substantially enhance the CCDPACK package. As its name implies, this package performs automated reduction of CCD data and the present aim is to extend this automation to the generation of mosaics from CCD images of the sky.

The process being developed starts with automatic star identification and matching, followed by image alignment, then normalization and mosaic formation. A user has only to specify the set of frames to be combined in order to obtain a final mosaic in which each contributing frame is correctly registered and normalized with its neighbours.

Each stage in this process uses optimal weighting techniques based on statistical error information (variances) which form a standard part of the STARLINK Extensible N-Dimensional Data Format (NDF). Variance estimates for the final mosaic may also be produced and can be used by subsequent applications such as aperture photometry routines.

5. Position, Intensity and Shape Analysis (PISA)

The PISA package locates, de-blends, fits and parameterizes objects in a 2-dimensional image. It can perform de-blending using either iso-photal or profile-fitting techniques, and measures intensities using iso-photal, profile-fitting or curve-of-growth analysis. It also contains facilities for producing model data, and for performing non-parametric KNN (k'*th* nearest neighbour) multivariate discriminant analysis to assist in object classification.

6. Spectral Data Reduction (SPECDRE)

SPECDRE is a newly-developed package for spectral data reduction which has been designed to support multi-wavelength working and to exploit features of the STARLINK NDF data format. One of its main purposes is spectral fitting, particularly of infra-red spectra.

A novel feature of SPECDRE is its ability to processes statistical error information (variances) associated with the data. In addition, it is also able to maintain information about inter-pixel *co-variances* produced during re-binning. This means that optimal weighting and accurate significance estimates for fitted parameters are possible even after re-binning or merging of spectra. SPECDRE is also able to handle missing data ("bad" pixels) and up to 8 bits of *pixel-quality* information.

SPECDRE stores information (e.g., about fitted parameters) in an *extension* within the NDF data file. This extensibility feature of the NDF format means that relevant information is retained with the dataset and may be used, for example, to re-create fitted functions at a later date. In fact, SPECDRE goes further by storing fitted parameters within an NDF data object *nested inside* the original one. This means that "images" of the fitted parameters and their errors are available for analysis by other NDF-based utilities.

7. Graphical User Interface Development (PHOTOM)

Work is also underway on a prototype Motif-based graphical user interface for the PHOTOM interactive aperture photometry routine.

8. How To Obtain STARLINK Software

The work highlighted above represents only the most active areas of development in a software collection which currently comprises some 122 items. Almost all of this collection is available for VAX/VMS systems and a large (and growing) fraction is also available for SUN Sparcstations running SunOs and for DECstations running Ultrix. Support for further operating systems is under consideration.

Further information about the Starlink Software Collection, additional software and hardware requirements, and copies of the software itself may be obtained on request from the STARLINK Software Librarian at the following address:

Room 1-28, R68,
Rutherford Appleton Laboratory,
Chilton, DIDCOT, OXON, OX11 0QX
United Kingdom.

Phone: +44 235 821900 x5363

E-mail contact (UNIX):

ussc@star.rl.ac.uk (Internet)
RLVAD::USSC / 19457::USSC (SPAN)
USSC@UK.AC.RL.STAR (JANET)

E-mail contact (VMS):

star@star.rl.ac.uk (Internet)
RLVAD::STAR / 19457::STAR (SPAN)
STAR@UK.AC.RL.STAR (JANET)

Acknowledgments. Thanks are due to Roger Stapleton, David Berry, Malcolm Currie, Paul Rees, Paul Harrison, Peter Draper, Horst Meyerdierks, Nick Eaton, Martin Bly and the many others who are primarily responsible for the work reported here.

Astronomical Data Analysis Software and Systems II
ASP Conference Series, Vol. 52, 1993
R. J. Hanisch, R. J. V. Brissenden, and J. Barnes, eds.

ROSAT Data Analysis with EXSAS

H. U. Zimmermann, T. Belloni, C. Izzo, P. Kahabka, O. Schwentker

Max-Planck-Institut für Extraterrestrische Physik Giessenbachstrasse D-8046 Garching, Federal Republic of Germany

Abstract. EXSAS, based on the ESO–MIDAS image processing system, is a large software system for interactive analysis of X-ray and XUV data from the ROSAT satellite. A multitude of typical applications and utilities in the areas of data preparation, correction of instrumental effects, spatial, spectral and timing analysis have been implemented in a coherent context. The package is portable and may be requested from the ROSAT Scientific Data Center at Garching.

1. The EXSAS Environment

EXSAS—the EXtended Scientific Analysis System—has been designed and developed with an effort of 20 staff years by the ROSAT Scientific Data Center (RSDC) at the Max-Planck Institut für Extraterrestrische Physik (MPE) for the reduction of X-ray data in general, with particular emphasis on the analysis of data from the ROSAT X-ray and XUV instruments. EXSAS comprises a large collection of application modules as typically required in analyzing data of this wavelength regime. The package is embedded in the well-known astronomical image processing system ESO–MIDAS, developed, maintained and distributed by the European Southern Observatory (ESO). EXSAS will normally be run interactively on workstations, but may also be used in batch mode.

In the following we list some of the basic features and design criteria of the EXSAS/MIDAS environment that enable observers to effectively work with the system.

- EXSAS/MIDAS is portable; the package runs both on different UNIX systems and VAX/VMS systems.
- ESO-MIDAS has a fairly wide distribution and acceptance within the astronomical community in Europe, but can also be found on other continents. The "Deutsches Astronetz" selected it as its Standard Image Processing System some years ago. Thus, EXSAS can be installed with minimum effort at many astronomical sites.
- EXSAS/MIDAS uses only standard file formats (tables, images and in a few cases ASCII files) and strongly standardized and documented sets of interface routines for all internal and external input/output operations.
- All EXSAS applications are written and documented in Standard Fortran 77, a language that is well known to astronomers. This enables the user

to understand existing software modules and, if necessary, change them according to specific needs.

- The ESO–MIDAS environment contains a number of packages that are directly used in EXSAS like extended image and graphic presentation/manipulation packages and general fitting and statistical applications.
- A special EXSAS header, read and updated by each application, maintains general information on the origin, the history and the parameter space of each dataset.
- All application packages have an instrument-independent structure. Therefore the adaptation of further instruments is fairly easy to apply.
- Besides on-line help facilities and tutorials, a comprehensive EXSAS User's Guide supports the usage of the system.
- On the basis of these standards, it is easy to implement the user's own software into the system.

2. Transferring Data into EXSAS

To maintain data distribution independent from the specifics of different operating systems, all ROSAT data distributed by both the ROSAT Scientific Data Center and the German XUV Data Center at Tübingen (AIT) are delivered as a so-called ROSAT Observation Dataset (ROD) in FITS format (Flexible Image Transport System: a data transfer standard that is operating system independent and has been widely accepted by the astronomical community). A ROD consists of more than a dozen tables and images containing all primary and auxiliary data needed for detailed analysis: photon event data, instrument housekeeping and quality parameters, attitude and orbit files, calibration data and also selected results from a Standard Analysis performed on the whole dataset at the Data Centers.

3. EXSAS Application Packages

EXSAS applications have been grouped into 4 main packages:

Data Preparation

The package comprises

- Selection and binning of photon event data with the following properties
  - Selection with respect to basic spatial constraints (box, ring, sector) which can be combined in defining any spatial region of interest
  - Selection with respect to arrival time and pulse height; time selection according to the actual count rate or other time dependent quantities
  - Update of selection history in file header

  - Counting of selected events
  - Binning in time (lightcurves), in pulseheight (spectra) or spatially (profiles, images)
  - Any combination of selection and binning criteria via a simple command and executed in a single step; simultaneous output of several products

- Instrument Correction including
  - Vignetting, or any other similar detector position dependent instrument response
  - Dead time, filter transmission, photon loss due to point spread function effects
  - Computation of exposure maps
  - Application to spectra, light curves, profiles and images, with different degrees of precision

- Spectral manipulations allowing
  - Spectral rebinning by explicit definition or automatic algorithms
  - Background offset corrections; removal of particle background
  - Count rates and error evaluation; algebraic operations

- Utilities for data presentation
  - Display of final (corrected) results; true colour images; spatial distribution of photon energies

Spatial Analysis

- A sophisticated package for detection of point-like and moderately extended sources in images or photon event data has been implemented providing
  - Source detection in image data using a local background
  - Source detection in image data using a smoothed background
  - A maximum likelihood method to evaluate best estimates for the position, intensity, extent and corresponding errors using the information of all individual photon events involved

- Ring and box integrated profiles from image and photon event data, including vignetting correction

- Utilities for coordinate manipulation

- Utilities for data presentation

Spectral Analysis

Functionality of this instrument independent analysis package includes

- Construction of a wide range of model spectra (by free combination of standard and/or user defined models) and comparison or fit to an observed spectrum
- Calculation of corresponding errors using either the covariance matrix or a chi-square grid search method
- Determination of photon and energy fluxes, luminosities, etc.
- Standard plotting of results on fits, error ellipses, chi-square contours

Timing Analysis

The package contains

- Power spectra calculation including automatic peak detection
- Auto-Correlation and Cross-Correlation methods
- Period folding with barycentric correction of photon arrival times
- Different statistical tests on source variability
- Standard plotting and display facilities

4. General Support for EXSAS

The EXSAS software can be requested from the ROSAT Scientific Data Center. The only prerequisite is the access to an ESO–MIDAS installation (MIDAS will be distributed only by ESO). At the data center EXSAS is presently used by about 80 scientists. In addition the EXSAS software has been distributed to more than 50 institutes all over the world. The ROSAT Data Center provides full support and service for all EXSAS packages, updating and extending the functionality in regular intervals. An EXSAS Users Guide complements the internal documentation (help facilities, tutorials). ROSAT observers may also come to the ROSAT Scientific Data Center to evaluate their data there. For this purpose a number of image processing workstations and some visitor service is provided by MPE.

Discussion

Dorland: Can EXSAS use GSFC data?

Zimmermann: Yes.

Bloch: How easy is it to add other instruments' characteristics to EXSAS?

Zimmermann: The following tables must be provided.

- for spectral analysis – a detector response matrix and effective area table are necessary
- for spatial analysis (only imaging telescopes) – a subroutine or table for the point response function is necessary and some default settings in the parameter files for source detection have to be adapted

Astronomical Data Analysis Software and Systems II
ASP Conference Series, Vol. 52, 1993
R. J. Hanisch, R. J. V. Brissenden, and J. Barnes, eds.

PROS: An IRAF Based System for Analysis of X-ray Data

M. A. Conroy, J. DePonte, J. F. Moran, J. S. Orszak, W. P. Roberts, and D. Schmidt

Smithsonian Astrophysical Observatory, Cambridge, MA 02138

Abstract. PROS is an IRAF based software package for the reduction and analysis of X-ray data. The use of a standard, portable, integrated environment provides for both multi-frequency and multi-mission analysis. The analysis of X-ray data differs from that of optical data due to the nature of the X-ray data. The scarcity of data, the low signal-to-noise ratio and the large gaps in exposure time make data screening and masking an important part of the process.

1. Introduction

IRAF/PROS provides a complete end-to-end system for X-ray data analysis (Tody 1986; Worrall et al. 1992). PROS was developed for the analysis of data from the ROSAT and *Einstein* X-ray missions, but many of the tasks can be used on data from other missions. The AXAF Science Center (ASC) will extend this system to support the AXAF mission. In addition to the PROS software features, we will discuss the design philosophy, development environment and architecture used by PROS to generate a portable, multi-mission software analysis environment.

2. Development Environment

PROS is developed and supported by the SAO/ROSAT Science Data Center (RSDC). The project team is required to provide support for current users, to develop new tasks and updated calibration data, and to generalize the system for additional missions. All of these considerations contributed to the organization of the PROS project.

PROS has been maintaining a semi-annual build cycle, producing major releases in April and October of each year. Between releases, patch updates are issued as necessary to provide up-to- date calibration information, or to address serious bugs. The group provides *hotline* support for all users. Contact is most frequently via e-mail, but conventional mail and telephone contact are also supported. Each problem report is investigated and an official bug report is made to the *prosbug* database. The report is reviewed, assigned a priority, and ultimately scheduled for a software upgrade. Meanwhile, when possible, a work-around solution is provided to the user. Problems that have a significant impact on other users are collected and reported to users via the *Hints and*

Pointers e-mail service. Bug fixes and new tasks are made available in-house in a *beta-test* account. This test system is also available to users of the RSDC.

Each task in PROS is created by a development team consisting of one programmer, one technical staff member and one scientist. The scientist is responsible for providing the scientific specification for the task. The programmer implements the task. The technical staff assists the programmer and the scientist in testing by developing an automated test script. The test script for each task is incorporated into the system of automated test scripts which are exercised on the *beta-test* version of each software release.

To simultaneously support users, testers and developers, PROS maintains three separate IRAF/PROS accounts in-house. The primary account is the official IRAF and PROS current release. The second account is the *beta-test* account for use by the technical staff and the RSDC scientists. The third account is the development account used by the programming staff to implement bug fixes and develop new tasks. The programming account uses the public domain software, Revision Control Software (RCS), to log all modifications to existing code and on-line help files.

This architecture is designed to provide a stable environment for data analysis, while at the same time allowing rapid turn-around of problems and quick feedback on new tasks. It also provides a comprehensive record of both user-feedback and programmer updates and changes. These features should ensure a smooth continuation of the IRAF/PROS system into the ASC, which has chosen it as the basis for their analysis system.

3. FITS Support

IRAF/PROS has a complete set of FITS readers/writers to ensure that data be importable into PROS and that results be exportable to other systems. The event-lists for both the US ROSAT data and the *Einstein* data are provided in FITS bintable format. The European ROSAT data are provided in FITS table format. The **fits2qp** task converts any of these formats into an IRAF/QPOE file for use within PROS. Additional data, in FITS image and FITS table format, can be read with the TABLES **strfits** task. An attempt to define standard FITS templates for X-ray data has been undertaken to simplify this process in the future (Corcoran et al. 1993).

PROS output files are in QPOE, *image* or *table* format, each of which is convertible to FITS. The task **qp2fits** will write the QPOE file to a FITS bintable file, while the TABLES task **stwfits** will convert the other formats to FITS for export to other systems.

4. Data Structures

X-ray data have several properties that require special software support. The data files have large dimensions, but very sparse data. Users need to analyze only the data which are of interest, often limited to a small section of the observed field. Also, the data are acquired under constantly varying observing conditions. Therefore, it is important to record the precise observing time and to allow users to define additional intervals for exclusion (Conroy et al. 1992).

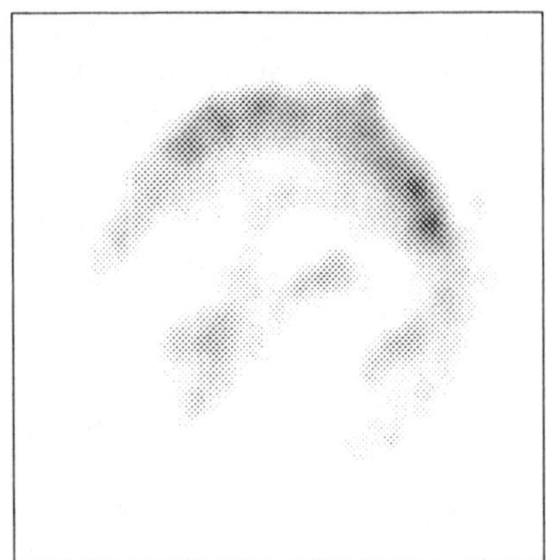

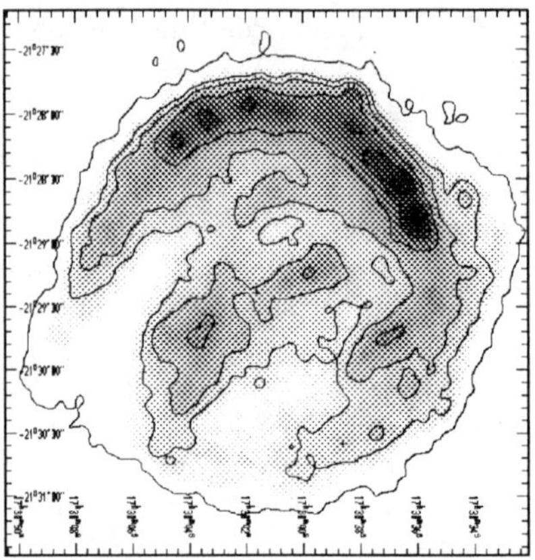

Figure 1. (left) A ROSAT/HRI observation of the Kepler Supernova Remnant, restored with the **lucy** deconvolution task, courtesy of J. P. Hughes and P. O. Slane. (right) A VLA 21 cm. observation of the Kepler Supernova Remnant with the ROSAT contours superposed, courtesy of John Dickel (Dickel et al. 1988).

4.1. Event List Data

PROS is designed to work on event-list data. However, it is essential that this event-list also be interpretable as an *image* array. The IRAF/QPOE file was designed to meet this need. It stores the complete event-list data, but it is recognized by all IRAF tasks that accept an input *image*. The events are automatically converted to an *image* array within the IRAF environment, with a user-specified resolution. The *event* format for the QPOE file is self-defining, allowing the QPOE file to be mission independent. The PROS QPOE file is compatible with the Einstein, ROSAT, ASTRO-D and EUVE data files.

Using the same IRAF tasks with QPOE files as with *images* means that multi-wavelength analysis is easy within the IRAF environment. Optical *images* from Hubble Space Telescope (HST), radio *images* from VLA radio observations or X-ray data can all be displayed using exactly the same procedures. Figure 1 shows the use of IRAF/PROS to display results from two different wavelengths.

4.2. Filtering

PROS supports two types of data filtering. The first is the built-in IRAF QPOE filtering that allows users to automatically filter on any of the event attributes. PROS has implemented a second filtering scheme, that extends the allowed attributes to the Temporal Status Conditions. These include most of the instrument and satellite housekeeping parameters, as well as several statuses such as aspect quality, background count rate, and viewing geometry.

4.3. Masking

PROS makes use of the IRAF Pixel Mask (PLIO) to do the spatial selection of photons. Like the QPOE file, the IRAF PLIO file is accepted as an *image* by all

IRAF tasks. The PROS team has developed the *regions* interface to facilitate the specification of the desired mask (Mandel et al. 1993). Other *image* utilities, such as **isoreg** and **imcalc**, can also be used to facilitate spatial mask creation.

5. Scientific Analysis Tasks

In addition to the environmental and system features, the PROS system consists of scientific analysis tasks specific to X-ray data. Many of these tasks rely on calibration information that is mission and instrument specific. PROS explicitly supports the ROSAT and *Einstein* missions by supplying the necessary calibration files for their analysis. As PROS is extended to support other missions, such as AXAF, the design will evolve to make extension to new instruments definable by the user.

All display and graphics tasks will accept *image* files as well as QPOE and PLIO files. In addition all graphics can be superposed on the TV display (Eisenhamer 1992). A brief overview of the PROS analysis tasks follows.

tv display Tasks **display** and **xdisplay** will produce a TV display of the data.

sky grids The **imcontour** task (DePonte & Worrall 1992) calculates and graphs the iso-intensity areas of the images and displays them on a skygrid.

coordinates Support for the World Coordinate System (WCS) is provided in all the IRAF and PROS tasks. PROS provides additional interfaces to facilitate conversions, including an interactive mode from the *image* display.

graphics All non-image output data files from PROS analysis are produced in TABLE format which can be graphed either with the TABLES **sgraph** task or with the Interactive Graphics Interpreter **igi** (Levay 1992).

source detection The **detect** package is designed to perform Maximum Likelihood Source detection on data exhibiting Poisson statistics. It uses a signal-to-noise threshold calculation (DePonte & Primini 1993).

PRF modeling The **imcalc, immodel** and **imsmooth** tasks provide the ability to generate complex Point Response Function model images that can be convolved with observations.

data extraction The **imcnts** task is a utilitarian tool used to extract background subtracted counts from complex regions.

timing corrections The **timcor** package provides the conversions from spacecraft clock to UTC and calculation of the barycenter timing correction.

periodic analysis The tasks **ltcurv** and **fft** provide general capabilities to examine periodic data. The **period** and **fold** tasks include a provision for a decaying period. The **qpphase** task generates a QPOE files with an additional event attribute, *phase*, that then allows the data to be split according to phase (Manning et al. 1993).

spectrum extraction The **qpspec** task allows users to extract a background corrected spectrum from a QPOE file for use in PROS or for export to other analysis systems.

model specification and fitting PROS has a flexible spectral model specification language which allows multi-component model fitting. Also, the **fit** task allows fitting of multiple data sets.

flux conversion Fluxes for any object can be calculated from the **xflux** task.

Acknowledgments. We are grateful to Belinda Wilkes, Martin Elvis, Pepi Fabbiano, and Chris Fassnacht, for ROSAT, HST and VLA data files provided for illustrations. We are also very grateful to Paul Martenis, Kathy Manning, David Alexander, Steven Guimond and Dave Borden for their expert assistance in all aspects of PROS development and support. PROS is partially supported by NASA contracts NAS5-30934 (RSDC) and NAS5-30751 (*Einstein*). PROS is available at no charge by contacting rsdc@cfa.harvard.edu, (6699::RSDC, rsdc@cfa) or 617-495-7148.

References

Corcoran, M.F., Pence, W., White, R., & Conroy, M. 1993, this volume

DePonte, J., & Worrall, D. 1992, in Astronomical Data Analysis Software and Systems I, A.S.P. Conf. Ser., Vol. 25, eds. D.M. Worrall, C. Biemesderfer & J. Barnes, 334

Dickel, Sault, Arendt, Matsui, & Korista 1988, ApJ, 330, 254

Eisenhamer, J.D. 1992, in Astronomical Data Analysis Software and Systems I, A.S.P. Conf. Ser., Vol. 25, eds. D.M. Worrall, C. Biemesderfer & J. Barnes, 331

Levay, Z.G. 1992, in Astronomical Data Analysis Software and Systems I, A.S.P. Conf. Ser., Vol. 25, eds. D.M. Worrall, C. Biemesderfer & J. Barnes, 337

Mandel, E., Roll, J., Schmidt, D., VanHilst, M., & Burg, R. 1993, this volume

Manning, K.R., Conroy, M.A., DePonte, J., Moran, J.F., Primini, F.A., Seward, F.D., & Aschenbach, B. 1993, this volume

Tody, D. 1986, in Instrumentation in Astronomy VI, SPIE, 627, part 2

Worrall, D.M., Conroy, M., DePonte, J., Harnden, Jr., F.R., Mandel, E., Murray, S.S, Trinchieri, G., VanHilst, M., Wilkes, B.J., et al. 1992, in Data Analysis in Astronomy IV, eds. V. Di Gesu et al. (Plenum Press), 145

Discussion

Unknown: How many sites have PROS? How large is the package?

Conroy: PROS currently has 113 registered sites, running under several machine architectures. PROS has approximately 225,000 lines of code.

Astronomical Data Analysis Software and Systems II
ASP Conference Series, Vol. 52, 1993
R. J. Hanisch, R. J. V. Brissenden, and J. Barnes, eds.

The ALEXIS Data Processing Package: An Update

J. J. Bloch, B. W. Smith, and B. C. Edwards

Astrophysics and Radiation Measurement Group, Los Alamos National Laboratory, Los Alamos, NM, 87545

1. Introduction

The ALEXIS experiment (Array of Low Energy X-ray Imaging Sensors; Priedhorsky et al. 1990) is a mini-satellite containing six wide angle EUV/ultrasoft x-ray telescopes. Its purpose is to map out the sky in three narrow (5%) bandpasses around 66, 71, and 93 eV. The 66 and 71 eV bandpasses are centered on intense Fe emission lines which are characteristic of million-degree plasmas such as the one thought to produce the soft x-ray background. The 93 eV bandpass is not near any strong emission lines and is more sensitive to continuum sources. The mission will be launched on the Pegasus Air-Launched Vehicle in early 1993 into a 400-nautical-mile, high-inclination orbit and will be controlled entirely from a small ground station located at Los Alamos. The project is a collaborative effort between Los Alamos National Laboratory, Sandia National Laboratory, and the University of California-Berkeley Space Sciences Laboratory.

The six telescopes are arranged in three pairs. As the satellite spins twice a minute they scan the entire anti-solar hemisphere. Each f/1 telescope consists of a spherical, multilayer-coated mirror with a curved, microchannel plate detector located at the prime focus. The multilayer coatings determine the bandpasses of the telescopes. The field of view of each telescope is 30 degrees with a spatial resolution of 0.5 degree, limited by spherical aberration.

The data processing requirements for ALEXIS are large. Each event in one of the six telescopes is telemetered to the ground with its time of arrival and position on the detector. This information must be folded with the aspect solution for the satellite to reconstruct the direction on the sky from which the photon came. Because of the way the six telescopes scan the sky, the effective exposure calculation is also very compute-intensive. ALEXIS may generate up to 100 megabytes of raw data per day, which are converted into a gigabyte per day of processed data. The entire analysis system is built on a set of SPARCstation platforms.

2. Software Overview

While the processing job for ALEXIS is sizable, the programming staff is small. We chose Research Systems Incorporated's IDL package as our software development platform because it allowed us to maximize our programming efficiency. IDL was used from the start of instrument development through flight. We use IDL as a top-level executive for the processing tasks (replacing Unix shell scripts), as a device-independent graphics engine, as a database manager, and as a final data manipulator. IDL routines spawn special-purpose C programs to perform detailed telemetry deconvolution and other specialized functions.

2.1. ALEXIS Data Streams and the End-to-End Philosophy

Early in the ALEXIS project, a uniform standard for all binary data files and streams was adopted. This uniform data standard allowed us to adopt an end-to-end system test and development philosophy. Simulation software for the ALEXIS instrument produced data streams just as the flight system does. When we tested individual ALEXIS telescopes in the laboratory, the ground test equipment (GSE) also generated data streams that looked as they would in flight. In this way, software developed to analyze experiment simulations could later be used for instrument testing, and then for flight operations, with little or no modifications. This greatly aided our software development efforts. We did not have to re-write existing software to match new data formats that could have arisen at each phase of experiment planning and integration.

Figure 1 shows the result of a 12 hour simulation of ALEXIS experiment operations. The raw photon events from one of the six telescopes are binned onto a Hammer-Aitoff map projection. The scale shows counts. The map was produced with IDL software designed for flight use.

2.2. Use of IDL

The ALEXIS experiment will generate up to 100 Megabytes of data per day that must be automatically processed and reduced. At the project's start, we began writing analysis software from the bottom up. We produced Unix C language "filter" programs, each doing a small piece of the processing, that were designed to be linked together in large Unix pipeline processes. We debated for some time how to tie all the programs together into an automated system for processing, archiving, and plotting the flight data. Our initial choice was to write Unix shell scripts, and use a plotting package such as MONGO for graphics. After a demonstration of the IDL data processing/graphics package, we came to the conclusion that IDL could exceed Unix shell scripts in versatility and provide a device-independent graphics capability as well. IDL would also provide a high-level, array-oriented, data manipulation functionality. Another advantage of IDL was an existing library of astronomy routines that was available from the UIT project at Goddard Spaceflight Center, an effort funded by the NASA Astrophysics Data Program. Almost no work was lost in the transition to IDL. All of the effort had been spent to that point writing the C processing routines, and they were used unchanged in the IDL environment.

Currently, our software design has control or "Glue" routines in IDL at the top level. These, in turn, call specialized IDL functions which perform specific data processing tasks. These functions in turn spawn Unix pipelines of C-language data filters to do the majority of the telemetry data manipulation. Temporary files or output Unix pipes produced by these spawned processes can be accessed easily within IDL to obtain the transformed data for producing the final results.

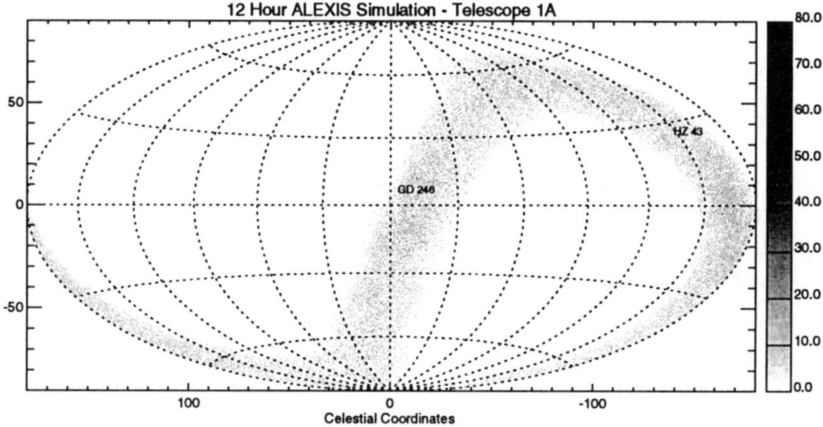

Figure 1. 12 hour simulation for an ALEXIS telescope displayed with IDL software. The greyscale legend is marked in raw counts. Map pixels are 0.5 degree wide. The map has no effective exposure corrections. Two point sources evident in the simulation results are marked.

2.3. Current Status

Our IDL and C software package for ALEXIS continues to evolve as we gain experience from analyzing simulated data sets and supporting mission simulations. Our experience with mission simulations, (where the spacecraft is fooled into thinking it is on orbit and communicates via an RF link to the ALEXIS ground station), has shown that it is easy to generate too much paper as part of the production data analysis. We have been striving to make the production software smarter, so that plots of housekeeping values are only generated when they are absolutely necessary. We have found the combination of IDL and C to be flexible in re-configuring analysis codes on a very short timescale.

Acknowledgments. This work was supported by the Department of Energy.

References

Priedhorsky, W.C., Bloch, J.J., Cordova, F., Smith, B.W., Ulibarri, M., Chavez, J., Evans, E., Siegmund, O.H.W., Marshall, H., Vallerga, J., & Vedder, P. 1990, in Extreme Ultraviolet Astronomy, eds. R.F. Malina & S Bowyer (New York: Pergammon Press)

Astronomical Data Analysis Software and Systems II
ASP Conference Series, Vol. 52, 1993
R. J. Hanisch, R. J. V. Brissenden, and J. Barnes, eds.

The IDL Astronomy User's Library

W. B. Landsman

Hughes STX Co., Code 681, NASA/GSFC, Greenbelt, MD 20771

Abstract. The IDL Astronomy User's Library is an anonymous FTP site containing over 400 astronomy related IDL procedures. I provide a brief overview of the history and philosophy of the Library, as well as a summary of the available procedures.

1. Introduction and History

IDL (Interactive Data Language) is a commercial[1] programming, plotting, and image display language, which is available for most Unix workstations, Vax/VMS, Microsoft Windows, and Convex supercomputers. IDL has found fairly wide use in astronomy, especially in the fields of solar and planetary astronomy, and in the analysis of data from space-based telescopes. IDL is used by astronomers who prefer its "hands-on" approach to data, and who find it far easier to program than either FORTRAN or C. Unlike many other interpreted languages, the programming ease of IDL usually does not come at the expense of computing speed or versatility.

Several design features of IDL contribute to its combination of power and ease of use. Every statement at the command level is also a valid programming statement and vice-versa, and the user always has immediate access to print or plot any IDL variable. A vectorized compiler allows IDL code to often be considerably shorter as well as easier to read than the equivalent algorithm implemented in FORTRAN. The most common spectral and image processing operations, such as convolution or median filtering, are built-in functions. Other features of IDL include an OpenWindows/Motif widget interface, Z-buffer graphics, dynamically allocated structure variables, and dynamic linking to C or Fortran executables.

Although IDL is a natural language for spectra and image processing, it does not contain any software specific to astronomy. As more astronomers began to use IDL for their data analysis, an obvious need developed for a central location of general astronomical software written in IDL, to avoid duplication of effort. In May 1990, I obtained a 3 year grant under the NASA Astrophysics and Software Aids program to develop an IDL Astronomy User's Library (AUL). The co-investigators on the original proposal were scientists working on several different space missions—GRO, IUE, ASTRO/UIT, COBE, and HST—which include IDL among their data analysis tools.

---

[1]Research Systems Inc., Boulder, CO

2. Astronomy Library Configuration

The AUL is not meant to be an integrated package but rather is a collection of procedures from which users can pick and choose (and possibly modify) for their own use. The procedures of the AUL are available as either individual ASCII files or collected in a .tar file from the anonymous FTP account on the machine idlastro.gsfc.nasa.gov. A duplicate version of the Library exists on a VMS machine accessible via DecNet.

All aspects of the AUL—including system management, programming, and user support—are run by Frank Városi (HSTX) and myself on a part-time basis. One consequence of this limited manpower is that the AUL is very dependent on the user community for program submission and for notification of programming or documentation errors. More than half the procedures currently in the AUL were written by outside users.

Another consequence of the limited manpower is that the AUL is continuously updated and does not have "versions". For example, bug corrections are implemented in all files as soon as they are found. Users can monitor the additions and modifications to the AUL by examining a "news.txt" file which lists all changes to the AUL in chronological order. In addition, news about major changes to the Library is sent every 3–4 months to over 240 users on an e-mail distribution list.

The AUL has a set of minimal programming and documentation standards but no attempt is made to enforce a uniform programming style. We do try to avoid procedure conflicts with existing instrument-specific IDL packages, such as those for IUE (Bonnell 1991) or ROSAT (Reichert 1992).

3. Astronomy Library Contents

The procedures available in the AUL fall under the general categories of (1) FITS I/O, (2) interfaces with other astronomy data reduction systems such as IRAF and STSDAS, (3) astronomical utilities such as coordinate conversion and precession, (4) database software, and (5) general spectra and image processing tools.

AUL procedures exist to convert between disk FITS files and IDL variables, which allows the use of FITS as a native data format. Other procedures will convert between FITS and STSDAS files, which are more appropriate for non IEEE-based machines. The AUL includes a complete FITS binary table package, which was written by Bill Thompson (ARC) in support of the SOHO mission. AUL procedures will also recognize the proposed world coordinate system representation for FITS (Hanisch & Wells 1992), which can be used for coordinate display or overlays, or for image alignment.

AUL procedures are available to read and write STSDAS images and tables, and IRAF OIF (.pix and .imh) files. These procedures allow users to combine the best features of IDL and IRAF/STSDAS while working on the same data files. Limited AUL support is also available to read and update MIDAS images and tables, and to read AIPS images under VMS.

The AUL includes photometry package based on the DAOPHOT algorithms of Stetson (1987). The IDL procedures are neither as sophisticated nor as easy to use as the recent version of DAOPHOT available in IRAF 2.10. However, the IDL procedures are useful for adapting the DAOPHOT algorithms to non-standard uses.

The AUL also includes IDL database software originally developed for the GHRS team (Lindler & Feggans 1988), but also used by the ASTRO/UIT and ALEXIS data analysis groups (Bloch et al. 1991). This software allows the powerful data manipulations tools of IDL to be used on databases derived either from standard catalogs or from the user's own data. Several popular astronomical catalogs from the Astronomical Data Center CD-ROM (Gessner et al. 1991) are included in the AUL formatted as IDL databases.

Additions to the AUL that are planned for the near future include a deconvolution package (Városi & Landsman 1993) and image mosaic software (Városi & Gezari 1993).

Additional information about the IDL Astronomy User's Library can be obtained by contacting the author at landsman@stars.gsfc.nasa.gov.

Acknowledgments. The IDL Astronomy Library is funded under NASA grant NASW-4509 to Hughes/STX.

References

Bloch, J.J., Smith, B.W., & Edwards, B.C. 1991, in Astronomical Data Analysis Software and Systems, A.S.P. Conf. Ser., Vol. 25, eds. D.M. Worrall, C. Biemesderfer & J. Barnes, 502

Bonnell, J. 1991, IUE Newsletter No. 45, 76

Gessner, S.E., Brotzman, L.E., Mead, J.M., & van Steenberg, M.E. 1991, BAAS, 23, 908

Hanisch, R.J., & Wells, D.G. 1992, preprint

Lindler, D., & Feggans, J.K. 1988, BAAS, 20, 709

Reichert, G. 1992, The Rosat IDL Recipes Cookbook

Stetson, P. 1988, PASP, 99, 191

Városi, F., & Gezari, D.Y. 1993, this volume

Városi, F., & Landsman, W.B. 1993, this volume

Astronomical Data Analysis Software and Systems II
ASP Conference Series, Vol. 52, 1993
R. J. Hanisch, R. J. V. Brissenden, and J. Barnes, eds.

GRO/EGRET Data Analysis Software: An Integrated System of Custom and Commercial Software Using Standard Interfaces

N. A. Laubenthal, L. McDonald (Hughes/STX), P. Sreekumar (USRA), D. Bertsch, A. Etienne (Hughes/STX), N. Lal, J. Mattox (Computer Sciences Corporation)

NASA/Goddard Space Flight Center, Code 660, Greenbelt, MD 20771

P. Nolan, J. Fierro

Stanford University, HEPL, Stanford, CA 94305-4085

1. Introduction

The Energetic Gamma Ray Experiment Telescope (EGRET) is the high-energy instrument on the Compton Gamma Ray Observatory (CGRO), which was launched in April 1991. It is designed to be a very low background instrument and provides more than an order of magnitude improvement in sensitivity over the previous experiments such as SAS-2 and COS-B. EGRET views more than half a steradian of the sky at a time, but measures the arrival directions of individual gamma ray photons to an accuracy of a few degrees or better. EGRET's primary goal is to map the gamma ray sky in the energy range of 30–20,000 MeV.

To accomplish this mission, a viewing plan has been established that spans the time of activation through November 1992. The Gamma Ray Observatory Phase 1 observing plan has been divided into approximately 50 viewing periods. During each viewing period, the instrument is pointed in one direction for up to three weeks. The instrument gathers data within approximately 45 degrees of the pointing direction.

Results from data analyzed to date have already provided a wealth of data with numerous surprises. Before the launch of CGRO, there was only one quasar known to be visible at gamma ray energies, namely 3C 273. So far EGRET has seen 15 additional quasars, many characterized by maximum power output at gamma ray energies with a few showing strong time variability of the order of a few days to weeks (Fichtel et al. 1993). This has raised more questions than answers towards our understanding of the processes that sustain the energy output from these distant sources. EGRET has also detected 4 pulsars (Kniffen et al. 1993) with the most exciting one being Geminga. Geminga was discovered first by the SAS-2 experiment in the mid-seventies as a very bright object at MeV energies with no detectable emission at any other wavelengths and had confounded astronomers until the recent discovery by ROSAT (Halpern & Holt 1992) and subsequently by EGRET (Bertsch et al. 1992) that it is a pulsar spinning at a rate of 4.22 rotations per second. This is the only gamma ray pulsar with no radio counterpart and is yet not understood. In addition, data from EGRET has yielded a variety of information on gamma ray bursts (Schneid et al. 1992), solar flares (Kanbach et al. 1992), diffuse continuum emission from our galaxy as well as the nearby Large Magellanic Clouds (Sreekumar et al. 1992). Further details can be found in the references provided.

2. The Spark Chamber Instrument

EGRET is a spark chamber instrument containing layers of grids of wires that are interspersed with thin metal foils that serve as an interaction medium. The entire assembly is enclosed in a pressure vessel containing mostly neon gas. A gamma ray that enters the detector may interact and produce an electron-positron pair. The interaction of these secondary particles with the gas causes sparks, whose positions are registered in an x-y readout system. The event spark data are made up of pairs of x-z and y-z coordinates, z being the direction along the detector axis. In each view the gamma ray appears as an inverted Y or V. From this information, three-dimensional pictures of the event can be constructed.

3. EGRET Data Base Generation

EGRET data are received at the Goddard Space Flight Center for processing. The initial telemetry processing takes the telemetry bit stream and separates the data packets into more logical and functional data quantities. All of the spark information for each event is written into a separate record that also includes the time of the event, data quality flags, spacecraft position, instrument pointing direction and relevant housekeeping information. Other records created at this stage include housekeeping records and up to 3 types of burst records.

A significant fraction of events telemetered to earth do not satisfy the strict requirements that are established for gamma rays of cosmic origin. A pattern recognition program, aptly named *SAGE* (Structuring and Analysis of Gamma ray Events) makes an association between sparks and the tracks of the electron and positron in two orthogonal views. *SAGE* then uses its many decision criteria to determine if the event is a good gamma ray, a spurious event to be rejected, or a questionable event. These questionable events (20%) are reviewed by trained analysts who decide whether or not the event is acceptable, and in a small fraction of cases, a manual analysis is done. Graphical software has been written by the EGRET software team to display the spark chamber event, to allow the analysts to change the spark classifications and to allow the analyst to classify an event as a gamma ray or not. These edited events are merged back into the EGRET event data base.

The next level data bases are then created from the edited gamma ray event data base. These data bases do not carry forward the individual spark information. They include only the event's arrival direction, time, energy and other analysis information. The Gamma Ray Summary Data Base (SMDB) holds all good gamma ray events. Subsets of the gamma ray data base can be generated using selection criteria to choose only events with specific characteristics such as time intervals, restricted arrival direction in either rectangular regions or a circular region centered on a source location, energy results, telemetry data quality, spacecraft position, analysis status and return codes, and other appropriate criteria. These data bases are called the SELECT data bases.

SMDB files and SELECT data base files contain records for each good gamma ray event with the following information: date and time to a resolution of 8 microseconds, spacecraft inertial coordinates (km), quality information, gamma ray angles, gamma ray arrival direction (earth coordinates), gamma ray arrival direction (in celestial and galactic coordinates), gamma ray energy, and gamma ray energy uncertainty. After a pulsar analysis has been done, information is added including the time correction to the solar system barycenter, barycenter direction, the pulsar binary phase, the pulsar phase at the event time, and the right ascension and declination of the pulsar.

Other ancillary data are needed for scientific data analysis. These data sets are the timeline file and the exposure history file. The timeline file is generated manually from the information about the timeline of the instrument. The time line file identifies: viewing periods, spacecraft pointing directions of the Z and X axes, viewing period start and stop times, times when the instrument is in a special mode not suited for normal gamma ray studies, calibration, earth albedo mode, test intervals, and any other excluded interval (e.g., gas refills, telemetry transmission errors).

The exposure history file summarizes the essential housekeeping information needed to calculate the exposure of a given viewing period. It contains livetimes integrated from a start time up to the point when a mode or aspect change occurs.

4. EGRET Data Analysis and Display

Once these ancillary data sets are available for the particular viewing period to be analyzed or for the particular target being observed over several viewing periods, standard analysis data products can be generated. These include a set of standard skymaps for each viewing period in each of 10 standard energy intervals. The user can also create additional SELECT files or SMDB files containing events for his or her specialized analysis. There is a set of mapping software which will create count maps (*MAPGEN*), intensity maps and exposure maps (*INTMAP*). These maps are stored in FITS format to take advantage of the full complement of tools available in several commercial analysis systems. The FITSIO library (Pence 1991) is used by custom-written analysis routines generated by the EGRET team. The maps can also be scaled and added together (*ADDMAP*). A standard naming convention for the EGRET FITS filenames is employed, for easy communication of the contents of the map: tttttt.vpxxxx.yzzz, where tttttt is "counts", "exposr" or "intens", where xxxx is the viewing period number with leading zeros, where y indicates the file location (s=Stanford University, m=Max Planck Institute for Extraterrestrial Physics, g=Goddard Space Flight Center), and where zzz is a unique count with leading zeros for that viewing period, with numbers 001 through 010 being reserved for standard products.

The mapping programs use a custom-built user interface, which we have called *xdialog*. *Xdialog*, based on the Xaw toolkit from X11, is mouse based, object oriented, and allows field types such as integers (with validation), real numbers (with validation), Boolean fields (with validation), strings (using a text editor), scrollable lists, and multiple choice selections.

The maps can then be displayed with an IDL *SKYMAP* program. This program has a user interface using IDL widgets. The upper left area of the screen has a text widget used to display a log of user commands, as well as messages and results from *SKYMAP*. The lower left area of the screen is used to display FITS filenames which can be read and displayed; this uses a scrollable bar. Along the right side of the screen the skymap color table is displayed. The central area of the screen displays the gamma ray skymap, using the user-defined color mapping. Along the bottom of the display are function buttons to zoom, change the color table, determine the maximum pixel, generate a hardcopy, display contours, change coordinate grids, and display the FITS file header. IDL has been an extremely flexible and powerful tool for the skymap display. Routines from the Goddard IDL Astronomy User's Library and the JHU/APL IDL User's Library were often used to develop 2-D and 3-D imaging programs.

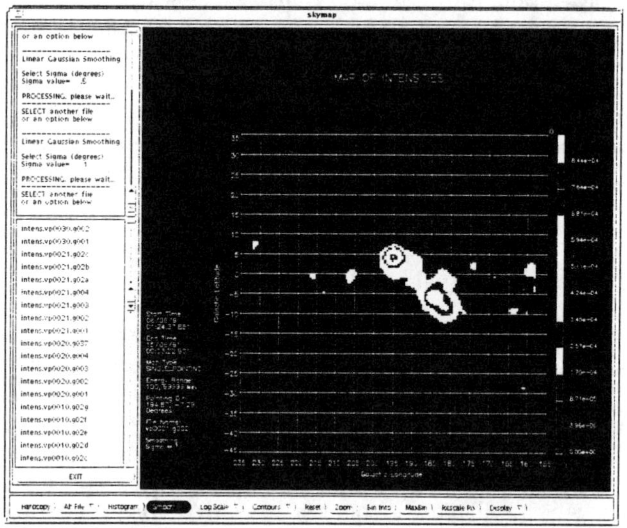

Figure 1. IDL *SKYMAP* display and user interface

The *DIFFUSE* program computes a map of the background in a region of the sky. Another analysis program used to find potential gamma ray sources is the maximum likelihood program, *LIKE*, which uses the diffuse background map and the EGRET FITS skymaps to evaluate the likelihood of a point source being present with the given background. It can also be used to determine the source location, subtract the background and subtract a strong source so that a weaker one can be seen. Often astronomical source data bases such as NED or SIMBAD are used to attempt to identify a gamma ray source candidate. *DIFFUSE* and *LIKE* provide FITS formatted output maps.

Other custom-built software, such as the Stanford University *spectral* and *pulsar* analyses, take advantage of the XView toolkit for display and Postscript output for the color hardcopy. *Spectral* generates an output spectra and *pulsar* generates a light curve.

5. Conclusion

With the use of standards like FITS files and IDL, new ways of displaying and visualizing data can be generated with a minimum of development effort. The following display combines output from various custom-written analysis routines and uses IDL to visualize pulsar activity. The combination of EGRET-specific software, powerful commercial tools for graphics (such as IDL) and data standards like FITS formats has proved to be a flexible and productive approach. Additionally, movies have been produced using the skymap images similar to those below to show pulsars flaring.

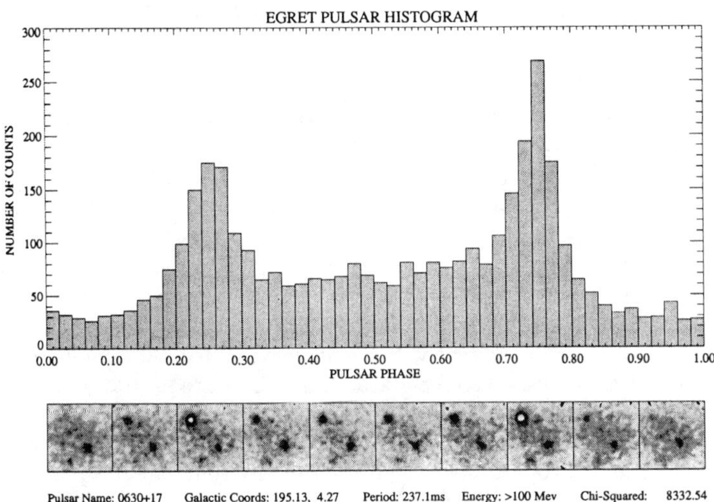

Figure 2. *Pulsar* spectrum display combined with 10 *skymap* displays

References

Bertsch, et al. 1992, Nature, 357, 306

Fichtel, et al. 1993, Proceedings of the Compton Observatory Symposium (Washington University)

Halpern, J.P., & Holt, S.S. 1992, Nature, 357, 222

Kanbach, et al. 1992, A&A, (in press)

Kniffen, et al. 1993, Proceedings of the Compton Observatory Symposium (Washington University).

Pence, W.D. 1991, FITSIO software package and documentation is available from the HEASARC at GSFC

Schneid, et al. 1992, A&A, 255, L13

Sreekumar, et al. 1992, ApJ, 400, L67

Astronomical Data Analysis Software and Systems II
ASP Conference Series, Vol. 52, 1993
R. J. Hanisch, R. J. V. Brissenden, and J. Barnes, eds.

The ISO-SWS Off-Line System

P. R. Roelfsema, D. J. M. Kester, P. R. Wesselius, N. Sym

Space Research Organisation of the Netherlands, Postbus 800, NL-9700 AV, Groningen, The Netherlands

K. Leech

ESTEC, Postbus 299, 2200 AG, Noordwijk, The Netherlands

E. Wieprech

Max-Planck Institut für Ekstraterrestrische Physik, Giessenbachstrasse, Garching bei München, Germany

Abstract. The software which is currently being developed for the Short Wavelength Spetrometer (SWS) of the Infrared Space Observatory (ISO) will be described. SWS is a 2–45 μm spectrometer containing two independent gratings and two Fabry-Perot's. A spectral resolution of ~1000 to ~20000 can be obtained.

Software is currently being developed for the acquisition, calibration and analysis of SWS data. This software is firstly required to run in a mode without human interaction in the "Pipeline," to process data as they are received from the telescope. In the Pipeline the software is required to work properly within the environment designed by the European Space Agency (ESA) for the spacecraft operations. For testing and calibration of the instrument as well as for evaluation of the planned operating procedures the software should also be suitable for use in an interactive environment. Thirdly the software will also be used to characterize the instrument. To be able to use the software modules for all these three purposes, interfaces have been designed such that the modules can be linked with both the Pipeline data interfaces as well as with MIDAS table interfaces for use in Interactive Analysis.

1. Capabilities of the ISO Payload

There are four focal plane instruments on board ISO, all having spectroscopic capabilities. The Short Wavelength Spectrograph (SWS[1]) and the Long Wavelength Spectrograph (LWS) form the core of these spectroscopic capabilities, covering the wavelength range 2.4–45 μm (SWS) and 45–180 μm (LWS, resolu-

[1]The SWS has been developed, built and tested by a consortium of scientists and engineers from the Netherlands (SRON/Groningen, SRON/Utrecht, the Technisch Fysische Dienst in Delft) and Germany (the Max Planck Institut für Extra-Terrestrische Physik in Garching).

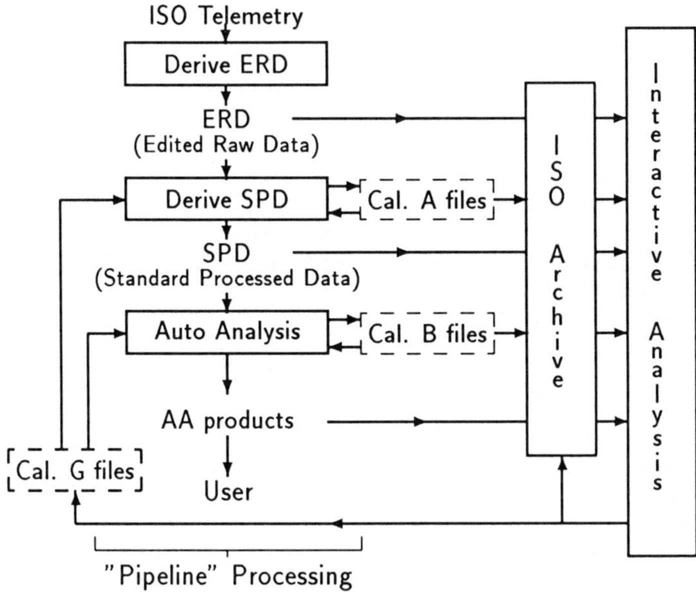

Figure 1. ISO data streams.

tion ~200 or 10^4 with a F.-P.). Furthermore ISO contains a photometer (2–5 and 6–12 μm, with a resolving power of 90) and a camera. A more extensive description can be found, e.g., in Kessler, 1991.

1.1. SWS, the ISO Short Wavelength Spectrometer

SWS consists of two nearly independent grating sections, the shortwave (SW, 2.4–13 μm) and longwave (LW, 11–45 μm) sections, each with its own collimator, grating and re-imaging optics. The gratings are used in the first three orders. The wavelengths are scanned by rotating a mirror, located close to the grating. Each grating section has two different detector arrays of 1×12 elements which are oriented along the dispersion direction. The accuracy of the alignment between the two sections and the four arrays allows simultaneous independent use of both grating sections.

The LW scanner not only selects the wavelength range to be directed to its two detector arrays, but also selects the wavelengths to be directed to the two input mirrors of the two F.-P. interferometers. The F.-P. reflectors are mounted onto a common pair of parallel plates whose distance and parallelism can be varied by adjusting the currents in the three electromagnetic actuators using the calibration tables given by the instrument microprocessor. These are obtained by a calibration procedure, using a SWS internal source located behind a fixed high resolution Fabry-Perot, thus generating a set of ultra-narrow emission lines. The two F.-P. interferometers have each a pair of detectors. Further details about SWS are given by Waters et al., 1993, and references therein.

2. The ISO Data Streams

Since all four instruments will be sending down data at various times which will come through the same channel, ESA has designed a data flow environment to which all instruments must adhere. The lower level architecture, file access, archiving, etc., routines are all provided for by ESA, the instrument groups must provide the instrument specific software and calibration files (e.g., calibration tables and routines). The data flow is indicated schematically in Figure 1.

In the figure three main bodies are indicated; the "Pipeline", the ISO data archive, and Interactive Analysis. The Pipeline is the system which takes telemetry data and generates products that can be shipped to the user. It runs continuously and should keep up with the incoming telemetry data. Many products from the Pipeline are saved in the ISO archive in the form of FITS files (mostly binary tables). In the "Interactive Analysis" (IA) environment *all* these products can be accessed and re-analysed.

In the pipeline the telemetry data are received and converted to the first standard data product, the so called Edited Raw Data (ERD). For SWS the ERD is almost the same as the down-linked data. Next a first series of calibrations is carried out on the ERD, and a next intermediate product, the Standard Processed Data (SPD) is generated. At what point the data are called SPD is left to the discretion of the instrument groups. In the process of converting the ERD to SPD so called "Calibration A" files *may* be used and/or generated. Also "Calibration G" files, derived from, e.g., standard system monitoring may be used. The distinction between these two sets of calibration files is subtle, but important; Cal. A files can be read *and* written by Derive SPD, Cal. G files can only be *read* in the Pipeline, they are *created* by Interactive Analysis. In the next phase the SPD is converted to something which is as close to a calibrated data set as is warranted through automatic processing. In this "Auto Analysis" flux calibration steps are carried out, but in principle for SWS also, e.g., line identification could be done here. Note that the entire Pipeline should run without human interaction, thus all its modules must be very robust.

2.1. SWS Reduction in the Pipeline

The reduction of the raw SWS data is intended to convert the detector output in mV as function of time to flux in Jansky as function of wavelength. As is shown in Figure 1, first an ERD is generated. Following this a number of steps are being carried out to generate an SPD in Derive SPD; 1) check grating and F.-P. setting, 2) identify out-of-range data values, 3) apply AC-correction, 4) apply detector crosstalk correction, 5) detect and correct for glitches, 6) determine the slope of the detector voltage in mV/sec, and 7) determine a wavelength for each sample using the grating parameters.

These wavelength calibrated data are then put into an SPD. From here Auto Analysis (AA) takes over; 1) determine observation type, 2) determine and subtract dark current, 3) determine flat-field correction, 4) apply flat-field, 5) use calibration data to convert from mV/sec to, e.g., Jansky's, and 6) interpolate onto a regular wavelength grid.

After this process the resulting spectrum is ready for further analysis by the astronomer. The output is a binary FITS table, and thus can be read

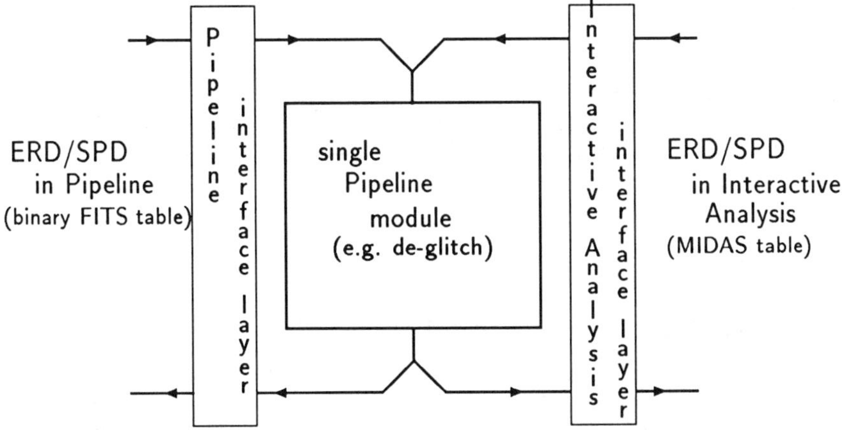

Figure 2. Re-use of Pipeline modules in Interactive Analysis.

into any astronomical reduction package. The different steps listed above are all implemented as different software modules. The actual algorithms can be changed without affecting the interface to the pipeline.

2.2. Interactive Analysis for SWS

Many of the calibration files must be generated and/or tested using appropriate software running under Interactive Analysis (IA). Furthermore IA will be used to evaluate the correctness and the quality of the Pipeline products. Finally IA will be the place for solving what could be called *any other business.* Although no definition is given of what IA looks like in detail, from the above some points are obvious:

- IA must be highly interactive,
- it must be easy to access *all* types of data products from within IA,
- all Pipeline functionalities must be present in IA.

The last requirement has two reasons; firstly to make intermediate Pipeline products which are not stored in the archive available to IA, and, secondly, to be able to test the Pipeline in an interactive environment. The most logical way to make the Pipeline modules available in IA is by building small shells around each module which take care of gathering all variables which are needed for input and output of the module. Some of these variables may be taken from other Pipeline products, others could be specified by the user. Thus by linking one of the Pipeline modules with either the IA interface and shell or with the Pipeline interface it can be used in both systems (see Figure 2).

When designing calibration software for a satellite like for ISO the algorithms will, almost by definition, prove not to be optimal once the satellite is flying. By using the same software modules in IA and the Pipeline in the way sketched above, it will be possible to upgrade them as experience is gained when the satellite is flying.

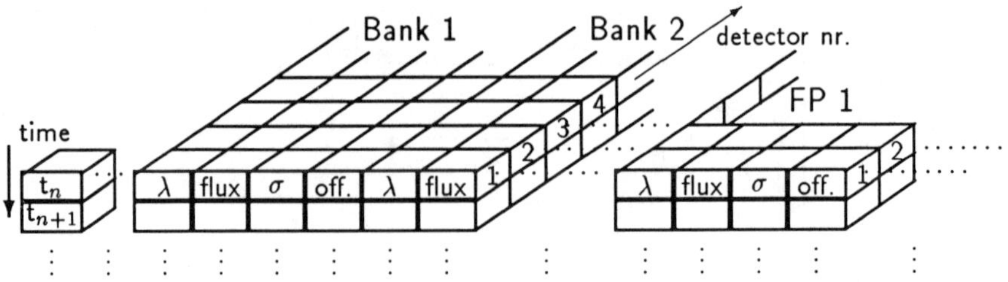

Figure 3. SPD as 3-D table in MIDAS.

2.3. The SWS IA Environment; MIDAS

The SWS software group has investigated a number of existing astronomical image processing packages to decide whether these could be used rather than investing effort in writing a new system. By using an existing package IA is automatically interactive, and the added advantage is obtained that it also allows access to other types of astronomical image processing. This will allow easy comparison of SWS data with data from other instruments. Since many of the ISO (SWS) products are tabular in nature, in the investigation much emphasis has been put on the possibilities of having tabular data in the system. It was found that the MIDAS table system is very suitable for the SWS data.

Given the SWS data as they are stored in ERD and SPD, the new 3-D extensions to the MIDAS table system (Peron, this volume) are found to be extremely useful. In Figure 3 a schematic representation of a SWS SPD as a MIDAS table is given. There are four detector banks for the two gratings (see above), each containing 12 detectors. Furthermore there are two Fabry-Perot detector banks each containing 2 detectors. During standard operations all 52 detectors are continuously read out. Due to the construction of SWS even if the system is setup such that the desired observation spectral range is transmitted only to, e.g., detector bank 1, the other banks may still serendipitously obtain astronomically useful information.

In each integration period for each of the detectors a unique wavelength, a detector output level (which will be converted to a flux density), the rms. noise in that level and a detector offset level are known. Thus for each integration period the data for one detector bank is a $4 \times N_{detectors}$ array (with $N_{detectors}$ 12 or 2). Apart from a time stamp, each "row" in the table will also contain additional information about the instrument, e.g., a status word, the positions of the gratings, etc. Typically there will be several thousand rows in such a table.

References

Kessler, M.F. 1991, Adv. Space Res., Vol. 11, No. 2, COSPAR, "Infrared and Radio Astronomy, and Astrometry", eds. J. Kovalevsky, M.A.C. Perryman, P.R. Wesselius, P.D. Barthel, G.F. Smoot & R.T. Schilizzi, 217

Waters, L.B.F.M., De Graauw, Th., Beintema, D.A., Loup, C., Roelfsema, P.R., Valentijn, E.A., & Wesselius, P.R. 1993, in Infrared Spectroscopy, in press

Astronomical Data Analysis Software and Systems II
ASP Conference Series, Vol. 52, 1993
R. J. Hanisch, R. J. V. Brissenden, and J. Barnes, eds.

PCIPS 2.0: Powerful Multiprofile Image Processing Implemented On PCs

O. M. Smirnov

Institute of Astronomy of the Russian Academy of Sciences, 48 Pyatnitskaya st., Moscow 109017 Russia

N. E. Piskunov

Observatory and Astrophysics Laboratory, University of Helsinki, Tähtitorninmäki, SF-00130 Helsinki, Finland

1. Introduction

Over the years, the processing power of personal computers has steadily increased. Modern 386- and 486-based PCs are fast enough for many serious applications, and inexpensive enough even for amateur astronomers. PCIPS is an image processing system based on these platforms that was designed to satisfy a broad range of data analysis needs, while providing maximum expandability, all on low-end hardware. It features an intuitive graphical use interface (GUI) and a flexible software development environment. Besides a basic processing package, the system has a CCD/Echelle reduction package and a spectral package, and others are in development.

2. General Capabilities

The minimum hardware configuration that PCIPS requires is a PC/AT or compatible with a 287/387 floating-point coprocessor, EGA graphics, 640K of memory and a hard disk. Such a bare-bones configuration is sufficient for elementary processing, especially if the images are relatively small. As tasks become more complicated and images bigger, faster CPUs and more memory can lead to enormous gains in performance. For example, when the Echelle reduction package described later on in this paper was run on a 33 MHz 486 with 8 Megabytes of memory, complex operations with 1200×770 16-bit images (1.8 Megabytes each) became nearly instantaneous.

All the actual image processing is performed by external modules, called *applications.* As new applications are developed, they are easily added to the system. There is also an extensive Application Program Interface (API) which lets users create their own custom applications in C or FORTRAN.

PCIPS supports one- and two-dimensional images of large formats (up to 2 048×2 048 or up to 16 384 pixels long) and a variety of data types: 8-, 16- and 32-bit signed and unsigned integers, 32- and 64-bit (double precision) floating point. The system can freely convert images between different types and sizes to satisfy the requirements of various applications. The images are kept in compressed format in the image database (IDB), which supports directories and storage of variables or other data, such as coordinate scales or color transfer functions. When images are accessed, compression and decompression is performed on the fly and invisibly to the user or application programs.

If expanded memory is available, it is used for intermediate storage of images. Once it is full, the least-recently used fragments are shifted out to the IDB. On a system with little memory, this disk swapping occurs constantly when several images or one big image is accessed. Large memories practically eliminate disk swapping and lead to drastic performance improvements.

3. The User Interface

The user interface for PCIPS is provided by the integrated environment (IE). The IE maintains a mouse- and keyboard-driven menu to load images and start applications, a context-sensitive help system, and a set of graphical buttons and scrollbars used in visualization of images. The basic idea was to implement standard visualization tools as a part of the IE that are always accessible to the user, so that the actual applications are responsible for image processing only.

Two-dimensional images are displayed in 16 colors using a linear or logarithmic color transfer function. To make up for the low color resolution (imposed by the standard PC hardware), the IE has a set of "color buttons" that may be used to dynamically change the color transfer function, zooming into any data range in seconds. One-dimensional images may be plotted by themselves or over other images. Another set of buttons provides such functions as zooming, shrinking (i.e., the averaging of several image points to make one screen pixel, used for displaying very large frames) and hard copy. In addition, PCIPS actually maintains two separate visualization screens which may be used to display different images or different parts of one image. The system can instantly flip the display between the two screens. A "Sync" feature can be turned on so that operations (zoom, etc.) with one screen are automatically duplicated on the other one.

A "mouse zoom" window constantly shows whatever is under the mouse cursor with selectable ($\times 2$ or $\times 4$) magnification. This is very useful for precisely marking points when certain applications require them. The window also tracks the coordinates of the cursor in the displayed image's local coordinate system.

An important feature of the PCIPS user interface is its *consistency*. All applications automatically receive the same GUI that looks like part of the IE, even though they are in fact external programs. This ensures that all of them have the same "look and feel", without any effort on their programmers' part. What's more, the IE's visualization tools remain accessible even while an application is running. This results in a very productive environment, because most components of the system retain their functions at all times, so the user practically never feels himself restrained.

4. The Application Interface

PCIPS has an API called AIS (Application Interface and Services) which allows users to develop their own applications in C or FORTRAN. The basic philosophy of AIS is that an application should be placed in an abstract environment, with no awareness of such entities as users or disk storage or operating systems. Instead, it must contend with images, parameters, points and the like.

This approach helps solve many problems. Its main virtue, though, is that it resolves what we call the user-programmer conflict (Smirnov et al. 1992) by placing both in a comfortable environment and letting PCIPS figure out what to do when a conflict of interests arises (by programmer we, of course, refer only to the application programmer, who reaps all the benefits at the expense of us, the PCIPS programmers). For example, the programmer doesn't want to waste time implementing a GUI which the user expects—fine, PCIPS provides the GUI for the application automatically. Another example: the programmer has cut a few corners and as a result his application supports a limited range of data types, while the user is trying to process an image of an unsupported type. PCIPS helps out by converting the user's image to a supported type before giving it to the application, and perhaps by converting the result back to the original type.

4.1. Requests

All applications obtain input or output images and parameters by generating *requests*. From the programmer's point of view, this is as simple as one function call—certainly no harder than getting them as input from the console, the way it was done long before anybody had heard of menus and mice. PCIPS takes care of the user interface: it shows the request in a window, obtains input from the user, and reports the results to the application. This way, all applications, even custom ones, have the same intuitive GUI without any efforts from their programmers. Some requests even get special treatment: for example, if the application requests a parameter of type WINDOW (coordinates of a window), PCIPS will display the appropriate image and let the user select a window with the mouse.

When requesting images, the programmer specifies what type and size of images he is prepared to handle. PCIPS allows a lot of flexibility in these specifications, which can range from "any type or size" and "two 8-bit integer 256×256 images" to even "two 32-bit floating point images of the same size; an image of any type, but of the same size as the first two; and two images of the same type with the same size" (the latter is much easier to program than to say, by the way). This is widely used in applications, for example, addition requires two input images and one output image (for the result) that are of the same size and type. PCIPS automatically performs temporary type/size conversion for images that do not meet the applications' specifications.

4.2. Image Access

AIS gives applications two ways to access images—point-by-point and direct access. Point-by-point access involves calling a function to obtain or set the value of each point of the image. It is very easy to use, and offers acceptable performance for most purposes. However, it can be slow when used with big images or intensive calculations, so AIS also supports direct access to the data in the image, which can significantly improve the performance of the application. No matter what method he employs, the programmer doesn't have to worry about memory constraints, because PCIPS implements disk swapping in a transparent way.

5. Applications

A basic package of applications is included with the system. It contains applications for:

- elementary mathematical operations over images and constants
- elementary statistics, filters and fitting
- geometric transformations, i.e., rotate, resample, cut fragment, transpose, etc.
- import/export of images in various formats: FITS, ASCII, binary, GIF, PostScript

Sophisticated and specialized reduction capabilities are available via additional application packages. The CCD/Echelle package includes all the necessary applications for reducing Echelle frames:

- quick look at frame
- search for Echelle orders
- detect and remove cosmic ray hits
- normalize by flat field
- extract Echelle orders into 1D spectra (includes geometric correction of each order)

The spectral package takes the processing a step further:

- build dispersion curve for spectrum
- fit the spectrum's continuum level
- convert spectrum to residual intensity/wavelength scale
- determine line characteristics
- compare to synthetic spectrum

Another package for photometry of CCD frames is also being assembled. Of course, the user need not stop at that—if he has his own packages for data reduction, integrating them into PCIPS is not a problem.

6. Future Development

The next version of PCIPS will be a completely different system, running under Windows and providing sophisticated programming capabilities (Smirnov & Piskunov 1993). The only thing that it will inherit from PCIPS will be the application interface, AIS, so that source code compatibility with existing applications is ensured. In fact, AIS was specifically designed for the *factory programming* metaphor which will be implemented in the new version.

7. Feedback

If you have any questions or comments about the system, we can be reached by E-mail at the following addresses: `piscounov@cc.helsinki.fi` (Nikolai Piskunov) and `oms@airas.msk.su` (Oleg Smirnov).

Acknowledgments. We are very grateful to the Smithsonian Astrophysical Observatory: this presentation would have been impossible without its financial support. The system was developed on hardware provided by the Institute of Astronomy, and partially funded by the Observatory and Astrophysics Laboratory.

References

Smirnov, O.M., & Piskunov, N.E. 1993, this volume

Smirnov, O.M., Piskunov, N.E., Afansyev, V.P., & Morozov, A.I. 1992, in Astronomical Data Analysis Software and Systems I, A.S.P. Conf. Ser., Vol. 25, eds. D.M. Worrall, C. Biemesderfer & J. Barnes, 344

Part 3. Data Acquisition

Section A. Real-Time Systems

Astronomical Data Analysis Software and Systems II
ASP Conference Series, Vol. 52, 1993
R. J. Hanisch, R. J. V. Brissenden, and J. Barnes, eds.

The VLBA Correlator—Real-Time in the Distributed Era

Donald C. Wells
National Radio Astronomy Observatory, Charlottesville, Virginia, 22903-2475 USA

Abstract. The Correlator is the signal processing engine of the Very Long Baseline Array [VLBA]. Radio signals are recorded on special wide-band digital recorders at the 10 VLBA antennas and are shipped to the Array Operations Center in Socorro, New Mexico, where they are played back simultaneously into the Correlator. Real-time software and firmware controls the playback drives to achieve synchronization, compute models of the wavefront delay, control the numerous modules of the Correlator, and record FITS files of the fringe visibilities at the back-end of the Correlator. The Correlator system contains a total of more than 100 programmable computers, which communicate by means of various protocols. The VLBA Correlator's dependence on network protocols is an example of the radical transformation of the real-time world over the past five years: real-time is becoming more like conventional computing.

1. Radio Interferometry

The angular resolution of radio telescopes is approximately given by $\frac{\lambda}{D}$, where λ is the wavelength and D is the aperture diameter. For example, a 100 m antenna at 10 cm (3 GHz) has an angular resolution of about 10^{-3} radian, which is about 200 arcsec, whereas typical optical ground-based imagery has about 1 arcsec resolution. In order for radio telescopes to obtain resolution comparable to optical telescopes D must become *much* larger. Radio astronomers achieve this with interferometers; pairs of separated antennas form two-slit interferometers, with D the separation instead of diameter, and a fringe pattern on the sky is seen. Radio interferometers measure the fringe visibility (spatial coherence function) by cross-correlating the pairs of noisy signals, and this gives samples of the 2-D Fourier Transform of the sky. A set of antennas forms $n(n-1)/2$ baselines, and Earth rotation over periods of hours scans the samples through the FT space. The FT can then be inverted to get images, and it is necessary to deconvolve them to remove sidelobe artifacts. The VLA in NM has 351 baselines with D up to 35 km, and at 10 cm it gives 0.5 arcsecond resolution. With this capability the VLA made a revolution in radio imaging in the 80s.

2. VLBI

VLBI [Very Long Baseline Interferometry] gives 100 or more times higher resolution than the VLA by using baselines up to 10,000 km (i.e., the diameter of

Earth). Baselines will reach 100,000 km by about 1995 using the VSOP and Radioastron satellites. Even though the wavelengths are long, the enormous baselines used in VLBI allow it to be the astronomical technique with the highest angular resolution.

An important category of astronomical sources observed with VLBI are the AGN (Active Galactic Nuclei) "engines" and the relativistic plasma jets associated with them. The radio galaxies and quasars which emit these jets are the brightest sources in the Universe. The energy source is believed to be gravity, via accretion disks around rotating black holes, with the jets being emitted along the spin axes. VLBI enables us to see the jets closer to the "engines" which make them. The hope is that such observations will provide constraints on feasible models of the engines.

A second category of VLBI sources are masers, usually seen around stars. Stellar radiation, especially UV, pumps various gas species in clouds around various types of stars, usually hot young stars embedded in gas clouds where they formed few million years ago. The most common maser types are OH, H_2O and methanol. In favorable cases the masers emit considerable beamed energy. The sources are almost points, almost monochromatic. Observations of these sources can give information about the kinematics of the gas clouds.

Finally, VLBI techniques can use the astronomical souces as references for geodetic measurements. Antenna locations are measurable with $\approx$1 cm RMS, which enables continental drifts (mm/year) to be detected with timespans of a few years. Many countries have VLBI geodetic programs today.

3. The Very Long Baseline Array [VLBA]

This NRAO project began circa 1983 and will begin making user observations in 1992. It consists of ten 25 m antennas located on US territory, capable of operating at up to 86 GHz (3 mm), equipped with hydrogen maser clocks plus GPS receivers for synchronization. Each antenna has two digital instrumentation tape recorders which record at up to 128 Mb/s, and can record up to 700 GB on one reel in 12 hours. The monitor and control of the antennas is done by Motorola 147s under VxWorks, commanded over the Internet[1] from NRAO's Array Operations Center [AOC] in Socorro, NM, following scripts created by observers. Logs are put into an Ingres DBMS, while data tapes are shipped to the AOC.

The Correlator is the signal processor of the VLBA. It plays back the data tapes after they arrive at the AOC, using 24 playback drives [PBDs] (20 plus 4 for staging), which enables it to perform two simultaneous ten-station VLBA runs or a single 20-station run. Currently 9 PBDs are available. The Correlator produces fringe visibilities at up to $\approx$500 KB/s, using a set of binary table (`BINTABLE`) extensions in one FITS file. The data will be archived on DAT tapes. AIPS has been enhanced to read the VLBA files and to support VLBI requirements.

---

[1]Security issues for RT systems are now a *serious* concern for our community.

4. Overview of the Correlator

The project began circa 1984 at CalTech with a design study; because of NSF funding limitations this effort was then suspended. The project was resumed by NRAO in 1987. Some team members have been on the project since the CalTech phase 8 years ago. The team consists of ten people: one manager, four engineers, one technician and four programmers. The author is the leader of the real-time programmer team, and joined the project in May 1988.

The Correlator hardware depends on 3000 custom VLSI chips. The VLSI chip fabrication by LSI Logic suffered an unexpected, and unfortunate, 13-month delay. The hardware is highly programmable; the engineers spend much of their time programming microcomputers in assembly language. Interfacing between the RT software and the hardware is via two custom protocols based on commercial VME interface cards. The Correlator hardware fills four wide racks in addition to the 24 PBDs, each of which is a full rack; the facility occupies considerable floor space at the AOC. As of November 1992 the hardware is operational for the most common observing modes, and work is continuing to debug the full set of operational modes. The RT software is sufficient to operate the hardware in a non-automated fashion, and work is continuing to add automatic scheduling, operator interfaces, additional observing modes and the archive and distribution subsystems.

5. The Correlator Software Project

The team uses Unix workstations, originally Sun-3s, now SPARCs. We use Sun's NSE [Network Software Environment] for configuration management. We use many imported software components, both commercial (multi-vendor) and PD; interactions and dependencies have sometimes inhibited software upgrades. The file server is also the database server; currently it is a SPARCstation-2 with 64 MB RAM and 3.2 GB disk (almost full now, as expected). The DBMS, a critical component of the Correlator system, is Ingres, with Sun's "Simplify" graphical DBMS tool. We have coded numerous embedded-SQL DBMS applications which run on the server. The RT OS is "VxWorks" from Wind River Systems, chosen because in Fall 1988 it had the lowest price among the RT OSes with acceptable network support and because of the prior favorable experience of CARA's Keck project and NOAO's GONG project. The programming language for both DBMS and RT is classical K&R C—we have not yet switched to ANSI C.

6. The RT Software

The observing logs are moved from the M&C Ingres server at the AOC to the Correlator Ingres server using Ingres/NET; until June 1992 they passed 3000 km from New Mexico to Virginia via the Internet. Embedded-SQL programs on SS-2 compute "jobs," which are tabular relations in third-normal-form computed in the DBMS from the logging data and expressed as ASCII-text files (see sample in Figure 1). The syntax of the script language is similar to the script language used to command the telescopes, the principal differences being that iteration

```
!* Model script file for VLBA Correlator Job *!
!* 3 stations, 8 baseband channels, 2 minute scans  *!
!* 30 July 1989 -- JMB  *!

!*----- Job Control Card ----------------*!
!table 'job'!
  jobid = 11021    program = 'VW23G'
  n_drives = 3  n_chans = 8   n_fft_pts = 512  n_stns = 3    fft_factor = 8
  date_start = 80Sep22  start = 12h00m00.0s
  date_stop  = 80Sep22  stop  = 12h30m00.0s
!row!
!endtable!

!*----- Observations Table -----------*!
!table 'observations'!
  date = 80Sep22  start = 12h00m00.0s  stop = 12h02m00.0s source ='BLLAC'
  name = 'HY'  array_id = 100 !row!
  date = 80Sep22  start = 12h02m00.0s  stop = 12h04m00.0s source ='VIRGO'
  name = 'HY'  array_id = 100 !row!
  date = 80Sep22  start = 12h04m00.0s  stop = 12h06m00.0s source ='BLLAC'
  name = 'HY'  array_id = 100 !row!
  date = 80Sep22  start = 12h00m00.0s  stop = 12h02m00.0s source ='BLLAC'
  name ='MPI'   array_id = 100 !row!
  date = 80Sep22  start = 12h02m00.0s  stop = 12h04m00.0s source ='VIRGO'
  name ='MPI'   array_id = 100 !row!
  date = 80Sep22  start = 12h04m00.0s  stop = 12h06m00.0s source ='BLLAC'
  name ='MPI'   array_id = 100 !row!
  date = 80Sep22  start = 12h00m00.0s  stop = 12h02m00.0s source ='BLLAC'
  name ='GB'    array_id = 100 !row!
  date = 80Sep22  start = 12h02m00.0s  stop = 12h04m00.0s source ='VIRGO'
  name ='GB'    array_id = 100 !row!
  date = 80Sep22  start = 12h04m00.0s  stop = 12h06m00.0s source ='BLLAC'
  name ='GB'    array_id = 100 !row!
!endtable!

!*----- Stations Table -------------------*!
!table 'stations'!
  name = 'MPI'
  x =  4.03394212e+6   y =  4.86993120e+05 z =  4.90043183e+06
  axistype = 'altaz'   axisoff = 0.00
!row!
  name = 'HY'
  x =  1.49240669e+06  y = -4.45726733e+06  z =  4.29688210e+06
  axistype = 'altaz'   axisoff  = 0.00
!row!
  name = 'GB'
  x = 8.882882548e+5 y = -4.92448405e+6 z = 3.94413087e+6
  axistype = 'polar'  axisoff  = 0.0
!row!
!endtable!

!QUIT!
```

Figure 1. Several Tables in Job Script Language

operators have been removed and table/row operators added. Note that the language does not tell the Correlator what to do explicitly, instead it describes sets of related objects and actions are implied.

The job scripts are copied from the SS-2 to the RT CPU using the NFS protocol—VxWorks mounts the SS-2 partition. During the loading process space is `malloc`-ed and the tables are transliterated to an equivalent set of arrays of C-structures linked by pointers (see Figure 2). Each row of a table in the script becomes a C `struct` and each table become an array of this `struct`. Each "job" is defined by tables-of-tables which contain pointers to the arrays of `struct`. A "queue" table contains pointers to the job tables. All information needed for processing observations is contained in the script/structures, and the RT tasks follow the pointers to find the schedule and parameters needed for processing.

In Figure 2 we see a mixing of table notation with C pointer notation; this equivalence of the DBMS relations for a job to the job script and to the job structures is one of the key ideas of the Correlator architecture. We have found that such a set of tables in "third-normal-form" is a powerful abstraction which is intuitively understood by everyone, and it forms a convenient language for communication between team members. I recommend that other designers seriously consider casting analogous problems into this form.

Figure 3 shows the association of the tasks with the hardware subsystems. Multiple `jobTasks` are spawned, each with a pointer to its job structures; startup overhead is overlapped. Each `jobTask` spawns `stnTasks`, one for each station, and each `stnTask` spawns one or more `tapeTasks`. Each `jobTask` spawns one or more `arrayTasks`, which write the FITS files to a local 1 GB SCSI disk. The individual tables (`BINTABLE` extensions) are written as separate files to permit concurrency (e.g., generation of FITS headers in the background at lower priority). The ticking frequency is 7.6 Hz (131 ms period); `jobTasks`, `stnTasks`, `tapeTasks`, `arrayTasks`, etc., operate as synchronous (ticking) state machines executing at various priority levels, sharing memory and interlocking critical resources with semaphores. The tasks `malloc()` RAM as needed and `free()` it before they terminate. The geometric delays are computed using the `CALC` package (in Fortran) which we have imported from the NASA Crustal Dynamics Program. The `tapeTask` positions tapes to within 10 ms while moving at 180 in/s (4.6 m/s), and a firmware servo then uses a buffer memory to synchronize the bit streams to 1 bit (30 ns) so that cross-correlations can be computed by the VLSI hardware. A number of permanent tasks (`tickTask`, `tapeMntTask`, `modelTask`, `xbrTask`, ...) manage the resources which are shared across jobs.

Note that the 24 PBDs are connected to the 24 signal processing engines via a crossbar. The extra four PBDs enable tape staging; in full operation the average interval between tape changes will be about $\frac{1}{2}$ hour. A crossbar task is used to coordinate and synchronize crossbar reconfigurations; the design is intended to enable a `tapeTask` to switch from one PBD to another without stopping and restarting. The geometric model computations are done by a model task acting as a server to stage the CPU-intensive computations.

The operator has a variety of window-based displays to monitor and control the Correlator software. These displays operate on workstations, communicating across the LAN, and using a custom window-management protocol layered on

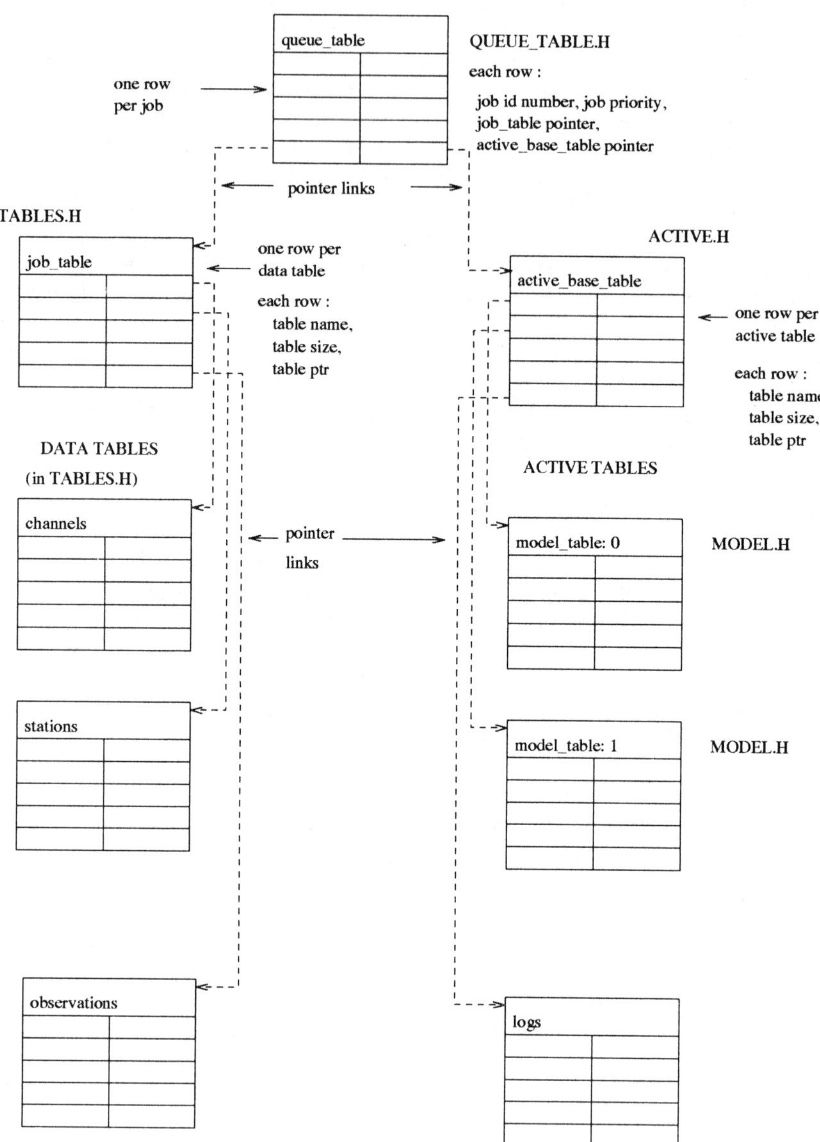

Figure 2. Table Hierarchy in the RT Environment

TCP/IP sockets.[2] Various logs and database transaction files flow back to the SS-2 file/DBMS server across the LAN via NFS.

The shading of the boxes in Figure 3 indicates the status of the subsystems: grey indicates substantially complete, striped indicates work-in-progress and white indicates the few modules which have been deferred until later. As of November 1992, all of the hardware is available and most of the software directly associated with the signal chain is operational.

7. RT Software Issues

In this section I will discuss a number of issues which arise in contemporary RT design, and which RT designers often debate.

7.1. Networked Versus Embedded

Networked systems facilitate development support on state-of-the-art platforms while enabling RT systems to support remote debuggers, remote login, X-clients (and servers), NFS clients (and servers), RPCs for custom services and for concurrent RT and sockets for custom monitor and control services. Embedded systems are fine for static applications that need to be robust and don't need to evolve. The VLBA Correlator contains more than 100 embedded CPUs which act as high speed front-end processors, almost dedicated at the one-task-equals-one-CPU level.

7.2. Synchronous Versus Asynchronous

Synchronous (ticking) RT systems can be designed with *proofs* that they will meet "hard" deadlines while managing resources. Asynchronous RT systems can achieve higher performance statistically, but often at the expense of poorer worst-case performance.

7.3. Lightweight Threads Versus Memory Protection

Lightweight threads are the task abstraction of VxWorks: multiple instances of any C function can be spawned as tasks, each with its own stack space for dynamic variables, but sharing static variables. The tasks all execute in the same memory address space, and a wild pointer in one task can trash the memory of another task or even overwrite the kernel. Each task executes in its own protected memory space in many older RT OSes. This is safer, but task startup is slower and it is harder to share code and data between tasks.

7.4. Semaphores-for-Everything Versus Interrupt-Lockout

VxWorks guards all critical regions with semaphores, and interrupts are almost never inhibited. Interrupts are disabled during kernel critical regions in many older RT OSes. This produces a variable latency in context switching, and usually statistically poorer responsiveness overall.

---

[2] A new implementation today could be X-based, because an X-windows client implementation is now available for VxWorks.

VLBA Correlator Status

as of 1992 November 1

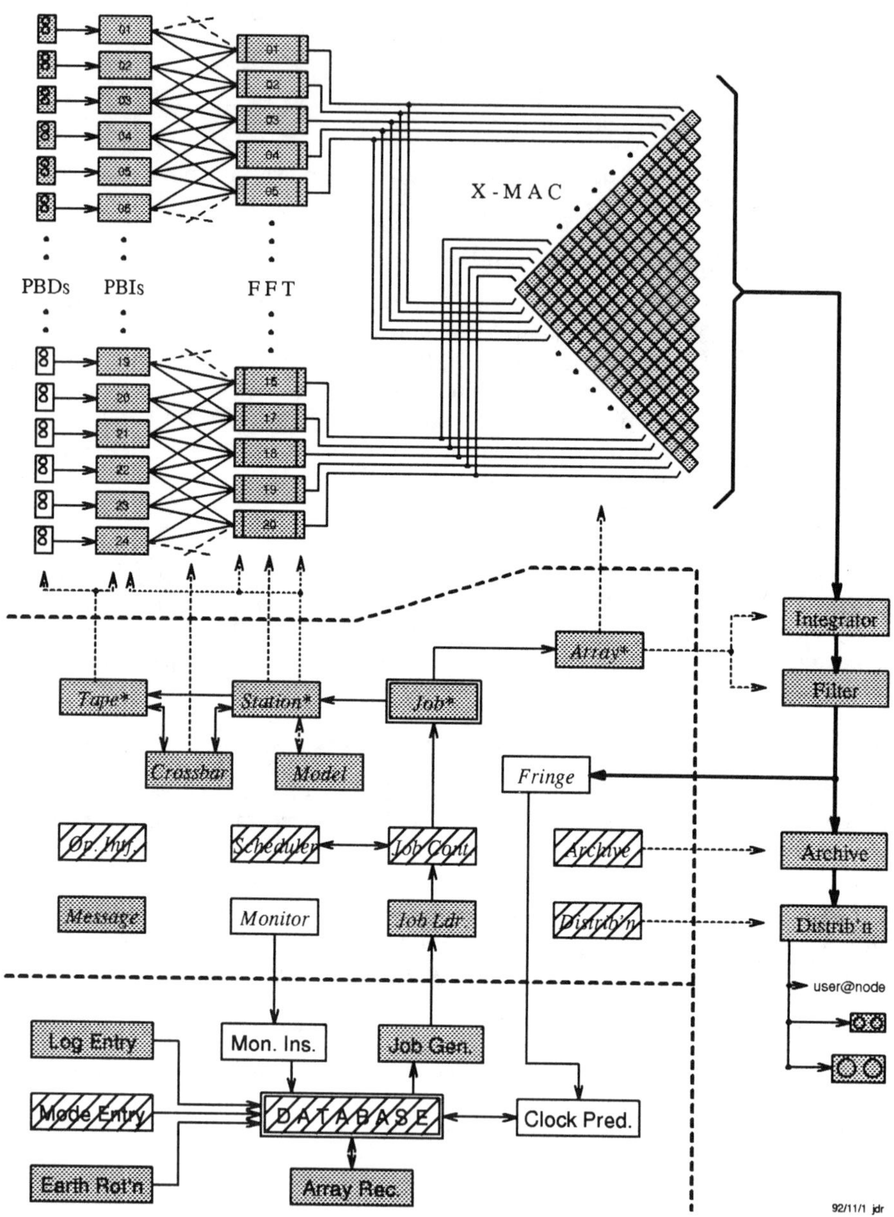

Figure 3. VLBA Correlator Block Diagram

I don't like the word "deterministic" in the RT context. It is misleading, because RT latencies are always variable. The real issue is the shape of the histogram of latencies. Older systems which inhibited interrupts in the kernel often claimed to be deterministic, but they never were.

7.5. Two Nasty Technical Problems

Cache memory is a frequent source of trouble, especially in the presence of intelligent device controllers. RT device driver availability is a perennial problem.

Concurrent CPUs are no longer a significant problem. Multiple VME-based CPUs can share RAMs, can even have one CPU run SunOS and other run VxWorks with backplane network driver as well as shared memory.

7.6. Is RT Unix in our Future?

Some RT Unix systems have high performance. Some "hard" RT kernels are Unix-like (e.g., VxWorks). Solaris 2.0 is expected to have a "hard" kernel, with shared memory, semaphores and lightweight threads. The Posix committees have draft standards for RT features, and probably these will become universal in Unix systems and "hard" RT kernels (like VxWorks) within a few years. The techniques used by RT programmers are good starting points for programming concurrent CPUs and highly responsive user interfaces.

8. A Final Word about this Paper

As Figure 3 shows, the project is not finished at this time (December 1992). Therefore, this paper cannot be regarded as the final word—probably there will be other papers about this major astronomical instrument system. I have not discussed the hardware design of the Correlator in anything like a proper fashion, and even in the software area I have omitted serious discussion of certain major subsystems which are still under development. Therefore, this paper must be regarded as only a preliminary introduction to the Correlator.

Acknowledgments. The VLBA Correlator project is a *team* project. The project manager is Jon Romney, who made major contributions to the software design. My fellow software team members are John Benson, Ray Gonzalez and Jim Horstkotte, each of whom has had responsibility for major subsystems. The hardware/firmware team is led by Ray Escoffier, and also includes Chuck Broadwell, Walter Brown, Joe Greenberg and Gene Runion. In addition to firmware, Broadwell also constructed device driver code to interface to the Correlator hardware. I am grateful to my fellow team members (hardware and software) for countless hours of camaraderie over more than four years. I regard them as *virtual co-authors* of this paper, but I take full responsibility for the inevitable mistakes of omission or commission in it. Each of them surely has his own story to tell—my version of the story is necessarily incomplete.

Figure 3 was graciously contributed by Jon Romney.

Discussion

Barg: Expand more on the issue of security with regards to real-time systems.

Wells: Dedicated subnet at private addresses. RT system only accepts command from a limited number of addresses.

Conrad: Can we have light weight processes and memory protection with VxWorks??

Wells: Yes. Some people have used the task schedule hook to use the Sparc MMV for home brew memory protection (see VxWorks exploder). Of course, this is the direction being taken by Solaris 2.0. The technology that is really driving this is the need to exploit multiple processors. We have a convergence of these two worlds:

1. Solaris/multiple processors
2. VxWorks/real-time

Blum: What about power failures?

Wells: Software has some restart/reboot capabilities—will have more later. In the Lab it's not so critical—remote would be more so. Can catch up pretty quickly via speedup option. The telescope can restart automatically.

Bloch: We at Los Alamos have had great success with VxWorks boxes attached to workstations as ground support equipment for space payloads. This separates the hardware dependence on the VX works box from the user interface on the workstation. We also are considering using VX works on spacecraft processors, hooking an ethernet card to the s/c and attaching a workstation for s/w development.

Astronomical Data Analysis Software and Systems II
ASP Conference Series, Vol. 52, 1993
R. J. Hanisch, R. J. V. Brissenden, and J. Barnes, eds.

CCD Data Acquisition Systems at Lick and Keck Observatories

R. I. Kibrick, R. J. Stover

UCO/Lick Observatory, University of California, Santa Cruz, CA 95064

A. R. Conrad

W. M. Keck Observatory, P.O. Box 220, Kamuela, HI 96743

Abstract. This paper describes the evolution of the existing CCD Data Acquisition System (DAS) at Lick Observatory, and its influence on the design of the CCD DAS currently under development for Keck Observatory. Tradeoffs between various types of interfaces between instrument computers and CCD controllers are also explored.

1. Introduction

Before discussing the design of CCD Data Acquisition Systems (DAS), it is useful to briefly review the major functions they perform. We omit those DAS functions which are instrument- or telescope-specific and not directly tied to the operation of the CCD.

1.1. Dewar and Chip-Level Functions

These include: generation of CCD clocking waveforms and bias voltages, signal processing and digitization of the video signal(s) from the CCD, transmission of the digitized data, exposure timing and shutter control, dewar temperature sensing and control, and optionally, dewar cryogen level sensing and control.

1.2. Image Recording Functions

These include: collecting (and, if needed, descrambling) the pixel data stream from the CCD, collecting all relevant observational parameters (e.g., exposure time, image dimension) for the FITS header, and recording the header and pixel data to disk and/or tape. Sufficient parameters should be recorded to allow subsequent data reduction and analysis. Additional parameters should be recorded to accurately describe all controllable aspects (e.g., CCD, instrument, and telescope) of an exposure so that a nearly-identical one can be recreated at a later time (Conrad & Lupton 1993). Ideally, the complete image file, or at least its FITS header, should be automatically entered into an archival data base.

1.3. Image Display Functions

These include both the real-time display of CCD images as they are being read out, and the display of previously recorded images. Such displays should at least provide adjustable intensity mapping and modes (e.g., linear, log), panning and zooming, interactive cursor readout, row and column plots of selected regions, and basic image statistics (e.g., image centroids, FWHM).

1.4. User Interface Functions

These include status display and control of: exposure state, time, and type (e.g., dark frame); CCD binning and windowing; readout modes (e.g., MPP); number and configuration of readout amplifiers; current image file directory and file name; frame number; recording modes (e.g., disk/tape); dewar temperature and cryogen level.

The user interface functions should be isolated into a separate series of processes, and loosely-coupled to the underlying hardware control processes by a well-defined interface protocol. This permits such user interfaces to be shared across multiple instruments (Lupton & Conrad 1993, Conrad & Lupton 1993).

1.5. Engineering Functions

These include: real-time modification and display of CCD waveforms and bias voltages, system performance measurements (e.g., readout noise), and subsystem diagnostics (e.g., data link loopback testing, power supply monitoring). For safety of the CCD, such functions would be inaccessible during routine observing, and would be reserved for qualified engineering and technical support personnel.

2. Evolution of the Existing CCD DAS at Lick Observatory

The first incarnation of the Lick CCD DAS was implemented in 1980. It established a basic model for the partitioning of CCD DAS functions which has endured for nearly 12 years. This model consists of 2 major hardware components: a CCD controller, located in close proximity to the CCD dewar, which performs the dewar/chip-level functions, and an instrument computer, remotely located in the control room, which performs the remaining CCD DAS functions. These two components are linked by the Lick serial multiplexer bus, a custom serial link developed at Lick during the early 1970s and which until recently formed the main communications and control network at the Lick telescopes. The custom hardware and associated software device drivers required to operate this link were embedded directly into both the CCD controller and the instrument computer, both for efficiency and economy.

2.1. PDP-8/I System, 1980–1981

In the initial 1980 version, the instrument computer was a DEC PDP-8/I minicomputer (a 1970s vintage machine acquired for running the Lick IDS scanner), with 32K 12-bit words of physical memory plus floppy disk and 9-track magnetic tape (Lauer 1981). The 12-bit word size was well matched to the fast 12-bit analog to digital converters (ADCs) of that era. The CCD controller (custom built at Lick) employed a Rockwell 6502 8-bit microprocessor, was programmed in assembly language, and utilized 2KB UV-erasable EPROM chips for program and waveform table storage (Robinson 1987). This overall system was barely adequate during our first few years of CCD observing, and was quickly overwhelmed as CCD dimensions and ADC resolutions increased.

2.2. LSI-11 Systems, 1982–1987

In 1982, the PDP-8/I instrument computer was replaced by DEC LSI-11/23 and LSI-11/73 Qbus-based systems, programmed in C and running a derivative of Bell Labs Version 7 Unix. The instrument computer side of the CCD DAS was re-written by Stover to utilize the multi-tasking capabilities of Unix, and the user interface functions were split off into a separate process which communicated with the underlying hardware-control processes by means of pipes. The custom PDP-8-based hardware and software drivers for the Lick serial multiplexer had to be re-designed and re-implemented for the LSI-11/Q-bus/Unix system. The CCD controller remained much the same, except that the 12-bit ADC was replaced with a 15-bit device to take advantage of the increased word size provided by the LSI-11 systems.

2.3. ISI 680x0 Systems, 1987–Present

Once again, advances in both CCD and computer technology overtook the capabilities of the instrument computer, and the LSI-11s were replaced with Integrated Solutions (ISI) computers in 1987. These ISI systems initially contained dual Motorola 68020-series CPUs (upgraded to 68030 CPUs in 1989) sharing a common VME-bus backplane. The first CPU is master of most of the peripherals such as disks, tapes, and image displays, and runs the non-time-critical user interface, data display, and data reduction processes. The second CPU, which is diskless, is dedicated to real-time control of the lower-level hardware, including processes for operating the CCD controller, the instrument controller, and the telescope controller. The two CPUs communicate via shared VME-bus memory, via sockets over the backplane, and via ISI's transparent remote file system which allows access to the disk from the second CPU.

Most of the C code from the LSI-11-based DAS ported directly to the ISI systems, which run a derivative of Berkeley 4.3 Unix. The separate user interface was retained in much the same form, but pipes were replaced by sockets. The MUSIC system was created (Stover 1989), which introduced three additional DAS processes (traffic, infoman, and runner) to coordinate distributed processing between the two CPUs and to provide for multiple instances of the user interface process. This latter feature made possible coordinated local/remote observing. Yet again, the custom hardware and software drivers for the Lick serial multiplexer were re-implemented to accommodate the change from the Qbus to the VME-bus and from Bell to Berkeley Unix. During this transition, the CCD controller remained virtually unchanged.

The ISI-based systems are still the primary ones used both at the Lick telescopes on Mount Hamilton and at the Lick CCD Detector Development laboratory on the UC Santa Cruz campus. The important trend to note in this 12-year evolution is that while the instrument computer side of the model has gone through multiple changes of processor architecture, hardware bus, and operating system, the CCD controller side has remained virtually unchanged. Also note that a non-negligible portion of the cost and delay associated with each instrument computer upgrade was the re-design and re-implementation of the custom hardware and device drivers required to support the embedded interface to the CCD controller.

2.4. New Architectural Model, SPARC System, 1992

In mid-1992, Stover implemented a CCD DAS at Lick which represents the first significant departure from the two-component 'instrument computer/CCD controller' model. In this new system (developed for the Coudé Auxiliary Telescope attached to the Shane 3-meter Telescope at Lick), a Sun Sparcstation (which has no VME-bus) is used as the instrument computer. However, rather than attempting to redesign the custom hardware for the Lick multiplexer (in this case, to work with the Sparc's S-bus) and embed it within the instrument computer, the custom hardware (and software) instead was isolated to a separate, diskless, VME-bus chassis. The Sparc instrument computer thus remains a standard 'off-the-shelf' system, containing no custom hardware or device drivers.

The separate VME-bus chassis (which we refer to as the data capture computer) represents a third component of the model, and logically sits between the instrument computer and the CCD controller. It contains basically the same VME-bus hardware as used on the ISI systems for interfacing to the Lick multiplexer. It also contains a Sparc-1E processor board, running SunOS 4.1e, which assumes most of the functions previously performed by the second CPU in the ISI systems. The Sparc-1E board boots from the Sparc instrument computer via Ethernet, and the custom device driver required to operate the Lick multiplexer is emulated on the Sparc-1E using the Unix mmap() system function. The data capture computer receives the pixel stream from the CCD controller, combines this data with appropriate FITS header data, and records, via NFS, the resulting image file on a remotely-mounted disk partition of the instrument computer. By isolating the custom hardware and software into the data capture computer, and using only a well-accepted industry standard interface to connect it to the instrument computer, future upgrades and/or replacement of the instrument computer hardware and software should be greatly simplified.

3. The Keck Observatory CCD DAS

An overview of the hardware architecture for the Keck CCD DAS show 3 major components: an 'off-the-shelf' instrument computer, a semi-custom data capture computer, and a CCD controller. At this level, the architecture is quite similar to that of the recently-implemented system at Lick described in the previous section. The similarity extends to a lower level, in that the Keck system (at least currently) employs a Sparc-based instrument computer, and a Sparc-1E-based VME-bus crate for the data capture computer. There, however, the similarity ends, since the Keck CCD DAS utilizes a completely different CCD controller, which connects to the data capture computer through an entirely new interface.

3.1. The Leach CCD Controller

Although the existing Lick CCD controller has proved to be a reliable workhorse over the past decade, it is clearly obsolete and does not provide all the features desired for the Keck system, such as dynamically programmable CCD waveforms or support for large numbers of readout amplifiers. Although some consideration was initially given to developing a completely new CCD controller, the system developed by Bob Leach at San Diego State University was found to meet the requirements for the Keck system (Leach 1990).

The Leach system is based on Motorola DSP56001 digital signal processors (DSPs) which are highly-parallel processors capable of performing multiple simultaneous functions at a 10 MHz rate. These processors are programmed in DSP assembler, and the software and CCD waveform tables are stored in an on-board EEROM which can either be programmed off-line in a PROM burner, or in-place at runtime via a 40 MHz dual-fiber optic link which connects the Leach CCD controller to the data capture computer. The Leach CCD controller provides all of the dewar and chip-level functions, and supports interleaved readout from large numbers (16 or more) of amplifiers.

The Leach controller is gaining fairly wide acceptance, and is currently in use at San Diego State University (SDSU), UCO/Lick Observatory, Caltech, Michigan Dartmouth MIT Observatory (MDM), University of Chicago, Universitie de Paris-Meudon, and Steward Observatory. It is also becoming somewhat of a standard on Mauna Kea, where it is also used at the Canada-France-Hawaii Telescope (CFHT), University of Hawaii Institute for Astronomy (UH-IfA), and Keck Observatory.

3.2. Interfacing to the Leach CCD Controller

The 40 MHz fiber optic link from the Leach CCD controller represents much the same problem as the 4 MHz serial multiplexer link from the existing Lick CCD controllers. Since both are non-standard, they require some sort of custom hardware and software interface. Unfortunately, while the Leach CCD controllers are becoming somewhat of a standard, most institutions have developed their own unique interface. Many of these interfaces were developed in parallel, and some of these are described in detail elsewhere in these proceedings.

In the UH-IfA system (Jim et al. 1993), the Leach CCD controller is interfaced via a small, external, custom fiber-optic adapter card to a DSP-input port on a NeXT computer, which contains an integral DSP56001 as standard equipment. The NeXT serves as the data capture computer and also provides an excellent software development environment for the DSP-software for the Leach controllers. The NeXT also replicates some of the functionality of the instrument computer, although images can also be uploaded via Ethernet to a SPARC-2 host for more extensive processing. The UH-IfA system is relatively inexpensive, efficient, and compact.

In the MDM system (Metzger et al. 1993), the Leach CCD controller is interfaced to a Sun SPARC-2 via an external custom hardware interface which connects to a DR-11W parallel interface emulator attached to the Sparc's internal S-bus. In this case, the SPARC-2 may be viewed as the instrument computer with an embedded custom interface to the CCD controller, as was the case under the original Lick 2-component CCD DAS model. This system involves minimal hardware, and is also quite inexpensive and compact. Since the pixel data stream is transferred directly from the CCD controller to the SPARC-2, any potential Ethernet bandwidth limitations are avoided.

Other permutations exist, involving possible combinations of the above two systems using a DSP56001 S-bus board manufactured by Ariel. In addition, Leach has recently released his own custom-designed VME-bus DMA controller board, which employs an on-board DSP56001 both to implement the VME-bus DMA-handshaking as well as to provide the interface to the fiber optic link from the Leach CCD controller.

More recently, a proposal to the Keck project has been advanced by McCarthy at Caltech and Stubbs at UCSB to design a SCSI-based interface to the Leach CCD controller. This proposed interface would emulate a SCSI device on the instrument computer side, and provide an internal memory buffer and fiber optic interface on the Leach CCD controller side. Since SCSI is a widely-accepted industry standard, such an interface could greatly simplify the task of interfacing the Leach controllers to a variety of both new and existing computer equipment.

Since none of these systems existed at the time that the Keck CCD DAS was specified, it employs yet another flavor of interface to the Leach CCD controller. As mentioned earlier, the data capture computer in the Keck system consists of a VME-bus crate containing a SPARC-1E board connected to the instrument computer via Ethernet. The VME-bus crate also contains a 32 MB VME-bus memory board, for local buffering of complete CCD images, and an IKON DR11-W (emulator) parallel interface DMA controller board, which transfers the pixel data stream into the VME-bus memory via DMA. All of this is 'off-the-shelf' hardware. The IKON DR11-W board in turn connects (via two DR11-W-style ribbon cables) to a custom-built wire-wrapped board which provides the interface to the the fiber optic link from the Leach CCD controller.

This custom interface board can buffer 4 K pixels of data in an on-board FIFO, and provides dual pathways to the VME-bus memory. Data can be transferred between the board and the VME-bus memory under DMA control via the IKON board, or directly via the VME-bus, using programmed I/O from the SPARC-1E. Extensive internal and external hardware loopback modes are provided to facilitate maintenance of this custom board.

During CCD readout, pixels are received by the data capture computer via the fiber link from the Leach CCD controller and transferred (via DMA) into the VME-bus memory. As each successive group of pixels (typically several thousand pixels) is received, it is packaged into UDP-datagrams by vxWorks-based software running on the Sparc-1E, and sent via Ethernet to the instrument computer, using a reliable, high-performance image-transmission protocol. There the image data are received, displayed in real-time, combined with FITS header data, and recorded. As soon as the last pixel of an image has been read out, sent down the fiber, and transferred via DMA into the VME-bus memory on the data capture computer, the next exposure can immediately be started, thus maximizing observing time.

The CCD read out, transmission down the fiber, DMA into VME-bus memory in the crate, re-transmission down the Ethernet, and display on the instrument computer are all overlapped into one big pipeline. Provided that the aggregate pixel data rate from the Leach controller remains below the effective Ethernet bandwidth (currently about 450 kilopixels/second), the Ethernet does not introduce any significant delay in the receipt of data by the instrument computer. For both Keck first-generation optical instruments (HIRES and LRIS), the aggregate pixel data rate (about 30 to 120 kilopixels/second) will remain well below that bandwidth limitation.

Compared to many of the interfaces to the Leach system described earlier in this section, the interface used by the Keck CCD DAS appears very expensive, bulky, and complex. Its total bandwidth is constrained by the Ethernet, and this could become a problem for future instruments involving large mosaics or more rapid readout-amplifier sampling rates. Nonetheless, the existing Keck interface to the Leach CCD controller offers some important advantages in terms of its potential longevity and its ability to prevent data loss and minimize loss of observing time in the event of instrument computer problems that occur during CCD readout.

3.3. Minimizing Loss of Data and Observing Time from Instrument Computer Crashes

It is an unfortunate reality of observing that instrument computers can and do crash. If this occurs during a CCD readout, it often means the loss of part or all of the data from that exposure. From the viewpoint of an observatory, given the nightly cost of operating a telescope like the Keck 10-m, such a loss may translate into a significant sum. From the viewpoint of the observer whose data is lost, such a loss may represent catastrophe.

Instrument computers crash for a variety of reasons, many of which are hardware-related. Since instrument computers are often burdened with lots of power-hungry peripherals, it is sometimes impractical to connect them to battery-backup power, and they are thus susceptible to power glitches and failures. In addition, these peripherals often involve lots of moving parts (e.g., disk drives, tape drives), and are subject to various mechanical failures (e.g., head crashes, tape jams). These failures often require the system to be power-cycled to swap components, or at least rebooted. Also, since users are in constant contact with the instrument computer, it is susceptible to derangement by electrostatic discharge (ESD), which is especially a problem in the dry mountain air where these computers are typically operated.

But instrument computers also crash (or at least require rebooting) for a variety of software reasons. They typically run very complex multi-user operating systems which are never stable nor fully-debugged. New releases of operating system software and major application packages (e.g., IRAF), while solving some old bugs, usually introduce many new ones. In addition, it is difficult to predict or control what other unknown software users will introduce onto the machine, either home-grown or public-domain software obtained off the network, and any of these can introduce subtle bugs. While such bugs may not crash the system outright, they may result in various resource exhaustion problems (e.g., filled disk partitions, consumption of all process slots) which so paralyze the system that it must be rebooted anyway. Usually, such system reboots not only lose any data which may be in the process of being read out, but they also result in significant loss of observing time, since subsequent exposures cannot be initiated until the reboot completes. If the instrument computer is configured with lots of disk drives whose contents must undergo file system consistency checks, this may take tens of minutes.

The Keck interface to the Leach CCD controller chassis avoids these problems by buffering the CCD image in the VME-bus memory of the data capture computer. Since both the Leach controller and the data capture computer will be on battery-backup power, they should be immune to most power glitches and outages. Because both are typically out of reach of the observer, they should also be less susceptible to ESD. Neither has any peripherals with moving parts, so they should be immune to disk head crashes and tape jams. Furthermore, the software in both is significantly less complex than the multi-user systems and massive application packages which run on the instrument computer. That software should also be considerably more stable, and immune to user perturbation, since users do not log in directly to either the Leach CCD controller or the data capture computer.

In the event that an instrument computer either temporarily hangs, crashes, or needs to be forcibly rebooted while the CCD is being readout, no data will be lost, because it is buffered in the data capture computer. Once the instrument computer either recovers or reboots, the data capture computer will resend as much data as needed to insure that the complete image is successfully received. In the case of a hard failure of a given instrument computer, or if the time required to reboot it is excessive, it is a simple matter to have the data capture computer send the image to any backup instrument computer on the network. In fact, the entire Keck CCD DAS can be switched over to a backup machine in less than a minute, without the need for any recabling, rebooting, or power-cycling of equipment. Thus, data loss is avoided, and loss of observing time is minimized.

In the case of instrument computer problems, most of the previously described interfaces to the Leach CCD controller cannot avoid either a partial data loss, or a significant loss of observing time if the instrument computer must be rebooted. The proposed SCSI interface holds some promise in this regard if it can be designed to buffer a complete image, and retain it across a reboot of the instrument computer to which it is attached. However, in the event of a hard failure of the instrument computer, data loss would most likely occur, since the SCSI interface unit would need to be power-cycled before it could be moved to a backup instrument computer. Alternatively, this SCSI interface could be attached directly to the SCSI port on the Sparc-1E in the Keck data capture computer, simply replacing the existing custom interface board and IKON DR11-W DMA controller. In this way, the existing crash recovery capabilities of the Keck CCD DAS could be retained, at the expense of losing the increased bandwidth that would otherwise be achieved were the SCSI interface attached directly to the instrument computer.

4. Next Generation CCD DAS at Lick Observatory

While those portions of the Lick CCD DAS which are most visible to the observer (e.g., instrument computers, user interfaces, data displays) will probably change very little in the near term, it is fairly certain that over the next year the Leach

CCD controllers will begin to replace the existing Lick CCD controllers, both at the Lick telescopes and in our CCD Detector Laboratory. What is less clear is the particular type of interface that will be used to interface the Leach CCD controllers to our instrument computers.

One option that is currently being evaluated is Leach's recently-released VME-bus DMA controller board. (A SunOS 4.1e device driver for the Leach DMA board has been provided by Steve Smith at CFHT, where the Leach DMA board is also being used.) We are testing this board in a data capture computer similar to the one employed in the Sparc-based system recently installed at the CAT telescope and which was previously described in section 2.4. In this test system, Leach's VME-bus DMA controller board replaces the VME board used to operate the Lick serial multiplexer link that connects to the existing Lick CCD controllers. Should these tests prove successful in the lab, the CAT telescope will provide the next likely test bed to determine how the overall system performs under actual observing conditions. Another option which will likely be evaluated is the proposed SCSI interface, provided that funding for that development effort is approved.

In making the transition from the Lick CCD controllers to the Leach controllers, an initial concern was whether a convenient, safe, and efficient means could be developed to allow CCD waveforms to be dynamically adjusted in real-time using the Leach controller, since adjustment of these waveforms is a frequent operation in our CCD detector lab. With the existing Lick controllers, the various CCD clocking and bias voltage levels are set by potentiometers, and can quickly be adjusted using a screwdriver while using an oscilloscope to watch the resulting effect on the CCD video signal. However, changes to waveform timing require recompilation, burning new UV-erasable EPROM chips, and installing these chips into the controller, which typically requires at least 10–15 minutes. Since the waveforms generated by the Leach controllers are set by programmable DACs, and since new waveform tables can be quickly downloaded via the fiber optic link, the potential exists for fully software controlled, dynamically adjustable waveforms.

To realize this potential, Stover has implemented a CCD waveform generation tool, wavegen (Stover 1991). Wavegen provides an X11/Motif-based graphical user interface which allows the user to draw new CCD waveforms, and to easily manipulate both the timing and voltage levels of existing waveforms. Wavegen also enforces CCD-specific voltage limits, to prevent the user from accidentally generating levels which might damage the chip, and also enforces timing constraints imposed by the Leach controller hardware. Once the user is satisfied with a given set of waveforms, wavegen produces an ASCII source file in DSP56001 assembler format which can be assembled to generate the appropriate waveform tables. While this makes the generation of these tables simpler, safer, and better documented, it still does not result in real-time waveform adjustment. However, it has reduced the time required to change a waveform (in either timing, voltage, or both) to under one minute, and that reduction has already markedly increased our efficiency in using the Leach CCD controllers to characterize the performance of new detectors.

Wavegen also provides a mechanism by which it can broadcast (via the MUSIC system) change notification event messages whenever the user makes any adjustment to a waveform. These messages can be monitored by another process, wavemon, which collects the changes and downloads an updated set of waveforms to the Leach CCD controller whenever the user requests. A preliminary version of wavemon is now being tested, and should be fully operational early next year. Once that is finished, it will be possible to safely and efficiently adjust Leach CCD controller waveforms in real-time.

5. Conclusions

Although predicting future trends is always somewhat problematical, if we use our experience at Lick over the past 12 years as a guide, it is likely that instrument computer hardware and software will continue to change much more rapidly than CCD controller hardware and software. That trend should be seriously considered when selecting an overall architecture for a CCD DAS.

The need for frequent re-implementation of the interface between the instrument computer and the CCD controller can be reduced by isolating the custom parts of that interface into a separate data capture computer, and by insuring that the data capture computer connects to a basically 'off-the-shelf' instrument computer using a widely-accepted (and hopefully long-lived) industry standard interface such as Ethernet or SCSI. In some implementations, such as the Keck CCD DAS, this data capture computer should significantly reduce both the loss of data and observing time that currently results from instrument computer failures.

The Leach CCD controller is gaining considerable acceptance in Astronomy, and may become something of a standard. All users of this controller will benefit from continued discussions regarding the best means of interfacing to this system, and from sharing experience gained in the operational details of these controllers. Efforts are now in progress to organize an informal users group, and to continue such discussions via electronic mail.

Acknowledgments. The overall design of the Keck CCD DAS was the result of a lengthy collaboration between the authors and Cohen, Cromer, Harris, and Southard at Caltech, Lewis and Lupton at Keck, Allen, Atwood, Keane, Ricketts, Robinson, Tucker, Vogt, and Wei at UCO/Lick, and Leach at SDSU. Detailing all of the contributions by these people would be a lengthy report in itself.

References

Conrad, A.R., & Lupton, W.F. 1993, this volume

Jim, K.T.C., Yamada, H.T., Luppino, G.A., & Hlivak, R.J. 1993, this volume

Lauer, T. 1981, The CCD Data-Taking System, Lick Obs. Tech. Rept. 28

Leach, R.W., & Beale, F.L. 1990, Proc. SPIE, 1235, 284

Lupton, W.F., & Conrad, A.R. 1993, this volume

Metzger, M.R., Tonry, J.L., & Luppino, G.A. 1993, this volume

Robinson, L.B., Stover, R.J., Osborne, J., Miller, J.S., Vogt, S.S., & Allen, S.L. 1987, Optical Engineering, 26, 795

Stover, R.J. 1989, MUSIC, Lick Obs. Tech. Rept. 54

Stover, R.J. 1991, Wavegen User's Manual, Lick Obs. Tech. Rept. 60

Discussion

M. Davis: What is the level of effort going into this (person-years)?

Kibrick: I don't have a precise number for this. The Keck CCD data acquisition system represents the combined efforts of two programmers at Lick, two at Caltech, and one at CARA/Keck Observatory, working together over a two year period, plus an additional programmer at CARA during the second year. However, none of them was working on this system full-time, and the level of effort allocated to this portion of the project varied considerably during that period. A crude estimate would be 5 programmers working on-average 35%-time for two years, plus 5% the additional programmer 5% the second year, or about 3.9 programmer-years.

Christian: What interface/interaction do you have with the telescope control system? (The point of my question is: Is the interaction with the telescope control passive or active? Will users have access to telescope control?)

Kibrick: Like the CCD data acquisition system, there are two different types of "active" user interfaces to the Keck telescope control system:

1. A windows-based point-and-click style user interface, intended to provide easy interactive access for routine operations.

2. A keyword-based command language user interface, intended for batch-mode operations or for more unusual or "custom" operations not easily provided by the windows-based interface.

The window-based user interface to the telescope control system currently operates in parallel with that of the CCD data acquisition system, so that the functions provided by each are controlled from separate windows. However, there is nothing in the system architecture which precludes providing both CCD and telescope control functions from a common window, should that prove desirable.

The keyword-based command language for the telescope control system employs the same style, syntax, and synchronization mechanisms as that of the CCD data acquisition system, so that commands to both systems can be inter-mixed in the same command script. They can also be inter-mixed with commands to data reduction systems such as Figaro or IRAF, so that operations such as focusing the telescope and spectrograph can be easily automated.

There is also a "passive" interface between the CCD data acquisition system and the telescope control system which provides for the automatic logging of all relevant telescope-related parameters in the FITS header recorded with each image.

Shaw: As you know, optical, ground-based astronomical data are typically obtained by sleepy, sometimes inattentive astronomers at high-altitude observatories. To what extent is simplicity a design consideration for the user interface to the Keck data-taking system?

Kibrick: Providing a simple and consistent appearance to the user interfaces was a major consideration in the design of the both the windows-based and command language user interfaces for the Keck data taking systems.

In the case of the windows-based system, we have tried to keep the displays simple and uncluttered, and to provide most functions via simple point-and-click operations which minimize the need for additional keystrokes. We have also tried to minimize the use of walking-windows and other mouse operations which can demand more dexterity than sleepy astronomers often have.

In the command-language user interface, we have kept things simple by employing a keyword-based system with a flat namespace, and by minimizing the number of actual commands (show, modify, waitfor) and command options (wait, nowait, notify). This user interface is consistent across all of the Keck instruments and the telescope control system. The resulting interface is elegantly simple, yet extremely powerful.

In addition to simplicity, a major design consideration in all of our user interfaces was providing the capability to run multiple, simultaneous invocations of these interfaces, both locally and remotely, in order to support coordinated local/remote observing. For example, a local observer at the Mauna Kea summit might observe (in real-time) in cooperation with a less oxygen-deprived and hopefully more alert colleague at the Keck headquarters in Waimea, or perhaps even in California. Once the instruments and telescope complete their testing/debugging phase and commence routine operations, remote observing (at least from Waimea) may become quite common.

Astronomical Data Analysis Software and Systems II
ASP Conference Series, Vol. 52, 1993
R. J. Hanisch, R. J. V. Brissenden, and J. Barnes, eds.

The U. H. Institute for Astronomy CCD Camera Control System

K. T. C. Jim, H. T. Yamada, G. A. Luppino, and R. J. Hlivak

Institute for Astronomy, University of Hawaii, 2680 Woodlawn Drive, Honolulu, HI 96822

Abstract. The U. H. CCD Camera Control System consists of a NeXT workstation, a graphical user interface, and a fiber optics interface which is connected to a San Diego State University CCD controller. The U. H. system employs the NeXT-resident Motorola DSP 56001 as a real time hardware controller interfaced to the Mach-based UNIX of the NeXT workstation by DMA. Since the SDSU controller also uses the DSP 56001, the NeXT is used as a development platform for the embedded control software. The fiber optic interface links the two DSP 56001s through their Synchronous Serial Interfaces. The user interface is based on the NeXTStep windowing system. It is easy to use and features real-time display of image data and control over all camera functions. Both Loral and Tektronix 2048×2048 CCDs have been driven at full readout speeds, and the system is designed to readout four such CCDs simultaneously. The total hardware package is compact and portable, and has been used on five different telescopes on Mauna Kea. The complete CCD control system can be assembled for a very low cost. The hardware and software of the control system have proven to be reliable, well adapted to the needs of astronomers, and extensible to increasingly complicated control requirements.

1. Introduction

Over the past year and a half we have designed and built a new CCD camera system with our specific requirements in mind. We needed a new system that could replace our old CCD cameras to take advantage of the latest technological advances in CCDs and could improve on the previous generation of systems. In order to improve the reliability of the cameras at Mauna Kea Observatory (MKO), we wanted to create a modular system in which any component could be quickly replaced. Modules standardized with other CCD camera groups were clearly preferable. In addition, any module can be upgraded while still maintaining data-taking capability.

Since the camera system would be used on several telescopes at MKO, the system had to be portable. It had to be small and it had to be able to operate in various observing environments, from observing rooms with complete network connections to domes with no computers or network access available. Consequently, the camera system had to be able to operate independently of

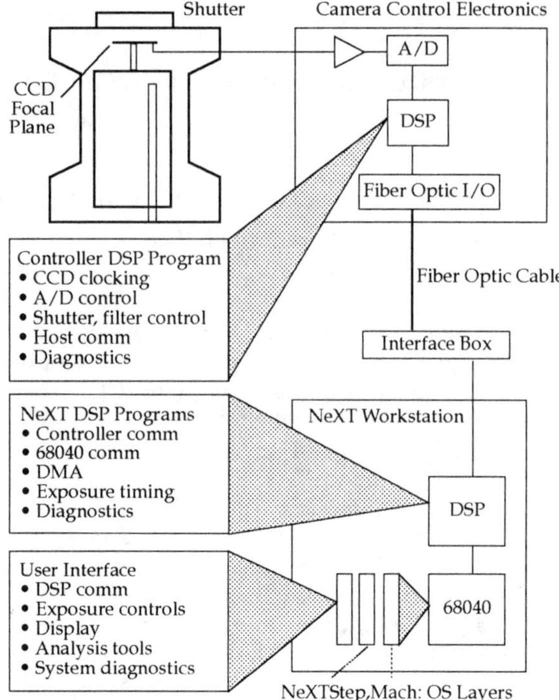

Figure 1. A schematic of the CCD camera system shows how the camera, control electronics, and workstation are connected. The functions of the three sets of software are indicated in the boxes.

any other resources, yet it had to be able to use other computing and network resources if they were available.

We needed a CCD controller which would be fast enough to clock our 2×2 mosaic camera of 2048×2048 CCDs at 50,000 pixels per second each (Luppino et al. 1992). For our new controller, we wanted the voltage levels of the CCD biases and clocks to be set by digital to analog converters (DACs) rather than the potentiometers of our previous generation system.

The workstation that controls the whole camera system would have to be easy to use, fast enough to display the CCD image as it was being read out, and, preferably, low cost.

2. The Hardware

A schematic of the system we designed and built is shown in Figure 1. The modular dewar, with the interchangeable camera head, is shown connected to the controller electronics which, in turn, is connected to the workstation through a fiber optic interface. Each of these components is described below.

At the heart of the camera system is the CCD, mounted in a modular dewar described extensively in Luppino and Miller 1992. The CCD controller electronics reside in a small box attached directly to the hermetic connector at the side of the dewar. We chose the SDSU controller (Leach 1988; Leach & Beale 1990) design since it met the requirements described in the introduction. It is based on the Motorola DSP56001 digital signal processor (DSP), and consists of a single digital timing board and an analog board for each CCD amplifier. The DSP and a two channel fiber optic link for receiving commands and transmitting data are mounted on the timing board. The analog boards have high-speed DACs which create the CCD biases and clock waveforms and contain the signal chain (preamp, dual-slope-integrator correlated-double-sampler, and 16-bit A/D converter). The controller also operates the shutter, with exposure timing determined either by the controller or by the workstation.

Commands and data are sent to and from the CCD controller through a simple fiber optic to serial interface to a NeXTstation computer (refer to Figure 1 for a schematic of the system.) The interface converts the high speed serial stream of the fiber optic link to a 4 MHz stream suitable for the serial interface of a DSP56001 resident on the motherboard of the NeXTStation. This approach offers several advantages. The NeXTstation DSP56001 serves as a development system for the CCD controller electronics which also use the Motorola DSP56001. The software provided by NeXT includes the 56001 assembler and a useful debugger from Ariel Corp (a C compiler is available from Motorola). Programming effort is reduced because both the controller and the workstation interface use the same programming environment, the DSP56001 assembler language. Finally, since all of the data must pass through both DSPs, it is possible to perform real-time filtering or preprocessing of the data at two stages in the data path, although that capability is not used now.

The control software for the CCD camera is distributed among three processors running simultaneously. Each performs a different task, as outlined in the boxes on the left side of Figure 1. The low-level CCD clocking waveforms, bias voltages, shutter control, and CCD readout are managed by a program resident in the DSP56001 of the SDSU controller. The NeXT DSP maintains real time control over the SDSU DSP through commands sent through the fiber optic interface. The NeXT DSP controls the timing of the shutter by sending appropriate interrupts to the SDSU DSP. The NeXT DSP communicates with the main CPU of the NeXT, a 33 MHz MC68040, through DMA over the host interface of the DSP. The NeXT DSP is loaded with a series of programs by a set of higher order functions executing on the MC68040. The user normally accesses these functions through the user interface, written in Objective C with a few lower level routines written in C for higher execution speed. In addition to the GUI, there is a command line interface and a low-level function library.

3. The User Interface

From the beginning, the user interface was designed with the needs of the observing astronomer in mind. The system is designed to provide the observer with the most important tools and operations needed for observing at the telescope. Experience showed that displaying the image as it was being read out was very important. The astronomer also needs to be able to quickly evaluate the data to make sure that the observations are proceeding correctly.

The user interface should allow all of the above operations without the need to use another computer, increasing reliability and portability, and making observing faster and more efficient. The user interface must be easy to use and to learn. To meet these requirements, we designed a system including only those functions really needed at the telescope: a real time display, the ability to zoom and pan that display, a tool for quickly evaluating the seeing, a tool to perform image statistics, and tools for quick look photometry and spectroscopy.

Our program can logically be divided into three areas: readout controls, displays, and analysis tools. The readout controls emphasize our philosophy of providing the minimum number of buttons that allow full control over CCD camera functions. The displays offer several ways of looking at the data, such as line cuts, zooming, and numeric data values. The analysis tools include the seeing evaluation, statistics, photometry, and spectroscopy tools.

The main windows of the user interface are displayed in Figure 2. Usually, the observer simply enters the exposure time and clicks on the Expose button. The CCD is exposed, the image is read out, and it is displayed in the Main Display window. For finer control over CCD functions the observer has several options, which are available as buttons and check boxes to the right of the Expose button. The Set Up window is where the configuration of the system is set. Through this window new controller software can be downloaded to the CCD controller and the configuration (display geometry and CCD controller software) can be selected as different cameras are attached to the system. From the Zoom window the other analysis tools, such as the seeing evaluation tool, are accessed by a set of buttons. The Plot window, which can display either column or row plots, is updated by mouse clicks on the Main Display, as are the other tools.

4. Conclusion

The CCD camera system described here has been used successfully at MKO since May 1992 with many different CCDs. We have successfully operated the system at five different telescopes at MKO, in environments ranging from the CFHT with full network support in a warm control room to the UH 0.6 m telescope with no network in a cold dome. As a testimony to the advantages of standardization, we were able to control a SDSU controller-based CFHT camera with only a day's worth of modifications. The interested reader should compare the design of our system with the others based on the SDSU controller described elsewhere in these proceedings (Kibrick et al. 1993; Metzger et al. 1993).

References

Kibrick, R.I., Stover, R.J., & Conrad, A.R. 1993, this volume

Leach, R. 1988, PASP, 100, 1287

Leach, R.W., & Beale, F.L. 1990, Proc. SPIE, 1235, 284

Luppino, G.A., & Miller, K. 1992, PASP, 104, 215

Luppino, G.A., Jim, K.T.C., Hlivak, R.J., & Yamada, H. 1992, Proc. SPIE, 1656, 414

Metzger, M.R., Tonry, J.L., & Luppino, G.A. 1993, this volume

Discussion

Kibrick: Do you have a safety mechanism (including a dewar ID that is computer readable) that prevents a user from downloading a set of waveforms that is inappropriate (i.e., unsafe) for the CCD which is in fact connected?

Jim: We currently do not have such a safety mechanism, although it could easily be accomplished by using an ID code in the EPROM of the Leach controller for each dewar (as CFH and probably others have done). However, we have never had a problem occur in six months of use on the telescope. By default, when the user interface is started, the correct code for the particular camera is downloaded. Since the code name is similar to the name of the camera, a problem is unlikely to occur even if the observer where to download the controller code manually.

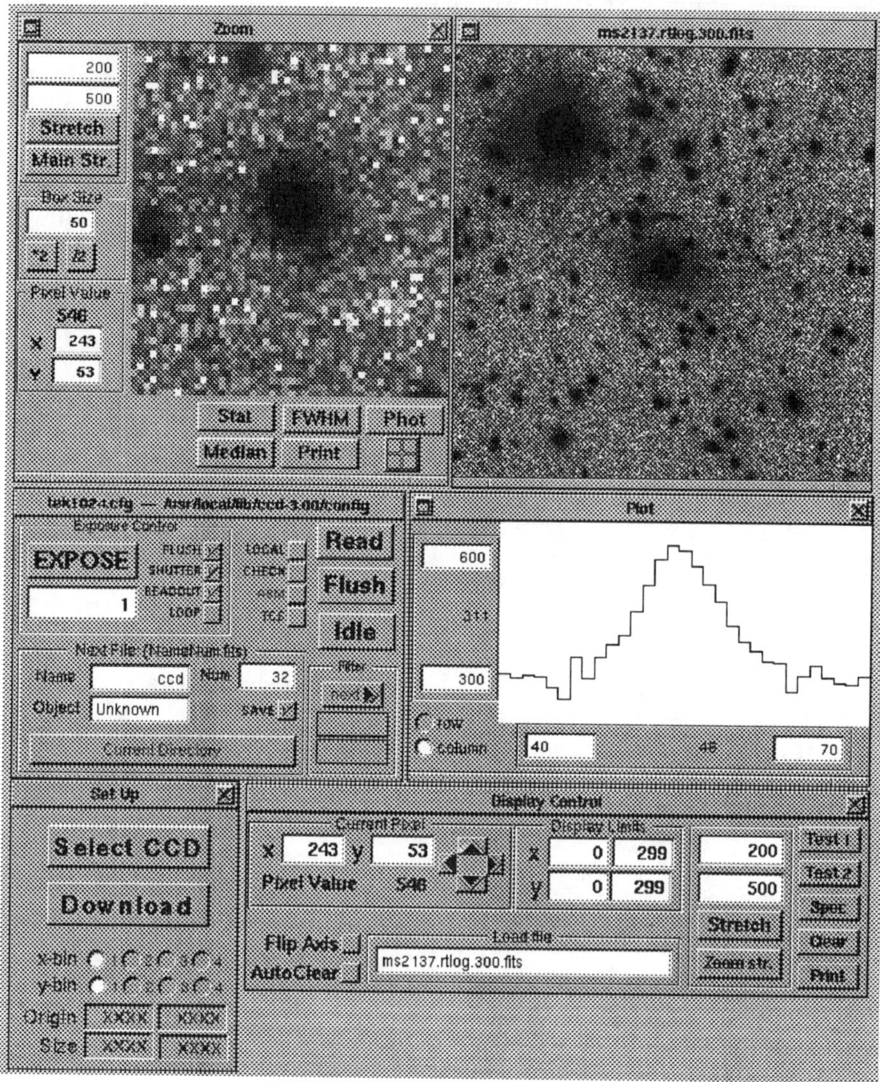

Figure 2. The main windows of the user interface of the CCD camera system. At the top left is the Zoom window, to its right is the Main Display window, below both are the Exposure Control and the Plot windows, and at the bottom are the Set Up and Display Control windows. The Main Display shows the image using a linear stretch set by the observer in the Display Control window. The astronomer can also control the region of the CCD that is displayed. The Main Display is linked to the analysis tools by mouse clicks. Double clicking on the Main Display, for instance, will update the Zoom window to show the the region around the cursor.

Astronomical Data Analysis Software and Systems II
ASP Conference Series, Vol. 52, 1993
R. J. Hanisch, R. J. V. Brissenden, and J. Barnes, eds.

The Data Acquisition System for the AAO 2-Degree Field Project

K. Shortridge, T. J. Farrell, and J. A. Bailey

Anglo-Australian Observatory, P.O. Box 296, Epping, NSW 2121, Australia

Abstract. The software system being produced by AAO to control the new 2-degree field fibre positioner and spectrographs is described. The system has to mesh cleanly with the ADAM systems used at AAO for CCD data acquisition, and has to run on a network of disparate machines including VMS Vaxes, UNIX workstations, and VME systems running VxWorks. The basis of the new system is a task control layer that operates by sending self-defining hierarchically-structured and machine-independent messages.

1. Introduction

The Anglo-Australian Observatory (AAO) is building a system that will provide a two-degree field of view at prime focus. A robot positioner will be used to locate up to 400 optical fibres at pre-determined positions in this field. While observations are being made using one set of 400 fibres, the robot will be positioning a second set of fibres in a background field that can be moved in to replace the first when the telescope is moved to a new position. The fibres feed two spectrographs each with a 1024 square CCD detector. The software for this new system has to control both the new positioner and the new spectrographs, and also has to fit comfortably into the existing observatory software systems.

2. Existing AAO Systems, and the Use of ADAM

Recent AAO data acquisition systems have been ADAM-based systems. ADAM (Astronomical Data Acquisition Monitor, Kelly 1992) allows data acquisition systems to be built out of individual modules (tasks) communicating through a message system and responding to named actions each controlled by a number of named parameters. An important part of the ADAM design is that an ADAM task is supposed to remain constantly receptive to new messages; it is not supposed to respond to an action request by performing that action to completion to the exclusion of everything else. Complying with this requirement means that a task can handle a number of concurrent actions, and also that it can respond to requests to cancel a current action. This is achieved by splitting actions that would normally take some time to complete into 'stages'.

An ADAM task comprises a 'fixed part' which is responsible for handling any message traffic and an 'action' routine that the fixed part calls to handle any action requests that it receives. This action routine should complete the requested action immediately, if this is possible. Otherwise, it should perform the first stage of the action, then return to the fixed part, indicating that it would like to be rescheduled, either immediately, or after a specified delay, or in response to some external event. The next time the action routine is called, it may be either to perform the next stage of a continuing action, or to begin or continue (or cancel) a quite different action. For example, the action routine for a task controlling a spectrograph may respond to a 'change filter' action by starting a filter wheel moving and then rescheduling itself at regular intervals to check whether the wheel is yet in the required position. If the interface is capable of generating a message or an interrupt (if the spectrograph is controlled by a separate microprocessor, for example) then the action routine can simply request to be rescheduled when that interrupt or message is received.

AAO has had a lot of success with ADAM. In particular, its CCD observing system, ('OBSERVER') has proven a very flexible system. The modular nature of the system, coupled with a 'generic' approach to its original design, has resulted in a system that can be reconfigured to control a number of different combinations of instruments. For example, there is a 'detector control' module, whose actions are defined in a general purpose way, such as 'EXPOSE', 'READOUT', etc. When the system is being used with a conventional CCD, this module is the normal CCD control task. When the IRIS infra-red detector was introduced, it was incorporated into the system by writing a new detector control task for it and adding a new spectrograph task that could control the new IRIS spectrograph. In principle, the 2dF, for all its complexity, can be thought of as just a new spectrograph, and all that is required is a 2dF spectrograph control task that can replace the usual spectrograph control task in the OBSERVER system.

3. New Requirements of the 2dF

However, the 2dF system—even if 'just a spectrograph' in principle—is in practice a complex system involving the control of a robotic positioner that, amongst other things, has to position 400 fibres to 10 micron accuracy in a target time of 20 minutes. The control of the mechanics of this is an exercise in the precision real-time control of a robotic system, including the use of cameras to check the position of the fibres. Also, the output from those fibres used for guiding has to be fed back into the telescope control system. Calculating the configurations—the mapping of fibre number onto required position in such a way that the fibres do not interfere or overlap unacceptably—and the configuration changes—moving from one configuration to another without producing a disastrous tangle of fibres—both need a significant application of CPU power, and the data has to be reduced at least to some extent as it is taken in order to keep the amount of data manageable. The computation required for the configurations and the data reduction mandates the use of UNIX workstations, while the real-time control of the positioner will be done using a VME system running the VxWorks real-time kernel.

The existing ADAM system runs on the AAO's VAXes, and as yet the real-time parts of ADAM have not been ported to other systems other than

VMS. Moreover, ADAM as it stands at present is not strong in its networking capabilities, even between VAXes. We felt, as a result of our experience of ADAM, that we wanted to use an ADAM-type system for as much of the new system as possible. We wanted the various positioner modules to behave like ADAM tasks, able to respond to commands sent through the sort of general purpose control interfaces we have already developed for ADAM tasks. Being able to talk to any part of the system through a standard interface has a great attraction, both for the control and diagnosing of the final system and also because it means that the general structure of any part of the system will be familiar even to those of our programmers who were not connected with its development.

So for 2dF we have taken the opportunity to extend the ideas of ADAM to produce a system that can operate well in conjunction with ADAM, but that is inherently networked and portable. Some of the ideas used in this system have been discussed within the ADAM community in the context of an ADAM II system, and we regard this project to some extent as a test-bed for concepts that might eventually be incorporated into an ADAM II design. In particular, the original ADAM system provided a limited message size with a restricted message format and a message throughput that was not only limited to a single machine but was also less efficient than one would like.

4. Components of the New 2dF Software System

The system we are producing provides a layer called DITS (Distributed Instrumentation Tasking System) that looks very similar to the existing ADAM task library, although with a more flexible and powerful repertoire of routine calls. It is built upon two main sub-systems: a message system (IMP) that handles the sending of messages between the distributed tasks on the system, and a hierarchical data system (SDS) that allows complex self-defining, machine independent, structures to be encoded into messages without the need for the receiving and sending tasks to share precise byte by byte message format definitions. So the fixed part of a 2dF system task will rely on DITS routines to monitor the message queues of the task, interpret the structured messages (to the extent required to know what part of the system to pass them on to) and then to reschedule the various routines in the task accordingly.

The message system (IMP) is designed to be able to handle inter-task and inter-processor messages of any length, efficiently, and without ever requiring a task to 'block' (i.e., be unresponsive to 'cancel' messages, enquiry messages, etc.) other than when deliberately waiting for external input—all of which will be through such messages. The essential requirement here is that a message 'send' operation should never be able to block. An initial version of this system is now running on VMS, some flavours of UNIX (Solaris 1.0 and ULTRIX), and on VxWorks. On a VAX 4000 (the machine of those tried so far on which it runs fastest) a task can send a short message to another task and receive a reply in about 320 microseconds. This seems an adequate performance, given that it will be slowed down somewhat by the higher levels of the software.

The messages will be encoded in structures using the SDS routines. For example, a DITS message may contain a structure of a type that could be declared in C as:

```
typedef struct {
    unsigned long int Seq;          /* Sequence number expected */
    long int Actptr;                /* Action pointer */
    long int Obeyid;                /* Unique to each invocation */
    long int ReasonStat;            /* Status of the message */
    long int Flags;                 /* Message flags */
    Dits___MsgType Type;            /* Message type */
    Dits___NetTrIDType TransID;     /* Transaction id */
    } Dits___MessageType;
```

While a structure like this could be sent, raw, as a message containing `sizeof(struct DITS__MessageType)` bytes, this could only be easily interpreted if received by a program that had the same structure definition built into it and was running on an identical machine. By adding the structure definition information and encoding details (byte order, floating point format, etc.) that SDS uses, we can send messages that are larger, because of the additional information, but that can be decoded easily by any program on any machine. For example, a monitor program that intercepts message traffic can list the contents of any message, showing the full hierarchy, including item names, of the message structure and the data it contains, while knowing nothing about the detailed content of the message other than that it is an SDS structure. We feel that for a distributed system, this huge flexibility is worth the overheads. (And it does not preclude a more efficient use—two cooperating programs could still encode a complex data structure using an agreed private format and send the whole thing as an SDS structure containing the encoded information as a single item. However, we shall try to avoid this.)

The DITS tasking layer that sits upon IMP and SDS provides a set of facilities similar to those of the ADAM task library. Individual tasks in the system will execute actions requested by tasks higher in the system hierarchy, often by requesting that actions be performed by lower level tasks. Tasks will all be required to remain receptive to new messages at all times, so that actions can be cancelled cleanly if necessary and so that concurrent actions can be handled by a single task. This also allows easy communication with all tasks for diagnostic purposes, since any task can respond to an enquiry as to its present state. In emergencies it also allows fine control to be exercised over any task in the system, something that is also made easier by all tasks having the same conventions about how actions are invoked.

5. The General Structure of a 2dF System Task

Figure 1 attempts to show the relationships between the various parts of a task in the 2dF system. Messages arrive on the message system and are decoded by the fixed part of the task. Messages connected with parameter values cause changes in the task parameters, other messages are handled by calling an appropriate action routine in the task-specific code. These action routines can make use

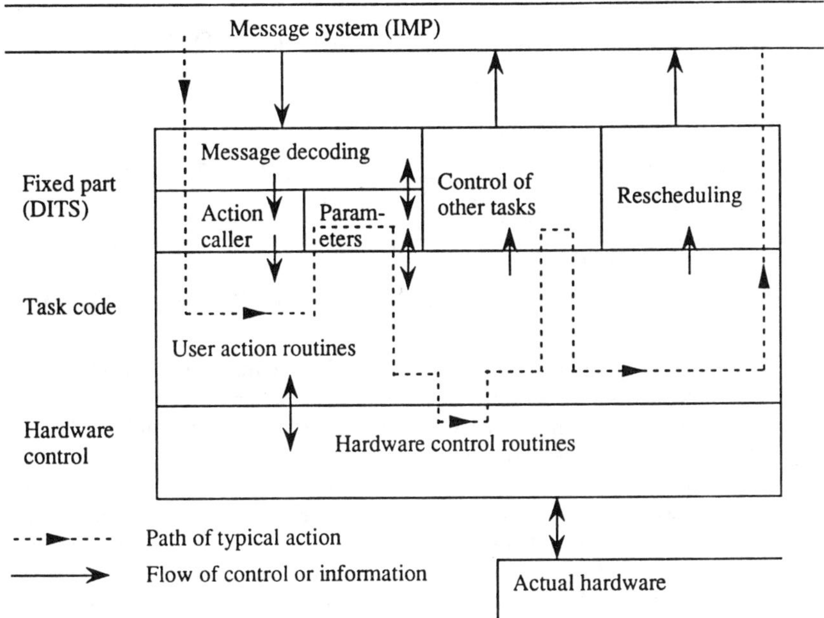

Figure 1. Structure of a typical 2dF system task

of the tasking routines in the fixed part to communicate with other tasks in the system, generally by sending them action requests. Or, the action routines, particularly those in the lower level tasks, may control the system hardware directly. Finally, an action routine returns to the fixed part, either indicating that it has finished or requesting that it be rescheduled.

The dashed line in the diagram indicates the flow of control for a typical action request. It is decoded by the fixed part, then an action routine is called. This routine will generally pick up parameter values for the action, then may perform some hardware control function or exercise control over some other task or tasks in the system (doing both, as shown, would in practice be unusual, since the higher level tasks will control other tasks and the lower level tasks will control hardware). Finally, it returns, possibly requesting that it be rescheduled. Rescheduling is handled by the fixed part arranging for a new message to be sent to the task after a specified delay, or will be as the result of a message being sent by another task. Either way, the whole task is driven solely by the arrival of messages on the message system. A message may be sent from interrupt level, so hardware interrupts can also be used to trigger new messages.

References

Kelly, B.D. 1992, ADAM — Guide to Writing Instrumentation Tasks, Starlink User Note 134

Astronomical Data Analysis Software and Systems II
ASP Conference Series, Vol. 52, 1993
R. J. Hanisch, R. J. V. Brissenden, and J. Barnes, eds.

Wilbur: A Low-Cost CCD System for MDM Observatory

Mark R. Metzger, John L. Tonry

Department of Physics, Massachusetts Institute of Technology, Cambridge, MA 02139

Gerard A. Luppino

Institute for Astronomy, University of Hawaii, Honolulu, HI 96822

Abstract. We describe "Wilbur", a CCD camera constructed for the Michigan-Dartmouth-MIT Observatory. The camera system hardware was constructed using existing designs for the dewar and control electronics and a commercially available control computer. The requirements for new hardware design was reduced to a simple interface, allowing us to keep the cost low and produce a working system on the telescope in under three months. New software written for operation of the camera consists of several individual components which provide data acquisition from the CCD, control of the telescope, and operation of auxiliary instruments. The hardware and software are modular, giving the flexibility to operate with other existing and future detectors at the observatory. The software also provides advanced CCD readout features such as shutterless video and drift scanning, and can be operated remotely from other computers over an IP-based network.

1. Introduction

The rapid advancement of CCD technology has made large-format detectors available at modest cost. To get a large-format detector working at the Michigan-Dartmouth-MIT Observatory (MDM), we had to address several issues. A significant problem was that the existing computers at MDM, Sun 2's, had been in place since 1984 and were already at the limit of their capability handling the smaller CCDs in use at MDM. Simply using a new computer for the new camera only is awkward, as observers would have to switch between different interfaces when using different detectors. It would also be difficult for the observer to control the telescope and other equipment using the old machine while taking data with another computer (there are no telescope operators at MDM).

We therefore chose the design for the new CCD camera in the context of an overall upgrade of the computer system at MDM. Telescope and auxiliary instrument (e.g., filter wheel, guider) control would eventually be migrated to the new system. The new system should be capable of running the existing CCDs at MDM that would not be retired, and be flexible enough to handle new CCDs we might acquire in the near future. And, of course, we wanted the system to be inexpensive and to operate on the telescope as soon as possible.

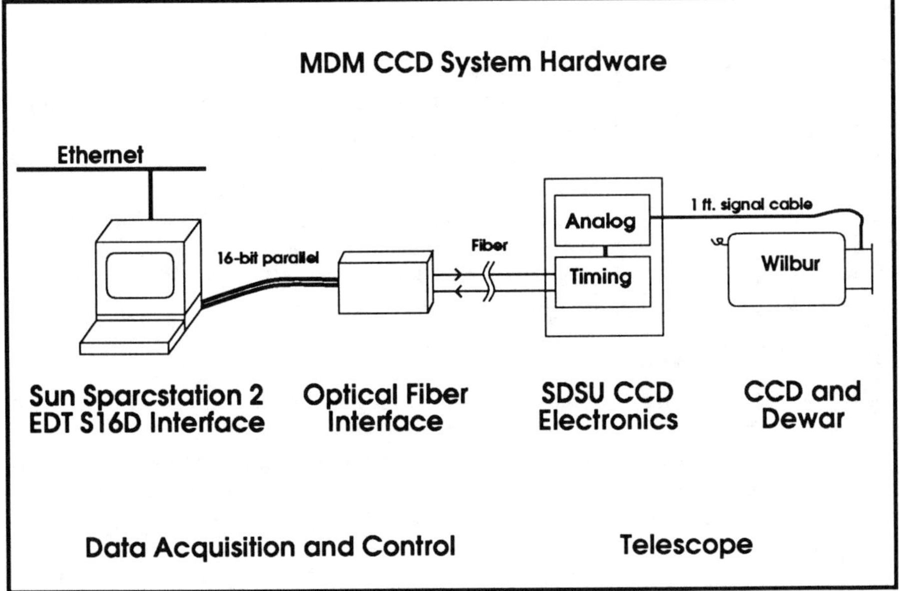

Figure 1. Diagram of the MDM CCD Camera System.

2. Hardware Overview

The hardware and characteristics of the CCD in Wilbur are described in detail by Metzger et al. (1993). The CCD clocking and A/D conversion are provided by a set of circuit boards from San Diego State University (Leach 1989). The versatility of the Leach electronics allows a wide range of CCDs to be operated with no additional hardware: the Leach system is controlled by a separate microprocessor (a Motorola DSP) which generates all of the bias voltages and clock signals. Thus most CCDs can be operated by simply writing code specific to the chip and downloading it to the DSP. Communication between a Sun Sparcstation 2 and the CCD electronics is via a pair of fiber-optic cables through a custom interface. The configuration is similar to that described in this volume by Jim et al. (1993), using a different front-end computer and interface to the Leach electronics. A diagram of the data system is shown in Figure 1.

The dewar chosen was a modification of the modular dewar described by Luppino and Miller (1992). The dewar consists of a cryogenic can attached to an interchangeable camera head, allowing different CCDs to be enclosed in separate camera heads and avoiding duplication of the entire dewar. The camera head was necked down to allow use on the instruments at MDM, and has a smaller window than the dewar described by Luppino and Miller. The CCD is a Loral 2048$\times$2048 device (Geary et al. 1991) of good cosmetic quality. The large (3 ℓ) capacity of the dewar provides a hold time of $>$ 24 hours with the Loral chip.

3. Software

The software consists of several independent components. CCD control, shutter operation, and data acquisition are controlled by a simple command-line style program. Other instrument control is provided by a suite of three window-based programs that operate under X Windows. One provides control of telescope pointing, the second provides control of an instrument containing the filter wheel, calibration lamps, and finder/guider cameras (the Multiple Instrument System; MIS), and the third is a window interface to control telescope guiding. The window programs also incorporate IP-based socket interfaces, which allows other software to control or inquire status of the telescope and instruments.

3.1. Data Acquisition

CCD control, exposure timing, data acquisition, and data storage are provided by a single program run in a terminal window of a Sparcstation. The observer controls the exposure through a set of simple commands. Data are stored on disk in FITS format (Wells et al. 1981), and header information is supplied automatically including exposure timing, telescope position, filterwheel status, and documentary information provided by the observer such as object name, comments, etc. Also supported are features such as on-chip binning and subarray readout, and overclock regions. The command interface is kept concise, so that the observer can set up the most frequently changed information, exposure type, exposure time, and object name, with a single command.

The acquisition software will operate with many different CCDs using different code for the DSP in the Leach electronics. The code is written in a standard fashion so that once DSP code for a particular CCD is downloaded, the acquisition software can query the electronics for information on the CCD format, binning, and overscan. We are currently using the system for both the Loral chip in Wilbur and a smaller Texas Instruments virtual phase device in a different dewar assembly (Luppino 1989). The software will also operate with both the 2.4 m and 1.3 m telescopes, using a file to identify different configurations.

The software currently supports two special operating modes. The first is a shutterless video mode: the shutter is opened and a subarray is read out and displayed at up to 10 Hz. The video mode is extremely useful for adjusting the telescope focus: one can see the image change as the focus is adjusted, and a real-time readout of the stellar profile width is provided. Since the fast readout can track a good portion of the image motion, a better focus can be obtained with this method than ordinary focus frames. This becomes especially important when the seeing is good and minor variations in temperature must be tracked carefully (the 2.4 m telescope frequently produces images having seeing of 0.7 arcsecond or better). A drift-scan or TDI mode (Boroson et al. 1983) is also provided for clocking out long strips of data while tracking the telescope slower than sidereal rate.

3.2. Telescope Control

Control of the 2.4 m telescope is performed through a window displayed on the Sparcstation under the X Window System ("Xtcs"). The layout of the window is shown in Figure 2. The telescope can be controlled by manually entering

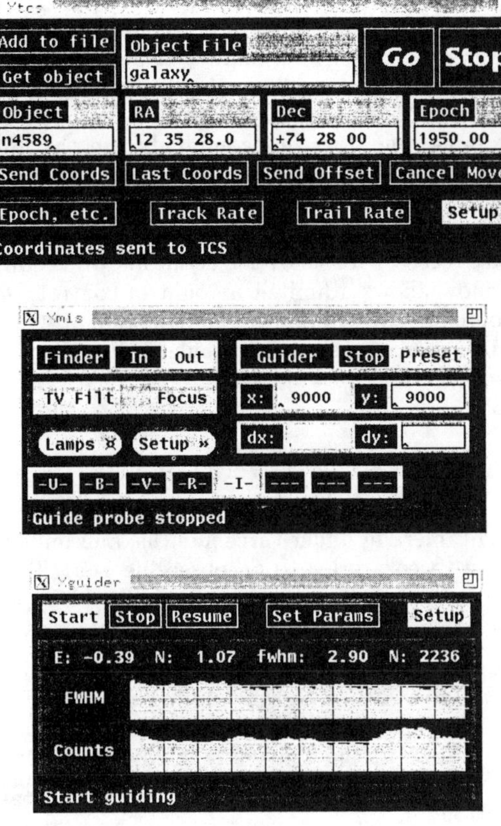

Figure 2. The TCS, MIS, and guider control windows, from top to bottom.

coordinates or by using a text file containing a list of coordinates accessed by name. Other features include coordinate offsets, tracking rate adjustment, and setting encoders. Less frequently used features are provided in separate pop-up windows.

The telescope control window operates by sending commands to the Telescope Control Computer (TCS) through a serial line. The TCS was reprogrammed to allow full control of the telescope over this line, and to provide complete telescope status as needed. Status information can be obtained by other software (e.g., the data acquisition program) via messages exchanged through Internet Protocol-based sockets. The program can be run on an entirely separate machine, if needed: the window can be displayed remotely on the observer's screen and the data acquisition software can obtain the telescope's position over the Ethernet. The socket-based interface can also allow automated or remote control of the telescope, though this feature is not currently used.

3.3. MIS and Guider Control

The filter wheel, calibration lamps for spectroscopy, finder mirror, and guider probe are all controlled by the MIS window ("Xmis") on both the 2.4 m and 1.3 m telescopes (see Figure 2). A configuration file specifies the setup, including filter and lamp names and whether the various components of the MIS are being used (for example, one does not commonly use the filter wheel with a spectrograph). Status of the MIS can be obtained via a socket interface similar to Xtcs; the data acquisition program queries for the status with each exposure and includes the information in the FITS header. The MIS window software sends commands and receives status over a serial line to the MIS.

The guider window allows a limited amount of control over the guider. It allows starting, stopping, and resuming guiding, and provides a sliding strip chart of relative guide star intensity and profile width. Parameters specific to the guider can be set in a pop-up window.

4. Conclusion

While many useful functions are left to be implemented, we were able to produce a working CCD camera in under three months, and implement most of the control software in two weeks prior to an observing run. The system is modular, low cost (in both materials and labor), and should be flexible enough to accommodate the needs of MDM Observatory for several years.

Acknowledgments. MM and JT acknowledge the hospitality of the Institute for Astronomy, University of Hawaii, during construction of the camera. Bob Barr's tireless help throughout the project was invaluable. This work was supported by many sources at the University of Michigan, Dartmouth College, and M.I.T., and by NSF grants AST89-58065 (JT), AST90-20680, AST90-15920 (MM), and AST92-01394.

References

Boroson, T.A., Thompson, I.B., & Shectman, S.A. 1983, AJ, 88, 1707

Geary, J., Luppino, G.A., Bredthauer, R., Hlivak, R., & Robinson, L. 1991, Proc. SPIE, 1447, 264

Jim, K.T.C., Yamada, H.T., Luppino, G.A., & Hlivak, R.J. 1993, this volume

Leach, R. 1989, PASP, 100, 1287

Luppino, G.A. 1989, PASP, 100, 931

Luppino, G.A., & Miller, K.R. 1992, PASP, 104, 215

Metzger, M.R., et al. 1993, in preparation

Wells D.C., Greisen, E.W., & Harten, R.H. 1981, A&AS, 44, 363

Astronomical Data Analysis Software and Systems II
ASP Conference Series, Vol. 52, 1993
R. J. Hanisch, R. J. V. Brissenden, and J. Barnes, eds.

The ADAM Environment and Transputers

B. D. Kelly, B. V. McNally, and J. M. Stewart

Royal Observatory, Blackford Hill, Edinburgh EH9

Abstract. The ADAM environment is used by Starlink for data analysis and by UK-involved observatories in Australia, Hawaii and the Canary Islands for instrumentation. The Royal Observatory Edinburgh has produced a Transputer version of the ADAM kernel to allow instruments which make use of Transputers via an Ethernet-TCP/IP connection for data acquisition/control to integrate more closely with the ADAM systems running at the telescopes.

1. Introduction

ADAM is the Starlink software environment (Allan 1992), providing the usual facilities to data analysis applications, such as a data system (HDS), a parameter system, error and message reporting facilities, graphics, etc. In its other use, as an environment for instrumentation (Stewart et al. 1992), ADAM has other properties which become important—in particular a tasking model and intertask communication.

As instrumentation systems become more network-based it has proved necessary for ADAM intertask communication to be networked as well. The next step involves heterogeneous networks, where cooperating ADAM applications are not only running on separate physical machines but are also running on different types of machine. Starting from a VAX/VMS implementation, Starlink is at an advanced stage in rendering ADAM portable and available on Unix systems. Meanwhile, ROE have produced an ADAM subset to run on Transputer systems.

2. Tasking and Communications Overview

An ADAM instrumentation application is built into a task. A typical instrumentation system consists of many such tasks. The tasks are loaded as separate processes into one or more machines and hook themselves into the intertask communication system. ADAM user interfaces behave similarly, leading to the situation summarised in Figure 1. This shows that more than one user interface can be present, and that tasks can communicate with one another as well as with user interfaces.

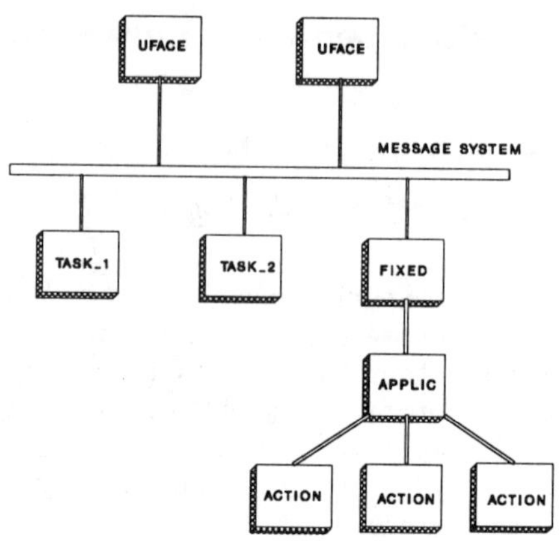

Figure 1. ADAM task and communications overview

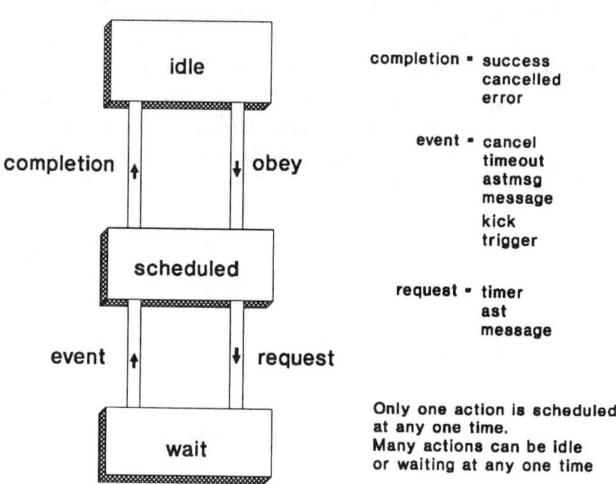

Figure 2. State transitions of an action

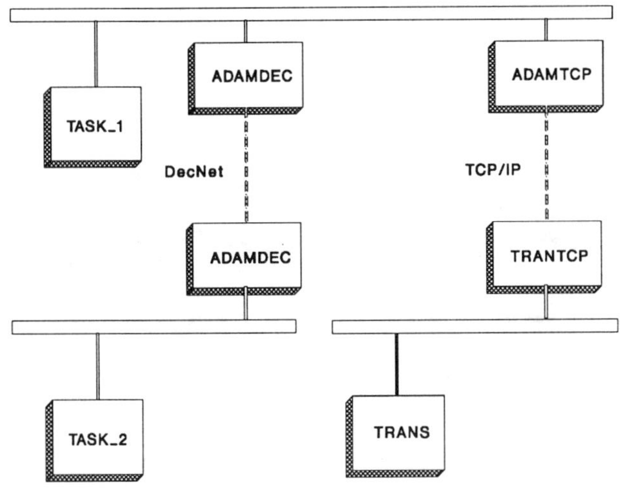

Figure 3. Networking the communications system

3. Structure of a Task

Figure 1 shows the general structure of a single task. The "fixed-part" is provided by the ADAM system and its function is to receive and interpret messages. When necessary it calls the application. The application consists of a number of "actions", corresponding to commands which can be sent to the task. The fixed-part is aware of the actions in the application, and keeps track of the state of each action. Figure 2 shows the state transitions possible for an individual action. Note that only one action can be scheduled at a time (i.e., the different actions are in the same process), but a number of actions can be simultaneously in the WAIT state pending a message arriving from a parallel thread or AST handler.

4. ADAM Networking

ADAM networking is achieved by having a network process interfacing the network to the intertask communications system. Figure 3 shows a VAX/VMS system containing an ADAM task, TASK_1, ADAMDEC which interfaces to DECnet and ADAMTCP which interfaces to TCP/IP using sockets. On the other end of the DECnet connection is a VAX/VMS system containing another copy of ADAMDEC plus TASK_2. On the other end of the TCP/IP connection is a Transputer system running TRANTCP, interfaced to ethernet via an Inmos B300 module. TRANTCP uses a socket interface and acts as a gateway process to the Transputer system network. Figure 4 shows two Transputers, the first containing the gateway interfacing to TCP/IP. Both Transputers are shown with a Router task which interfaces the individual Transputers to their internal network, plus an ADAM task marked "T". The current implementation

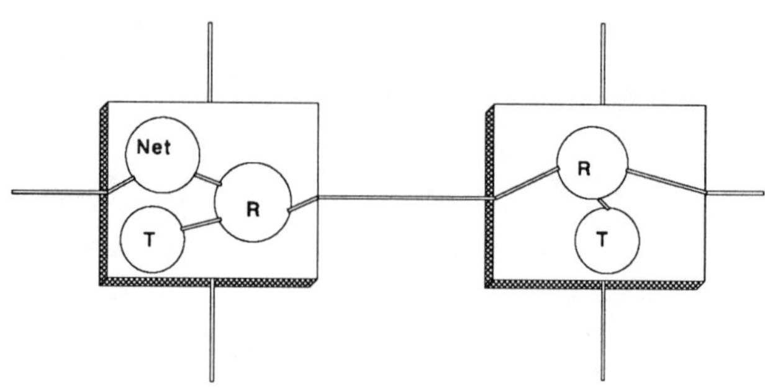

Figure 4. Inside the Transputer network

assumes the Transputers are organised as a chain carrying the ADAM network protocols. The two remaining links of each Transputer are available for use by the applications.

5. Transputer ADAM Subset

Transputer ADAM implements the full instrumentation tasking model as shown in Figure 2, but provides only a subset of the ADAM libraries to the application programmer. These are a cut-down parameter system, simplified error and message systems, and full support for intertask communications. The interfaces are provided only for the "C" language. It will be noticed that this list excludes support for HDS and the noticeboard systems, so bulk data has to be returned from the Transputer system by the application making sockets calls associated with a free Transputer link connected to the B300 module.

6. Current Applications

Two systems are currently under development at ROE which make use of the Transputer-ADAM system.

- ALICE, a project to upgrade the main infrared cameras and spectrometers at UKIRT to use 256×256 arrays, with the potential to use larger arrays in the future. The first phase of this project will be completed in mid 1993. At a maximum frame rate of 120 Hz, necessary to prevent saturation at 5 microns, the system generates a raw data rate of 7.9 Mpixels/s. Realtime data processing, being implemented on a network of 25 transputers, includes coadding data, calculating variances, image display and image sharpening (Sylvester & Paterson 1992).

- SCUBA (Submillimetre Common-user Bolometer Array), a bolometer array which is being built for the James Clerk Maxwell Telescope, and which uses a network of 12 Transputers. Data are collected from 131 bolometers at 128 samples/sec, despiked, digitally demodulated and resampled into 2-D images (Cunningham & Gear 1990).

References

Allan, P.M. 1992, in Astronomical Data Analysis Software and Systems I, A.S.P. Conf. Ser., Vol. 25, eds. D.M. Worrall, C. Biemesderfer, & J. Barnes, 126

Cunningham, C.R., & Gear, W.K. 1990, in Instrumentation in Astronomy VII, SPIE Vol. 1235

Stewart, J.M., Beard, S.M., Mountain, C.M., Pickup, D.A., & Bridger, A. 1992, in Astronomical Data Analysis Software and Systems I, A.S.P. Conf. Ser., Vol. 25, eds. D.M. Worrall, C. Biemesderfer, & J. Barnes, 479

Sylvester, J., & Paterson, M.J. 1992, in Proc. of ESO Conference on Progress in Telescope and Instrument Technologies, held in Garching, Germany 27–30 April 1992

Astronomical Data Analysis Software and Systems II
ASP Conference Series, Vol. 52, 1993
R. J. Hanisch, R. J. V. Brissenden, and J. Barnes, eds.

AXAF VETA X-ray Data Acquisition and Control System

R. J. V. Brissenden, M. T. Jones, M. Ljungberg, D. T. Nguyen, and J. B. Roll, Jr.

Smithsonian Astrophysical Observatory 60 Garden Street, Cambridge, MA *02138*

Abstract. We describe the X-ray Data Acquisition and Control System (XDACS) used together with the X-ray Detection System (XDS) to characterize the X-ray image during testing of the largest pair of AXAF-I mirrors. Features of the system include layered UNIX tools, real-time X-window displays and IRAF compatible parameter files. The system operation is illustrated by describing a mirror alignment test.

1. Introduction

The Advanced X-ray Astrophysics Facility – Imaging (AXAF I) will be launched in 1998 and consists of an X-ray telescope, high and low resolution gratings and two imaging detectors: the High Resolution Camera and the AXAF CCD Imaging Spectrometer. The X-ray telescope is comprised of four paraboloid and hyperboloid pairs of Wolter type I grazing incident mirrors, the largest of which were completed during the summer of 1991. The mirror elements were assembled into the Verification Engineering Test Article (VETA) and aligned and tested in X-rays at the X-ray Calibration Facility (XRCF) at Marshall Space Flight Center during September and October, 1991.

The XRCF consists of an X-ray source located 518 m from a $23 \times 7.6 \times 7.6$ m vacuum test chamber and connected by a pipe of approximately 1.2 m diameter. The test chamber housed the VETA and the VETA X-ray Detector System (VXDS) which consisted of a focal plane assembly of detectors mounted on stages and apertures (XDA), a data acquisition and control system (XDACS), a set of quadrant shutters located in front (the source end) of the VETA and normalization detectors located in front of the shutter assembly. In addition, the X-ray Flux Monitor (XFM) detector system monitored the flux in the X-ray pipe 45 m from the source and the Motion Detection System (MDS) measured the relative vibration of the focal plane instruments, VETA and source.

Tests were conducted at five energies in the range 0.277 keV (C)–2.7 keV (Mo) (the mirrors were uncoated and had little response above 2.5 keV). During testing, X-rays were generated at the source, detected by the XFM, entered the chamber, detected by the normalization detectors, focussed by the VETA and detected by instruments at the XDA. In this paper we describe the XDACS component of the VXDS and provide an example of test operation. For a more detailed description of the VXDS see Podgorski et al. (1992) and for the XDACS see Brissenden et al. (1992).

2. System Description

The X-ray Data Acquisition and Control System (XDACS) consisted of the software, computers and controllers used to perform the control, data acquisition, monitoring, logging and analysis functions during mirror testing. The central operator workstation (a SUN 4/330) interfaced with subsystems via TCP/IP over ethernet or IEEE 488 bus as shown in Figure 1.

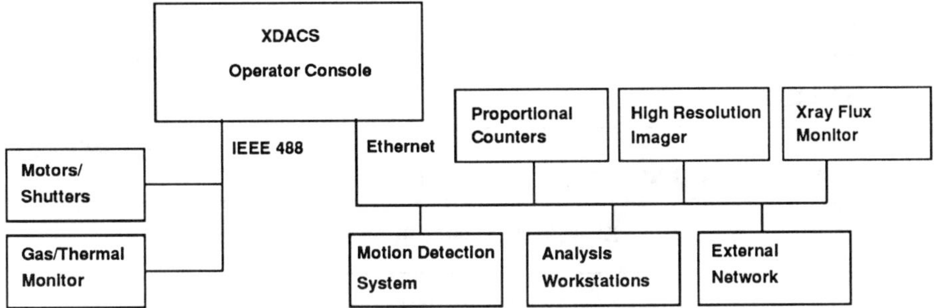

Figure 1. XDACS network architecture.

The subsystem functions are described briefly in Table 1. In addition, detailed analysis was performed during testing with four workstations and data access was available from remote sites through the external network.

Subsystem	Description
Motors/Shutters	Motor stages moved detectors and apertures in the focal plane with 2 μm accuracy. Shutters were opened and closed during focus and alignment tests.
Gas	Gas supply system controlled flow rate and type of gas through the focal plane and normalization flow proportional counters.
Thermal	Thermal monitoring system measured temperatures at various places of interest on and around the equipment.
High Resolution Imager	Imaging detector with 20 μm spatial resolution and 0.1–3.5 keV energy response.
Proportional Counters	Flow and sealed counters were provided as focal plane and normalization detectors.
X-ray Flux Monitor	Proportional counter with gas and thermal system located near the source to monitor beam intensity.
Motion Detection System	Detected relative motion between the focal plane, VETA and X-ray source with 50 Hz resolution.

Table 1. Description of the major XDACS subsystems.

The software architecture (Figure 2) was layered with a common UNIX and X-windows user interface, a central archive function and independent low level

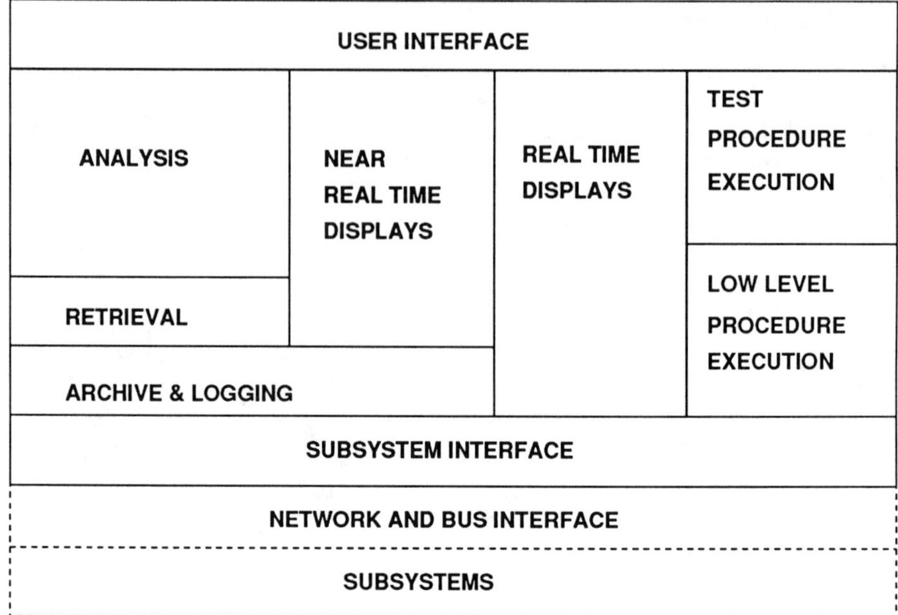

Figure 2. VETA software architecture. Dashed lines indicate network or hardware components.

subsystem interface programs. Subsystem functions were available as UNIX style commands and higher level procedural commands were built as shell scripts.

High level commands implemented complex procedures by combining lower level scripts that initiated subsystem commands. Data and status were returned to the Subsystem Interface level and made available to the archive and the real time displays. Quick-look analysis was performed with data retrieved from the archive and less time critical data were displayed in near real time through the archive.

Programs in the Test Procedure Execution and Low Level Procedure Execution layers were coded in korn shell. User and system parameters were stored in IRAF style parameter files and access was provided by a convenient set of interface tools available from both the command line and as library calls (Roll and Mandel 1992). As an integration language the korn shell provided highly flexible string manipulation, immediate access to UNIX tools, rapid development of new procedures and a familiar user environment which reduced operator training. The Subsystem Interface programs were coded in optimized C or C++ as were the display tools.

The real time display tools were developed as X Windows clients and displayed photon event data from the High Resolution Imaging (HRI) detector and the Proportional Counter subsystem. The HRI display had many of the features of SAOimage including pan, zoom, color map manipulation and pixel readout with updates every 0.1 sec. The Proportional Counter display updated every second and showed the cumulative spectra from the two focal plane and two normalization counters, and provided zoom on region of interest, statistics summary and selectable autoscale. Data from the motors subsystems were displayed in near real time as an ASCII table displayed in an X-window. The Gas and thermal displays were combined and showed pressures, flow rates, valve status and temperatures as green or red highlighted values. When a temperature of pressure went out of range the operator was notified and the displayed value changed from green to red.

3. Test Operation

The operator initiated procedure steps by setting test parameters and executing the specific test as a system command. The operator monitored data from all subsystems in real or near-real time during test execution and results were logged to the XDACS data archive either directly of after quick-look analysis. Examples of tests included: beam centering, (mis)alignment measurement, encircled energy, full-width-half-maximum and effective area. Typical parameters associated with these tests included: focal plane instrument of choice, integration time, scan step size, aperture type and size.

Data from multiple subsystems were combined during tests, for example, proportional counter spectra from the normalization detectors were used to normalize spectra acquired by moving a focal plane detector and aperture via the motorized stages and MDS data were continually monitored to determine if vibrations exceeded allowed limits.

3.1. Test Example: Alignment

The alignment of the mirror elements was accomplished by iteratively measuring the tilt (misalignment) errors and adjusting the position of the hyperboloid relative to the fixed paraboloid. The tilt errors were calculated by combining the centroids of the four images acquired by opening each of the quadrant shutters in turn. The focus error was also derived and an adjustment was made at each iteration to the focal plane detector focus position. For a detailed description of the alignment of the VETA see Brissenden et al. (1991).

A typical alignment test proceeded as follows (parameter values are representative). The operator entered the alignment command and was prompted for all test parameters prior to initiating the test. The first shutter was opened with the remaining three closed. The focal plane flow proportional counter was moved behind a 25 μm circular aperture and the counter/aperture pair moved to the first point in a 9×9 grid of 25 μm centers. Data were accumulated for 10 seconds simultaneously in the focal plane and normalization counter and the resulting spectra recorded. The normalized counts were calculated by integrating the counts under the line in both spectra and dividing. The resulting value was the first point in the 9×9 "image" and data were taken similarly at the other 80 points, moving the counter/aperture pair in a 2-D raster scan. An image was built in this manner with each shutter open in turn and the centroids calculated and combined to determine the tilt and focus errors.

During the scan, the operator examined proportional counter spectra in real-time, zoomed on regions of interest (typically the X-ray line), obtained cursor readout for channels of interest, and saw updated statistics associated with the chosen display parameters. Summary data were displayed in a console window after each scan point and included the normalized counts. Other subsystem data such as motor positions and status, gas system pressures, thermal system values and vibration levels were displayed in other X-window displays.

Proportional counter data were transmitted via TCP/IP in binary format displayed, and stored in the XDACS archive. Scan images were stored as ASCII files with associated parameters and converted into IRAF format files for analysis. Programs to perform the centroid and tilt calculations were written as IRAF tasks and made available to the shell for access by procedural scripts.

Once the VETA was aligned, characterization of the PRF and other tests were performed at all five energies. The majority of tests involved scanning the proportional counter behind an appropriate aperture and building 1-D or 2-D data sets. The tests included full-width-half-maximum, encircled energy and effective area. Other tests were performed to beam center, check focus, flat field, measure background and off axis response.

4. System Evolution

The detection system to be used to test the fully assembled AXAF flight mirror (the High Resolution Mirror Assembly—HRMA) is currently being developed at SAO. The control and data acquisition component of the HRMA X-ray Detector System (HXDS) will be based closely on the successful system architecture of the VXDS XDACS. The HXDS will generate the calibration data for the AXAF HRMA and provide monitoring and other data during the Science Instruments calibration. The calibration data will be available during the mission through the AXAF Science Center.

Acknowledgments. The authors extend thanks to Eric Mandel for enthusiastic support throughout the system development and test, and in particular for providing so much of the parameter interface library. This work is supported by NASA contract NAS8-36123.

References

Brissenden, R., Hughes, J.P., Kellogg, E., & Zhao, P., Internal SAO report SAO-AXAF-AR-91-080, 3, November 1991

Brissenden, R., Jones, M., Ljungberg, M., Nguyen, D., & Roll, J., Jr. 1992, SPIE Proceedings, "VETA X-ray Data Acquisition and Control System", 1742

Roll, J.B., Jr., & Mandel E. 1992, "The Parameter Interface" in NASA Workshop: User Interfaces for Astrophysical Software, GSFC, ed. E. Mandel

Podgorski, W., Flanagan, K., Freeman, M., Kellogg, E., Norton, T., Ouellette, P., Roy, A., & Schwartz, D.A. 1992, SPIE Proceedings, "VETA-I X-ray Detection System", 1742

Astronomical Data Analysis Software and Systems II
ASP Conference Series, Vol. 52, 1993
R. J. Hanisch, R. J. V. Brissenden, and J. Barnes, eds.

The Keck Task Library (KTL)

W. F. Lupton and A. R. Conrad

W. M. Keck Observatory, P. O. Box 220, Kamuela, HI 96743

Abstract. KTL provides a simple way of writing event-driven applications which interact with multiple underlying sub-systems. This paper explains what KTL is and why it was produced. Several examples are given, both of sequential and of event-driven applications, and KTL's generic sub-system interface is described.

1. Introduction

The Keck software group is responsible for Keck telescope and primary mirror control software. Instrument control software is the responsibility of the instrument builders. As can be seen from Table 1, this has resulted in many different approaches.

Table 1. Keck software environments.

system	environment	language	message system	other
DCS[a]	VMS / VxWorks	C	pipes / sockets	
ACS[b]	VMS / VxWorks	C	RPCs	
PCS[c]	VMS	Fortran / C	sockets	
HIRES / LRIS[d]	SunOS / VxWorks	C	MUSIC[g]	keywords[h]
LWIRC / LWS[e]	SunOS / VxWorks	C++ / C	RPCs	keywords[h]
NIRC[f]	SunOS / VxWorks	C	sockets / RPCs	

[a] Drive and Control System (telescope control, in-house)
[b] Active Control System (primary mirror control, LBL / in-house)
[c] Phasing Camera System (primary mirror characterization, UC Irvine)
[d] HIgh RESolution spectrograph (UC Lick) / Low Resolution Imaging Spectrograph (Caltech)
[e] Long Wave Infra-Red Camera (UC Berkeley) / Long Wave Spectrometer (UC San Diego)
[f] Near Infra-Red Camera (Caltech)
[g] Multi User System for Instrument Control (UC Lick, Stover 1989)
[h] Application code works in terms of keywords and values rather than explicit messages

In the short term, our job is to make all these systems work together. In the long term, as a gradual process, we aim to move towards a single software environment running under Unix (initially SunOS) and VxWorks.

KTL was conceived as a means of unifying the telescope's messaging model and the instruments' keyword model, the idea being that, by defining a single top layer, code duplication will be minimized and a better system will result. The other main requirement was that it should handle multiple event sources; for example, a single application should be able to handle message, keyword, X and file I/O events.

2. Services and Styles

A KTL application interacts with one or more underlying sub-systems (known as *services*). To connect to a service, application code specifies a service name—typically an instrument name—and a *style*, which defines the way in which the application will interact with the service. Two styles are currently supported: *keyword*, where the application reads and writes named keywords and the resulting inter-task message traffic is hidden; and *message*, where the application deals directly with messages. The keyword style is intended mainly for user interfaces, and the message style is intended mainly for lower-level applications.

New services and styles can be supported without changing KTL; refer to Section 5. for more details.

Figure 1 shows how to use KTL to read the current position of the telescope (as in all the examples, error checking is omitted for brevity).

```
#include "ktl.h"
int main ()
{
    KTL_HANDLE *handle;
    double ra, dec;

    ktl_open( "dcs", "keyword", 0, &handle );
    ktl_read( handle, KTL_WAIT, "RA", NULL, &ra, NULL );
    ktl_read( handle, KTL_WAIT, "DEC", NULL, &dec, NULL );
    ktl_close( handle );
    printf( "(RA,DEC) = (%g,%g)\n", ra, dec );
}
```

(a) The service is dcs and the style is `keyword`. (b) `ktl_open()` returns an opaque handle which is passed to `ktl_read()` and `ktl_close()`. (c) The two `NULL` parameters to `ktl_read()` are `client_data` and `context`, neither of which are used here because callbacks are not being used.

Figure 1. A sequential KTL application.

3. Events and Callbacks

Figure 1 illustrates that the simplest KTL applications are very simple indeed. Most KTL applications, however, are event driven: they wake up only when a keyword changes value or a message arrives. An application connects to all desired services, expresses interest in specified events, then enters a dispatch loop in which it waits for events and calls the appropriate service's event-handling routine. Each event is associated with a callback routine which is invoked when the event occurs.

Figure 2 illustrates the use of KTL callbacks (details of the event loop are omitted). The first three arguments of KTL callbacks are directly comparable to those of Xt callbacks (`widget`, `client_data` and `call_data`). `context` is discussed later.

```
#include "ktl.h"
int callback();
int main()
{
    KTL_HANDLE *handle;
    KTL_CONTEXT *context;
    fd_set fdset;

    ktl_open( "dcs", "keyword", 0,
        &handle );
    ktl_context_create( handle, callback,
        NULL, NULL, &context );
    ktl_read( handle, KTL_CONTINUOUS,
        "RA", NULL, NULL, context );
    ktl_ioctl( handle, KTL_FDSET, &fdset );
    forever {
        select( ... );
        handle other event types;
        ktl_dispatch( handle );
    }
}
```

```
int callback(
    char *keyword,
    void *client_data,
    double *value,
    KTL_CONTEXT *context )
{
    printf( "%s = %g\n",
        keyword, *value );
    return 1;
}
```

(a) `ktl_open()` is called as before. (b) `ktl_context_create()` creates a `context` which contains, among other things, `callback()`'s address. (c) `ktl_read()` issues a "continuous read" on the `RA` keyword (it doesn't really issue a read, but asks to be notified when the `RA` keyword changes its value). (d) `ktl_ioctl()` returns the appropriate fd set for `select()`. (e) The event loop blocks awaiting events. (f) When `RA` changes, `ktl_dispatch()` is called and it invokes `callback()`.

Figure 2. An event-driven KTL application.

4. Command Context

A callback will frequently want to interact with other tasks. Event-driven tasks should not block (and thus be unresponsive) during such an interaction. `ktl_read()` and `ktl_write()` support a `KTL_NOTIFY` flag which results in callback invocation when a response is received.

Figure 3 gives an example of such an interaction. Here the `context` parameter is used extensively; it contains sufficient information for KTL to route the response back to the appropriate callback routine (always the callback that initiated the transaction). This is done using unique message ids which allow responses to be matched to commands. KTL facilities in this area owe a lot to those of the ADAM "D-task fixed part" (Kelly 1992).

5. Sub-system Interface

As mentioned earlier, KTL services and styles can be added without changing KTL. This is possible because KTL defines a generic interface to service/style

```
int callback( char *keyword, void *client_data, double *value,
              KTL_CONTEXT *context )
{
    int filter = 8;

    switch( context->state ) {
      case KTL_INITIAL_STATE:
        printf( "%s = %g\n", keyword, *value );
        ktl_write( hires, KTL_NOTIFY, "FILTER", NULL, &filter, context );
        context->state = KTL_MESSAGE_STATE;
        break;
      case KTL_MESSAGE_STATE:
        printf( "Filter move response with status %d\n", context->status );
        if ( context->done )
            context->state = KTL_FINAL_STATE;
        break;
    }
    return 1;
}
```

(a) When the `RA` changes, the callback is invoked and the first case of the `switch` statement is taken. (b) `ktl_write()` with `KTL_NOTIFY` initiates a filter move, the callback changes `context->state` and it exits. (c) When a response is received from `HIRES`, the callback is again invoked and this time the second case is taken. (d) If the move is not complete, the callback reschedules awaiting a final response. (e) Otherwise the callback again changes `context->state` and KTL will delete the context.

Figure 3. An KTL callback that does a non-blocking keyword write.

specifics. All *styles* must provide seven routines: `open()`, `ioctl()`, `read()`, `write()`, `event()`, `respond()` and `close()`, which in turn branch to *service*-dependent routines.

Normally these seven routines and the routines that they call are linked into shareable libraries and activated at run-time. This is done using a set of three `ktl_load_*()` routines[1]. This means that generic applications can be written and then used with new services without re-linking.

Figure 4 illustrates the interactions between KTL routines and service/style-specific routines.

6. Current Status

Keyword-style services have been implemented for the telescope and four of the five instruments. They use a mixture of MUSIC (telescope and optical) and RPCs (infra-red) as their message systems. Message-style services have been implemented for the telescope, MUSIC and a telescope simulator. The latter is nice in that it keeps all the simulation code below the generic KTL sub-system interface.

---

[1] Under SunOS they use `dlopen()`, etc., and will have to be re-implemented for other operating systems.

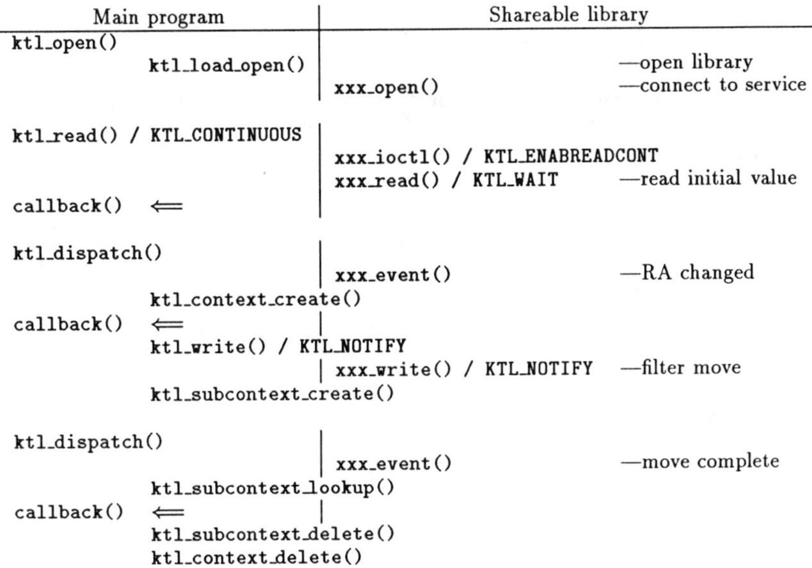

Main program
Shareable library
ktl_open()
ktl_load_open()
—open library
xxx_open()
—connect to service
ktl_read() / KTL_CONTINUOUS
xxx_ioctl() / KTL_ENABREADCONT
xxx_read() / KTL_WAIT
—read initial value
callback() ⟸
ktl_dispatch()
xxx_event()
—RA changed
ktl_context_create()
callback() ⟸
ktl_write() / KTL_NOTIFY
xxx_write() / KTL_NOTIFY
—filter move
ktl_subcontext_create()
ktl_dispatch()
xxx_event()
—move complete
ktl_subcontext_lookup()
callback() ⟸
ktl_subcontext_delete()
ktl_context_delete()

Figure 4. The KTL sub-system interface.

7. Conclusions

KTL is young but is already becoming an important part of the Keck software system. Its benefits will be felt particularly strongly as telescope control software is ported to Unix, since the KTL message style provides essentially the same facilities that are currently available to telescope control tasks.

Refer to Lupton (1992) for general information on KTL. Refer to Conrad and Lupton (1993) and Kibrick et al. (1993) for background information on Keck instrumentation software and the ways in which it uses KTL.

References

Conrad, A.R., & Lupton, W.F. 1993, this volume

Kelly, B.D. 1992, Starlink User Note 134, ADAM — Guide to Writing Instrumentation Tasks

Kibrick, R.I., Stover, R.J., & Conrad, A.R. 1993, this volume

Lupton, W.F. 1992, Keck Software Document 28, KTL Programming Manual

Stover, R.J. 1989, UCO/Lick Technical Report 54, MUSIC — a Multi-User System for Instrument Control

Discussion

M. Davis: What documentation is available?

Lupton: The KTL programmer's manual is available via anonymous FTP from kaula.keck.hawaii.edu (128.171.96.41). It is in docs/ktl_manual.ps. Other files in this directory with "ktl" in their names may also be of interest.

Astronomical Data Analysis Software and Systems II
ASP Conference Series, Vol. 52, 1993
R. J. Hanisch, R. J. V. Brissenden, and J. Barnes, eds.

Efficient Transfer of Images over Networks

J. W. Percival

Space Astronomy Laboratory, University of Wisconsin, Madison, WI 53706

R. L. White

Space Telescope Science Institute, 3700 San Martin Drive, Baltimore, MD 21218

Abstract. Effective remote observing requires sending large images over long distances. The usual approach to the transfer problem is to require high bandwidth transmission links, which are expensive to install and operate. An alternative approach is to use existing low-bandwidth connections, such as phone lines or the Internet, in a highly efficient manner by compressing the images. The combined use of existing low-cost infrastructure and standard networking software means that remote observing can be made practical even for small observatories with limited network resources.

We have implemented such a scheme based on the H-transform compression method developed by White (1992) for astronomical images, which are often resistant to compression because they are noisy. The H-transform can be used for either lossy or lossless compression, and compression factors of at least 10 can be achieved with no noticeable losses in the astrometric or photometric properties of the compressed images. The H-transform allows us to organize the information in an image so that the "useful" information can be sent first, followed by the noise, which makes up the bulk of the transmission. The receiver can invert a partially received set of H-coefficients, creating an image that improves with time. The H-transform is particularly well-suited to this style of incremental reconstruction, because the spatially localized nature of the basis functions of the H-transform prevents the appearance of artifacts such as ringing around point sources and edges.

Our implementation uses the WIYN Telescope Control System's TCP-based communications protocol. We sent an 800x800 16-bit astronomical image over a 2400 baud connection, which would normally take about 71 minutes; after only 60 *seconds*, the partially received H-transform produced an image that did not differ appreciably from the original. This paper presents a quantification of the efficiencies, as well as examples of images reconstructed from partial data.

1. Overview

We want to send images from the WIYN Observatory to Remote Observing sites for target acquisition, data quality monitoring, and quick-look analysis. We want to avoid expensive, private, high-speed data links, and we want to transmit the image in a way that allows the Remote Observer to see something quickly, with improvement as time goes on. This allows the Remote Observer to cancel the transmission as soon as the image is good enough.

We use a type of wavelet transform called the H-transform (White 1992) that organizes the image data for serial transmission.

We compare three methods:

- Sending the raw image in raster order (i.e., row by row)
- Sending the raw image in bitplane order (i.e., MSB first)
- Sending the H-transform in bitplane order (i.e., MSB first)

Sending images or H-transforms in bitplane order (MSB first) is good because the higher bitplanes are sparse and compress well. We use 1-dimensional or 2-dimensional quadtree encoding for each bitplane, with a final run-length encoding of the quadtree codes.

For a test image, we used an 800x800 pixel, 16-bit image of the Coma cluster digitized from the Palomar Sky Survey.

2. Transmission Software

We have developed a message protocol for the WIYN Observatory Control System that will be used for:

- Commands from the user interfaces
- Telemetry (engineering data) from the real-time system
- Data (images, spectra, catalogs, etc.)

This protocol uses TCP/IP sockets as a reliable byte stream, with a special message protocol for exchanging messages between applications.

We have written an image server application that sends images over the network in a machine-independent way, and an X-windows display client that allows the Remote Observer to select the transmission method and display style (linear, log, etc.), updates the display as more data are received, and allows the Remote Observer to cancel the transmission when the received image is good enough, or to wait until the received image has been reconstructed perfectly.

The H-transform can have negative pixels, so a special protocol must be used to avoid sending all the sign extension bits in a negative number. We store the sign bits, and transmit absolute values. After each bitplane is sent, a special "negation plane" is sent to negate any ultimately negative number whose absolute value has just had its most significant one-bit sent. Negating the H-transform coefficients "as you go" keeps the H-transform inversion well behaved on the remote end.

3. Comparison of Results

Figures 1, 2, 3, and 4 show the image after receiving the highest four non-zero bitplanes of the original raw image. After receiving one bitplane, the image is simply a bitmap with two intensities, black or white. After receiving two bitplanes, four intensities are shown, and so on. The full intensity resolution of the original is recovered only after receiving all the raw bitplanes.

Figures 5, 6, 7, and 8 show the image after receiving the highest seven non-zero bitplanes of the H-transform. (The highest seven bitplanes of the H-transform are compared to the highest four bitplanes of the raw image because these represent roughly comparable amounts of data.) Note that a variety of intensity variations is visible in even the earliest of the H-transform reconstructions.

Our quality measure Q is the "reduced χ^2":

$$Q = \frac{1}{\sigma_0^2 N} \sum (I_1 - I_0)^2 \tag{1}$$

where I_1 represents the partially received image, I_0 represents the original image, σ_0 is the standard deviation of the noise in the original image, and the sum is over all N pixels in the image. $Q \approx 1$ occurs when the images are globally consistent to within the noise (although the differences may be highly spatially correlated especially when the image is sent in raster order).

Figure 9 shows Q as a function of the number of transmitted 4-byte words (including protocol overhead). The raster mode converges most slowly, as expected. $Q < 1$ occurs after 326000 words (98.7%) have been transmitted. Sending the image in bitplane order causes convergence after 46861 words (20.1%). Sending the H-transform in bitplane order causes convergence in only 4564 words (1.59%).

Figures 10, 11, and 12 show the intensity histograms of the original image, and the images in Figure 4 and Figure 8 respectively. Figure 4 shows good "visual" convergence after receiving plane 10 of the raw image, but Figure 11 shows that the data are still heavily quantized. Figure 12 shows a much less quantized intensity histogram after receiving plane 10 of the H-transform, more closely approximating that of the original image.

References

White, R.L. 1992, High Performance Compression of Astronomical Images, available via FTP from stsci.edu

Figure 1. Raw image, bitplane order, planes 13–15. 0.313% of the data have been received. Plane 13 is the highest non-zero plane, so the image at this point is only 1-bit deep.

Figure 2. Raw image, bitplane order, planes 12–15. 1.36% of the data have been received. Two non-zero planes have been received, giving 4 intensity levels.

Figure 3. Raw image, bitplane order, planes 11–15. 2.81% of the data have been received. Three non-zero planes have been received, giving 8 intensity levels.

Figure 4. Raw image, bitplane order, planes 10–15. 11.3% of the data have been received. Four non-zero planes have been received, giving 16 intensity levels.

Figure 5. H-transform, bitplane order, planes 13–31. 0.746% of the data have been received. Note the various spatial and intensity scales that are already present in the image.

Figure 6. H-transform, bitplane order, planes 12–31. 1.59% of the data have been received.

Figure 7. H-transform, bitplane order, planes 11–31. 3.07% of the data have been received.

Figure 8. H-transform, bitplane order, planes 10–31. 7.9% of the data have been received.

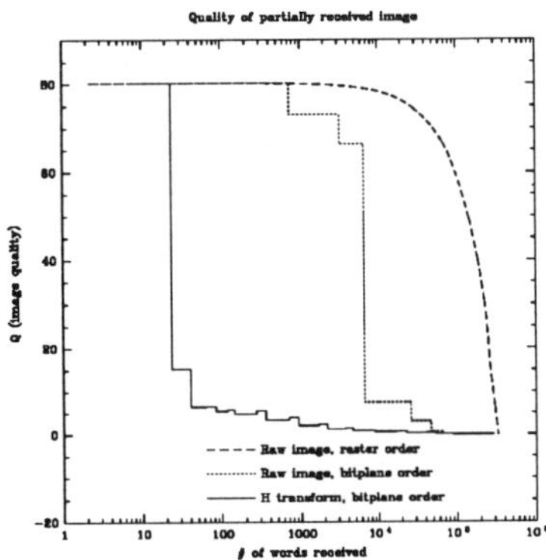

Figure 9. Quantitative improvement: reduced χ^2 measure as a function of the number of transmitted 4-bytes words (including protocol overhead). Note that the H-transform, bitplane method reaches the asymptotic part of its quality curve when fewer than 400 *bytes* of data have been received!

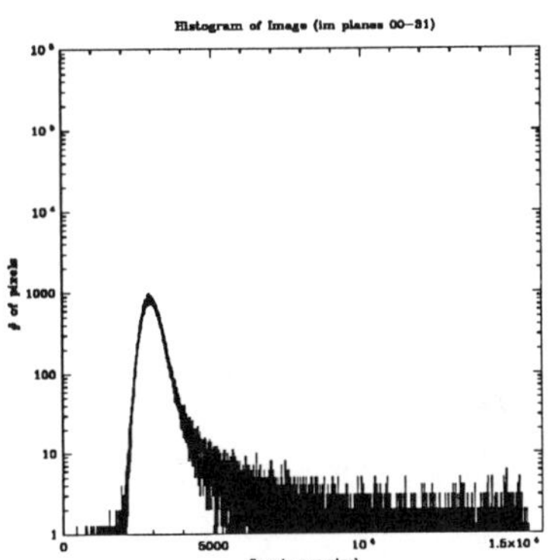

Figure 10. Intensity histogram of original image.

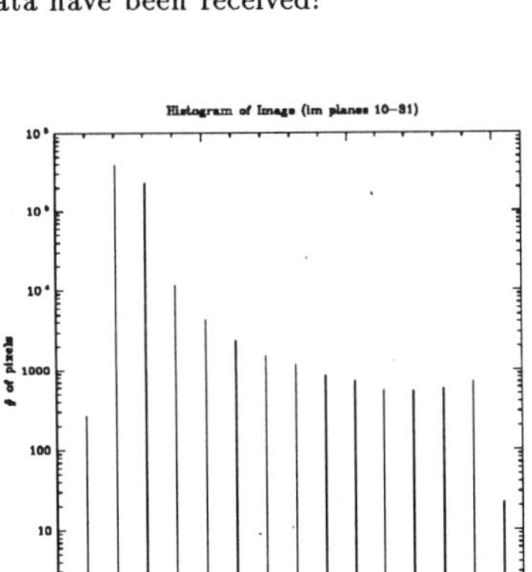

Figure 11. Intensity histogram of image after planes 10–15 of the raw image have been received. Moving through the raw image in bitplane order quantizes the received image according to the number of received bitplanes. The quantization shown here makes it impossible to determine the sky brightness accurately.

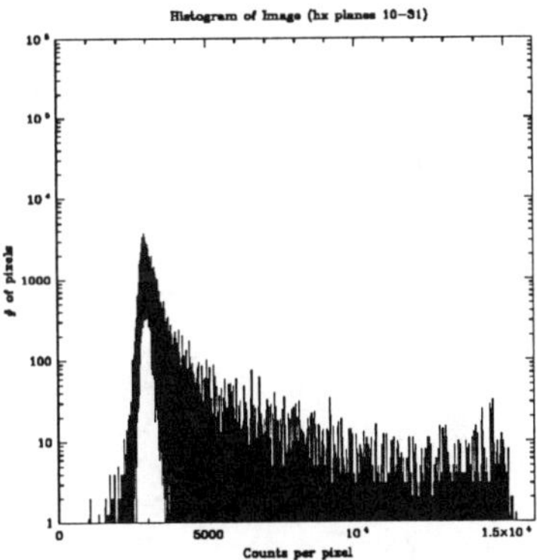

Figure 12. Intensity histogram of image after planes 10–31 of the H-transform have been received. Sending the wavelet coefficients in bitplane order greatly reduces intensity quantization in the received image. Note the improved accuracy possible in estimating the sky brightness.

Part 3. Data Acquisition

Section B. Scheduling Systems

Astronomical Data Analysis Software and Systems II
ASP Conference Series, Vol. 52, 1993
R. J. Hanisch, R. J. V. Brissenden, and J. Barnes, eds.

The Application of Artificial Intelligence to Astronomical Scheduling Problems

Mark D. Johnston
Space Telescope Science Institute, 3700 San Martin Drive, Baltimore MD 21218

Abstract. As artificial intelligence (AI) technology has moved from the research laboratory into more and more widespread use, one of the leading applications in astronomy has been to high-profile observation scheduling. The SPIKE scheduling system was developed by the Space Telescope Science Institute (STScI) for the purpose of long-range scheduling of Hubble Space Telescope (HST). SPIKE has been in daily operational use at STScI since well before HST launch in April 1990. The system has also been adapted to schedule other missions: one of these missions (EUVE) is currently operational, while another (ASTRO-D) will be launched in February 1993. Some other future space astronomy missions (XTE, SWAS, and AXAF) are making tentative plans to use SPIKE. SPIKE has proven to be a powerful and flexible scheduling framework with applicability to a wide variety of problems.

1. Introduction

Efficient utilization of costly space- and ground-based observatories is an important goal for the astronomical community: the cost of modern observing facilities is enormous, and the available observing time cannot accommodate the demand from astronomers around the world. The complexity and variety of scheduling constraints and goals has led several groups to investigate how artificial intelligence (AI) techniques might help solve these kinds of problems. The earliest and most successful of these projects was started at Space Telescope Science Institute in 1987 and has led to the development of the SPIKE scheduling system to support the scheduling of Hubble Space Telescope (HST).

The aim of SPIKE at STScI is to allocate observations to timescales of days to a week, observing all scheduling constraints, and maximizing preferences that help ensure that observations are made at optimal times. SPIKE has been in use operationally for HST since shortly after the observatory was launched in April 1990. Although developed specifically for HST scheduling, SPIKE was carefully designed to provide a general framework for similar (activity-based) scheduling problems. In particular, the tasks to be scheduled are defined in the system in general terms, and no assumptions about the scheduling timescale are built in. The mechanisms for describing, combining, and propagating temporal and other constraints and preferences are quite general. The success of this approach has been demonstrated by the application of SPIKE to the scheduling of other satellite observatories: changes to the system are required only in the specific constraints that apply, and not in the framework itself. In particular, the SPIKE framework is sufficiently flexible to handle both long-term and short-term scheduling, on timescales of years down to minutes or less.

This paper briefly discusses the role of automated scheduling in astronomical applications, the SPIKE system in the context of Hubble Space Telescope scheduling, and the application of SPIKE to other problem domains.

2. Role and Purpose of Scheduling in Astronomical Applications

One most often thinks of applying scheduling software during *mission operations* where the purpose is to meet such goals as: maximizing observing efficiency and throughput; maximizing the quality of data taken (e.g., by scheduling observations at times which minimize background contributions or proximity to relevant calibrations); improving the capability to react effectively to changes (in the schedule or in external conditions); and minimizing the burden of work on the operations staff. However, the use of scheduling tools during *mission development* phases can play a very important role in the design of a successful mission. For example:

- **scenario evaluation:** develop baseline and contingency operations scenarios ("Design Reference Mission") using realistic system models
- **performance metrics:** encourage identification of quantitative goals by which observatory performance can be measured
- **design trade-offs:** evaluate tradeoffs (as early as possible) during mission design and development

The use of scheduling tools as "mission simulators" can provide early and quantitative insight into the scientific impact of different cost choices, thus maximizing the science return from the mission.

3. Scheduling Technology Overview

Scheduling has been the subject of much work over the years: even simplified unrealistic problems are known to be NP-complete, i.e., there are no efficient algorithms. In recent years there have emerged two main avenues of attack:

- Operations research: These techniques have been under development since WWII (e.g., French 1982). They typically center on branch-and-bound, integer programming, or mixed integer-linear programming problem specifications. Unfortunately, the difficulty of casting realistic scheduling constraints into the mathematical framework means that these methods are often limited to finding exact solutions to a very simplified problem. Some good search heuristics are known (e.g., Adams, Balas & Zawack 1988), but many are strongly problem-specific.
- Constraint-based reasoning: This refers to a family of methods derived from artificial intelligence (AI) research, based on the pioneering work by

Mark Fox and others from CMU from late 1970s (see Fox 1987). These methods exploit constraint knowledge to guide the search for feasible schedules. They can can handle much more realistic classes of constraints and goals, but often at the expense of problem size. Within the constraint-based approach there is significant variation from one specific application to another, depending on various possible choices for such aspects as scheduling search (constructive vs. repair), the degree of lookahead during search, and the role of stochasticity. *Integrated planning and scheduling* has so far been limited to a research effort, but it is being applied to astronomical scheduling (Muscettola 1992). This approach may have a bright future, provided that performance issues can be adequately addressed.

The only currently fielded AI-based system for astronomical scheduling is SPIKE, developed at STScI for Hubble Space Telescope scheduling. SPIKE has its origins in the constraint-based approach pioneered by Fox and colleagues, but SPIKE also integrates other AI techniques (especially evidence combination) and incorporates novel search algorithms that have recently stimulated significant computer science research (Minton et al. 1990, 1992). In the following we first give an overview of the HST scheduling problem and process, and then describe the SPIKE system.

4. HST Experience and the SPIKE System

4.1. HST Science Scheduling

Figure 1 shows the high-level scheduling flow for the HST ground system, described in more detail in Miller (1989), Adorf (1990), and Miller and Johnston (1991). The major elements are:

- Proposal Preparation: An astronomer who plans to use HST must create an observing proposal which specifies the observations to be made. This proposal is the primary input to the planning and scheduling process. The Remote Proposal Submission System (RPSS) and Proposal Entry Processor (PEP) handle observing proposals, including electronic submission by astronomers.
- Planning: The TRANSFORMATION expert system converts the proposal from a high-level specification into detailed task descriptions for scheduling.
- Long-term Scheduling: Since the HST scheduling problem covers such a wide range of timescales and consists of tens of thousands of tasks, a two-tiered approach to scheduling was adopted. A long term plan spans approximately one year and allocates tasks (as defined by TRANSFORMATION) to specific weeks or parts of a week. From the long-term plan, week-long segments are extracted for short-term scheduling. Long-range scheduling is done with the SPIKE system, which was developed at the STScI.
- Short-term Scheduling: Short-term scheduling with the Science Planning and Scheduling System (SPSS) performs the final sequencing of groups

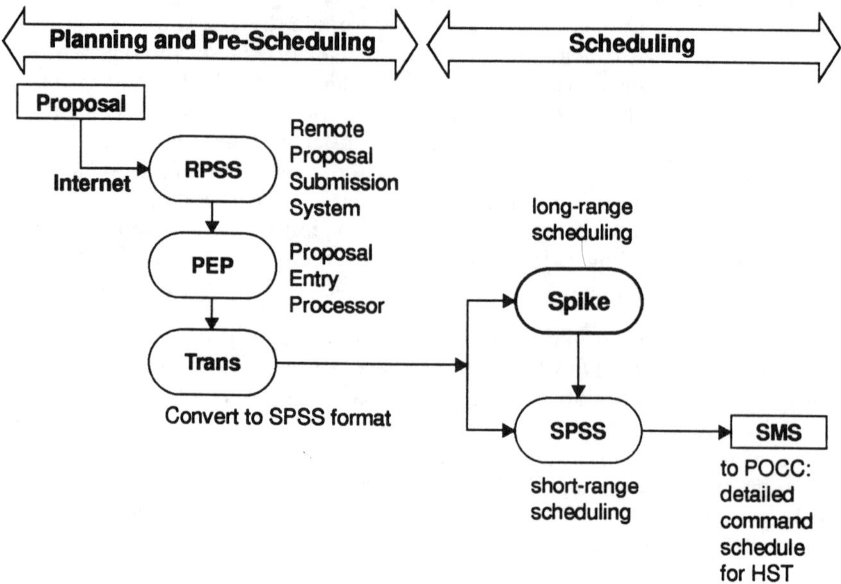

Figure 1. The high-level scheduling operations flow for HST.

of observations within a week, generates the detailed command list, and transmits the results as the Science Mission Specification (SMS) to the HST Payload Operations Control Center. SPSS was originally developed by TRW and now maintained by the STScI.

4.2. SPIKE System Architecture

The SPIKE architecture consists of two conceptual levels. A lower "constraint representation and reasoning" level captures descriptions of the tasks or activities to be scheduled, along with absolute or relative timing constraints and preferences. An upper "strategic scheduling" level consists of the schedule generation (search) modules as well as the components which interact with the user via the graphical user interface. These are briefly described in the following, and are covered in more detail in Miller et al. 1987, 1988; Sponsler et al. 1991; and Johnston and Miller 1993.

4.3. Constraint Representation and Reasoning

Scheduling constraints convey two major types of information to the scheduler:

- **Feasibility constraints** specify conditions or times when activities may or may not be scheduled. These constraints are often termed "strict" or "hard" constraints, since they may not be violated under normal circumstances. A few examples in the HST scheduling context are:
  - provide a minimum separation of two months between a specific observation and a repeat observation on the same target

- never schedule an observation when the Sun is within 50° of the target
- do not roll the spacecraft more than 30° from its nominal orientation

- **Preference constraints** specify quality judgments on scheduling conditions which are preferred but not required. These may be based on objective or subjective factors. In HST scheduling, for example, it is desirable to schedule at times which:
  - minimize scattered light from the bright limb of the Earth
  - maximize the chance of successfully acquiring guide stars
  - place an observation as close to nominal orientation as possible

The SPIKE constraint representation level exploits the "suitability function" framework which is one of the new technology developments incorporated in SPIKE (see Johnston 1989, 1990; Johnston & Adorf 1992). Suitability functions represent the degree of preference for scheduling tasks as a function of time. They can be based on absolute time, or relative to the scheduled time of other activities, and can be quantified by expert human schedulers so that they properly represent the combined preferences due to a variety of factors. A few examples best illustrate the types of constraints that can be handled:

- schedule task A_1 only during intervals when coordinated observing with another observatory can be conducted
- schedule a calibration task C_1 within 24 hours of the start time of task A_1
- schedule task A_2 after task A_1, but with at least 30 days of time separation between them
- group tasks $A_1 \ldots A_n$ within 3 days
- schedule task A_1 at time t with preference $P_1(t)$
- schedule task A_2 at time t_2, given that task A_1 has been scheduled at time t_1, with preference $P_{12}(t_2 - t_1)$. This allows for such preference constraints as "schedule A_2 as soon as possible after A_1, but no sooner then 2 days and no later than 10 days."

An arbitrary number of constraints like these can be stated and handled by SPIKE. Note in particular that the last two formulations provide for very general constraint specifications.

In addition to these types of constraints and preferences, the constraint representation level can also represent scheduling decisions (e.g., to schedule an activity at a specific time or within a range of times), and can propagate their combined implications. This allows the strategic scheduling level to maintain a high degree of awareness of the state of the schedule and of the significance of the constraints and preferences.

It is important to note that the SPIKE constraint representation level has no built-in time granularity. Temporal relationships can be specified with whatever resolution is appropriate to the problem. For example, it is possible to define solar constraints with 1-year periodicity, and at the same time define constraints which change on a timescale of seconds (e.g., target occultations, command execution times).

4.4. Strategic Scheduling—Search and Optimization

SPIKE's strategic scheduling level is responsible for generating and modifying schedules as a whole. SPIKE contains some extremely efficient scheduling search techniques that represent new contributions to the state-of-the-art and which have shown to be widely applicable (Johnston & Adorf 1992; Johnston & Miller 1993; Minton et al. 1990, 1992; Johnston & Minton 1993). The most effective techniques developed for and used by SPIKE are known as "stochastic multistart repair" algorithms. They are based on a repeated three-stage process which is simply stated as follows:

1. **Trial assignment:** Using special-purpose optimized heuristics, construct an initial schedule—which may contain constraint violations. The best heuristics for this stage include stochastic elements and have been developed through extensive experimentation.

2. **Repair:** Repair the schedule using additional heuristics. The best repair methods derive from a very successful artificial neural network developed for use in SPIKE (see Johnston & Adorf 1992).

3. **Deconflict:** Remove any remaining scheduling conflicts or constraint violations. This step is heuristically-guided as well, and is optimized for particular scheduling problems or stages.

The strategic scheduling level also provides the interfaces to the SPIKE graphical user interface (GUI). This is an X-window interface which gives the SPIKE user visibility into the schedule and into the constraints and preferences which apply to individual tasks. Through the GUI the SPIKE user can run the automatic scheduler, focus on groups of tasks (e.g., high priority tasks, or those with only a few potential scheduling times), make manual scheduling decisions, generate text and graphical output products, and perform other schedule manipulations.

One of the main goals of SPIKE is to maximize schedule efficiency. Since SPIKE uses the task suitability function as a guide to time preference choices, this clearly means that observing efficiency will be a major component of the suitability function for each task. In addition, SPIKE can represent the fact that tasks can have variable duration depending on when they are scheduled (for example, because the background level is varying to a degree that exposure times must be shortened or lengthened). This allows SPIKE to optimize the schedule for efficiency by maximizing the suitability of each task, and at the same time to accurately represent the fact that efficient scheduling can make observing time available for use by other tasks.

4.5. SPIKE Implementation

SPIKE is implemented in Common Lisp (ANSI standard X3J13, pending) and makes use of the Common Lisp Object System (CLOS), also part of the standard. SPIKE was designed from the outset (1987) as an object-oriented system and has benefitted from the many known advantages of this approach. The SPIKE graphical user interface is implemented using the Common Lisp Interface Manager (CLIM), which is a proposed standard and which is presently available on a large number of platforms (from Macintoshes to Crays). CLIM generates window displays via X-windows and thus fully supports the operation of SPIKE in a distributed workstation environment.

The operational use of SPIKE for HST and other missions currently makes use of the Allegro Common Lisp (CL) environment from Franz, Inc. The Allegro CL system is widely available on Unix workstations, minis, and mainframes. Although not part of the Common Lisp ANSI standard, Allegro CL (and nearly all other commercial Lisp implementations) provide for "external function calls" so that code written in FORTRAN, C, or other languages can be used from Lisp.

HST SPIKE and its associated software represents approximately 100K lines of code. However, it should be remembered that Lisp is a particularly expressive language: in most other languages the total volume of code would be much higher for equivalent functionality.

5. Adapting SPIKE to Other Missions

As noted above, SPIKE was developed by STScI for long-term scheduling of Hubble Space Telescope. However, the design and implementation of SPIKE considered from the outset the more general aspects of astronomical scheduling. As a result, the SPIKE core system is entirely independent of the specifics of the HST scheduling environment, policies, constraints, etc. This has made it straightforward to adapt SPIKE to other types of scheduling problems by incremental extensions. Such extensions draw on common elements from HST or other problem classes (via object inheritance), but at the same time provide easily for new and customized behavior.

The success of this approach has been demonstrated by the large number of successful adaptations of SPIKE that have been made over the past several years (see, e.g., Johnston 1988a, 1988b; Johnston & Rosenthal 1991). A few comments on the types of changes required will help explain the general nature of the work required to customize SPIKE for new problem.

5.1. Extreme Ultraviolet Explorer (EUVE)

This mission starts a pointed General Observer (GO) mode in January 1993 which will be scheduled by SPIKE on a one-year cycle. Most EUVE observations are quite long (1–3 days), so it is feasible for the scheduler to work with the detailed schedule for a whole year at once. Modifications to SPIKE for EUVE included new constraint calculation software to determine whether one or two star trackers had clear fields of view during earth shadow passages, and variable exposure durations, to account for the fact that exposure times could change significantly depending on target visibility and the SAA. A comparison of SPIKE's best scheduling heuristic against a standard "best next" scheduling dispatch rule showed that SPIKE could schedule as much as *20 additional days of observing in a year.* This is a dramatic indication of the kind of efficiency improvements that can be obtained using the technology embodied in SPIKE.

5.2. ASTRO-D

The size of the ASTRO-D scheduling problem is similar to that of HST, but for this mission SPIKE will be used for both long-term and short-term schedul-

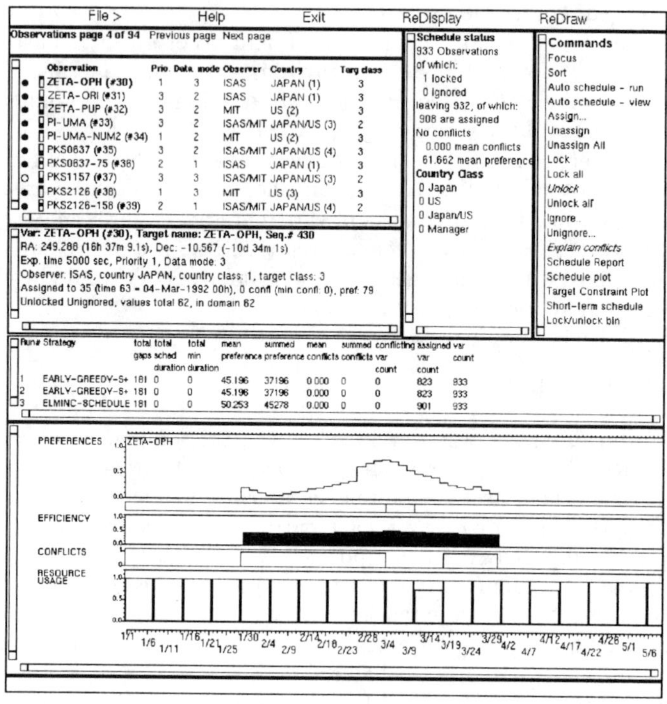

Figure 2. An example of the SPIKE user interface for ASTRO-D displaying a portion of a long-term schedule.

ing (Isobe et al. 1993). The same SPIKE core software is used in both situations: the difference is entirely in the task and constraint definitions. For long-term scheduling, most constraints are treated "statistically" and have applicable timescales of days or longer (see Figure 2 for an example). This mode is similar to HST long-term scheduling in that observations are allocated to week-long segments which are subsequently scheduled in detail. For short-term scheduling, constraints are treated with much finer time resolution (minutes or less) In addition to such basic constraints as target visibility (due to Sun, Moon, and Earth avoidance constraints), SAA passage, and star tracker availability, several new classes of constraints were implemented for SPIKE/ASTRO-D, including slewing to align the detector axis along the earth's magnetic field during SAA passages, ground station pass availability, and the capacity of the onboard data recorder (which limits the number of high data rate observations during periods of sparse ground station contacts). These additional ASTRO-D constraints are accessible through the customized X-window user interface. For example, the SPIKE user can select with the mouse a specific ground station pass, display its detailed timing, and then toggle its availability (which will be immediately reflected in the schedule).

5.3. X-ray Timing Explorer (XTE)

Like ASTRO-D, the adaption of SPIKE for XTE provides for both long-term (years) and short-term (minutes) scheduling. Very few constraint modifications were required for XTE. Following the adaptation, an evaluation of SPIKE was conducted which included a scheduling test of a large number of observations. The result, documented by Morgan (1992), showed that SPIKE could achieve observing efficiencies of 70% or better, which, for a low-earth orbiter mission, is extremely high: cf. Benvenuti and Pirenne 1991, Petro, Stockman, and Whitmore 1992, and Benvenuti 1992.

Acknowledgments. The author thanks the SPIKE development team at STScI: Glenn Miller, Jeff Sponsler, Mark Giuliano, Tony Krueger, and Mike Lucks. Space Telescope Science Institute is operated by the Association of Universities for Research in Astronomy for NASA.

References

Adams, J., Balas, E., & Zawack, D. 1988, "The Shifting Bottleneck Procedure for Job Shop Scheduling", Management Science, 34, 391

Adorf, H.-M., 1990, "The Processing of HST Observing Programs", Space Telescope European Coordinating Facility Newsletter 13, 12Q15

Benvenuti, P., & Pirenne, B. 1991, "HST Observing Efficiency", ST-ECF Newsletter Aug. 1991, No. 16, 18

Benvenuti, P. 1992, "HST Efficiency — Is It Good or Bad?", ST-ECF Newsletter Jan. 1992, No. 17, 26

Isobe, T., Johnston, M., Morgan, E., & Clark, G. 1993, this volume

Fox, M. 1987, Constraint-Directed Search: A Case Study of Job Shop Scheduling (Morgan Kaufmann: San Mateo, CA)

French, S. 1982, Sequencing and Scheduling (Ellis Horwood: Chichester)

Johnston, M. 1988a, "Artificial Intelligence Approaches to Spacecraft Scheduling", Proc. ESA Workshop on Artificial Intelligence Applications for Space Projects, ESTEC (Noordwijk, Holland), 5

Johnston, M. 1988b, "Automated Observation Scheduling for the VLT", Proc. ESO Conf. on Very Large Telescopes and their Instrumentation, ESO (Garching, Germany), 1273

Johnston, M. 1989, "Reasoning with Scheduling Constraints and Preferences", Space Telescope Science Institute Spike Tech. Report 1989-2

Johnston, M. 1990, "Spike: AI Scheduling for NASA's Hubble Space Telescope", Proc. 6th IEEE Conf. on AI Applications, Santa Barbara, CA, 184

Johnston, M., & Rosenthal, D. 1991, "A Day in the Life of the Mars Rover: Capabilities of the Rover System Architecture in Realistic Simulations", Proc. Computing in Aerospace 8, Baltimore, Maryland

Johnston, M., & Adorf, H.-M. 1992, "Scheduling with Neural Networks — The Case of Hubble Space Telescope", Computers and Operations Research 19, 209

Johnston, M.D., & Minton, S. 1993, "Analyzing a Heuristic Strategy for Constraint-Satisfaction and Scheduling", in Intelligent Scheduling, eds. M. Fox & M. Zweben (Morgan Kaufmann: San Mateo), in press

Johnston, M., & Miller, G. 1993, "Spike: Intelligent Scheduling of Hubble Space Telescope Observations", in Intelligent Scheduling, eds. M. Fox & M. Zweben (Morgan Kaufmann: San Mateo), in press

Miller, G. 1989. "Artificial Intelligence Applications for Hubble Space Telescope Operations", in Knowledge Based Systems in Astronomy, eds. F. Murtagh & A. Heck, (Springer Verlag: Berlin), 5

Miller, G., Rosenthal, D., Cohen, W., & Johnston, M. 1987, "Expert Systems Tools for Hubble Space Telescope Observation Scheduling", Telematics and Informatics, 4, 301

Miller, G., Johnston, M., Vick, S., Sponsler, J., & Lindenmayer, K. 1988, "Knowledge-Based Tools for Hubble Space Telescope Planning and Scheduling: Constraints and Strategies", Telematics and Informatics, 5, 197

Miller, G., & Johnston, M. 1991, "A Case Study of Hubble Space Telescope Proposal Processing, Planning and Long-Range Scheduling", Proc. Conf. Computing in Aerospace 8 (AIAA), 1

Minton, S., Johnston, M., Philips, A., & Laird, P. 1990, "Solving Large-Scale Constraint Satisfaction and Scheduling Using a Heuristic Repair Method", Proc. 8th Nat. Conf. on Artificial Intelligence, Boston, MA (Morgan Kaufmann), 17

Minton, S., Johnston, M., Philips, A., & Laird, P. 1992, "Minimizing Conflicts: A Heuristic Repair Method for Constraint Satisfaction and Scheduling", Artificial Intelligence, in press

Morgan, E. 1992, "Evaluation of Spike for XTE", MIT Center for Space Research Tech. Report, April 1992

Muscettola, N., Smith, S.F., Cesta, A., & D'Aloisi, D. 1992, "Coordinating Space Telescope Operations in an Integrated Planning and Scheduling Architecture", IEEE Control Systems Magazine 12(2)

Petro, L., Stockman, P., & Whitmore, B. 1991, "HST's Observing Efficiency", Space Telescope Science Institute Newsletter November 1991, 4

Sponsler, J., Johnston, M., Miller, G., Krueger, A., Lucks, M., & Giuliano, M. 1991, "An AI Scheduling Environment for Hubble Space Telescope", Proc. Computing in Aerospace 8, Baltimore, Maryland, 14

Discussion

S. Reynolds: On Sparc platforms, how much memory and disk space are required to run SPIKE?

Johnston: SPIKE runs on sparcs with 32 MB of memory. As far as disk space goes, approximately 10 MB are required for the software alone. Additional scheduling data associated with it requires somewhat more.

Kibrick: You mentioned the possibility of using SPIKE for software development scheduling. Has this actually been done?

Johnston: Yes, we used SPIKE for some software project scheduling for several years at STScI, but eventually decided that simple Macintosh project scheduler packages were adequate. SPIKE was particularly useful when there were lots of resources to consider simultaneously, and for projects with large number of pre-defined pieces.

Shaw: How easy is it to express a hardware or other limitation as a constraint in the SPIKE software?

Johnston: Generally it's easy to express a hardware or other constraint in SPIKE if it falls into a category that is similar to the constraints already there. So, for example, hardware limits on star trackers or computer storage space are trivial. Completely new kinds of constraints could require new code to model and interface to the system. How much work this would be depends on the details of the constraint.

M. Davis: Do you ever find instances when the detailed scheduling finds serious flaws in the long-term scheduling results?

Johnston: Usually not serious flaws, but it is certainly true that detailed scheduling can reveal problems that have to be dealt with by changing the long-term schedule. In general, if the constraints are well-modeled in the long-term problem, then the chances of a major upset are minimal when the short-term scheduling is performed.

Olson: Can SPIKE describe to its user why it made a particular decision in the automated scheduling processing?

Johnston: Generally not, since any one particular decision often depends on a whole chain of past decisions. However, SPIKE can describe what constraints and preferences went into the decision, and can separately describe the cumulative impact of past decisions as well, so that in practice it is often possible for a user to understand why a decision was made.

Astronomical Data Analysis Software and Systems II
ASP Conference Series, Vol. 52, 1993
R. J. Hanisch, R. J. V. Brissenden, and J. Barnes, eds.

The Application of SPIKE to ASTRO-D Mission Planning

T. Isobe

Center for Space Research, Massachusetts Institute of Technology, 77 Massachusetts Ave., Cambridge MA 02139

M. Johnston

Space Telescope Science Institute, 3700 San Martin Drive, Baltimore MD 21218

E. Morgan, G. Clark

Center for Space Research, Massachusetts Institute of Technology, 77 Massachusetts Ave., Cambridge MA 02139

Abstract. SPIKE is a mission planning software system developed by the Space Telescope Science Institute (STScI) for use with Hubble Space Telescope (HST). SPIKE has been developed for the purpose of automating observatory scheduling to increase their effective utilization and scientific return. In this paper, we describe the adaptation of SPIKE to the fourth Japanese X-ray astronomy satellite ASTRO-D. SPIKE has been successfully extended to handle both long-term and short-term scheduling for this mission.

1. Introduction

ASTRO-D is Japan's fourth X-ray astronomy mission and the first for which the US is providing a part of the scientific payload (Tanaka 1990). Scheduled to fly in February 1993, its four large-area telescopes will focus X-rays from a wide energy range onto a pair of charge coupled devices (CCDs) and imaging gas proportional counters (Arnaud, Day & White 1991). ASTRO-D will be the first X-ray imaging mission operating over the 0.5–12 keV band with high energy resolution. This combination of capabilities will enable a varied and exciting program of astronomical research to be conducted.

ASTRO-D will weigh 420 kg and will be launched from Kagoshima Space Center (KSC) by the Institute of Space and Astronautical Science (ISAS) into an approximately circular orbit at an altitude varying between 550–650 km. It will be three-axis stabilized to an accuracy of one arcminute; post facto attitude reconstruction will be accurate to about 0.1 arcminute. The mirror axis is parallel to the plane of the solar panels, which means that the observable region of the sky is a belt centered on a great circle 90° from the sun vector. Solar panel attitude restrictions limit the half-width of the belt to 30°. Direct contacts between the satellite and the ground stations will be possible for ten of the fifteen orbits per day; five from Japan and five from Australia (Tanaka 1990). Data taken between ground station passes are recorded onboard for later playback.

After an initial seven-month calibration and verification phase, observing time on ASTRO-D will be open to competitive proposals. Based on the previous X-ray missions (i.e., Einstein and ROSAT), it is expected that more than 1000 proposals will be submitted every half year. The expected throughput of ASTRO-D is in the range of 5 to 20 objects per day, which is much too large to schedule without computer assistance. Based on this, ISAS decided to use the SPIKE observation planning software system developed at Space Telescope Science Institute (STScI).

2. SPIKE

SPIKE is a mission planning software system developed by a team of programmers at the Space Telescope Science Institute (Johnston 1990, Sponsler et al. 1991, Johnston & Adorf 1992) to help solve the Hubble Space Telescope (HST) scheduling problem. SPIKE was developed for the purpose of automating astronomical observatory scheduling to increase their effective utilization and scientific return. SPIKE exploits artificial intelligence (AI) technology for representing constraints and for searching for feasible and optimal schedules. SPIKE's low-level constraint representation handles a wide variety of absolute and relative constraints and preferences. SPIKE's most successful high-level scheduling strategies are derived from a novel artificial neural network algorithm. Graphical display and manipulation of the schedule is an important feature of the system. Besides HST, several space missions are considering or planning to use SPIKE, including EUVE, XTE, and AXAF (see, e.g., Morgan 1992).

SPIKE accepts a list of targets and constraint specifications, checks constraints (such as sun avoidance, horizon angle limits, and South Atlantic Anomaly interference), and computes a *suitability function* for each observation. Suitability functions represent the degree of preference for scheduling an observation as a function of time. SPIKE also applies any relative timing constraints (e.g., A before B, group $A_1, \ldots, A_n$ within 10 days) and prior scheduling decisions, if any. It then schedules the observations during a specified time period to find a feasible observation plan. These plans tend towards optimal by maximizing the suitability of all scheduled observations. SPIKE does not impose any particular time granularity on the scheduling problem: for HST, Spike is used only for long-term scheduling (i.e., over a period of about one year to a resolution of a few days). The actual time granularity is dictated by the nature of the observations and the constraints.

Although SPIKE can give excellent schedules which do not violate specified constraints, users must realize that SPIKE can only assist them in constructing schedules that accurately embody the scientific judgment and priorities that may apply. The user has the freedom to change the schedule at any time (even by overriding constraints), and for generating text and graphical views of the schedule that can help in making policy decisions. SPIKE is not a proposal entry or evaluation system, although it could be used to determine whether proposed observations violate mission constraints. It could also be used by proposers to evaluate the feasibility of their requests.

3. ASTRO-D SPIKE

Although the version of SPIKE used for the ASTRO-D mission is almost identical to that used for the HST, there are a few differences. The most important differences center on the scheduling constraints: ASTRO-D has star tracker, ground station, and magnetic field alignment constraints which do not apply to HST, and so new constraint modeling software had to be implemented to represent these.

In addition, the ASTRO-D version of SPIKE is used both for long-term and short-term scheduling. There are several ASTRO-D specific functions that were developed for the the short-term scheduling mode. For example, ASTRO-D uses two ground stations for data downlinks, instead of the Tracking and Data Relay Satellite System (TDRSS) for data transmission. As a consequence, ASTRO-D is constrained by limited on-board data storage capacity to schedule high data-rate observations during periods of frequent ground contacts. The ASTRO-D version of SPIKE considers this constraint and schedules high bit rate observations accordingly.

A further difference between the ASTRO-D and HST versions of SPIKE is the user interface: the HST graphical user interface provides access to numerous functions that are not relevant to ASTRO-D. Conversely, there are necessary user interactions for ASTRO-D that are not supported by the HST user interface (e.g., modifying the ground station contact pass availability). For this reason a new ASTRO-D user interface was implemented. This will make it easier to customize the user interface when the occasion arises.

The procedures by which a user works with ASTRO-D SPIKE to build a schedule are illustrated in Figure 1. Processes shown as rectangles are steps taken by the user, while those represented by ovals are performed by the SPIKE software. The general concept is that long-term (six-month) schedules are constructed to allocate observations to 7-day intervals (with some oversubscription), then each 7-day interval is scheduled in detail using the SPIKE short-term mode.

4. SPIKE Implementation

SPIKE is written in ANSI standard Common Lisp. Since Lisp compilers are available for many computers and from several companies, it is possible to run SPIKE on many different platforms. Versions of SPIKE are currently running on Sun SPARCStations, DECStations, and Macintosh. SPIKE for ASTRO-D has been developed using Allegro Common Lisp from Franz, Inc., and will run on a Sun SPARCStation at ISAS for ASTRO-D operations.

SPIKE also uses Common Lisp Interface Manager (CLIM) as the user interface toolkit. This software is also available from Franz Inc. (and from other vendors for different Lisp compilers). On the SPARCStation, CLIM is based on the X-windows system, and is thus compatible with a distributed workstation environment. To run SPIKE efficiently requires 32MB or more of RAM and ~200MB of swap space (depending on the size of the scheduling problem).

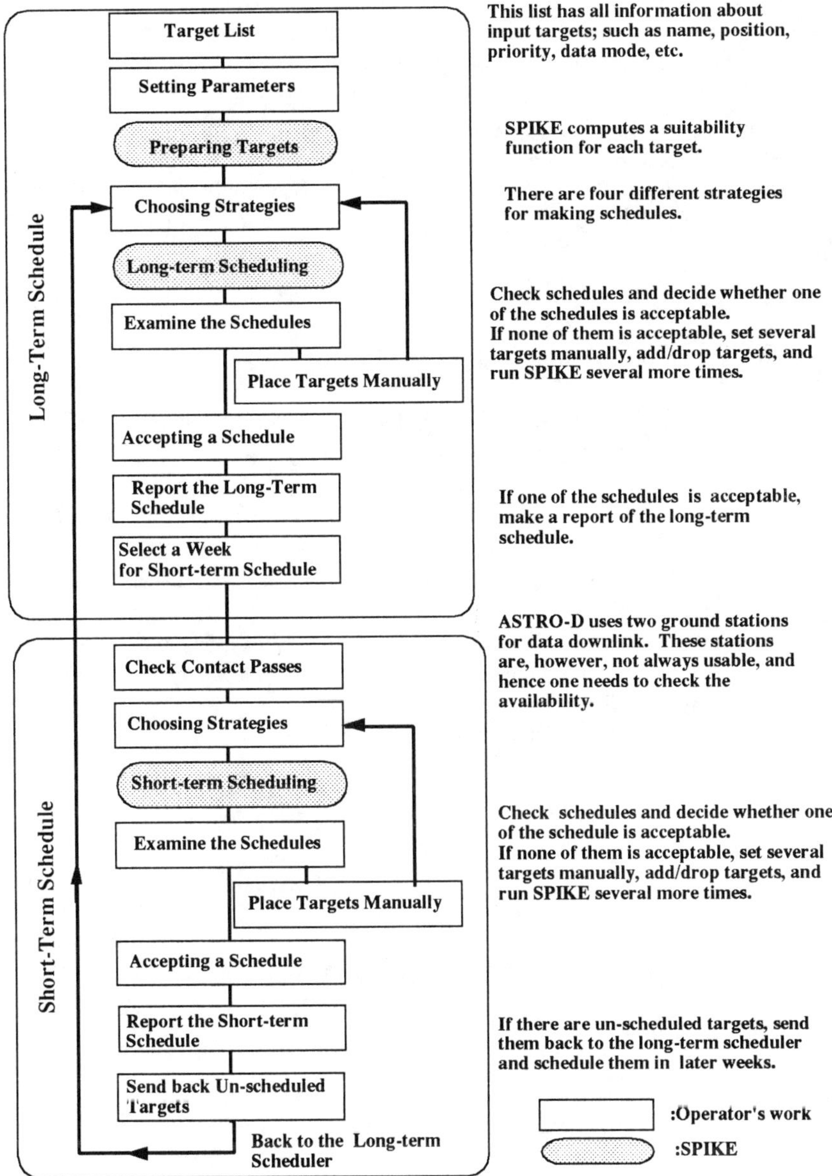

Figure 1. ASTRO-D SPIKE Schedule Process

5. Conclusion

We have successfully ported the SPIKE scheduling system from Hubble Space Telescope to ASTRO-D. The ASTRO-D version of SPIKE will schedule both long-term and short-term observations for this low-earth orbiting X-ray satellite observatory. The success of this effort helps demonstrate that efficient mission scheduling technology can be transferred from one mission to another.

Acknowledgments. This project is funded by NASA grant NASW-4372. We also thank ASTRO-D project teams at ISAS and at MIT, and SPIKE team at STScI for advice and assistance. Space Telescope Science Institute is operated by the Association of Universities for Research in Astronomy for NASA.

References

Arnaud, K., Day, C., & White, N. 1991, ASTRO-D Project Data Management Plan

Johnston, M.D. 1990, in Proc. Sixth IEEE Conf. on Artificial Intelligence Applications (Santa Barbara, March 5-9, 1990), Los Alamitos, CA: IEEE Computer Society Press 1990, 184

Johnston, M.D., & Adorf, H.-M. 1992, Int. J. Computers and Operations Research 19, 209

Morgan, E. 1992, Evaluation of SPIKE for XTE (MIT Technical Report, Center for Space Research).

Sponsler, J., Johnston, M., Miller, G., Krueger, A., Lucks, M, & Giuliano, M. 1991, in Proc. AIAA Conf. on Computers in Aerospace 8 (Baltimore, October 21-24 1991), 14

Tanaka, Y. 1990, Adv. Space Res. 10-2, 255

Part 4. User Interfaces

Astronomical Data Analysis Software and Systems II
ASP Conference Series, Vol. 52, 1993
R. J. Hanisch, R. J. V. Brissenden, and J. Barnes, eds.

Happy Families of AXAF Software

E. Mandel, R. J. V. Brissenden, M. Freeman, D. Nguyen, and J. Roll

Harvard-Smithsonian Center for Astrophysics, Cambridge, MA 02138

Abstract. We have developed an *X11* graphical user interface layered on top of an *IRAF*-compatible parameter interface that provides a common "look and feel" for diverse analysis systems. We describe representative *AXAF* applications that make use of this interface.

1. Introduction

> "All happy families resemble one another, each unhappy family is unhappy in its own way." (from *Anna Karenina*, by Leo Tolstoy)

At the Smithsonian Astrophysical Observatory (*SAO*), we are developing independent "families" of software to support various aspects of the Advanced X-ray Astrophysics Facility (*AXAF*) project, including telescope modeling and performance testing, detector development, calibration, and user data analysis. In general, software environments and methodologies are different for these different projects: in-house mirror modeling software uses a rapid prototyping approach, while the software used to run calibration tests at NASA's Marshall Space Flight Center requires a carefully designed architecture tuned to the hardware being tested. Still more stringent methodological requirements on portability, evolvability, and maintainability govern the development of software that will be utilized world-wide by the astronomy community to analyze *AXAF* data.

At the same time, we see an opportunity to make maximal use of resources by seeking commonality, since much of the *AXAF* pre-launch software can form the foundation for post-launch user analysis tools. Thus, we need to reconcile the independent software requirements of each of our *AXAF* groups with the advantages of having a common software approach, in order to develop the best possible software for the *AXAF* project as a whole.

Taking a hint from Tolstoy, our response to this challenge has been to try to make these software systems resemble one another as much as possible. Recognizing that user interface technology is an area in which we can attain commonality without imposing undue restrictions of any individual project, we have sought to build interfaces that will provide a common "look and feel" for different systems, while preserving direct access to the underlying tools. The result of our efforts thus far is the development of an *X11*-based graphical user interface (*GUI*) called *ASSIST* that is *layered* on top of an *IRAF*-compatible parameter interface.[1] The former provides commonality at the graphical level while the latter does so at the command level.

[1]The *ASSIST* is an acronym for "AXAF Science Support Interactive Software Tool".

We chose *IRAF* as the basis for our parameter interface because of its rich set of parameter-handling capabilities and also because of its widespread familiarity within the astronomical community. Our interface provides non-*IRAF* programs with the same "look and feel" as *IRAF* tasks, with full support for functions such as hidden, query, and learn modes, and range checking of inputs. It consists of an application programming interface (*API*) to set and retrieve parameter values, and a set of programs for manipulating parameters at the command line.

The *ASSIST* graphical user interface is based on the *AnswerGarden* system that is being developed by Mark Ackerman at MIT. It has sophisticated support for executing programs and *IRAF* tasks, examining documentation and help files, browsing tutorials, maintaining on-line user notes, asking questions of experts, and developing a knowledge base of frequently required information. These and other aspects of the parameter and graphical interfaces are described in detail elsewhere (Mandel et al. 1992). In this paper, we will focus our attention on the methods we use to connect the *ASSIST* interface to different analysis systems. Use of *ASSIST* (as well as the underlying *AnswerGarden* system) in a wider range of *SAO* applications, including access to a document data base and the organization and management of user questions, is described elsewhere (Brissenden, 1992).

2. Interfacing the GUI to Analysis Systems

Viewed simply as a graphical interface to an analysis system, the *ASSIST* consists of three types of windows (see Figure 1):

- **Grapher windows** organize tasks and programs in hierarchical "trees". Each "leaf" of a tree can represent an *IRAF* task, a script or program that uses the *IRAF*-compatible parameter interface, or another branch of the tree. When a leaf is activated using the mouse, a separate window is created to display the appropriate task, program, or branch.

- **Parameter editor windows** represent programs and tasks. They consist of a command strip containing command buttons for running tasks, viewing help, etc., as well as a tabular display of parameters that can be edited using *emacs*-style key strokes. The parameter editor window is built on top of the *IRAF*-compatible parameter interface, so it can be utilized by any *IRAF* task, or any script or program that uses the parameter *API*.

- A **target analysis window** runs the underlying analysis system (e.g., an xterm window that runs either *IRAF* or a command shell).

The parameter editor and the underlying parameter *API* play a central role in interfacing the *ASSIST* to an analysis system. When a parameter editor window is created for a particular program, the parameter file for that program is read using the parameter *API* and displayed as a graphical table. Parameters can be edited in the tabular display, or the task can be executed using the command buttons. While it is active, the parameter editor will check periodically to see if its parameter file has been changed on disk, in which case it re-reads the parameter values into the table. This mechanism allows command-line parameter changes in the target window to be fed back into *ASSIST*.

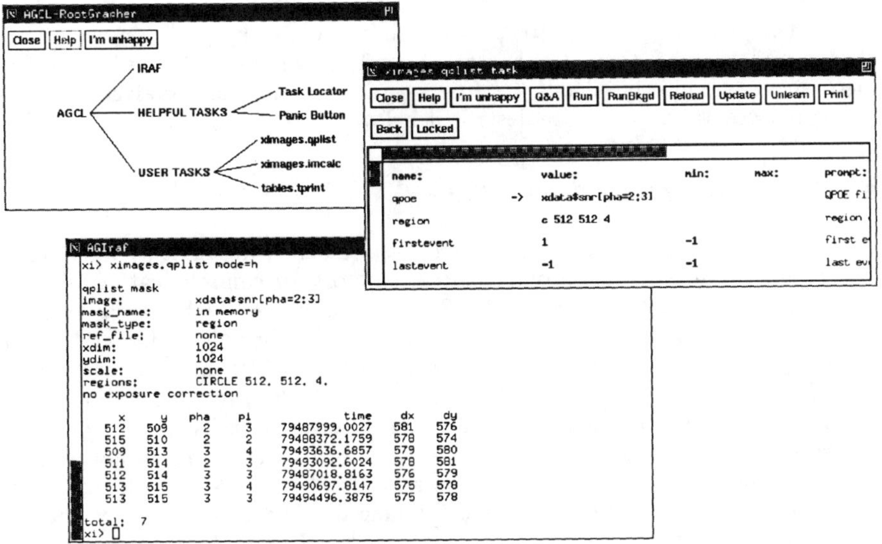

Figure 1. Grapher "trees" organize parameter editors which execute tasks and programs by sending commands to the analysis window.

When a parameter value is modified in the graphical parameter table, an update command is sent from the parameter editor to the target window to change the stored value of the parameter. When sending commands, the *ASSIST* uses its knowledge of the distinction between *IRAF* tasks and non-*IRAF* programs to send the appropriate type of command to the target window. Thus, to modify a parameter value in an *IRAF* task, the *IRAF* parameter assignment statement is sent to the window running *IRAF*. For a non-*IRAF* program, the *pset* parameter setting command is sent. Similarly, when a task or program is executed by the *Run* command in the parameter editor command strip, the appropriate execution string is sent to the target window. In all cases, the actual work is done by the target analysis system itself: the parameter editor simply sends the correct command to the underlying system.

Commands are sent from the *ASSIST* to the target analysis window by means of the *Xkibitzer* facility.[2] The *Xkibitzer* is an application programming interface (along with the high-level *Xkib* program built on top of the *API*) that converts an *ASCII* string to a series of *X11 KeyPress* events and then sends these events to another window using the *X11 XSendEvent* subroutine. The target analysis window receives these events exactly as if the user had typed them at the keyboard. Note that this design implies that the underlying analysis window is always available to the user.

[2] A "kibitzer" is someone who stands on the sideline and gives advice or meddles in the affairs of others.

Thus, the *ASSIST* organizes and displays analysis tools in a hierarchical tree structure and runs these tools on behalf of the user by packaging command strings and sending them to the target analysis window. Using this simple mechanism, we interfaced our *GUI* to a variety of different analysis systems. Three representative cases are described below.

3. Interfacing ASSIST with IRAF

The *AXAF* Science Center at *SAO* will base much of its analysis software on *IRAF*, and so the *ASSIST* provides direct support for communicating with the *IRAF* environment. When the *ASSIST* is started, the *IRAF* hierarchy of packages and tasks is automatically added to the *ASSIST* graphical hierarchy. The hierarchy is dynamically determined by reading the *IRAF* help data base, so that each *IRAF* site will have its own configuration of *IRAF* packages represented in the *ASSIST*. In addition, users can add often-used tasks to their personal top-level Grapher window, so that these tasks can be accessed without having to search the *IRAF* tree. A special function is also provided to locate *IRAF* tasks using a keyword search through the help data base. In this way, the *ASSIST* provides a graphical interface that matches *IRAF*'s hierarchical package structure, while also supporting methods for accessing the tasks directly.

4. Interfacing ASSIST with Programs Using the Parameter Interface

The power of our two-layered interface is illustrated by the software development efforts of the *AXAF* mirror performance analysis group. They have developed a suite of spectral fitting programs in the Unix environment, using the *IRAF*-compatible parameter interface to manage the dozens of parameter options associated with these programs. The system is designed as a pipe-line processing system, in which a given program writes its parameter results into the parameter file of the next program in the sequence. The pipe-line is typically run on hundreds of data sets at a time, with the action being managed by the parameter interface.

During development of this system, it became clear that these fitting programs would have great value when used interactively. To provide the system with a full graphical parameter interface, it was only necessary to include the names of the fitting programs in a given user's *ASSIST* initial load file of often-used programs. The *ASSIST* automatically adds these programs to the top-level *ASSIST* grapher, so that they become available alongside the standard *IRAF* tasks. And because each parameter editor updates itself when a change is made to its parameter file, the display of parameters is always kept current as one program writes its return values into the parameter space of another. It is therefore easy to set the parameters in the one program, run that program, inspect the results in the parameter display of the next program, execute that program, etc.

5. Interfacing ASSIST with Other Systems

Another group at *SAO* is responsible for validating Kodak's mirror performance predictions by duplicating portions of the latter's finite element analysis (*FEA*) of the mirror optics and support structure. An *FEA* to raytrace data interface has been developed using the *PV-Wave* interactive environment. The *ASSIST* interface was desired for these *FEA* tools, and so a simple technique was developed to layer the *ASSIST* on top of *PV-Wave* or any command-driven system.

The technique uses simple shell scripts to mediate between the *ASSIST* and *PV-Wave*. Each *PV-Wave* tool is represented by a parameter editor which, in turn, is connected to a script that manages a parameter file for that tool. Users set parameter values using the tabular display, and these are stored in the parameter file. The *Run* command starts the script in the xterm window. The script then uses the *Xkibitzer* to send commands to the *PV-Wave* window. These commands cause *PV-Wave* to read the script's parameter file to get user-specified parameter values and perform the required actions. Upon completion, *PV-Wave* uses the *pset* parameter program to write return values back into the parameter file. These are then redisplayed by the parameter editor. Thus, all of the benefits of the *GUI* are available to the user, even though the target analysis system is not directly supported by the interface.

6. Conclusions

In principle, the *ASSIST GUI* can be layered on top of any command-driven system, either by supporting that system's command interface directly, (as is the case for *IRAF* tasks and programs that use the parameter interface), or by means of intermediate scripts (as is the case for *PV-Wave)*. Our near-term plans will center on extending the scope of the *GUI* to support such diverse tools as image display and data base access. Ultimately, our aim is to develop a complete *AXAF* environment for running programs, accessing information, examining documents, and getting assistance, an environment in which the different families of *AXAF* services can happily resemble one another.

Acknowledgments. Special thanks go to Ralph Swick (DEC) for many helpful suggestions concerning the *ASSIST* in general and the *Xkibitzer* in particular, and to Mark Ackerman, who enthusiastically extended the *AnswerGarden* to help make this all possible. This work was supported under NASA contracts to the *IRAF* Technical Working Group (NAGW-1921), the *AXAF* High Resolution Camera (NAS8-38248), the *AXAF* Support Team (NAS8-36123) and the *AXAF* Science Center (NAS8-39073).

References

Mandel, E., Roll, J., Murray, S.S., & Ackerman, M.S. 1992, to appear in Proceedings of the Conference on Astronomy from Large Databases-II

Brissenden, R.J. 1992, in Proceedings of the Workshop on User Interfaces For Astrophysical Software, produced for NASA by the Harvard-Smithsonian Center for Astrophysics

Discussion

Silberberg: As the task tree becomes large, the task-tree window becomes cluttered. Also, the user does not always know the task name desired. How does ASSIST help in such a scenario?

Mandel: ASSIST is planning an "Outline" window which makes navigating large task trees easier.

Pollizzi: Does ASSIST abandon the CRT user? I.e., what do you do from home?

Mandel: The ASSIST is layered on top of analysis systems such as IRAF. It sends commands to the "target" analysis window on behalf of the user. Thus, the underlying system is still available to the user who only has access to a CRT.

Astronomical Data Analysis Software and Systems II
ASP Conference Series, Vol. 52, 1993
R. J. Hanisch, R. J. V. Brissenden, and J. Barnes, eds.

Tools from the *IDL* Widget Set within the *X Windows* Environment

Benoit Turgeon[1]

Space Astrophysics Laboratory, Institute for Space and Terrestrial Science, 4850 Keele Street, North York, ON, M3J 3K1, CANADA

Abstract. This paper describes two new tools that were developed from the IDL widget set in the *X Windows* environment: XSm, a slide-making utility, and XCc, a widget application allowing the user to change the shape of the cursor. Although simple-minded, these utilities were developed as a stimulant to other authors to use the new IDL widget programming techniques. In addition to widget examples, this paper includes instructions about retrieving the documentation related to the IDL Astronomy User's Library that was created at the Space Astrophysics Laboratory.

1. Introduction

The introduction of Graphical User Interfaces (GUIs) in the software industry has both advantages and disadvantages. Among the large number of advantages is the facility with which the user is able to perform a given task. Among the disadvantages is the difficulty for the programmer to implement such interfaces.

In the *X Window System* (the *X Window System* is a trademark of the Massachusetts Institute of Technology), simple GUIs can be difficult to write, especially for people who are not familiar with programming in modern languages such as C, or simply do not have the time to invest in reading the rather large documentation pertinent to GUIs. A tremendous amount of work can be required to achieve only the simplest of applications. The main difficulty when writing *X Windows* applications resides in the flexibility of the windowing system itself. In effect, *X Windows* is perhaps the most advanced and flexible windowing system currently available on platforms used in the astronomical community, and imposes large constraints on the programmer to allow an almost limitless variety of applications to be written. Flexibility, in general, is directly proportional to programming difficulty for system designers.

The Interactive Data Language (IDL, *IDL* is a registered trademark of Research Systems, Inc.) widget interface simplifies the writing of GUIs dramatically and even makes writing them a pleasant experience. The tradeoff in using a more limited set of widgets is not a big issue since most astronomical applications are rather rigid in format and fairly standard in their requirements.

[1]Also with the Department of Physics & Astronomy, York University, Toronto, CANADA.

Research scientists and students at SAL are currently starting to use the widget capabilities of IDL to write easy-to-use GUI applications. With the release of the latest version (V3.0) of the IDL documentation, the gap that existed in the widget documentation has been filled and writing successful GUIs require now a minor time investment.

This paper will present two of the first GUIs written at SAL: XSlideManager and XChangeCursor in Sections 2 and 3. These should be seen as development tools and their purpose is to act as a stimulant to other authors to make use of the IDL widget programming techniques.

In addition, the author has produced an upgraded version of the documentation pertinent to the IDL Astronomy User's Library developed by W. B. Landsman and F. Városi at Hughes/STX (Landsman 1993). The description of the manual as well as its availability are outlined in Section 4. Section 5 is a "call for comments" on a proposal to store astronomical widget applications written in IDL in a publicly available directory at SAL.

2. XSlideManager (*XSm*)

A large fraction of our work at SAL involves images reduction and analysis of images coming from ground-based observatories or the Hubble Space Telescope (HST). Once the reduction process is completed, a final output for publication is needed, most often in a slide or 35 mm negative format. XSlideManager was developed to shorten the slide-making process and transform it to a point-and-click simple GUI.

XSlideManager assumes that the input FITS images have been processed and are ready for display. The first task for the user is to select one of the ten possible layouts for the arrangement of the images. If the user requires a 2×2 image slide, for example, the template button can be cycled through until the 2×2 layout appears on the screen. The following step is to import the images from disk and display them into the various sections of the template (in this case, 4 sections). The user can then add various element to the slide such as surrounding display boxes, annotations, color intensity scale, and titles.

Once the layout and all the necessary annotations have been declared, the user lets XSm take over. XSm will remove the cursor from the screen, eliminate the widgets and produce a screen-wide window containing the images and their annotations. The following step is to `screendump` the image into a file, transfer it to diskette and take it to the local photo-finishers for output on a print, a 35 mm negative or on a slide.

On-line help is available, as well as a tutorial and a few examples of slides produced with XSm. It should be mentioned that XSm was written specifically for our Sun workstations. Although we coded as generally as possible, some parts of the application are specific to the *X Window* system or the Sun Operating System. Porting to other platforms will not be done at SAL, but all prospective users are free to examine the code and modify it to match the requirements of their own systems. XSlideManager is available through `anonymous` FTP at the following site:

```
nereid.sal.ists.ca (132.251.40.2)
pub/idl_widgets/xsm1.tar.Z
```

Please note that the file is in "tar-compressed" format. If you can't read the file, contact the author through electronic-mail (the e-mail addresses appear at the beginning of this book). Instructions and a small introduction manual appear in the distribution.

Finally, the reader can get an idea of the quality of the software by looking at the images of the northern ultraviolet aurora on Jupiter taken by the HST (Caldwell et al. 1992). Those three images were created using XSlideManager.

3. XChangeCursor (*XCc*)

This other utility was developed mainly to get an easy access to the standard choice of cursors offered through *X Windows*. The other advantage of XCc over the IDL User's Library routine `chcurs` is the ability to look at the cursor choices before actually changing the cursor shape.

XCc will also support user-defined cursors in the standard 16×16 bitmap format, appended in the machine and human readable cursor database.

Although simple-minded, the coding of XCc was challenging, mainly because of the various methods of bitmap storage and the requirement that bitmaps should be easily editable with a simple text editor. However, this same application written entirely in the *X Windows* programming style would have been much more difficult.

This utility is also available through `anonymous` FTP at the same site that was described in Section 2. The filename is `xcc1.tar.Z`.

4. The IDL Astronomy User's Library Documentation

Over 400 IDL-based astronomy procedures are available through anonymous FTP at the Goddard Space Flight Center (GSFC, see Landsman 1993). On-line documentation for every procedure is available but no hardcopy or PostScript outputs are available. We have integrated the on-line help file into a TeX file that can be retrieved electronically from SAL:

```
nereid.sal.ists.ca (132.251.40.2)
pub/idl_astro/
```

The available files are: `contents_dec91.tex.Z`, `contents_dec91.ps.Z`, and `contents_dec91.dvi.Z`. The first file is the original TeX file, the second is the PostScript output and the last one is the DVI file for those who would like to modify the output format. All files are compressed by using the standard UNIX utility `compress`. As later versions of the library documentation are converted, they will be made available in the same directory, with the filename suffix `mmmyy`, where `mmm` is the month and `yy` is the year of the IDL Astronomy User's Library release.

5. Call for Comments

Beginning in early 1993, `nereid.sal.ists.ca (132.251.40.2)` will become more involved in the archival and distribution of IDL-related widget software. If you have any IDL widget applications that you wish to distribute, please contact the author through electronic mail. Arrangements will be made to import your applications to our site and include them in the publicly accessible directory `pub/idl_widgets`.

In addition, readers are encouraged to subscribe to the USENET newsgroup `comp.lang.idl-pvwave` for extended discussions on the IDL and PV-WAVE software systems.

Acknowledgments. I wish to underline the work of Aaron Aston at the Department of Physics & Astronomy at York University, Toronto, CANADA, for his contribution to the IDL Astronomy User's Library documentation.

References

Caldwell, J., Turgeon, B., & Hua, X.-M. 1992, Science, 257, 1512

Landsman, W.B. 1993, this volume

Astronomical Data Analysis Software and Systems II
ASP Conference Series, Vol. 52, 1993
R. J. Hanisch, R. J. V. Brissenden, and J. Barnes, eds.

GUIs in the ESO-MIDAS Environment

P. Ballester and K. Banse

European Southern Observatory, Karl-Schwarzschild-str.2, D-8046 Garching, Germany

Abstract. Graphical user interfaces are developed at ESO mainly for instrument and telescope control, archives and data reduction. The ESO GUI Common Conventions are based on the MOTIF graphic standard and secure the compatibility and similarity of the interfaces. Different interfaces have been developed in MIDAS using GUI builders: XHelp for accessing the on-line documentation, XEchelle and XSpectra for the reduction of echelle and long-slit spectra.

1. Introduction

ESO-MIDAS (Munich Image Data Analysis System) is the image processing system developed at ESO for astronomical data reduction. MIDAS is used for off-line data reduction at ESO and many astronomical institutes all over Europe. In addition to a set of general commands, enabling to process and analyze images, catalogs, graphics and tables, MIDAS includes specialized packages dedicated to astronomical applications or to specific ESO instruments.

Different graphical user interfaces are already in use at ESO or will be developed for the following applications:

- Instrument and telescope control, either locally or remotely.
- Archive operations and database access (STARCAT and others)
- On-line and off-line data reduction with MIDAS

A detailed description of the interface XEchelle is available in Ballester (1992). The present paper describes the options and solutions retained for the development of GUIs in MIDAS. Different viewpoints must be considered, consisting of the coding of interfaces and design aspects. Compatibility of "Look and Feel" makes necessary to opt for a given widget set and define a corporate style.

2. Developing Interfaces

2.1. Toolkits

Three toolkits are widely available in Unix (MIT Athena, OSF/MOTIF, Open Look), which differ in style and completeness, but also by their distribution policy. Some GUI builders provide interoperability: applications are developed with components common to the main toolkits and the code is generated differently for each toolkit (Raney 1991, Raney 1992). However they do not allow to exploit fully the possibilities of any particular toolkit. OSF/MOTIF, for its wide availability and its completeness, has been chosen as the ESO graphic standard.

2.2. GUI Builders

GUI builders are commonly recognized as a very successful approach for the development of graphical interfaces, since they allow developers to concentrate on design issues and spend less time in the low-level implementation problems. GUI builders can be found in two classes:

- design systems generate an exportable source code for each application.
- management systems interpret a project file to generate the interfaces.

UIDS (User Interface Design Systems) are very powerful development tools, although some possible drawbacks can be expected, like code defaults or differences of behavior between the designed and generated interface.

It is interesting to note that many public-domain GUI builders are available, like Dirt, Serpent or xgen. Graphical user interfaces for MIDAS have been developed first with such public domain softwares, and now with commercial UIDS.

2.3. Portability

MIDAS is distributed in about 170 astronomical institutes, 2/3 of them located in the member countries of the organization. 1/3 of the installations are performed on VMS systems and the Unix version of MIDAS is installed on almost all existing platforms. Differences between compilers, directory structures, versions of MOTIF or X11 are the main causes of portability problems.

2.4. Defining Guidelines

Design consistency is an essential issue since it not only helps users to identify quickly the structural elements of an interface, but also provides designers with ready-to-use solutions and avoid them to reinvent the same graphical wheels in too many different ways. Guidelines must constrain the realizations of interfaces without preventing good design solutions to be implemented. The ESO GUIs Common Conventions are based on the MOTIF Style Guide (OSF 1992), but also define ESO specific components and interfaces.

3. GUIs in MIDAS

Interfaces in MIDAS have been developed first at the application level, for specialized processes like reduction of spectra. GUIs are meant to facilitate the learning of specialized software and to provide to expert users a better visibility on complex applications (See Figure 1). Benefits are expected from interfaces in the area of application packages and for very frequently used facilities, like help commands, graphics or displays.

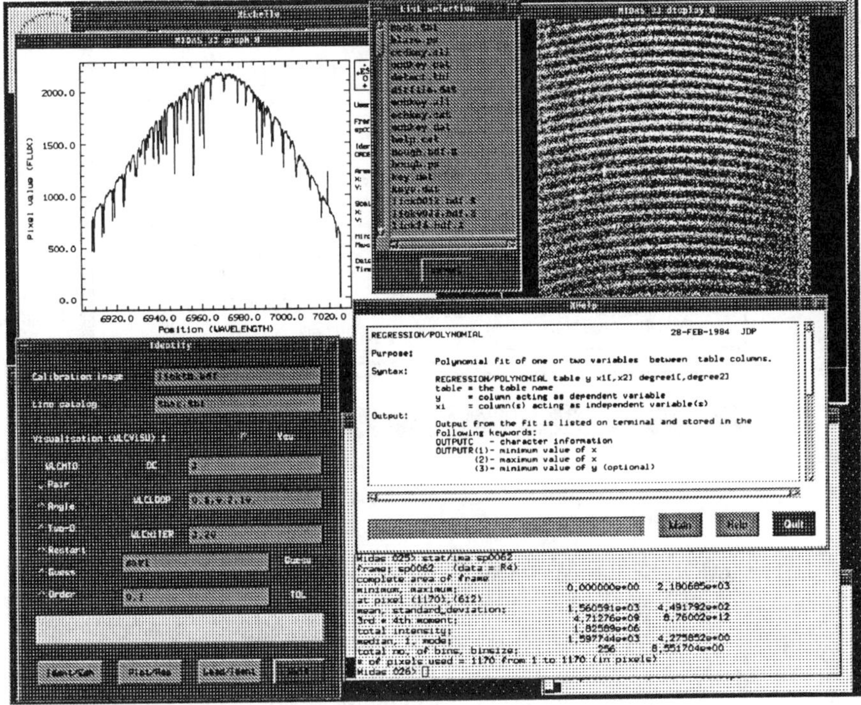

Figure 1. The MIDAS Environment

One of the advantages of GUIs is that they eliminate syntax problems by generating the adequate MIDAS commands corresponding to the user actions. The principle of complete independence between applications and interfaces is respected: MIDAS applications can be used without the interfaces by typing commands directly in the terminal window.

In spectroscopy, two interfaces are widely used: XSpectra and XEchelle. These interfaces were originally based on Athena widgets. XEchelle has been generated with Dirt. A first MOTIF interface, XHelp, has been realized with the UIM/X GUI builder from Visual Edge Corp. (Mikes 1991). This interface gives access to the on-line MIDAS documentation.

3.1. Interface XEchelle

The calibration and reduction process for echelle spectra is organized as a series of steps, some of them being optional. For each step, several methods may be available. Each method is controlled by a different set of parameters. The interface represents the status of the calibration and reduction process by displaying the options, methods and parameters values set by the user. Buttons and parameter fields not relevant to the selected options and methods are insensitive. During the design, emphasis has been given to optimize the usage of screen space, so that the interface fits between the already required display, graphics and terminal windows. The interface is independent from the MIDAS monitor

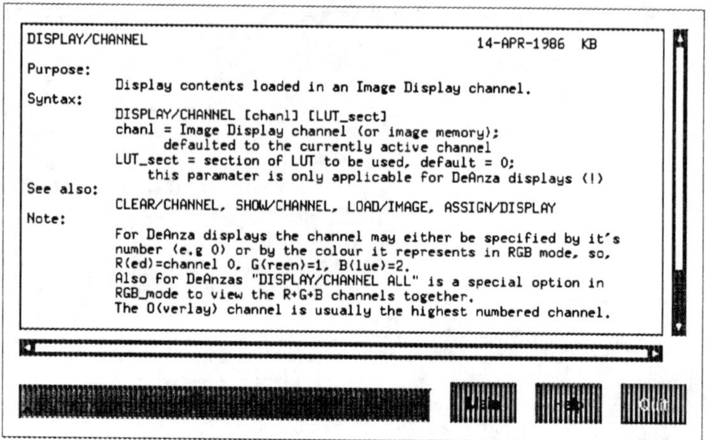

Figure 2. The Graphical User Interface XHelp

and from the Echelle package. Commands are sent to the MIDAS monitor, as if they were typed by the user in the terminal window.

3.2. Interface XHelp

The interface XHelp provides access to the on-line MIDAS help files and documentation. It is based on an hypertext-like behavior: the user clicks on command names to get the documentation. The general implementation of a 'See Also' entry (see Figure 2) allows navigation throughout the documentation. The interface is independent from the MIDAS monitor: documentation can be accessed even if the monitor is active.

References

Ballester, P. 1992, GUIs for MIDAS Applications, in Proceedings of the NASA Headquarters Astrophysics Division Workshop: User Interfaces For Astrophysical Software, Goddard Space Flight Center, Greenbelt, MD

Raney, S. 1991, Pick a GUI, Any GUI, in Unix World, May issue

Raney, S. 1992, Porting Your Way To Interoperability, in Unix World, June issue

Open Software Foundation 1992, OSF/Motif Style Guide, (Prentice Hall)

Mikes, S. 1991, Two Gooey Builders That Stick, in Unix World, January issue

Discussion

Misra: How much programmer time was involved in adapting MIDAS to MOTIF?

Ballester: Each package is adapted independently. About one week per package.

Hanisch: Do experienced MIDAS users actually use your GUIs, or do they continue to use the command-line interface they are already familiar with?

Ballester: Compared to the traditional command-line interface, GUIs offer many advantages which make them attractive also for experienced users:

- Better visibility and control over specialized applications,
- Minimized syntax problems and risks of mistyping,
- Easier access to the on-line help.

M. Albrecht: How much development time was saved by using a development tool (Unix)?

Ballester: Evaluating precisely the gain of development time can be the subject of an experimental study, but we did no such test. Our experience however, is that GUI builders reduce dramatically development efforts. Particularly, they allow concentration on the design rather than on the actual implementation, and generate a reasonably well optimized source code.

Astronomical Data Analysis Software and Systems II
ASP Conference Series, Vol. 52, 1993
R. J. Hanisch, R. J. V. Brissenden, and J. Barnes, eds.

The MIDAS Table File System and the Data Organizer

M. Peron and P. Grosbøl

European Southern Observatory, Karl-Schwarzschild-Straße 2, D-8046 Garching, Germany

Abstract. Before being able to actually reduce and analyse a new set of observations, the observer has to prepare, sort out and arrange the data, that is for instance, make a first quality check, classify data according to a set of rules and associate with each science frame a set of relevant calibration frames. This task can be cumbersome because of the complexity of the instruments and the large number of data files they produce. In the first part of the paper, the MIDAS Table File System (TFS) as such is introduced and utilities for relational database operations are described in detail. The prototype of a Data Organizer based on the TFS is presented. It uses only existing tools in a very straightforward and user-friendly way.

1. The Table File System

A table in ESO-MIDAS (ESO-IPG 1992) is a data structure for handling collections of heterogeneous data arranged in rows and columns and is constructed as a simple relational database where a row of the table is a tuple of the relation and a column an attribute (Peron et al. 1992). The supported data types are numerical data (8/16/32-bit integers or 32/64-bit reals) and character strings, all elements of a given column being necessarily of the same type. The entry at a given row and column may be either a single value or an array. The MIDAS table structure is therefore fully compatible with the Binary Table Extension proposed by FITS.

A header containing a set of descriptors is associated with each table. It defines the internal structure of the table and makes it possible to support different disk storage strategies (e.g., row by row, column by column).

1.1. Input/Output

The exchange of data to and from the Table File System is done either through standard ASCII files or FITS. This makes it easy for users to transfer data between the TFS and other environments. Thus, output files from text editors and database systems containing tabular data in any fixed format can be directly transferred into the TFS.

Table values can be listed out and the output formatting will be done using the display format associated with each column. Supported formats are Fortran-77 standard formats (e.g., E15.6; I4) and special display formats to accommodate sexagesimal and time values. Finally, facilities exist to plot/overplot table data.

2. Relational Database Operations on Tables

Relational operations can be carried out on tables: one can *sort* a table in ascending or descending order according to one or several columns, *project* one or several columns of an input table into a new table, *merge* two or more tables into an output one. The *select* and the *join* operations are discussed here in detail.

2.1. Selection of objects

The *select* operation is used to select a subset of rows which satisfy a selection criteria formulated as a logical expression. This expression may include reference to other columns, logical operators, mathematical functions for numerical columns and pattern matching expressions when columns containing character strings are involved. Table 1 illustrates the supported operators and functions.

OPERATOR	APPLICABLE ON	EXAMPLES
Relational Operator .LE. .LT. .GE. .GT. .NE. .EQ.	Numeric Column Character Column	:MAGNITUDE.LT.7 :IDENT.EQ."~*FF*"
Logical Operator .OR. .AND. .NOT.	Numeric Column Character Column	SELECT.AND.:RV.GE.4000.0 :DESCR.EQ."S[abO]*"..AND.:B_TOT.GT.15..AND.:B_TOT.LT.16
Arithmetic Operator + - * / **	Numeric Column	
Mathematical Functions sqrt(:a) ln(:a) log10(:a) exp(:a) sin(:a) cos(:a) tan(:a) abs(:a) int(:a) min(:a,:b) max(:a,:b) mod(:a,:b)	Numeric Column	(:BT-25.-5.*LOG10(:RV/50)).GT.-14
Character Functions concat(:a,:b) collapse(:a) tolower(:a) toupper(:a)	Character Column	CONCAT(:NAME,TOUPPER(:SEQ)).EQ."NGC3*"

Table 1. The *select* Operation on MIDAS tables

2.2. Join of Tables

TFS allows one to perform cross-matching of tables, that is, to find common objects in two tables by comparing one or two attributes (columns) of the objects (rows) from both files. Using *Equi-Join*, only objects with identical attributes will be matched while *Fuzzy-Join* matches objects having similar attributes within an error box.

3. The Data Organizer

To illustrate the basic functionalities of the Data Organizer, an observing run of three nights obtained with the ESO EMMI instrument has been extracted from the ESO archive using STARCAT (ST-ECF, ESO 1992). EMMI (Melnick et al. 1992) is a very flexible instrument which is permanently mounted on ESO's New Technology Telescope on La Silla, Chile. It offers many observing modes ranging from wide-field imaging to high-dispersion echelle spectroscopy. For this reason it is a typical example of an instrument producing a large quantity of very heterogeneous data files which are difficult to handle without proper tools.

3.1. Creation of the Observation Summary table

Each image descriptor which is considered relevant for the reduction process (e.g., exposure time, telescope mode) is mapped into one column of a table that is called the Observation Summary Table (OST), and the corresponding information for a given input frame is stored into one of its row.

The information is mainly collected from the header of the images or computed automatically, such as low order statistics of the image data, e.g., mean and root mean squares.

3.2. Definition of Classification Rules

Next, the data files need to be classified into groups according to some rules: One needs for instance to put together all the files observed in a given instrument mode (e.g., imaging, spectroscopy), or one wants to group the images according to the exposure time. A instruction for grouping Flat Field exposures could in natural language be: Select all files which have a descriptor IDENT matching one of the substrings 'FF', 'SKYFL' or 'FLAT'.

A first approach for establishing such a set of criteria would be to adopt a rule based expert system. But this doesn't take into account the fact that the information a user enters in the observing log file does not usually follow any standard conventions. Furthermore, every data set is different and only the user knows what reduction procedure is optimal for a given purpose. The rules which have to be established can actually be thought of as a logical expression supported by the *select* command, but writing such a expression can be tedious unless the user is provided with a utility which facilitates the formulation of the query.

An interface using the Table editor has been designed. It enables the user to enter interactively constraints on the existing fields of the OST. Relational operators (e.g., >, <, != or =) as well as values or ranges of values may be used.

The given constraints are ANDed and translated into a expression understandable by the SELECT command. This expression is stored in the header of the table. In addition, a command exists to edit it interactively.

3.3. Classification of the Input Images

The Image files satisfying a given selection rule are flagged with a string provided by the user. This result is stored in a column of the OST. Three different sets of rules have been written to classify our EMMI observations: the first set groups the files according to the exposure type (Table 2A), the second groups them according to the detector mode (Table 2B) and the last one according to the optical path. The rules of a given set are applied sequentially and the resulting strings are stored into one column of the OST (Table 2C), e.g., four different classes of exposure type have been generated: DK, SC, FF and TT. At this stage, it is already possible to make a preliminary statistical analysis of the observing run and this without having to interactively inspect individual files. One can, for instance, plot the sequence number of the images against the date and use different symbols according to the exposure type in order to have an overview of what has been observed. Or one may want to plot the rms noise of the flat field exposures against their mean pixel value to check the general consistency of the frames and whether there are any outliers (Figure 1).

3.4. Association of Images

The last task of the Data Organizer, which is still under development, consists of the association of science frames with suitable calibrations frames. This can be achieved by using the same rule generating interface as referred to above even though the rules to be applied are different: One may want to look for the flat fields which have been taken within a certain time interval of a given science exposure. One may also wish to plot parameters, for instance read-out noise versus time, in order to recognize trends or other peculiarities and to interactively reject frames which are deficient.

TABLE 2 (A)

COLUMN	RULES
FILENAME	
NPIX_1	
NPIX_2	
IDENT	= *FF*|*FLAT*|*SKYFL*
UT	
EXPTIME	
INSTRUMENT	= *EMMI #1*
_EI_MODE	
_ED_MODE	

TABLE 2 (B)

COLUMN	RULES
FILENAME	
NPIX_1	= 1700
NPIX_2	= 1700
IDENT	
UT	
EXPTIME	
INSTRUMENT	
_EI_MODE	
_ED_MODE	= *S*

TABLE 2 (C)

FILENAME	EXPTYPE	DETMODE	OPATH
ntt0211	SC	D1501	F610
ntt0212	DK		
ntt0213	DK	**D1700**	
ntt0214	**FF**	**D1700**	F610
ntt0215	**FF**	**D1700**	F610
ntt0216	TT		
ntt0217	SC	D1501	F610
ntt0218	DK	D1501	
ntt0219	SC	D1501	F610
ntt0220	SC	D1501	F610
ntt0221	SC	D1501	F610
ntt0222	SC	D1501	F610
ntt0223	SC	D1501	F610
ntt0224	**FF**	D1501	F610
ntt0225	SC	D1501	F610
ntt0226	SC	D1501	F610
ntt0227	**FF**	D1501	F610
ntt0228	**FF**	D1501	F610
ntt0229	**FF**	D1501	F610
ntt0230	**FF**	D1501	F610

Table 2. Definition of classification rules and classification of images

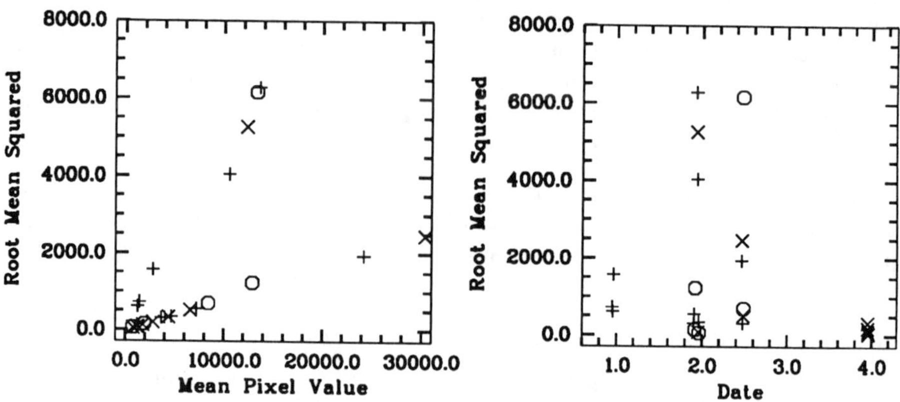

Figure 1. Flat Field Exposures (Symbols represent different filters)

3.5. Conclusions

We have presented a first implementation of a conceptually new tool which enables observers to organize the reduction of their data in an optimally customizeable way. Normally the operation is fully interactive, but the usage in batch mode and the definition of customized defaults are trivial. The Data Organizer is built entirely on existing capabilities of the MIDAS Table File System. Therefore the astronomer does not have to learn any new computer jargon, change environments, or convert data formats. The entire spectrum of utilities of an advanced image processing system can be incorporated without restriction. For applications which are (initially) based only on header information, it is possible to use the Data Organizer by just reading that information but without actually dumping the data to disk. This is of great value if a strategy has to be worked out how to cope with limited disk space. Finally, the Data Organizer will also be extremely valuable at the telescope because it offers practically unlimited options to monitor the ongoing observing program, e.g., performance of the detector, efficiency of the observer, status of the calibration data base, exposure levels, effective image quality, etc.

Acknowledgments. We are very grateful to M. Albrecht and D. Baade for very useful discussions and suggestions.

References

Peron, M., Ochsenbein, F., & Grosbøl, P. 1992, in Proc. of the Astronomy from Large Databases II

ST-ECF, ESO 1992, in Documentation for on-line catalogs

Melnick, J., Dekker, H., & D'Odorico, S. 1992, in EMMI & SUSI, ESO Operating manual

ESO-IPG 1992, in MIDAS Users Guide

Astronomical Data Analysis Software and Systems II
ASP Conference Series, Vol. 52, 1993
R. J. Hanisch, R. J. V. Brissenden, and J. Barnes, eds.

An IDL-Based Analysis Package for COBE[1] and Other Skycube-Formatted Astronomical Data

John A. Ewing
Applied Research Corporation, Goddard Space Flight Center, Code 685.3, Greenbelt, MD 20771

Richard Isaacman
General Sciences Corporation, Goddard Space Flight Center, Code 685.3, Greenbelt, MD 20771

Joel M. Gales
Applied Research Corporation, Goddard Space Flight Center, Code 685.3, Greenbelt, MD 20771

Sarada Chintala
Hughes STX Corporation, Goddard Space Flight Center, Code 685.9, Greenbelt, MD 20771

Peter Kryszak-Servin
General Sciences Corporation, Goddard Space Flight Center, Code 685.3, Greenbelt, MD 20771

Kevin G. Galuk
NYMA, Inc, Goddard Space Flight Center, Code 685.9, Greenbelt, MD 20771

Abstract. UIMAGE is an IDL-based software package designed to support the analysis of data from NASA's Cosmic Background Explorer (COBE). The package will also work with sky-map data-sets from other experiments which are also in the unfolded skycube format. The COBE project is using data in this form in order to avoid losses in the data's resolution which would occur if it was translated into another projection. Though UIMAGE will reproject data and allow users to manipulate the projections, any actual calculations are always performed by using the data in the original skycube format.

UIMAGE is a menu-based system, and can run on both X-window terminals and on traditional Tektronix-compatible terminals (certain functionality however is only accessible from X-windows). One advantage of a menu-based system is that guest investigators can get a quick acquaintance with the available data and functionality without being required to immediately become familiar with the syntax required to invoke local software.

[1]COBE is supported by NASA's Astrophysics Division. Goddard Space Flight Center (GSFC), under the scientific guidance of the COBE Science Working Group, is responsible for its development and operation.

1. Introduction

UIMAGE is a data analysis package written in IDL for the Cosmic Background Explorer (COBE) project. COBE has extraordinarily stringent accuracy requirements: 1% mid-infrared absolute photometry, 0.01% submillimeter absolute spectrometry, and 0.0001% submillimeter relative photometry. Thus, many of the transformations and image enhancements common to analysis of large data sets must be done with special care. UIMAGE is unusual in this sense in that it performs as many of its operations as possible on the data in its native format and projection, which in the case of COBE is the quadrilateralized spherical cube ("skycube"). That is, after reprojecting the data, e.g., onto an Aitoff map, the user who performs an operation such as taking a crosscut or extracting data from a pixel is transparently acting upon the skycube data from which the projection was made, thereby preserving the accuracy of the result. (Please refer to White & Stemwedel for more details about the skycube projection).

Current plans call for formatting external data bases such as CO maps into the skycube format with a high-accuracy transformation, thereby allowing Guest Investigators to use UIMAGE for direct comparison of the COBE maps with those at other wavelengths from other instruments. UIMAGE is completely menu-driven so that its use requires no knowledge of IDL. Its functionality includes I/O from the COBE archives, FITS files, and IDL save sets as well as standard analysis operations such as smoothing, reprojection, zooming, statistics of areas, spectral analysis, etc.

One of UIMAGE's more advanced and attractive features is its terminal independence. Most of the operations (e.g., menu-item selection or pixel selection) that are driven by the mouse on an X-windows terminal are also available using arrow keys and keyboard entry (e.g., pixel coordinates) on VT200 and Tektronix-class terminals. Even limited grey scales of images are available this way. Obviously, image processing is very limited on this type of terminal, but it is nonetheless surprising how much analysis can be done there. Such flexibility has the virtue of expanding the user community to those who must work remotely on non-image terminals, e.g., via modem.

UIMAGE stores each data array internally within a structure. Besides the data array, each structure contains information about the data which may be needed for certain operations. Some of this additional information includes the name of the projection, the name of the coordinate-system, the title, the bad-pixel indicator value, etc. The design puts no software-imposed limit on the number of data-objects which can be manipulated. The practical limit is a function of the available memory. The realm of data-objects which can be manipulated within UIMAGE include skycubes, individual faces of skycubes, reprojected maps, graphs, zoomed-sections of images, and 3-D objects containing skycube maps at different frequencies.

Additional software has been developed which transports a data array of an appropriate size from the user-level IDL environment into the UIMAGE internal data environment and vice versa. This allows users to do a partial analysis of data within UIMAGE, and then to transfer the data to the IDL environment in which they can do additional analyses that are not supported within UIMAGE itself.

2. Additional Features

UIMAGE's main menu (Figure 1) lists classes of operations. For each of these classes there corresponds a sub-menu (Figure 2) in which the actual operations may be accessed. On each menu there is a HELP option which, if selected, will result in the presentation of scrollable text which describes the significance of the various options in the current menu. (Scrollable text is supported for both X-window and non-X-window terminals).

```
IDL>
IDL> uimage

        UIMAGE [MAIN MENU]
 Data I/O and Management...

 Display Manipulation...
 Image Enhancement...
 Algebraic Operations...
 Spectrum Operations...
 Line Plots and Statistics...
 Modeling and Fitting...

 Journal Enable/Disable    OFF
 Report Comments or Problems

 HELP
 Exit UIMAGE
```

Figure 1. UIMAGE's main menu, as it appears on a non-X-windows terminal.

There are about 40 operations available within the package (please refer to Figure 2 to see the available functionality). Many of the operations are uniquely tailored for data in the skycube projection. As an example, smoothing a skycube image must be done in a special manner since some physically neighboring pixels may not be neighboring pixels within the image. This is the case with pixels that lie along the edge of a cube face. Our smoothing algorithm was designed to account properly for this geometry, and also to account for that fact that some pixels do not have a valid value due to some form of contamination.

The routines which implement the core functionality (locally referred to as "Science Analysis Tools") have been designed so as to be accessible outside of the UIMAGE package, i.e., directly callable from the IDL command line. This promotes re-use of code for analysts who will be working with skycube data, but who may not be actually using the UIMAGE package itself.

Whenever the user selects an operation within UIMAGE, he will get a menu which contains the titles of potential operands. For X-window users, the result of the operation will automatically be shown in a new window placed on the screen in such a way as to minimize overlapping with other windows.

DATA I/O AND MANAGEMENT

Read a data set...
Write a data file...
Create a PostScript file

Report object attributes
Remove an object

HELP
Return to MAIN MENU

SPECTRUM OPERATIONS

Extract spectrum from a pixel
Average spectra in an area

Extract a frequency slice
Integrate over frequency
Display a frequency table

HELP
Return to MAIN MENU

DISPLAY MANIPULATION

Refresh an image
Resize/redraw all windows
Change color table
Stretch contrast
Change a title

Resize a graph
Change X and Y axis ranges
Change graph axis labels

HELP
Return to MAIN MENU

ALGEBRAIC OPERATIONS

X1 + X2
X1 - X2
X1 * X2
X1 / X2

SQRT(X)
LOG10(X)
ABS(X)

C0 + C1*X1 + C2*X2 +...+ Cn*Xn
Average(X1, X2, X3, ..., Xn)

HELP
Return to MAIN MENU

IMAGE ENHANCEMENT

Smooth an image
Change array resolution
Histogram equalization
Edge enhancement

Zoom an image
3-D surface plot
Reprojection
Plot coordinate grid

HELP
Return to MAIN MENU

LINE PLOTS AND STATISTICS

Tek or VT200 grey scale plot

Cross sections, Sky cuts
Scatter plot (2 maps)
Contour plots

Single pixel information
Statistics & Histogram

HELP
Return to MAIN MENU

Figure 2. Sub-menus directly beneath UIMAGE's main menu.

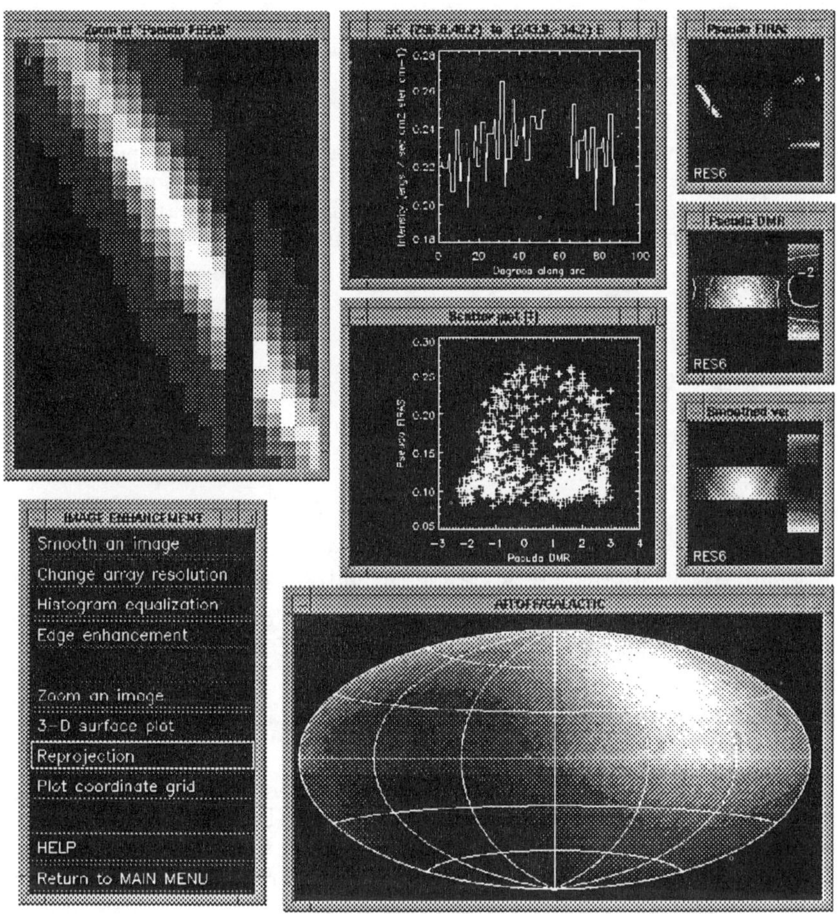

Figure 3. An example of how UIMAGE looks when run in an X-windows environment

Figure 3 shows how a typical X-window screen looks while UIMAGE is running. After having read in two sample data-objects, various operations (smoothing, contouring, reprojecting, cross-sectional plotting, scatter plotting, and zooming) were applied in order to generate additional objects. The menu that is shown in the lower left corner gives an example of how the menus appear in an X-window environment.

A package that can run on a variety of terminal types is accessible to a broad range of users. Although people on non-X-terminals won't be able to see color images, they can still do I/O, draw contours, make plots (histograms, geometric cross sections, intensity vs. frequency plots, or scatter plots), perform algebraic operations, generate grey-scale image-plots, generate PostScript files, etc. Though the name "UIMAGE" emphasizes the package's ability to assist with image-based analysis on X-terminals, the package nevertheless provides a fair amount of functionality for non-imaging environments.

UIMAGE currently works in IDL under the following operating systems: VMS, ULTRIX, SunOS, and PC/windows.

Acknowledgments. We would like to thank the following people for contributions to the development of UIMAGE: Ralf Petrich, Tom Piper, Kevin Turpie, Vidya Sagar, Ed Kaita, and Celine Groden. Software used to create menus on X-terminals was based on similar software from the STAR system.

References

White, R., & Stemwedel, S. 1992, in Astronomical Data Analysis Software and Systems I, A.S.P. Conf. Ser., Vol. 25, eds. D.M. Worrall, C. Biemesderfer & J. Barnes, 379

Discussion

Landsman: What are the advantages of the data-cube format?

Ewing: Each pixel in the Sky-cube represents an equivalent amount of steradians. The Sky-cube format has the data from COBE in its full photometric accuracy.

Shaw: Could you tell us how errors and exceptions are trapped and handled in your package?

Ewing: We try to check for arithmetical and I/O errors. When an error condition occurs and is appropriately identified, then we call IDL's MESSAGE command in order to inform the user of the condition, the current operation is aborted, and we put up the previous menu again. If the user wants, he can then select an option to perform another operation, or he can exit.

Astronomical Data Analysis Software and Systems II
ASP Conference Series, Vol. 52, 1993
R. J. Hanisch, R. J. V. Brissenden, and J. Barnes, eds.

GammaCore: The Compton Observatory Research Environment

T. McGlynn, J. Jordan, D. Jennings, N. Ruggiero, and T. Serlemitsos

Compton Observatory Science Support Center[1]*, Code 668.1, Goddard Space Flight Center, Greenbelt, Maryland 20771*

Abstract. The Compton Observatory Science Support Center (COSSC) is developing a coherent analysis environment for the analysis of Compton and other gamma-ray astronomy data. This environment, *GammaCore*, allows the astronomer to access the data analysis systems developed at the instrument team sites for the four Compton Observatory instruments. In addition users have access to standard astronomical tools, an archive system and other software developed at the COSSC. This paper describes the development of the *GammaCore*, particularly its user interface which is based on the AGCL, an early version of the AXAF Science Center's ASSIST system. The discussion focuses on the problems that were encountered in integrating extremely diverse software packages, and how these problems were overcome. Information on how to get access to the *GammaCore* is also given.

1. Introduction

The Compton Gamma Ray Observatory was launched in April 1991 and for the first time has provided the astronomical community with guest investigator (GI) opportunities in the gamma-ray regime from 30 keV to 30 GeV (Kniffen 1989). This paper describes the effort to build a common user interface to all software systems requisite for the analysis of Compton data. This is a challenging task since the necessary elements are extremely heterogeneous.

The Compton Observatory comprises four instruments: a low-energy all-sky monitor, the Burst and Transient Source Experiment (BATSE); a pointable spectrometer, The Orientable Scintillation Spectrometer Experiment (OSSE); a medium energy (roughly 1 MeV) imaging Compton telescope (COMPTEL); and a high-energy spark-chamber detector, the Energetic Gamma Ray Experiment Telescope (EGRET). Papers from *The Compton Observatory Science Workshop* (Shrader et al. 1992) give detailed information on these instruments and other aspects of the Compton Observatory mission.

Analysis of the data from these instruments is extremely complex and each of the instrument teams (ITs) which built the instruments has developed a sophisticated set of analysis tools. However, the Compton Observatory was

[1]Operated by Computer Sciences Corporation under contract to the National Aeronautics and Space Administration.

not originally to have a guest investigator component, so these systems were developed quite independently with little thought given to supporting non-team members. These systems are not compatible, are often hardware specific, and frequently use proprietary software.

The Compton Observatory Science Support Center (COSSC) was created to assist guest investigators is developing a software envelope which includes a common interface to the IT software systems. In addition, this research environment will allow users to access publicly available data from the Compton Observatory archive, provides help on how to use the observatory and software systems, and includes facilities to translate data between internal data formats and FITS. Also, it will allow astronomers to use tools from other software systems, notably IRAF, XANADU and IDL. Over the past year the COSSC has begun developing this environment, *GammaCore*. In the remainder of this paper we discuss how we have attempted to surmount the problems faced in accomplishing this task.

2. Elements of the *GammaCore*

To understand the requirements for the *GammaCore* user interface one needs to understand the diversity of the software that will be incorporated within it. This includes the data analysis systems developed by the instrument teams as well as various software elements created at the COSSC.

The instrument team analysis software forms the heart of the *GammaCore*. Each of these analysis systems represents tens to hundreds of years of work which would be virtually impossible to duplicate.

The BATSE analysis systems comprises a set of loosely coupled large analysis packages to do spectral, temporal and spatial analysis. The packages run only on VAX/VMS architectures and the data is tightly controlled using a INGRES database. Some elements of the BATSE analysis use standalone tasks. TAE and INGRES ABF forms are used to provide the user interface.

OSSE analysis is performed in IDL on VAX/VMS architectures. It uses very system-specific data formats and system specific code within IDL. The OSSE system uses the standard IDL user interface.

The COMPTEL analysis system was developed to work on IBM and Prime mainframe computers. To make it possible to port software among these machines the team developed their own user interface. This is currently tightly coupled to the native operating environments and an Oracle database. Work is underway to port the system to UNIX. The COMPTEL system uses a locally developed forms interface for most user interactions.

EGRET analysis is performed primarily on UNIX systems with a number of largely standalone tasks which communicate only via data files. Some database functions are performed on IBM mainframes. Specialized X11 interfaces have been developed for the individual tasks.

The great diversity of these instrument team software systems poses the greatest challenge for the *GammaCore* user interface: How can we integrate these systems into a package which allows the user the full power of the underlying systems, but which does not does not require the user to adapt to the idiosyncrasies of four very different systems, and how can this be done with very limited resources?

In addition to the instrument team systems, the COSSC is responsible for several software areas that must be integrated into the *GammaCore* as well as many small tasks that astronomers will find userful.

The most important COSSC responsibility is the data archive. The COSSC has developed a user catalog/data archive which allows the user to interrogate the catalog using INGRES ABF forms and then electronically retrieve data from the archive. The archive has been operational since August 1992. The archive software is kept independent of the underlying hardware using the GRASP software developed at the COSSC (Jordan et al. 1992).

Another major area developed at the COSSC are data conversion utilities. These provide the capability of converting instrument team format to FITS and the reverse. All data in the archive is currently stored in FITS. This software is described in Jennings et al. (1992), and is fully integrated into the *GammaCore.*

The third critical area which the COSSC has been integrating into the *GammaCore* is user help. This includes a standalone help package which describes all aspects of the Compton Observatory mission as well as context sensitive help within the *GammaCore* user interface.

3. The User Interface

The tool we chose in early 1992 to test the feasibility of unifying all of these disparate software systems was the AGCL, an early version of the AXAF Science Center's ASSIST program (see Mandel 1992). Two features of the AGCL were key to our decision. The AGCL supports a parameter editor which allowed us to isolate the underlying software systems from the user interface. Users set parameters in a single GUI used for all underlying tasks. The tasks were then run by scripts which read the parameter using utilities included with the AGCL, converted these parameters to the formats needed in the underlying software, and then started the appropriate executables in these systems. Thus, at least for task initiation, users see only a single interface.

The second critical feature of the AGCL was the *Xkibitzer.* This is a program which performs the simple task of emulating keystrokes in specified window on a given X console. The *Xkibitzer* provides a very general capability for running tasks on remote machines or in proprietary environments. For example suppose that the AGCL was running on a Unix system (as indeed it does) and that we wished to run a task on a VMS machine within the INGRES environment. To do this we have the AGCL spawn a new XTERM session then use the *Xkibitzer* to send commands to that window to log into our VMS machine. Once the login is complete the *Xkibitzer* is used to send commands to start up the INGRES environment, and finally the particular task desired is sent. In conjunction with the parameter interface which is used to modify the contents of the messages sent to the other screen, we can work in essentially any environment so long as we can establish a terminal session there.

One side benefit of this *Xkibitzer* approach is that the user actually can see the internals of the underlying software as it runs and if interested can begin to learn that system. Alternatively he or she can iconify the window and ignore the lower level software entirely.

Over the past several months we have integrated a number of the software elements described above into the AGCL to form the nucleus of the *GammaCore*. With some exceptions, discussed below, the process has been quite smooth. The AGCL naturally allows one to integrate software piece by piece. Perhaps more importantly, it allows several levels of integration starting with simple task initiation and providing progressively more detailed control as more of the task is extracted into a dialog with the parameter interface. The resulting system has been demonstrated both locally and to users at scientific meetings.

Our experience of the past year has taught us several lessons. Basically we feel that the overall approach is correct: the abstraction of the user interface from the software with the parameter interface is vital. The AGCL has provided us with a flexible tool which has proven to be relatively robust even though it is relatively new. The AGCL itself is small and readily maintainable.

We have encountered some problems with the system, including a crippling slowness in the *Xkibitzer* when it is used on an Xterminal with only modest connectivity to its server. The system is dependent upon using an X11 interface and the use of the *Xkibitzer* makes it difficult to see how the system could be extended to character-cell devices. There is no really good and consistent fashion that we found for handling errors withing the AGCL. The AGCL tends to use up real estate on the Xterminal screen very rapidly with new windows popping up all the time. In developing our interface we found these and other features we would hope could be added or changed in the AGCL but no real showstoppers.

4. *GammaCore* Status

During this past year the *GammaCore* has been developed as a prototype to evaluate the feasibility of accomplishing all of the objectives that we had set in creating our research environment. During this period substantial portions of the BATSE, OSSE and EGRET software systems; the archive catalog and retrieval software; and data conversion and help facilities have been brought within the *GammaCore*'s envelope. The AGCL provides IRAF support *ab initio*, and IDL has been used to support OSSE and BATSE analysis tools. While many of pieces have only been partially integrated, we feel that we have adequately demonstrated that this approach can work, and we are now engaged upon making a version of the *GammaCore* publicly available by the end of 1992.

In the next few years we anticipate extending the *GammaCore* to include all major analysis tools developed by the Instrument teams, and to more completely integrate these tools to generate a more uniform environment for the users. Anyone interested in using the *GammaCore* should contact the authors or the Compton Observatory Science Support Center.

5. Critical Integration Issues

We have learned a few things that we believe are not related specifically to our approach. In the AGCL and other similar interfaces the user is often faced with making a selection out of a large number of choices. We found that this is fine when the choices are things in the user's domain, e.g., different analysis tools, but tends to confuse the user (at least the novice) when the choices are the interfaces domain, e.g., the AGCL may simultaneously show a "`Help`", "`I'm unhappy`" and a "`Q&A`" buttons.

It turns out to be far easier to characterize and abstract the inputs to a program into a parameter interface than to deal with the outputs in any consistent fashion. In general we are not able to bring any real coherence to how programs display their results or final status.

The whole concept of a standard interface can cause problems when one tries to incorporate software which already has a specialized interface. In these cases the cost of standardization can be the elegance of the original interface. In these cases one may wish to a least have the option of running the software through that interface.

One of our hopes had been that we could generalize software by adopting common data formats so that one set of software could deal with multiple instruments. Our experience has shown that getting common data formats is at least as hard as getting a common interface.

Finally, the real problems that we have encountered in bringing up the *GammaCore* are not the software issues. Rather they were the human factors of establishing a consensus on what we were going to do: what should be included, which versions of software should be used, what data formats and standards should be adopted, and who was responsible. Our experience of trying to standardize after groups had gone their own ways for nearly a decade emphasizes the importance of cooperation in large projects. Nonetheless, we feel we have successfully integrated very diverse software elements into a reasonably coherent structure so that even in this extreme case, it is feasible and practical to incorporate existing software rather than rebuilding from scratch.

Acknowledgments. We would like to thank Eric Mandel and John Roll of the Smithsonian Astrophysical Observatory for providing an early copy of the AGCL and for their assistance in helping us to use it.

References

Kniffen, D. 1989, in Proceedings of the GRO Science Workshop, ed. W.N. Johnson, 1

Jennings, D.G., Jordan, J.M., McGlynn, T.A., Ruggiero, N.G., & Serlemitsos, T.A. 1993, this volume

Jordan, J.M., Jennings, D.G., McGlynn, T.A., Ruggiero, N.G., & Serlemitsos, T.A. 1993, this volume

Mandel, E., Brissenden, R.J.V., Freeman, M., Nguyen, D., & Roll, J. 1993, this volume

Shrader, C.R., Gehrels, N., & Dennis, B. 1992, The Compton Observatory Science Workshop, NASA Conference Publication 3137

Discussion

Adorf: I am full of admiration for your work addressing the problem of simultaneous use of heterogeneous systems. But in the user-interfaces I see that something is missing.

1. Where is the language-sensitive editor?
2. Where is the hierarchical MATHEMATICA-like editor, out of which I can execute whole procedures or fractions thereof?
3. Where is the "Meta-point" command that allows me to click on a function-name and that then takes me to the definition of that function, e.g., for inspection?

McGlynn: In some sense this question addresses the fundamental capabilities of the AGCL (or ASSIST) rather than our specific application of it to the Compton Observatory analysis systems. Our goal has been to make certain services available in a reasonably consistent and coherent fashion to users but we do not kid ourselves that the AGCL seamlessly integrates all of our software. It is unclear to me it would be possible to perform such an integration, certainly not without very considerable effort by both the designers of the integrating system and the integrators of the pre-existing software. The point of our talk is that the AGCL (and ASSIST) exist and can integrate diverse software in a rational fashion in a way that is feasible even for a small software group.

Valdes: The parameter editor gives a common parameter interface but what about parameter name/function conventions across diverse applications?

McGlynn: The parameter names used that the user sees using the parameter interface in the *GammaCore* need not be the same as the names used within the included software packages. We use this in an attempt to regularize the nomenclature of the GRO analysis systems by using common parameter names amongst the systems we are incorporating. This ability is one of the strengths of the parameter interface.

Astronomical Data Analysis Software and Systems II
ASP Conference Series, Vol. 52, 1993
R. J. Hanisch, R. J. V. Brissenden, and J. Barnes, eds.

Writing Instrument Interfaces with Xf/Tk/Tcl

Arne A. Henden

The Ohio State University, 174 W. 18th Ave., Columbus, OH 43210

Abstract. This paper describes the Xf/Tk/Tcl environment, currently being used to develop a user interface for the OSU OSIRIS near-infrared camera. Some important features of the development process are discussed.

1. Introduction

Tcl (Tool Command Language) was developed by J. K. Ousterhout (Ousterhout 1990a). It is a simple interpretive programming language (implemented as a library of C functions) that can be added to applications to provide a user scripting shell, a debugging tool, or for interprocess communication. Tcl is public domain, and can be obtained though anonymous ftp from sprite.berkeley.edu.

Tk was also developed by Ousterhout (Ousterhout 1990b). It is a toolkit for the X11 window system, implemented using Tcl. Like the Xt toolkit, Tk provides a higher level of control than Xlib and transparently performs many of the mundane tasks of widget creation and control. Like Tcl, Tk is interpretive, providing dynamic control of the user interface. Tk also provides a special command, *send*, which allows any Tk based application to send Tcl commands to any other Tk-based application. Tk is public domain, and can be obtained from sprite.

Xf is an interface builder for Tcl/Tk. It was written by S. Delmas (TU Berlin). It is a convenient method of creating the look-and-feel of the user interface. Xf can be obtained from barkley.berkeley.edu.

2. The OSIRIS Application

OSIRIS (Ohio State InfraRed Imaging Spectrometer) is a state-of-the-art system employing a Rockwell 256 × 256 2.5μ detector. OSIRIS can be used for direct filter imaging, grism spectroscopy, or slit spectroscopy using a grating. IR systems must contend with a higher sky background, thermal radiation, and more detector defects than CCD systems. Therefore, IR systems require considerable data reduction before data quality can be determined. Image processing systems such as IRAF and VISTA already exist on Unix workstations. Rather than reinventing the wheel, our approach has been to concentrate on the data acquisition and then to use a canned package for the quick look analysis.

We are currently using IRAF on a Sun SparcStation for the data processing. Data collection is performed by a 486 AT clone computer (the Instrument Computer, IC), transferred over a high-speed parallel link to another 486 computer (the Workstation Computer,WC), and then transferred to the Sun (the Reduction Computer, RC) over Ethernet using PC/NFS. This three-tiered approach has historical import and is not necessarily the best solution for OSIRIS. Our present plans are to remove the intermediate WC and do most of the instrument control from the Sun. We are developing our own Sun application program for instrument control, rather than using an existing wheel such as ICE, because of the unique aspects of remote observing required by OSU astronomers.

Our approach emulates others presented at this meeting: a graphical user interface (GUI) shell layered on top of a command-line interface to the instrument. Novice users can remain at the GUI level with help and prompts to collect data with a minimum of training. Those more familiar with the instrument can optionally move to the command-line window and directly enter commands for the instrument.

Towards this end, we are experimenting with Tk to generate the Sun GUI. The xyimage widget is used for the image display, the xygraph widget is used for plots, and the built-in widgets for radio buttons, pull-down menus, etc., are used for command input. The screen layout emulates the IC layout, with a command-line entry region at top, image display at the lower right, and status information at the lower left. Each status line is actually a button widget; if the widget is selected, the user can modify the parameter (for example, changing a filter). The Sun application program then builds an appropriate IC command, displays it in the command window and transmits it to the IC.

We will initially use an RS232 serial line for commands and status to/from the IC, and an Ethernet connection for the high-speed data transfer. We are investigating the replacement of the Ethernet connection with a high-speed parallel link for higher transfer rates.

3. Discussion

Tk is an easy package to use. While we do not normally recommend a public domain package for anything other than in-house use, we feel differently about Tk. It has sufficient documentation with its man pages and Usenix papers. Ousterhout will be releasing an Addison-Wesley textbook on Tcl/Tk within the next few months, and draft copies of the text are already available. Tcl/Tk have a good user base and a Usenet newsgroup. Full source code is provided, and many examples are available on barkley. Development was much faster than with equivalent Xt coding. Since there was one author for both Tcl and Tk, naming and use conventions are consistent.

One major benefit to using Tcl is that ports exist on both Sun/Unix and DOS. This means that one can use the same environment for hardware control as for the user interface. Since Tcl is interactive, it is ideal for real-time hardware diagnostics. Since both Tcl and Tk are written in standard C, software is maintainable. The system is public domain, so no royalties are necessary and the system is easily ported to other institutions.

The disadvantages of using Tcl/Tk include the large size of the compiled task (since you must include all elements of the language for even the simplest of applications); no instrument-specific application widgets (such as meters, 7-

segment displays, etc.); and its slow speed (since all arguments are passed as strings, then converted to numerics if needed). Since Tcl/Tk are not part of a commercial package, no one is accountable, there is no guarantee of longevity, and it is easy to lose version control. Xf is still rough and buggy, and does not have sufficient documentation to use effectively.

4. Status

The Sun GUI high-level shell has been written. The new IC interface has been added to the OSIRIS system, but the WC computer remains in place as it is needed for other instruments. OSIRIS is primarily used on the Perkins 1.8m telescope on Anderson Mesa (Flagstaff, AZ). An Internet node is being placed at Anderson Mesa in early 1993. This will give us direct connection to the Sun for software development as well as for low-cost remote observing. We anticipate a full-up demonstration of the new OSIRIS software within a few months after the Internet connection is made.

References

Ousterhout, J.K. 1990a, Proc. 1990 Winter USENIX Conference, p. 133

Ousterhout, J.K. 1990b, Proc. 1990 Winter USENIX Conference, p. 105

Astronomical Data Analysis Software and Systems II
ASP Conference Series, Vol. 52, 1993
R. J. Hanisch, R. J. V. Brissenden, and J. Barnes, eds.

An Object-Oriented Approach for Supporting Both Terminal and X Interfaces

J. Johnson

Science & Engineering Systems Division, Space Telescope Science Institute, Baltimore, MD 21218

Abstract. This paper describes a practical approach for designing user interface software which supports both character cell and X Window System displays. Application-level objects are used to encapsulate calls to the display library, allowing the full capabilities of each system to be used while minimizing and isolating the device-dependent code.

1. Introduction

The scientific community has a large installed base of character cell terminals. Due to their modest cost and the widespread availability of terminal emulators, terminals are readily accessible to virtually the entire user community. Therefore, terminals constitute the "lowest common denominator" of display devices. Making an application accessible to the greatest number of users requires an interface that supports character cell displays.

On the other hand, bitmap display devices such as workstations, X terminals, and personal computers are rapidly becoming the norm. Applications should be able to take full advantage of the improved technology which these machines provide: higher performance, direct manipulation using a mouse or other pointing device, high-resolution graphics, and increased display area.

What is needed is a way to develop an application that can run on either type of display device without doubling the effort involved or compromising usability on one platform or the other. The approach presented here was developed to provide a practical solution to these needs.

2. Overview of the Object-Oriented Approach

User interface applications consist of two parts: the components used to display information and accept user input, and the processing of the interaction between these components. Most graphical user interfaces are built around a core set of components such as menus, text entry fields, and forms. For a given application, the interaction between these components is the same regardless of the display system used.

Our approach is conceptually simple: treat each component as an object. All of the code necessary to implement the component for the desired display system is completely encapsulated within the object. The goal is to combine one or more primitives into a single, higher-level object which forms a useful architectural element of the application. This is in contrast to developing a thin "wrapper" around each primitive, which provides no additional functionality. Similar object-oriented approaches have been suggested for encapsulating X (Smith 1991) and Motif (Young 1992), but in each case are only designed to support a single display library.

Objects are only accessible via their public interface, which describes the set of functions which the object can perform. Ideally, all of the public interface functions are device-independent. This allows applications to be written as a collection of objects interacting in a wholly device-independent manner.

Supporting multiple display environments requires modifying each of the objects which contains device-dependent code. In practice, the number of objects requiring modification may be significantly less than the total number of objects in the system. Many objects, particularly those derived from more primitive ones, add only application-level behavior and do not call the display library directly.

When moving to a display environment which provides additional capabilities, the application can be extended to take advantage of these capabilities by adding additional objects. For example, an image display object could be added for X.

3. Benefits

Two critical issues to be consider when developing a large-scale user interface application are:

- Coping with the variety and continual evolution of display hardware and environments
- Maintaining the software over a long project lifetime

The object-oriented approach presented here addresses these concerns in two ways:

1. By resulting in a system that can be ported to a new display environment without affecting the entire application. Applications can be ported to any display system capable of implementing the various components. Of course, the closer the system matches the set of components, the less code will be necessary to implement each object. For example, implementing a text entry field using Motif is fairly trivial since a TextField widget is already defined, whereas implementing the same object in Curses requires considerably more effort. Testing and maintenance costs are also reduced since only the implementation of the individual objects has been changed, not the interaction between them.

2. By providing a foundation that can be extended with minimal impact to other portions of the system. All objects are application-level components, not low-level primitives as provided by most display systems. This allows arbitrarily complex components to be developed as a collection of objects, as a specialized object which is derived from a more primitive one, or as a combination of the two.

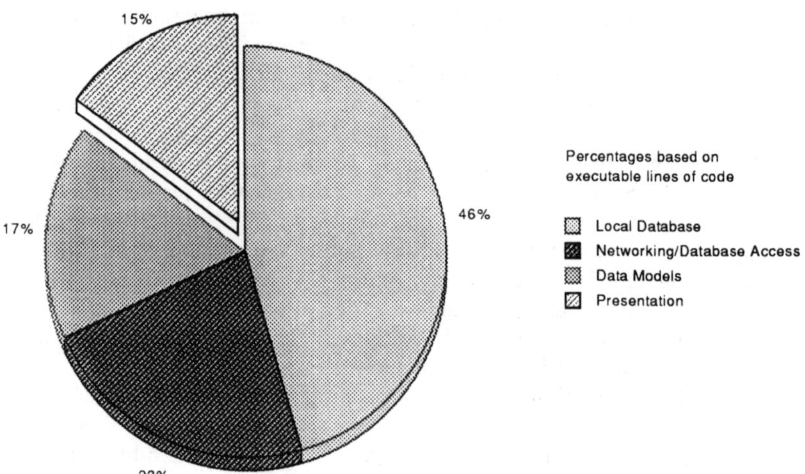

Figure 1. Breakdown of StarView Subsystems.

For example, consider a graphical skymap object which displays individual observations as markers on a rectangular portion of the screen. Such an object could be derived from a basic scatterplot object to inherit the functionality needed to create a rectangular display, traverse a list of points, draw the markers, and allow the user to zoom and pan the display. In addition, it could contain pushbutton and text entry field objects to allow user selection of the coordinate system, equinox, and epoch. Finally, the skymap object could provide specialized methods for changing the coordinate system, precessing coordinates, and drawing grids and labels.

4. Project Status

We are currently using an object-oriented approach to implement StarView, the user interface to the Hubble Space Telescope science archive (known as ST DADS). StarView allows the user to browse the HST catalog using both predefined and ad hoc queries, and to retrieve datasets from the archive. StarView is written in C++, and uses Vermont Views for the character cell interface and OSF/Motif for the X Window interface. Vermont Views is a commercial functional library for developing graphical user interfaces for terminals.

StarView is composed of four major subsystems: Presentation, Local Database, Data Models, and Networking/Database Access. All interaction with the user, including screen displays and accepting user input, is isolated within the Presentation subsystem. This subsystem will account for approximately 15% of the total StarView code upon completion, as shown in Figure 1.

The Presentation subsystem is organized as a collection of the following component classes: Form, Field, Menu, Dialog, and FormManager. Each of these classes (except for FormManager) consists of a base class and one or more derived classes. The C++ inheritance mechanism allows the derived classes to

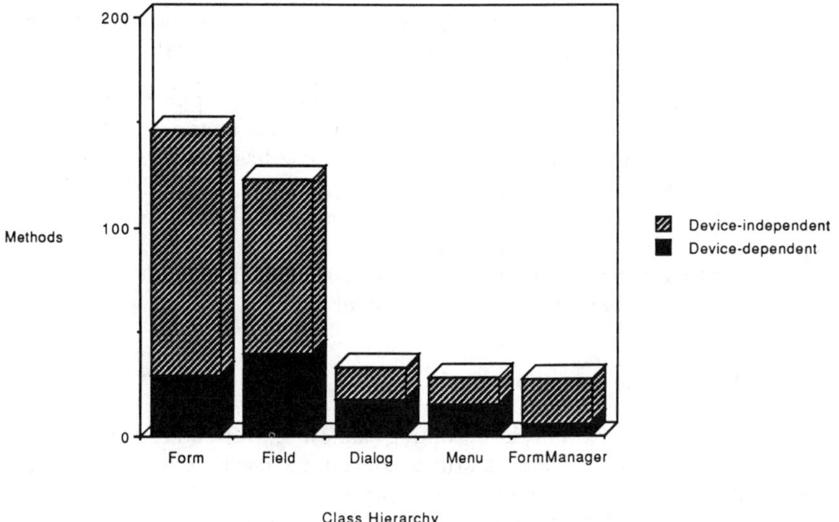

Figure 2. StarView Presentation subsystem classes.

inherit all of the functionality of the base class, then add their own specialized functionality. These classes can then serve as the base class for other derived classes, allowing for extensibility of the system. Figure 2 shows the number of device-dependent methods compared to the total number of methods used to implement each of these class hierarchies. Roughly 30% of the Presentation subsystem methods contain device-dependent code. We estimate that less than 5% of all StarView code is dependent on the particular display library used.

5. Inadequacies of Existing Vendor Packages

Several available commercial packages support both character cell and X displays (e.g., TAE+, JAM, XVT, and Open Interface). Using such a package can result in substantial cost savings over in-house development. You should evaluate these packages to determine if any of them meet your particular requirements. Most of these packages are quite sophisticated and reflect considerable development effort. However, during our research we discovered a variety of inadequacies which you should consider before deciding to use one of these packages:

- *Lack of support for VMS.* Admittedly, this is not an important factor for many developers. However, many astronomy departments and research facilities use VMS machines, and these sites should not be ignored. To be easily accessible to these users, StarView is required to support VMS.

- *Lack of programmatic interfaces.* Of the packages which include screen designers, the screens are constructed by the developer, and a binary (usually proprietary) definition file is read by the application at run-time. StarView requires that both the the user and the system be able to construct screens at run-time. This capability requires that functions be available for creating and manipulating screen components programmatically.

- *Restrictive licensing policies.* Some packages require purchasing run-time licenses, the cost of which can be prohibitive when supporting hundreds or thousands of users. A major goal of StarView has been to allow users to integrate its database access and data manipulation capabilities with their existing data analysis software. This requires that StarView source code and executables be delivered to end users without restrictions. To help achieve this goal, StarView will not require users to purchase third-party software or to complete license agreements.

6. Conclusion

While the X Window System is becoming the system of choice for developing graphical user interfaces, there is still a continuing need to support character cell terminals. Applications which must support both displays should be able to use both to their fullest extent. This is preferable to simply providing a character-based display within an X window. The approach presented here encapsulates the device-dependent code within application-level objects, which facilitates the support of multiple display environments. This approach allows the full capabilities of each system to be used while minimizing and isolating the device-dependent code.

References

Smith, J.D. 1991, Object-Oriented Programming with the X Window System Toolkits (John Wiley & Sons, Inc.)

Young, D.A. 1992, Object-Oriented Programming with C++ and OSF/Motif (Prentice Hall)

Astronomical Data Analysis Software and Systems II
ASP Conference Series, Vol. 52, 1993
R. J. Hanisch, R. J. V. Brissenden, and J. Barnes, eds.

The HEASARC Graphical User Interface

N. White, P. Barrett, P. Jacobs, B. O'Neel

NASA/Goddard Space Flight Center-HEASARC

1. Introduction

An OSF/Motif based graphical user interface has been developed to facilitate the use of the database and data analysis software packages available from the High Energy Astrophysics Science Archive Research Center. It can also be used as an interface to other, similar, routines. A small number of tables are constructed to specify the possible commands and command parameters for a given set of analysis routines. These tables can be modified by a designer to change the appearance of the interface screens. They can also be dynamically changed in response to parameter adjustments made while the underlying program is running. Additionally, a communication protocol has been designed so that the interface can operate locally or across a network. It is intended that this software be able to run using a variety of terminal types.

2. GUI Objectives

- Generic Design
- Table Driven
- Ability to Run Over a Network
- Dynamic Operation
- Independent of Application Program
- Machine Independent
- Provide Distributable User Interface for HEASARC On-Line Service
- Ability to Access Remote Archives

3. GUI Implementation

- Uses OSF/Motif Toolkit and Widgets
- Uses Sockets for Connection to Application Programs
- Uses PERL for Optimization of Program Tables
- Uses XPI for Implementation of Communication Protocol
  (Xanadu Parameter Interface - B. O'Neel)

4. GUI Tables

- Key Word Table (lists all commands and their parent groups)
- Key Table (lists the parameters associated with each command)
- Parameter Table (lists the characteristics of each parameter)
- Command Table (lists commands, command descriptions, and window classes)
- Applications Name Table (lists the application programs available to the GUI)

5. GUI Table Examples

```
          Key Word Table:
          --------------
   (group name, command name)

          display, dall
          display, dat
          display, dclass
          display, dcoord
          list, ldb
          list, lsam
          list, lind

          Key Table:
          ---------
  (command name, parameter name,
     positional specifier)

          sc, ra, 1
          sc, dec, 2
          sc, radius, 3
          sname, name, 1
          sname, full, 0

          Command Table:
          -------------
  (command name, command description,
    window type, window definition)

mind  ,make index,w,def
index ,index sample on a parameter,w,def
sort  ,sort the sub_sample,w,def
mdb   ,make user database,w,def
```

```
          Parameter Table:
          ---------------
 (parameter name, parameter type,mode
  default value, minimum value,
  maximum value, prompt, indirect command,
  indirect command position)

PAGEWIDTH,i,a,80,1,256,">Page Width"
INFILE,s,a,"index fits",,,">Name of FITS file"
PRDATA,b,a,"yes",,,"> Flag for data"
DB,s,a," ",,,"> ",ldb,2
```

6. GUI/Application Program System

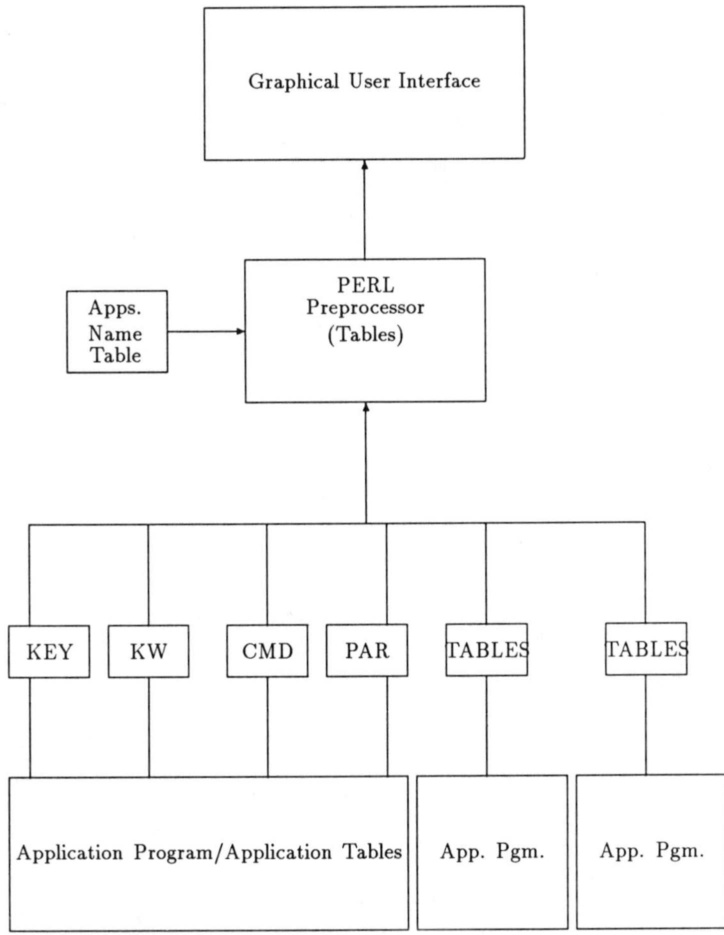

7. GUI Future Work

- Release Prototype for Evaluation (November, 1992)
- Release Version 1.0 (January, 1993)
- Distribute as Astrophysics Data System Service (Spring, 1993)
- Provide Multiple Window System Support (using XVT) (Summer, 1993)

Part 5. Data Analysis Applications

Section A. Imaging Algorithms and Techniques

Astronomical Data Analysis Software and Systems II
ASP Conference Series, Vol. 52, 1993
R. J. Hanisch, R. J. V. Brissenden, and J. Barnes, eds.

MOSAIC: an IDL Software Package for Manipulating Collections of Images

F. Városi[1] and D. Y. Gezari

Infrared Astrophysics Branch, Code 685, NASA/Goddard Space Flight Center, Greenbelt, MD 20771

Abstract. We have developed a powerful, versatile image processing and analysis software package called MOSAIC, designed specifically for the manipulation of digital astronomical image data obtained with two-dimensional array detectors. The software package is implemented using the Interactive Data Language (IDL), and incorporates new methods for processing, calibration, analysis, and visualization of astronomical image data, stressing effective methods for the creation of mosaics from collections of individual exposures, while at the same time preserving the photometric integrity of the original data. Since IDL is available on many computers, the MOSAIC software runs on most UNIX and VAX workstations with the X-Windows or SunView graphics interface.

1. Introduction

The MOSAIC software was written for the purpose of processing and analyzing images from the 58 × 62 pixel Goddard 5–18 μm infrared array camera system (Gezari et al. 1988, 1992). However, there is no restriction on the size or type of images, and the software has been used to create mosaics of images from other camera systems, such as 512 × 512 pixel CCD images. For examples of MOSAIC results see Gezari et al. 1992, Gezari 1992, Telesco and Gezari 1992.

2. Capabilities and Methods

The major tasks performed by MOSAIC are: input of a collection of images, creation of mosaics of the images, analysis, display and hardcopy of the results. Each of these major functions provides an extensive set of operations which the user can interactively apply to the data. The user interface is three-button mouse controlled, menu driven, and window based, so that the user can naturally follow a logical sequence of required processing steps. Future plans for the software include taking full advantage of X-Windows widgets for the user interface. For the more experienced user, the MOSAIC procedures and functions can be invoked directly at the IDL command level, permitting specialized operations that may not be in the presented list of options.

[1]Hughes STX Co.

The creation of a mosaic image consists of four basic steps: 1) pre-processing, which includes formatting and flat-fielding of the individual images, residual background offset corrections, bad pixel detection and masking, and other corrections (e.g., de-striping), 2) alignment of images by correlation of common spatial features, then referred to as a "raw mosaic", 3) matching of pixel intensities in overlapping image areas, 4) averaging the overlapping areas (or splicing images) to form a final mosaic image with improved signal-to-noise (SNR), called the "averaged mosaic".

The spatial arrangement or alignment of images is accomplished by the following methods, selected as options by the user. If accurate relative coordinates of the images are known they can be immediately arranged, by entering the coordinates interactively or reading a file containing the coordinates. If relative coordinates are not known, there must be some overlap between images in order to proceed. The user can choose between interactive or automatic creation of the raw mosaic of images, or a combined approach. One interactive method allows the user to align images at a common source feature by mouse controlled cursor selections. This method has an automatic counterpart which can be used when the alignment point is a peaked source in the images. In such cases the centroid of the source is automatically computed in each image and then used to automatically align the images.

Another interactive method allows direct manipulation of the images, similar to a desktop graphics environment, and this visually corresponds to "dragging" the image with the mouse. While being dragged, a blinking image is displayed, alternating with the other images in the window, so that the user can visually correlate the overlaps. The automatic counterpart to this image blinking-dragging option is the use of image cross-correlations to determine the optimal relative positions of the overlaps. In this automatic mode, the computer maximizes the cross-correlations of the image overlaps, thereby achieving the same or better results as the interactive approach. All operations for creating the raw mosaic occur on a fractional pixel grid.

The MOSAIC system also allows the user to choose image scaling, smoothing and magnification for the display of the raw mosaic. Other interactive options include the ability to "pop" and "push" displayed images in a raw mosaic, highlight the borders of images, display header information, or determine the noise (SNR) and background (sky) levels. The pixel values of individual images or all overlapping images can be displayed per user selects, as either arrays of numbers, mean and standard deviation in a box, graphs of profile cuts at any angle through the images, or as histograms showing the pixel value distribution.

After the raw mosaic is arranged spatially, the pixel intensity values in the overlapping image regions are often not the same. The pixel values in overlapping image areas can in most cases be matched effectively with the constraint of applying only linear transformations to the intensities (a factor plus a constant). This matching of intensity levels can be performed either in a manual/interactive or automatic approach. The user may specify two points in the image stack, usually one point on a source the other on sky, to use in the computation of the linear transform. The average of pixel values in the two specified areas of a chosen size then defines the linear transformation to be applied to all the images which overlap at the two points. An automatic method uses linear-least-squares fits between corresponding pixels to compute the linear transformations. However, such pixel to pixel matching can cause erroneous fits when image noise levels are high. A more robust approach is to automatically match the means and variances of the pixel intensity distributions in the overlapping regions. This approach recognizes the fact that the images are already spatially arranged so that the matching of pixel values can be accomplished statistically as a group. Carrying this approach further leads to the method of matching by histograms. Since histograms of the overlapping regions give the full distribution of pixel values, the act of matching the cumulative histogram graphs accomplishes the desired matching of image levels. A record of what linear transformations have been applied is kept in the IDL structured array of data, with the pixel intensities, relative image locations in mosaic space, positional coordinates and other parameters. The results of all processing can be saved at any stage of the effort and later restored to continue processing.

After the raw mosaic is spatial arranged and matched, the images are combined by averaging or splicing, on a whole or fractional pixel grid, to form the averaged/spliced mosaic image. As a user selected option, bad pixels can be ignored during averaging, such as at the edges of images. All of the information resulting in the successful creation of the final image is retained with the averaged mosaics. The system also allows averaged mosaics or any other images to be combined and manipulated together, creating another raw mosaic to form an even larger averaged mosaic, or just a group of images for analysis.

The display of an averaged mosaic and generation of hard copy is a module offering many user selectable options. Contour levels can be overlaid on the color image, or the data displayed with contours alone. Overlaid contours can even be from other image data of different resolution. The display format for mosaic images include positional coordinates of three types: relative pixels, relative arcseconds, or absolute R.A. and Dec. Other display options include: linear, geometric, or logarithmic scaling of data, selection of color palates and adjustment of color tables, insertion of hardcopy titles, labeling and marking of sources. All specifications for the display of an image are saved in an IDL structure, and so they can be restored to redisplay the image exactly as it was specified. Color hardcopy of the displayed mosaic image is produced by output of PostScript graphics files, in 32 level grey-scale or 256 pseudo-colors mode.

Final mosaic images obtained at different wavelengths can be aligned to form a multi-wavelength mosaic image stack, creating a "data cube" which can then be used for further analysis, such as formation of true-color images, deriving source spectra at any point in the image, or to derive model results displayed in the form of images (e.g., color temperatures, dust column densities, line-of sight extinction, etc.). The creation of a multi-wavelength mosaic is performed in three steps: 1) spatial alignment of images, 2) relative calibration of overlapping regions, 3) extracting the intersection of all images. Arithmetic and function operations can then be applied to the intersection region in the stack of images. The MOSAIC software also has an interactive interface to the DeConv_Tool package (Városi & Landsman 1993) for the purpose of instrument point spread function deconvolution.

3. Summary of MOSAIC Functions and Capabilities

PRE-PROCESSING:

- Cataloging and retrieval of observational image data files
- Background subtraction
- Flat-fielding
- Residual sky (background) subtraction
    - sky level matching
        - linear transform
        - histogram matching
    - synthetic sky frame
- De-striping and other corrections
- Bad pixel masking and removal

MOSAIC CREATION:

- Create raw mosaic of images by relative offsets
    - define coordinate system
    - enter relative coordinate offsets
    - read coordinate offsets from file
    - define scanning pattern of offsets
- Interactive raw mosaic creation:
    - aligning at common point
    - blink/drag image to position
- Automatic raw mosaic creation:
    - align at point source centroids
    - offsets by optimal cross-correlation of image overlaps
- Match image intensity levels
    - match using two points
    - linear least squares of overlap pixel values
    - match means and variances of pixels in overlap
    - match cumulative histograms of overlaps
- Manipulate raw mosaic
    - move (blink-drag) image or group of images
    - pop/push images
    - add/remove images
    - save/restore raw mosaic
    - scale and display images (linear/log/smoothing)
    - magnify or reduce display of images
    - display image borders
    - display header information
    - estimate statistics: noise level, FWHM, etc.
    - show fluxes in aperture
    - plot histograms of images and overlaps
    - plot profile cuts of overlapping images
    - average/splice all or subset of images
    - save averaged/spliced mosaic

DISPLAY of FINAL MOSAIC IMAGE:
- Display image and contour plots
  - overlay contour levels
  - overlay contours from a different image
  - relative/absolute coordinates
  - change scaling (linear/log , min/max)
  - zoom subregion
  - smooth data (low pass filter, Gaussian convolution, etc.)
  - mark/label sources
  - save/restore all display specifications
  - save/restore final mosaic image
- Show fluxes (in aperture) and statistics
- Determine source centroids and FWHM
- Plot profile cuts
- Hardcopy (PostScript)
  - contour maps
  - 32-level grey scale images
  - 256-level pseudo-color images

COLOR MANIPULATION:
- Select, create and save color tables
- Adjust pixel value to color table mapping

MOSAIC ANALYSIS:
- Create multi-wavelength mosaic image stack
- Normalization/calibration
- Plot spectrum at any point
- Plot profiles at any angle
- Compute arithmetic and functions of images in stack
- Output multi-wavelength data cube
- Deconvolution

Acknowledgments. The MOSAIC software was developed under NASA/HQ OSSA Astrophysics Division RTOP 188-44-23-08, with significant additional internal funding provided by the Laboratory for Astronomy and Solar Physics (Code 680) at NASA/Goddard.

References

Gezari, D.Y., Folz, W., Woods, L., & Wooldridge, J.B. 1988, SPIE, 973, 287

Gezari, D.Y., Folz, W., Woods, L., & Városi, F. 1992, PASP, 104, 191

Gezari, D.Y. 1992, ApJ, 396, L43

Telesco, C.M., & Gezari, D.Y. 1992, ApJ, 395, 461

Városi, F., & Landsman, W.B. 1993, this volume

Astronomical Data Analysis Software and Systems II
ASP Conference Series, Vol. 52, 1993
R. J. Hanisch, R. J. V. Brissenden, and J. Barnes, eds.

A Technique for Stacking Digitized Photographic Plates

Jonathan Bland-Hawthorn and Patrick L. Shopbell

Dept. of Space Physics & Astronomy, Rice University, Houston, TX 77251

Abstract. With the advent of fast scanning microphotometers and inexpensive digital mass storage, there has been a resurgence of interest in performing deep (B $\leq$ 25) panoramic surveys by co-adding large numbers ($\sim 10^2$) of digitized photographic plates. While the Kodak IIIa emulsions are highly linear recorders of photographic grain density, we demonstrate that the threshold and saturation levels which restrict the dynamic range of the emulsion can distort the higher statistical moments of the grain density fluctuations (variance, skewness, etc.) along the linear part of the characteristic curve. The variance of the grain fluctuations is only additive between digitized plates that preserve the Poissonian grain noise. In order to compensate for the statistical distortion, we derive the necessary pixel weighting for five scanning aperture sizes (2μm, 4μm, 8μm, 16μm, 32μm) as a function of the grain density.

1. Introduction

Wide-field photographic plates continue to play a crucial role in astronomical surveys (West 1991, Irwin 1992). In an age of rapidly evolving electronic detectors, plates still retain the unique ability to record high-density photometric information over fields of view exceeding a few degrees with a spatial resolution limited only by the seeing. Recent improvements in the performance of fast scanning microphotometers (e.g., COSMOS) now make it possible to extract $\sim 10^9$ data pixels from a plate in a matter of a few hours. This has allowed various groups to compile objective catalogues of very large numbers of galaxies. In the Edinburgh/Durham Galaxy Survey, Heydon-Dumbleton et al. (1988) scanned 60 overlapping plates that cover almost one steradian of the southern sky in order to compile a catalogue of roughly one million galaxies down to a limiting magnitude of $B_J \approx 20$. Rapid plate scanning also affords the opportunity to co-add or 'stack' a large number of photographic plates in order to increase detection limit sensitivity over a wide field. Tsvetkov (1992) reports that there are at least 1.3 million wide-field plates in archives throughout the astronomical community. Roughly speaking, this means that each square degree is covered by $30f^2$ plates, where f is the average field size in degrees, although these archives are heavily biased towards certain regions of the sky largely in the northern hemisphere. Malin (1988) has demonstrated that the plate limit improves by about two magnitudes when 36 UK Schmidt plates are co-added using traditional laboratory techniques. More recently, Hawkins (1991) used the COSMOS microphotometer to scan 58 plates of a common field; these were added in density space to form an image whose completeness limit was $B_J \approx 24$ mag with some images going a magnitude fainter. The importance of this result is exemplified by the number-magnitude relation for field galaxies: an improvement in the limiting sensitivity from $B_J = 22$ mag to $B_J = 24$ mag increases the detection rate of field galaxies by an order of magnitude.

2. The Statistics of Photographic Plates

For a given aperture size, the central limit theorem ensures that the density fluctuations obey Poissonian or Gaussian statistics (Marriage & Pitts 1956), even in the limit that the scanning aperture size is of order the emulsion grain size (Trabka 1969). However, it is clear from Figure 1*a* that the width of the grain noise statistics does not conform to a simple Poissonian ($\sqrt{N}$) model. This is underscored by Figure 1*b* which shows that there is a systematic trend in the skewness of the density fluctuations with increasing grain density. In the following section, we present a statistical treatment that attempts to explain the basic behavior in Figures 1*a* and *b*. Any departure from Poissonian statistics has important consequences for co-adding digitized photographic plates.

When we add together separate images with mean signal s_1 and s_2, it is normally assumed that the noise in both images, n_1 and n_2, obeys Poissonian statistics. If $n_1 = \epsilon_1\sqrt{s_1}$ and $n_2 = \epsilon_2\sqrt{s_2}$, the variance of the noise in the summed image is simply the sum in quadrature of the original variances, i.e., $\sqrt{n_1^2 + n_2^2} = \sqrt{\epsilon_1^2 s_1 + \epsilon_2^2 s_2}$. Therefore, in summing images, we are making at least two general assumptions: n_1 and n_2 are statistically independent and are governed by Poissonian noise such that $\epsilon_1 = \epsilon_2 = 1$. The first assumption is discussed by Shopbell, Bland-Hawthorn & Malin (1992) and reduces to showing that the weight of the covariance term, w_3, is negligible, viz.

$$\sigma_S^2 = w_1 n_1^2 + w_2 n_2^2 + w_3 n_1 n_2 \tag{1}$$

where $w_1 = \epsilon_1^{-2}$, $w_2 = \epsilon_2^{-2}$, and σ_S^2 is the variance of the summed image. For the second assumption, if the noise distributions deviate from Poisson statistics, the correction factors w_1 and w_2 need to be applied to the individual variances. These weights are derived below.

3. The Restricted Poissonian Distribution

The photographic plate is a highly linear recorder of photographic density. After Dainty & Shaw (1974), we demonstrate that the measured density fluctuations in Figures 1*a* and *b* can be understood with a relatively straightforward model. For ease of illustration, consider a linear *photon* detector comprising an infinite number of recording cells that is uniformly exposed with an average number of p photons per cell. The photon distribution over the cells will be governed by Poisson statistics such that the proportion of cells receiving r photons will be $p^r e^{-p}/r!$. In practice, there are minimum (threshold) and maximum (saturation) exposure levels that a detector can register. Assume that the first $(t-1)$ photons do not generate counts whereas the tth photon registers one count. Assume

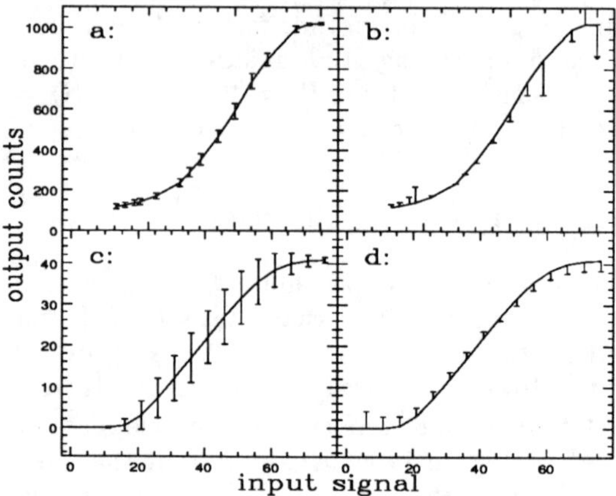

Figure 1. *a*,*b*.– The characteristic curve (or μ) for a scanned IIIa-F photographic plate measured from the density step wedges. In *a*, the error bars correspond to $\pm 1\sigma$, and in *b*, the vertical bars indicate the magnitude and sign of the skewness, κ. *c*,*d*.– The output response to a linear detector in the presence of saturation and threshold levels. When the linear response is restricted to a finite dynamic range, both moments deviate from their true Poissonian values.

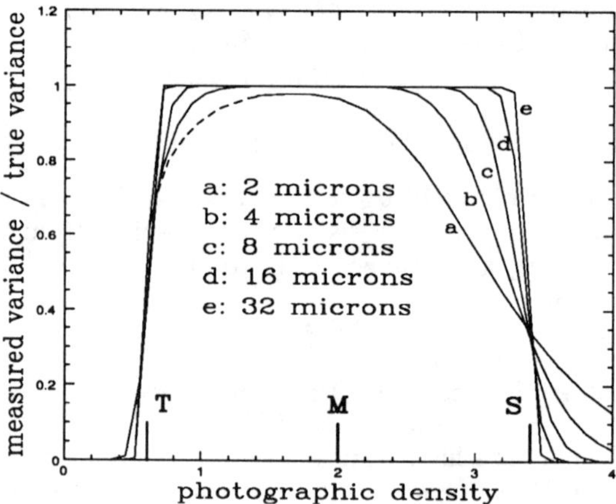

Figure 2. A plot that shows how the measured variance differs from the Poissonian variance as a function of photographic density and scanning aperture size. The values along the vertical axis are inversely related to the corrective weights (see text). The statistical distortion is clearly more important at high density and for small scanning apertures. The dashed line is an extrapolation as the numerical method fails in this region.

further that s or a higher number of photons register exactly s counts. The mean number of counts over all cells is then

$$\mu = \mathcal{E}(p) = \alpha(1 - \beta e^{-p}) \tag{2}$$

where $\alpha = s - t + 1$ and

$$\beta = \frac{1}{\alpha} \sum_{k=t-1}^{s-1} \sum_{r=0}^{k} \frac{p^r}{r!} \tag{3}$$

We have used the form of the expectation, $\mathcal{E}$, to determine higher moments of the distribution, viz.

$$\mathcal{E}(p^2) = \alpha^2(1 - \gamma e^{-p}) \tag{4}$$

$$\mathcal{E}(p^3) = \alpha^3(1 - \delta e^{-p}) \tag{5}$$

where the summation terms are given by

$$\gamma = \frac{1}{\alpha^2} \sum_{k=t-1}^{s-1} \sum_{r=0}^{k} (2[k-t+2]-1)\frac{p^r}{r!} \tag{6}$$

$$\delta = \frac{1}{\alpha^3} \sum_{k=t-1}^{s-1} \sum_{r=0}^{k} (3[k-t+2][(k-t+2)-1]+1)\frac{p^r}{r!} \tag{7}$$

Equations (4) and (5) provide information on the variance, σ^2, and the skewness, κ, of the distribution since

$$\sigma^2 = \mathcal{E}(p^2) - \mu^2 \tag{8}$$

$$\kappa^3 = \mathcal{E}(p^3) - 3\mu\sigma^2 - \mu^3 \tag{9}$$

For an ideal linear detector with infinite dynamic range, the expected values of the mean, standard deviation and skewness for a Poisson distribution are easily derived. In the presence of detector limits, all three statistical moments deviate from their true Poissonian values; the induced 'statistical distortion' is illustrated in Figures 1*c* and 1*d*. Intuitively, as the value of p approaches either detector limit from the linear section of the curve, one tail of the distribution measuring the statistical uncertainty in p starts to disappear. A comparison of Figures 1*a* and *b* with Figures 1*c* and *d* respectively reveals that the restricted Poissonian model explains the major trends in the scanned density step wedges.

4. Discussion

In order to simulate the response of the grain density fluctuations to the presence of threshold and saturation levels, we determine the linear dynamic range that produces the fluctuation level at some intermediate (midpoint) value along the straight section of the characteristic curve. We note in Figure 1 that the maximum statistical dispersion occurs roughly at the midpoint of the linear ramp. Since the threshold and saturation levels can only reduce the statistical dispersion, it follows that the grain density for which the statistical response is most likely to exhibit Poissonian behavior is that which produces the largest grain dispersion. Furenlid, Schoening & Carder (1977) have calibrated the grain density fluctuations, σ_D, on the linear part of the characteristic curve for a range of photographic emulsions. They find that under a wide range of conditions, there exists a unique relationship between σ_D and grain density, D: for the Kodak IIIa emulsions, we adopt $(\sigma_D/D) = (0.64/a)$ where a is the square aperture dimension in microns. We compute statistical models which attempt to simulate the grain density response for five aperture sizes: 2, 4, 8, 16, 32μm. The expected grain density fluctuations are 32%, 16%, 8%, 4%, 2% respectively. For all five models, the positions of the threshold (T), midpoint (M) and saturation (S) levels remain in a fixed ratio with respect to the linear scale defined by a zero density (Z) and a peak density (P). We adopt the following representative density values in each of the simulations: Z=0.0, T=0.6, M=2.0, S=3.4, P=4.0. An important boundary condition in our model is that the density fluctuation falls to zero at zero density. However, in reality, we note from Figure 1*a* that there will be finite background noise at both extremes of the density range. The results are summarized in Figure 2. It is clear that the statistical distortion is particularly severe for the smaller apertures and tends to restrict the linear density response, particularly to higher densities. For a scanning aperture of 4μm, notice that the higher statistical moments are truly Poissonian for only about half of the straight section of the characteristic curve.

In practice, one corrects for the statistical distortion by dividing out the form of the measured to true variance in Figure 2. The necessary pixel weights, w_1 and w_2, are functions of density and are essentially the reciprocal of the values plotted in Figure 2. The first statistical moment is handled by the density-to-intensity transformation; the third and higher moments are unlikely to have much influence. In stacking N digitized plates, with the proposed corrections, we would expect the magnitude limit to improve in proportion to $2.5 \log \sqrt{N}$. Thus, a hundred co-added plates could in principle increase the sensitivity limit by an order of magnitude everywhere within the full field.

References

Dainty, J.C., & Shaw, R. 1974, Image Science (Academic Press: London)

Furenlid, I., Schoening, W.E., & Carder, B.E. 1977, AAS Photo-Bulletin, 16, 14

Hawkins, M.R.S. 1991, IAU Working Group on Wide-Field Imaging, 1, 23

Heydon-Dumbleton, N.H., Collins, C.A., & MacGillivray, H.T., 1989, MNRAS, 238, 379

Irwin, M.J. 1992, in Digitised Optical Sky Surveys, eds. MacGillivray, H.T. & Thomson, E.B. (Kluwer: Holland), 43

Malin, D.F. 1988, in Astrophotography, Proc. IAU (Springer: Berlin), 125

Marriage, A., & Pitts, E. 1956, J. Opt. Soc. Am., 46, 1019

Shopbell, P.L., Bland-Hawthorn, J., & Malin, D.F. 1992, AJ, in press

Trabka, E.A. 1969, J. Opt. Soc. Am., 59, 662

Tsvetkov, M.K. 1992, IAU Working Group on Wide-Field Imaging, 2, 51

West, R.M. 1991, ESO Messenger, 65, 45

Astronomical Data Analysis Software and Systems II
ASP Conference Series, Vol. 52, 1993
R. J. Hanisch, R. J. V. Brissenden, and J. Barnes, eds.

Registering and Resampling Images in STSDAS

Ramon L. Williamson II

Space Telescope Science Institute, 3700 San Martin Dr., Baltimore, MD. 21218

Abstract. Registering images can be difficult, especially if the images to be registered are images at different wavelengths, where features in one image may look entirely different or be absent from the second image. The implementation and use of two new packages soon to be added to the STSDAS package, REGISTER and RESAMPLE, are discussed, and an example of the use of these packages is shown.

1. The REGISTER Package

The REGISTER package allows the user to determine the amount of translation, rotation, and/or magnification needed to make two images, spectra, or time series congruent.

The methods implemented to compute the registration parameters use:

- a set of the pixel coordinates of the same features identified in two files, or
- the FITS coordinate transformation parameters in the headers of two data files, or
- a single feature identified as the peak of a cross-correlation between two vectors.

The coefficients describing the registration are defined by the equations (for a two-dimensional image):

$$X = a + bx + cy \quad (1)$$

$$Y = d + ex + fy \quad (2)$$

where (X,Y) are the pixel coordinates of a feature in the Reference image, (x,y) are the pixel coordinates of a feature in the Secondary image, and the computed coefficients are a, b, c, d, e, and f.

Results may be produced by linking the output of REGISTER to RESAMPLE in a command language procedure. The output from REGISTER and the input to RESAMPLE consists of a matrix of coefficients (a through f above) fully specifying the registration.

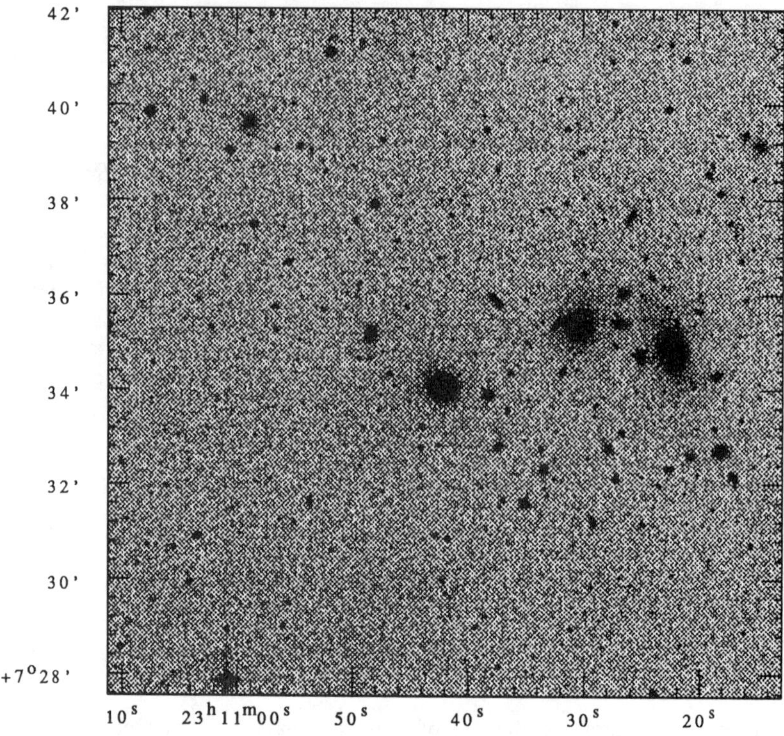

Figure 1. Guide Star image of NGC 7503.

2. The RESAMPLE Package

The RESAMPLE package resamples simple vector or image data for a given amount of translation, rotation (images only), and magnification, or reflection of the science data. Specific options included are:

- image rotation about the FITS reference pixel
- scale changes, i.e., magnification or demagnification (for images, independently on both axes)
- simple translation
- reflection (for images, about one or both axes)
- resampling and registration to a reference dataset

Output from the RESAMPLE task is the resampled image which may then be displayed and compared with the reference image.

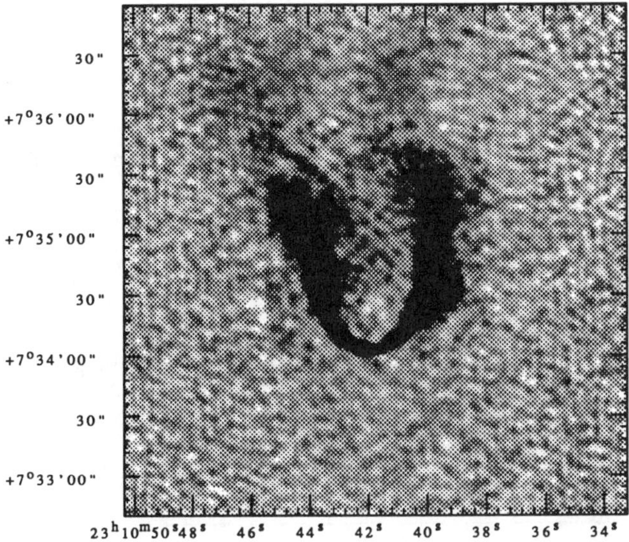

Figure 2. VLA image of NGC 7503.

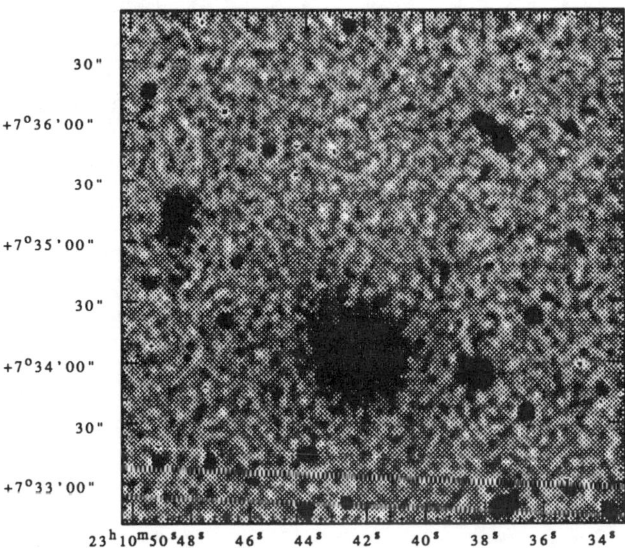

Figure 3. Resampled Guide Star image at scale of the VLA image.

3. An Example: Registrating and Resampling Optical and Radio Images of NGC 7503

This example shows the actual output from the REGISTER and RESAMPLE packages described above.

Figures 1–4 and Tables 1, 2 show the steps involved in registering and resampling an image from the Guide Star Catalogue and a VLA image of NGC 7503.

Figure 1 shows the original Guide Star image of the region around NGC7503. The labeling shows the scale.

Figure 2 shows the VLA image of NGC 7503. Again, the labeling shows the scale.

Table 1 shows the coefficients for the linear equation to convert any pixel in the secondary image to the same scale as the reference image. This table was found using the images' FITS parameters. The elements shown here have been rounded. Actual output is double precision.

coeff1	coeff2	coeff3	coeff4	coeff5	coeff6
−305.476	1.701	0.004	−305.456	−0.004	1.701

Table 1. The output registration parameters table from the REGISTER package.

deltax	deltay	xmag	ymag	rotang
−125.082	−127.000	1.701	1.701	−0.126

Table 2. Remaining output from the REGISTER package

Table 2 shows the remainder of the output table from the REGISTER package. These parameters show how much offset in X and Y as well as magnification and rotation are required to register the secondary image with the reference image.

Figure 3 shows the output from the RESAMPLE package, the resampled optical image from the Guide Star Catalogue at the same scale as the VLA image.

Figure 4 shows the output optical image with the input radio image overlayed using the STSDAS task newcont.

Acknowledgments. Many thanks to Dr. Bob Hanisch for the use of the VLA and GSC data shown in the example.

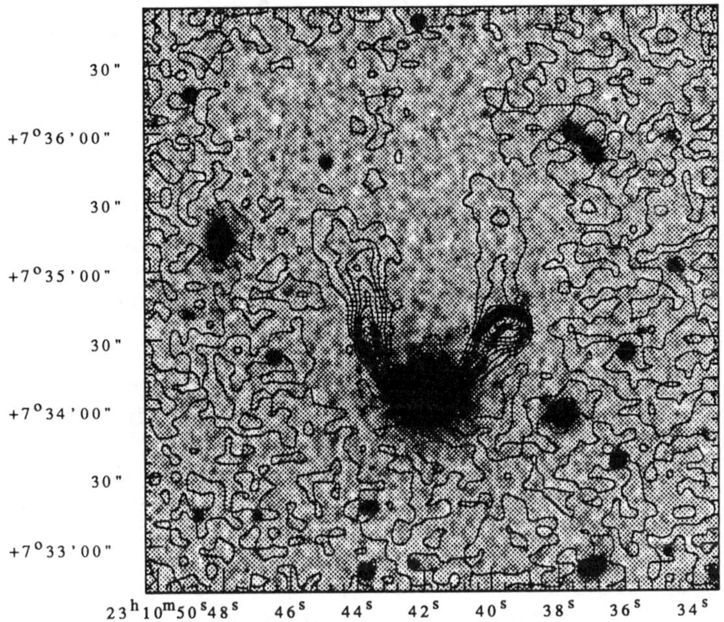

Figure 4. Contoured VLA image overlaid onto Guide Star image.

Astronomical Data Analysis Software and Systems II
ASP Conference Series, Vol. 52, 1993
R. J. Hanisch, R. J. V. Brissenden, and J. Barnes, eds.

Experiments with Recursive Estimation in Astronomical Image Processing

I. Busko

Divisão de Astrofísica, Instituto Nacional de Pesquisas Espaciais, C.P. 515, CEP 12201.970, São José dos Campos, SP, BRAZIL

Abstract. Recursive estimation (Kalman filtering) is used to adaptively decrease the noise level in astronomical images. Experiments show that the technique may be incorporated in an iterative image restoration algorithm, in order to control high-frequency noise buildup.

1. The Problem

Recursive estimation concepts were applied to image enhancement problems since the 70's. However, very few applications in the field of astronomical image processing are known (see e.g., Richter 1978, Capaccioli & Caon 1989). Those concepts were derived, for 2-dimensional images, from the well-known theory of Kalman filtering in one dimension. The historic reasons for application of these techniques to digital images are related to the images' *scanned* nature, in which the temporal output of a scanner device can be processed on-line by techniques borrowed directly from 1-dimensional recursive signal analysis (Nahi 1972, Habibi 1972, Woods & Radewan 1977, Woods & Ingle 1981).

An important aspect of recursive estimation is that the theory can be extended to include *non-stationary* phenomena, that is, phenomena which have their statistical properties variable in time (or position in a 2-D image). Many image processing methods are based on underlying stationarity assumptions either for the stochastic field being imaged, for the imaging system properties, or both. They will underperform, or even fail, when applied to images that deviate significantly from stationarity. Recursive methods, on the contrary, turn it feasible to perform *adaptive* processing, that is, to process the image by a variable processor with properties tuned to the images's *local* statistical properties (Rajala & Figueiredo 1981; Biemond et al. 1983, Tekalp et al. 1989). Astronomical images have very specific structural and statistical properties, and deviate significantly from stationarity. These properties can be helpful for the design of powerful, custom recursive processors.

Recursive estimation can be used to build estimates of images degraded by such phenomena as noise and blur. Current experiments reported in this work explore noise damping, in particular in the context of noise regularization in image restoration algorithms.

2. The Processor

For these experiments, a minimal recursive processor was implemented. The state vector has the simplest possible nonsymmetric half-plane support: only one point back in x and y coordinates (Woods & Radewan 1977). The Kalman gain is adaptively controlled by the image local statistical properties. Experiments were conducted with two basic types of control:

(i) *Signal intensity or signal-to-noise ratio*: in this mode the Kalman gain is related to local intensity or signal-to-noise ratio. The filter does not take into account local spatial structure.

(ii) *Autocorrelation or "spatial activity"*: the local autocorrelation function can be approximated by the autocorrelation of a stationary Markov process with circular symmetry:

$$R(|\mathbf{r}|) = Ke^{-\rho|\mathbf{r}|}$$

and the Kalman gain set as a function of the correlation (ρ) and "energy" (K) parameters. The filter takes into account both intensity and spatial structure.

An alternate but equivalent form is to use H-transform coefficients (Fritze et al. 1977; Richter 1978) to estimate K and ρ. These coefficients are related to the Fourier transform coefficients, and can be used to estimate the local power spectrum with much less computational effort. Notice that this procedure is not the same as the one introduced by Richter (1978). There he first uses an H-transform filter to eliminate non-significant information in the image, and subsequently smooth it by a Kalman filter driven directly by the H-transform coefficients. Here I do just the second step, but modified to compute the filter gain in such a way as to relate it to the local K and ρ parameters. In this way, problems associated with the square shape of the 2-D H-transform filter impulse response function are eliminated.

3. Some Results

Figure 1 depicts a 256^2 section of a photographic image of NGC1316 (taken from Space Telescope Science Institute's Guide Star Survey image bank) processed by the recursive adaptive filter with two different filter gain controlling parameters. A non-adaptive convolution with a Gaussian kernel (FWHM = 2 pixels) is included for comparison. Notice the dramatic decrease in the sky standard deviation attained by the adaptive mechanism, without changing significantly the stellar image shapes. The adaptive filter linearity is demonstrated in Figure 2.

Figure 3 depicts results from application of adaptive recursive filtering to noise regularization in image restoration. The iterative Richardson-Lucy algorithm (Richardson 1972, Lucy 1974) was applied to Supernova 1987a image made by HST FOC $f/96$ with F501N filter. Experiments were run with both pre-restoration adaptive noise filtering, as well as adaptive noise filtering as an integral part of the restoration loop. The idea here is to abort sky background high-frequency noise buildup inside the algorithm loop itself, as soon as it starts to become larger than noise in the input image sky background. Notice how the adaptive filter preserves high spatial frequencies in the ring and star images, however keeping the sky fluctuations under control.

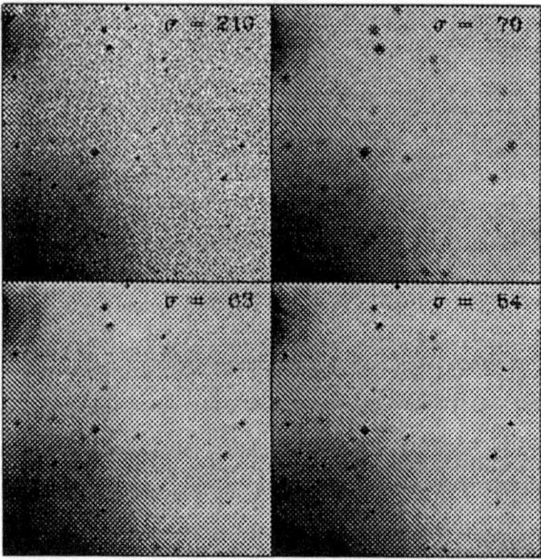

Figure 1. Photographic image of NGC1316 subjected to noise filtering. σ is the sky background standard deviation in each image. Upper left: original image. Upper right: non-adaptive filter with Gaussian kernel. Lower left: adaptive recursive filter with gain controlled by local signal-to-noise ratio. Lower right: adaptive recursive filter with gain controlled by local correlation, computed from H-transform coefficients.

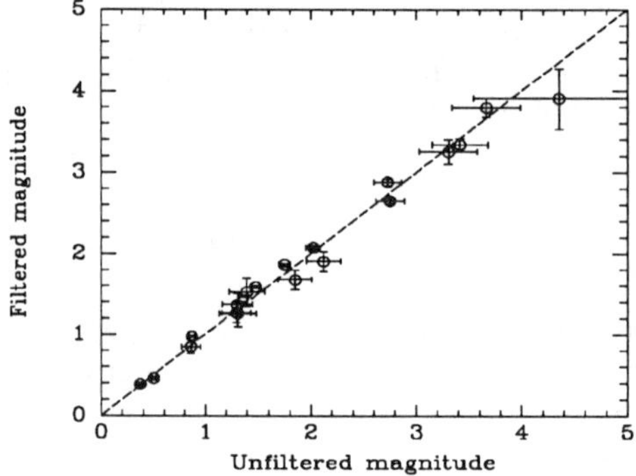

Figure 2. Adaptive filter linearity is demonstrated by stellar aperture photometry on NGC1316 image. Magnitudes (in arbitrary scale) were measured on both original, unfiltered image and on image filtered by adaptive filter in correlation mode. Same aperture parameters were used on both images. Dashed line is the locus $y \equiv x$. Notice that formal error bars from filtered image are smaller than from original image; this is due to smoother sky background in filtered image.

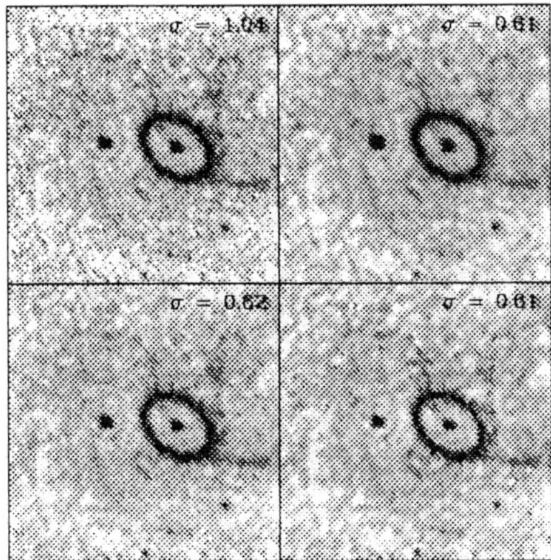

Figure 3. Adaptive noise filtering and image restoration. σ is the sky background standard deviation in each image. Upper left: test image restored by 35 Lucy iterations. Upper right: test image convolved with Gaussian kernel and restored by 35 Lucy iterations. Lower left: adaptive recursive noise filter with gain controlled by local correlation was applied to test image, subsequently restored by 35 Lucy iterations. Lower right: test image restored by 65 Lucy iterations; the same filter as in lower left panel was applied three times during the iteration loop. Test image has $\sigma = 0.60$; 65 Lucy iterations with no filter lead to $\sigma = 1.71$. Both Gaussian and adaptive filters were adjusted to generate, after restoration, the same sky standard deviation as in original test image.

4. The Future

The recursive algorithm will be expanded to handle more powerful and descriptive state vectors. As such, it will be possible to perform more complex operations. The first goal is to include a deblurring operation with fixed blur and stochastic image models. The next step will be the extension to positionally dependent blur, noise and image models, in a switching context as the one proposed by Tekalp et al. (1989). Image restoration under these conditions is not yet solved in a satisfactory way for terrestrial images, and mostly unexplored in the domain of astronomical images. An immediate application would be restoration of Hubble Space Telescope's Wide Field Camera images.

Acknowledgments. I am grateful to Urias Soares and Rodrigo Campos from CNPq/Laboratório Nacional de Astrofísica, for the photographic color reproductions. This work was partially supported by Fundação de Amparo à Pesquisa do Estado de São Paulo, under grant 92/3099-0.

References

Biemond, J., Rieske, J., & Gerbrands, J.J. 1983, IEEE Transactions on Acoustics, Speech, and Signal Processing, ASSP-31, 1248

Capaccioli, M., & Caon, N. 1989, in First ESO/ST-ECF Data Analysis Workshop, ESO Conference and Workshop Proceedings n. 31, eds. P.J. Grosbol, F. Murtagh & R.H. Warmels, p. 107

Fritze, K., Lange, M., Mostl, G., Oleak, H., & Richter, G.M. 1977, Astron.Nachr, 298, 189

Habibi, A. 1972, Proceedings of the IEEE, 60, 878

Lucy, L.B. 1974, AJ, 79, 745

Nahi, N.E. 1972, Proceedings of the IEEE, 60, 872

Rajala, S.A., & Figueiredo, R.J.P. 1981, IEEE Transactions on Acoustics, Speech, and Signal Processing, ASSP-29, 1033

Richardson, W.H. 1972, Journal of the Optical Society of America, 62, 55

Richter, G.M. 1978, Astron.Nachr, 299, 283

Tekalp, A.M., Kaufman, H., & Woods, J.W. 1989, IEEE Transactions on Acoustics, Speech, and Signal Processing, 37, 892

Woods, J.W., & Radewan, C.H. 1977, Transactions on Information Theory, IT-23, 473

Woods, J.W., & Ingle, V.K. 1981, IEEE Transactions on Acoustics, Speech, and Signal Processing, ASSP-29, 188

Astronomical Data Analysis Software and Systems II
ASP Conference Series, Vol. 52, 1993
R. J. Hanisch, R. J. V. Brissenden, and J. Barnes, eds.

Multiresolution Analysis in Two or More Dimensions

B. C. Bromley

Department of Physics and Astronomy, Dartmouth College, Hanover, NH 03755

Abstract. An algorithm is implemented to perform a multiresolution decomposition in $n \geq 2$ dimensions based on wavelets generated from products of 1-D wavelets and smoothing functions. The functions are chosen so that an n-D wavelet may be associated with a single resolution scale and orientation. This algorithm enables complete reconstruction of a high resolution signal from decomposition coefficients. The signal is oversampled to accommodate non-orthogonal wavelet systems and to provide approximate translational invariance in the decomposition arrays.

1. Introduction

In the analysis of signals in time or scalar fields in configuration space, transient or localized features may be isolated and characterized with the use of wavelets. As basis functions for representation, wavelets and their Fourier transforms are simultaneously localized in accordance with the uncertainty principle. Thus, they are indexed by position (configuration space localization) and scale (localization in Fourier space), a property that may be advantageous for the detection of specific objects or patterns in a signal.

The purpose of this work is to present an algorithm to obtain discrete wavelet transforms of multidimensional systems based on the multiresolution analysis of Mallat (1989). First, the multiresolution method in 1-D is reviewed; then, a simple extension to n-D systems is considered. In the algorithm discussed here, signals are oversampled as in the 2-D image analysis of Mallat and Zhong (1992), a feature that is beneficial in some applications including image compression and statistical analysis.

2. Multiresolution Analysis

To implement a wavelet decomposition, Mallat (1989) has designed a system in which a signal is examined at a hierarchy of resolution levels, with adjacent smoothing scales differing by a factor of 2. Each adjacent pair of smoothed signals in the hierarchy is connected by detail information which appears at the higher resolution level yet is lost at the lower resolution. This information can be used to completely reconstruct the higher resolution version from the lower level. Indeed, a signal of arbitrarily high resolution may be decomposed entirely into a hierarchy of details without loss of information.

In practice, a high resolution signal may be approximated by discrete samples taken with a smoothing function ϕ and its translates, as in a spline basis, for example. Wavelets are introduced to carry the detail information; dilation and translation of a single wavelet template ψ generate a basis for the details.

For a 1-D system, the multiresolution smoothing functions and wavelets are defined by

$$\phi(x) = \sum_{n=-\infty}^{\infty} 2h_n\phi(2x-n), \qquad \sum_{n=-\infty}^{\infty} h_n = 1; \tag{1}$$

$$\psi(x) = \sum_{n=-\infty}^{\infty} 2g_n\phi(2x-n), \qquad \sum_{n=-\infty}^{\infty} g_n = 0. \tag{2}$$

The coefficients h_n and g_n may be regarded as discrete elements of low-pass filter H and high-pass G, respectively, that define a particular multiresolution system. For example, $h_0 = h_1 = 1/2$ gives a tophat smoothing function (_⊓_), and $g_0 = -(g_1 = 1/2)$ gives the Haar wavelet (–⊔⊓–).

Equations (1) and (2) allow the smoothing samples and details of adjacent resolution levels to be directly related. With resolution index $j = 0$ indicating unit smoothing scale and samples taken at integer positions, lower resolution smoothed samples s and details w are obtained from

$$s_n^{j-1} = \sum_{k=-\infty}^{\infty} h_k s^j_{\lambda_j k+n}\ ; \quad w_n^{j-1} = \sum_{k=-\infty}^{\infty} g_k s^j_{\lambda_j k+n} \quad (j<0). \tag{3}$$

In the above expression, $\lambda_j = 2^{|j|}$ is the smoothing scale for resolution level j and the superscripts and subscripts of s and w refer to resolution level and position, respectively.

Equations (3) may be interpreted as convolutions with discrete filters H_j to smooth, and G_j to skim off the details. Successive application of such filters to a set of smoothed samples S^j yields a decomposition into wavelet mode amplitudes in the form of details $W^{j'<j}$, as shown schematically by

$$\begin{array}{ccccccccc} S^j & \rightarrow & [H_j] \rightarrow S^{j-1} & \rightarrow & [H_{j-1}] \rightarrow S^{j-2} & \rightarrow & \ldots \\ & \searrow & & \searrow & & \searrow & \\ & & [G_j] \rightarrow W^{j-1} & & [G_{j-1}] \rightarrow W^{j-2} & & \ldots. \end{array} \tag{4}$$

In this decomposition, the sampling density is constant at all resolution levels to enable non-orthogonal wavelet systems to pick up all detail information (Mallat & Zhong 1992). For orthogonal wavelet systems, the sampling density may be reduced by a factor of two at each drop in resolution without information loss.

3. Wavelet Decompositions in Multidimensional Systems

In the simplest extension of the multiresolution decomposition to systems of $n > 1$ dimensions, smoothing functions and wavelets are generated from n-tuple products of 1-D functions with identical resolution indices (Mallat 1989). For example, in 2-D, the unit resolution smoothing function is $\phi(x)\phi(y)$, while the

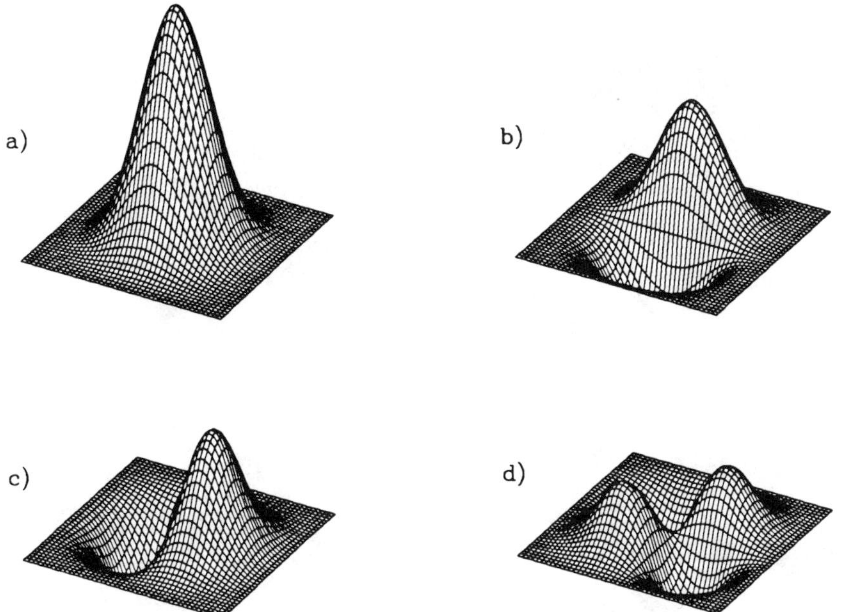

Figure 1. Separable 2-D cubic spline smoothing function (a) and wavelets (b-d), which couple strongly to edges.

corresponding wavelets are $\phi(x)\psi(y)$, $\psi(x)\phi(y)$, and $\psi(x)\psi(y)$, which couple to detail activity on the x-axis, y-axis, and both axes, respectively (e.g., Figure 1). Wavelet modes in this system are indexed by resolution, position, and direction of sensitivity or orientation.

To implement a separable n-D multiresolution decomposition, Eqs. (3) are applied along each axis to smoothing function samples defined on a regular grid. For a 2-D system, schematized in terms of the discrete filters H and G, the decomposing of a resolution j signal is

$$S^j \begin{cases} \nearrow [H_j(x)] \begin{cases} [H_j(y)] \to S^{j-1} \langle \cdots \\ [G_j(y)] \to W_y^{j-1} \end{cases} \\ \searrow [G_j(x)] \begin{cases} [H_j(y)] \to W_x^{j-1} \\ [G_j(y)] \to W_{xy}^{j-1} \, , \end{cases} \end{cases} \tag{5}$$

where the subscripts of W, refer to direction of sensitivity to detail variations. Figure 2 illustrates a smoothed sample set and the corresponding detail arrays obtained with Eq. (5).

This algorithm is coded for systems of arbitrary dimensionality. The decomposition of high resolution sampling coefficients is performed hierarchically

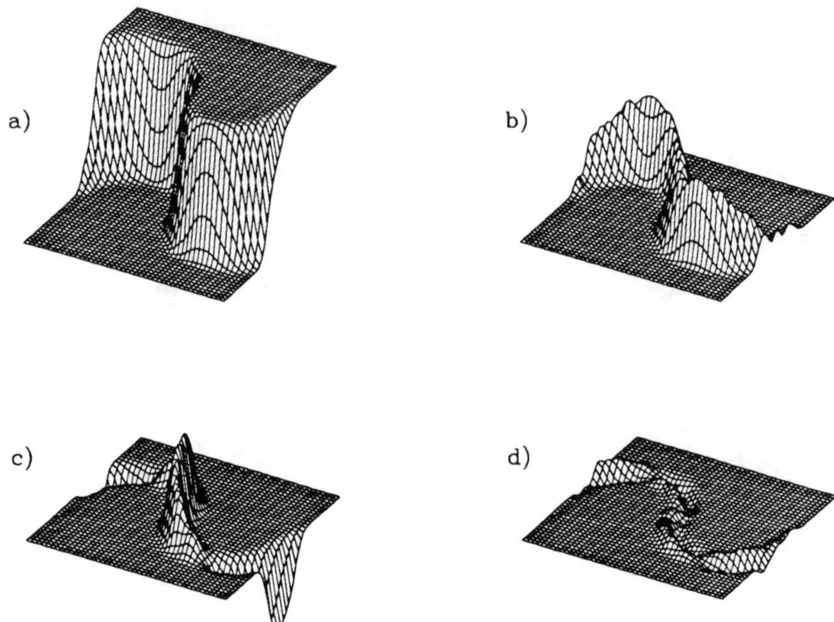

Figure 2. An example of smoothed sample coefficients on a 2-D grid (a) and details (b-d), associated with cubic spline functions in Figure 1. The detail arrays show the edge-detection capabilities of this particular wavelet system.

and yields 2^{n-1} detail arrays for each resolution level. The sampling density is held constant throughout, so that low resolution details are oversampled. This implementation has initializers for a variety of systems including Haar wavelets, compact functions of Daubechies (1988), Lemarié Battle wavelets (Battle 1987), and non-orthogonal splines. Complete reconstruction of high resolution samples from detail arrays is performed for all of these systems by inversion of Eqs. (3) (Mallat & Zhong 1992).

4. Discussion

Wavelet decompositions in multiple dimensions may be performed in a variety of ways. In addition to the separable multiresolution scheme considered here, other hierarchical decompositions have been implemented with tensor products of 1-D wavelets (e.g., Press 1992) and non-separable multiresolution functions (Kovacevic & Vetterli 1992). A general wavelet decomposition may be obtained by convolution with wavelets whose scale and position parameters are drawn from a continuum of values. Clearly, the choice of decomposition should reflect the nature of the application.

Here, the scheme for indexing multidimensional wavelet mode amplitudes by position, a single scale, and orientation is well suited to quantify compact objects whose characteristic scales are roughly the same in all dimensions. Alternately, the method may be used to isolate specific properties of an object regardless of its overall geometry, such as the sharpness of its boundary (e.g., Figure 2). The segregation of wavelets by scale also suggests that it may be useful for general representation and processing when noise has scale dependence.

In the original separable 2-D multiresolution method of Mallat, wavelets formed a complete orthonormal basis. In this case, the sampling density falls off with scale and the decomposition of a high resolution image can be done in place. In contrast, the decomposition discussed here produces 2^{n-1} oversampled detail arrays at each resolution level in n-D; for systems of even modest dimensionality ($n \geq 4$), the data storage requirements can be tremendous. This problem may be obviated by saving only local extrema of the wavelet coefficients corresponding to prominent detail features, a technique that has been used for data compression (Mallat & Zhong 1992, Maes 1992).

One reason for choosing this algorithm despite the awkward data production is that the detail arrays are approximately invariant to spatial translations of the original field. In the compact orthogonal decompositions, the detail arrays can change drastically when the original signal is translated, making quantitative interpretive processing difficult. A second reason for using this method is to enable lossless decompositions for non-orthogonal wavelets. This feature opens up the prospect of optimizing wavelet forms (as specified by filters H and G in Eqs. [1] and [2]) to couple with specific structures in a field.

In my astrophysical application, statistics of wavelet coefficients are sought for 3-D large-scale structure data. The organization of wavelets by scale simplifies interpretation in terms of models of clustering. Oversampling is desirable, as is the freedom to use optimized non-orthogonal wavelets without information loss.

Acknowledgments. I am very grateful to Dennis Healy for guidance regarding wavelet analysis and the multiresolution approach. Thanks are also extended to Isabel Dulfano and Gary Wegner.

References

Battle, G. 1987, Commun. Math. Phys., 110, 601

Daubechies, I. 1988, Commun. Pure Appl. Math., 41, 909

Kovacevic, J., & Vetterli, M. 1992, IEEE Trans. Info. Theory, 38, 533

Maes, S. 1992, in Proc. IEEE-SP Symp. on Time Frequency and Time-Scale Analysis, October 4-6, Victoria, B.C., p. 467

Mallat, S. 1989, IEEE Trans. Pattern Analysis and Machine Intellegence, 11, 647

Mallat, S., & Zhong S. 1992, in Wavelets and Applications, ed. Y Meyer (Masson/Springer-Verlag), p. 207

Press, W.H. 1992, in Astronomical Data Analysis Software and Systems I, A.S.P. Conf. Ser., Vol. 25, eds. D. Worrall, C. Biemesderfer & J. Barnes, 1

Astronomical Data Analysis Software and Systems II
ASP Conference Series, Vol. 52, 1993
R. J. Hanisch, R. J. V. Brissenden, and J. Barnes, eds.

Computation of Flat Fields for the HST Wide Field/Planetary Camera

J.-C. Hsu and C. E. Ritchie

Space Telescope Science Institute

Abstract. An algorithm developed by the WFPC IDT to generate flat fields using Earth streak exposures is now implemented in STSDAS. We explain in detail how this algorithm works and its possible deficiencies.

1. Introduction

The flat field of the wide-field/planetary camera (WFPC) on board the Hubble Space Telescope is obtained by taking exposures of the bright Earth. Because of the relative motion between the telescope and the Earth, these "Earth flats" show the pattern of streaks. An algorithm was developed by the WFPC Investigation Definition Team to use these exposures to generate flat fields of the CCD chips. A modified version of this algorithm has been implemented in STSDAS and is available to the general community for use in the reduction of WFPC data. The task is called STREAKFLAT.

2. Generating Flat Field from Earth Flats

The following steps are used to generate the WFPC flat field:

1. Calculate the mean flat by taking a straight average, pixel by pixel, of all input Earth flats (raw streak flats), but excluding pixels which are masked to be bad. The "raw" data here are data having minimal calibrations applied, i.e., A-to-D error correction, known bad pixels masking, and bias level correction.

2. Each Earth flat frame is divided by the mean flat obtained in step 1. The ratios from this division in the central 400 by 400 pixel region are averaged into a single value of ratio for that frame. This last step also involves exclusion of very large or very small ratios in this region by two passes of 2-sigmas rejection.

3. A first guess of the flat field is obtained from the scaled Earth flat for each pixel, i.e., at each pixel, every Earth flat frame is divided by its ratio from step 2 to generate the scaled Earth flat, and a median value is chosen among these scaled Earth flats.

4. Each of the Earth flats is divided by the median flat from step 3 to get an estimate of the streak pattern in that frame.

5. The estimate of the streak pattern from step 4 is then smoothed along the streak direction. The smoothing function used here is a box filter. The streak angle is determined from the spacecraft orbital velocity and the direction of its pointing.

6. Each Earth flat is divided by the smoothed streak pattern to give an estimate of the "chip flat" for that frame.

7. Repeat steps 1, 2, and 3 but using the "chip flats" generated in step 6, instead of the original Earth flats as input. The new median is the next approximation to the flat field. This new median flat is then used in place of the previous median flat and starts a new iteration by going through steps 4 to 7. Each new iteration will use a narrower filtering box.

3. Discussions

The number and widths of the filtering box along the streaks can be chosen by the user. From past experience, we set the default to be the following eight half-widths: 800, 600, 400, 250, 150, 90, 50, and 30 pixels.

The STREAKFLAT task under STSDAS does the iteration steps described above. The result is an unnormalized flat fields for each CCD chip. A separate task called NORMCLIP does the final normalization between chips and scales the flat field pixels such that the overall average of all four chips is unity. The task NORMCLIP also clips out extremely low or high flat field values.

When input Earth flats are taken within one or consecutive orbits without changing the space craft's roll angle, the streak angles will lie within a very small range. Our algorithm can not handle this situation with total satisfaction, i.e., some remnant streaky features may show up in the final flat field. In addition, the corners will have larger uncertainty in the final flat field. The solution to this problem is to plan the Earth flat exposures such that they cover as large a range of streak angles as possible.

Our algorithm is also based on the assumption that the streaks or deviations in each Earth flat is not large (not more than 10 to 20 per cent) compared to the general background. Visual inspection of each input file is recommended and Earth flats with abnormally large bright/dark features or steep gradients should be excluded from the final flat field calculation.

Running the STREAKFLAT task does use extensive computing resources. To calculate the flat fields of all four chips with 8 iterations needs roughly one hour (clock time) per input Earth flat file on a SUN SPARC 2 machine. Compared to STREAKFLAT, the task NORMCLIP uses insignificant computing resources.

Acknowledgments. We are grateful to Jeff Hester and John MacKenty for useful inputs and discussions. Much of this paper is based on Chapter 6 of the Final Orbital/Science Verification Report by the WFPC IDT.

Astronomical Data Analysis Software and Systems II
ASP Conference Series, Vol. 52, 1993
R. J. Hanisch, R. J. V. Brissenden, and J. Barnes, eds.

New Software For the IRAF Stellar Photometry Package

L. E. Davis

NOAO/IRAF Group, Tucson, AZ 85726

1. Introduction

Work on the IRAF stellar photometry package has, over the past year, focused on the following areas: 1) porting the DAOPHOT II algorithms and tasks (Stetson 1992) to IRAF/DAOPHOT, 2) testing the DAOPHOT II algorithms on KPNO data, 3) creating new, and improving existing interactive setup tools for IRAF/DAOPHOT, 4) creating new photometry analysis tools for both the APPHOT and IRAF/DAOPHOT packages, 5) improving interaction with the image display in the APPHOT package, and 6) making improvement to the sky fitting algorithms in APPHOT.

2. The DAOPHOT II Algorithms and Tasks

The DAOPHOT II tasks and algorithms (Stetson 1992) that have been ported to IRAF/DAOPHOT are: 1) a new task for selecting good candidate PSF stars, 2) a choice of 6 functions for defining the analytic component of the PSF function, 3) a choice of 0, 1, 3, or 6 look-up tables for defining corrections to the analytic PSF function, 4) use of all the PSF stars, weighted by brightness to compute the analytic component of the PSF, 5) optional automatic analytic PSF function selection, 6) optional down-weighting of deviant pixels in the look-up table computation, 7) optional use of saturated stars to fit the wings of the PSF, 8) optional iterative sky recomputation in the ALLSTAR task, and 9) user control of the flat fielding and profile errors.

In addition to the DAOPHOT II algorithm changes, the following algorithm options have been added to the IRAF/DAOPHOT tasks: 1) the PSF task can be driven by a starlist, a cursor command file or the image cursor, 2) recentering is now optional in PEAK and NSTAR as well as ALLSTAR, 3) sky refitting is optional in PEAK and NSTAR as well as ALLSTAR, and 4) user control of bad data rejection is implemented in PEAK and NSTAR as well as ALLSTAR.

3. A Test of the New Algorithms on KPNO Data

Some test results for a 450 second V band frame of the Horseshoe Nebula taken with the 2K square T2KA chip at the KPNO 0.9m telescope are presented. Figure 1 shows a set of contour plots of the PSF at nine equally spaced ($X = 256, 1024, 1792$ and $Y = 256, 1024, 1792$) grid points in the image, demonstrating that the PSF is strongly and asymmetrically variable across the frame. Figures 2(a) and 2(b) in the upper and lower left corners of the plot respectively, show the fitted magnitude versus error plots for stars in the frame, 1) fit with a constant

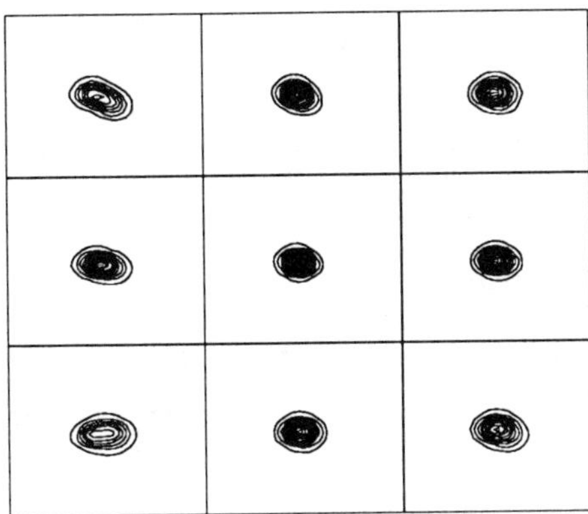

Figure 1. The PSF as a function of image position.

PSF and, 2) fit with a 2nd order variable PSF. The variable PSF fits clearly produce a lower mean error (0.015 versus 0.034 magnitudes) and lower scatter in the error (0.005 versus 0.012 magnitudes) for bright stars (mag <= 21.0 in the plots) than the constant PSF fits. Figures 2(c) and 2(d), in the upper and lower right corners of the plot respectively, show plots of the total magnitude minus the fitted magnitude versus radial distance from the center of the image for the 68 stars used to compute the PSF. There is a clear trend in the constant PSF plot, which is absent in the variable PSF plot, although some problems remain at small radial distances where there were very few PSF stars.

4. New Interactive Tools for IRAF/DAOPHOT

A new parameter editing task has been written for IRAF/DAOPHOT. The task DAOEDIT edits the IRAF/DAOPHOT parameters directly using the image display and a radial profile plot as shown in the Figure 3. A related task SETIMPARS saves and recalls the edited parameter sets by image name, making it simpler for the user to change images in a reduction session, or resume work on an image after a long delay.

Several new features have been added to the interactive PSF fitting task PSF. The PSF model and size can be changed, stars can be deleted from as well as added to the PSF star list, the function fit or refit, and the residuals viewed at any point in the fitting process.

5. New Photometry Catalog Tools

PEXAMINE is a task designed for interactively examining photometry catalogs produced by APPHOT or IRAF/DAOPHOT. PEXAMINE uses the photometry catalog, the graphics window, and optionally the image and image display, to

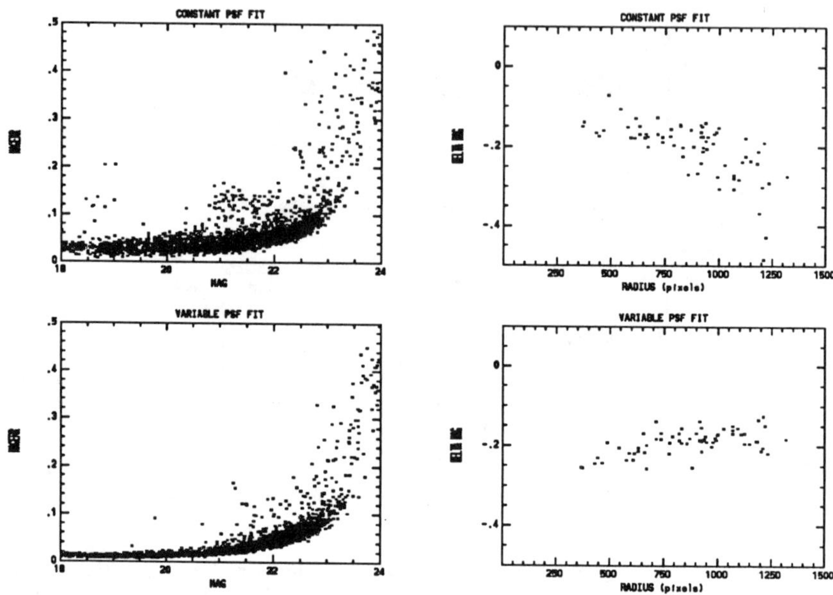

Figure 2. Comparison of the variable PSF fits with the constant PSF fits.

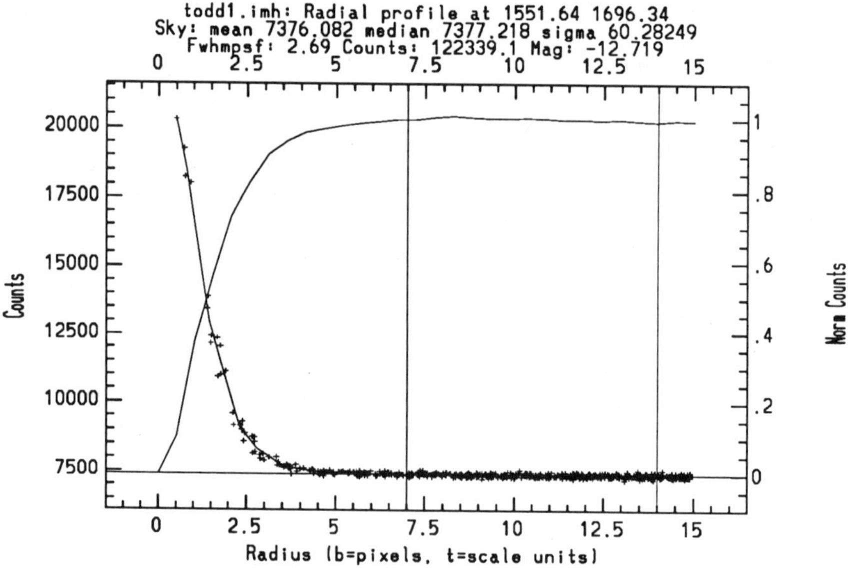

Figure 3. Sample IRAF/DAOPHOT interactive setup plot.

interactively examine and/or edit photometry catalogs. PEXAMINE complements the batch editing facilities of the PSELECT task.

PEXAMINE can be used to make x-y plots of columns in the photometry catalog, such as the sharpness versus chi plot shown in Figure 4, plot histograms of the columns, find records in the photometry catalog by pointing to the object on the image display or the appropriate x-y plot, make contour, surface, and radial profile plots of objects in the catalog, and interactively delete points from the catalog.

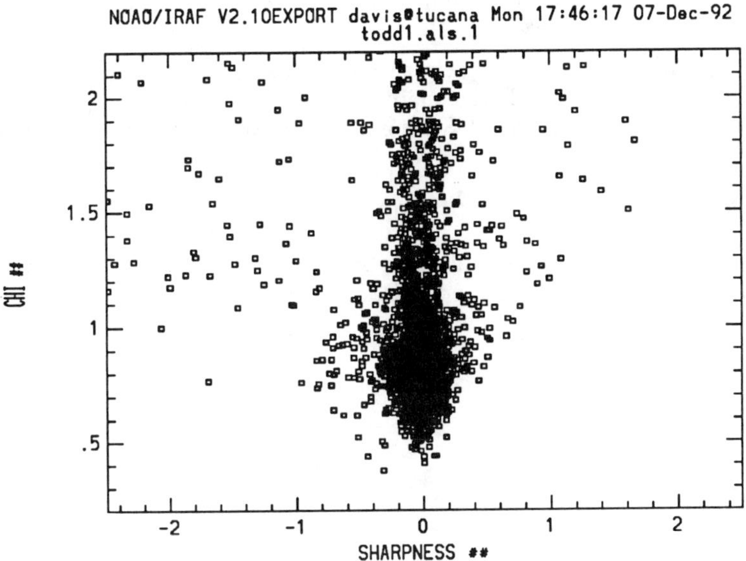

Figure 4. Sample IRAF/DAOPHOT interactive setup plot.

6. Improved Interaction With the Image Display

Limited image display marking facilities have been added to the APPHOT package tasks using the IRAF IMD graphics kernel. Object centers, sky annuli, and circular and polygonal apertures can now be marked on the displayed image. The IMD graphics kernel is not fully interactive so functions like erase are not implemented, and there are other limitations as well, but the facility is still useful.

7. Improvements to the Sky Fitting Routines

Several improvements have been made to the sky fitting routines in the APPHOT package. The new routines 1) provide better protection against very deviant pixels that are not removed by the datamin and datamax parameters,

2) permit the user to clip a certain percentage of the sky pixels from the low and high sides of the sky distribution before sky fitting, 3) permit the user to set different rejection criteria for the high and low side of the sky distribution, 4) plot sky pixel histograms in box style for greater ease of interpretation.

8. Current Status

The PEXAMINE task is already available in IRAF 2.10. The remaining new photometry software will become available as an external IRAF add-on package in early 1993.

Acknowledgments. The author wishes to thank P. Massey for his continuing interest in the project and for supplying interesting data sets with which to test the the new software, and P. Stetson for answering numerous questions.

References

Stetson, P.B. 1992, in Astronomical Data Analysis Software and Systems I, A.S.P. Conf. Ser., Vol. 25, eds. D.M. Worrall, C. Biemesderfer, & J. Barnes, 297

Astronomical Data Analysis Software and Systems II
ASP Conference Series, Vol. 52, 1993
R. J. Hanisch, R. J. V. Brissenden, and J. Barnes, eds.

Detection of X-ray Sources with PROS

Janet DePonte and Francis A. Primini

Smithsonian Astrophysical Observatory, Cambridge, MA 02138

1. Introduction

The problem of detecting discrete sources in x-ray images has much in common with the problem of automatic source detection at other wavelengths (see, for example, Stetson 1987). In all cases, one searches for positive brightness enhancements exceeding a certain threshold, which appear consistent with what one expects for a point source, in the presence of a (possibly) spatially variable background. Multidimensional point spread functions (e.g., dependent on detector position and photon energy) are also common. At the same time, the problem in x-ray astronomy has some unique aspects. For example, for typical x-ray exposures in current or recent observatories, the number of available pixels far exceeds the number of actual x-ray events, so Poisson, rather than Gaussian statistics apply. Further, extended cosmic x-ray sources are common, and one often desires to detect point sources in the vicinity or even within bright, diffuse x-ray emission. Finally, support structures in x-ray detectors often cast sharp shadows in x-ray images making it necessary to detect sources in a region of rapidly varying exposure.

We are developing a source detection package within the IRAF/PROS environment (Tody 1986; Worrall et al. 1992) which attempts to deal with some of the problems of x-ray source detection. We have patterned our package after the successful *Einstein* Observatory x-ray source detection programs. However, we have attempted to improve the flexibility and accessibility of the functions and to provide a graphical front-end for the user. Our philosophy has been to use standard IRAF tasks whenever possible for image manipulation and to separate general functions from mission-specific ones. In the current release, we provide a set of simple PROS tasks to detect discrete sources using the traditional (to x-ray astronomers) "Local Detect" technique. In this paper we discuss the (well-known) algorithms of the technique in the context of their IRAF/PROS implementation and present some preliminary results.

2. Local Detect

In this technique, an x-ray image is scanned with a square "detect cell" of side d, placed at locations separated by d/3 in each dimension. At each location, the significance (SNR) of a source assumed to be centered in the detect cell is calculated, with background estimated from a frame of width d/3 immediately surrounding the detect cell. Locations containing local peaks in SNR above a threshold are identified, and maximum likelihood positions for sources corresponding to each local peak are determined. A final SNR in a detect cell centered on each final position is calculated, and all locations with SNR greater than a second threshold are labeled as sources.

In computing source significance, we borrow heavily from the algebra derived for the *Einstein* IPC Rev. 1 **LDETECT** programs (Harnden et al. 1984). We define

- S = Expected total source counts,
- B = Expected background counts in detect cell,
- C = Total counts in detect cell of area A_C,
- Q = total counts in background frame of area A_Q,
- PRF(x, y) = Point Response Function,
- $\alpha = \int^{\text{detect cell}} \mathrm{PRF}(x,y)\, dx\, dy$,
- $\beta = \int^{\text{background frame}} \mathrm{PRF}(x,y)\, dx\, dy$.

S and SNR may be determined from the solution of the two simultaneous linear equations for C and Q, i.e.,

$$\mathrm{C} = \alpha\,\mathrm{S} + \mathrm{B} \quad \text{and} \quad \mathrm{Q} = \beta\,\mathrm{S} + \frac{A_Q}{A_C}\,\mathrm{B} \tag{1}$$

This yields

$$\mathrm{S} = \frac{\frac{A_Q}{A_C}\,\mathrm{C} - \mathrm{Q}}{\frac{A_Q}{A_C}\,\alpha - \beta} \tag{2}$$

and

$$\mathrm{SNR} = \frac{\mathrm{S}}{\sqrt{\mathrm{var(S)}}} \simeq \frac{\frac{A_Q}{A_C}\,\mathrm{C} - \mathrm{Q}}{\sqrt{\left(\frac{A_Q}{A_C}\right)^2 \mathrm{C} + \mathrm{Q}}} \tag{3}$$

As can be seen above, SNR is independent of α and β and is thus insensitive to uncertainties in the PRF. Calculation of S, however, is subject to such uncertainties. Further, the assumption of Gaussian statistics in computing SNR is not valid in the limit of small numbers of counts and will lead to overestimates of source significance in such cases. We will address this problem in a future release.

3. IRAF/PROS Implementation

We implement the above algebra with existing IRAF and PROS tasks rather than recoding it anew in Fortran or SPP. We generate an input image, I, by blocking an x-ray QPOE file at a resolution of 1/3 of a detect cell in each

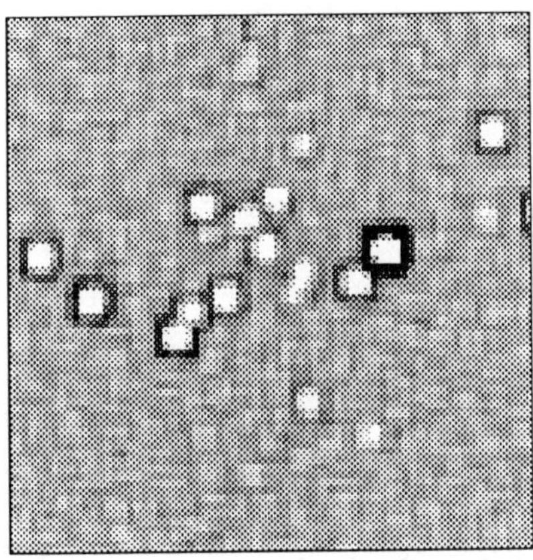

Figure 1. Source Significance Image, SNR, for M31.

dimension. The quantities C and Q may then be determined *as images* by convolution of I with simple masks, e.g.,

$$C = I \otimes \begin{bmatrix} 1 & 1 & 1 \\ 1 & 1 & 1 \\ 1 & 1 & 1 \end{bmatrix} \quad \text{and} \quad Q = I \otimes \begin{bmatrix} 1 & 1 & 1 & 1 & 1 \\ 1 & 0 & 0 & 0 & 1 \\ 1 & 0 & 0 & 0 & 1 \\ 1 & 0 & 0 & 0 & 1 \\ 1 & 1 & 1 & 1 & 1 \end{bmatrix} . \tag{4}$$

The **CONVOLVE** task in the IRAF images package performs the computation. The **IMCALC** task in the PROS ximages package is then used to calculate SNR *as an image* from the C and Q images. The quantities A_C and A_Q are simply the sums of all values in the C and Q convolution masks.

This approach offers significant advantages. Intermediate results are easily accessible and, since they are images, may be easily examined using the full set of IRAF image analysis and manipulation tools. In Figure 1 we display the SNR image for a ROSAT HRI observation of M31 (Primini, Forman & Jones 1993). More importantly, the approach is extremely flexible. Although the current implementation is confined to a traditional square detect cell of dimension d and contiguous background frame of width d/3, the user may easily modify the convolution masks and **IMCALC** statements to accommodate different detect cell geometries. Oversampling of the detect cell by more than a factor of 3, detached and/or larger background frames, or even different detect cell shapes, may all be accommodated. All that is required is that the convolution masks be symmetric about the center of the detect cell and that the detect cell and background frame be independent.

4. Determining Source Positions

The PROS task **BEPOS** is used to determine the best estimate and 90% confidence limit in source position from individual x-ray events in the *QPOE* file, using the minimum C statistic technique of Cash (1979). The C statistic in our case is defined as

$$C = 2B + 2S - 2\sum_{i=1}^{n} \ln(I_i) \tag{5}$$

where I_i is the probability of observing the ith event at position (x_i, y_i), assuming a source at (x_o, y_o)

$$I_i = \frac{B}{A_C} + \frac{S}{2\pi\sigma_{PRF}^2} e^{-\frac{(x_i - x_o)^2 + (y_i - y_o)^2}{2\sigma_{PRF}^2}} \tag{6}$$

Here, A_C is in units of instrument pixels. The C statistic is minimized with respect to (x_o, y_o) via an iterative technique, with an initial guess for B provided by another PROS task, **BKDEN**.

A Gaussian PRF is assumed. At present, σ_{PRF} may be input by the user if known. For *Einstein* and ROSAT instruments, PRF calibration data in TABLES format is accessed by the program, allowing estimation of σ_{PRF} from source position in the image. It is recognized that the Gaussian model is suitable for the *Einstein* and ROSAT HRI and ROSAT PSPC at best only in limited regions near the optical axes. It is used primarily because of ease of implementation. Further work is necessary to determine the effects of the model on position accuracy and uncertainty.

The 90% confidence limit in position is determined by incrementing the x,y positions about the maximum likelihood (minimum C statistic) position until the C-statistic has increased by 4.6, the value corresponding to 90% confidence for two interesting parameters. A calculation of source significance (SNR) at the final source position is made, using the counts in a detect cell centered on that position and the counts in the background frame surrounding that cell. Sources with SNR below a user-defined threshold are rejected. A figure of merit is also calculated to describe how well the photon distribution matches that expected for the PRF and sources which match the PRF poorly are identified in the output.

5. Results

We illustrate the performance of the current system in Figure 2. A ROSAT HRI observation of M31 (Primini, Forman & Jones 1993) and a ROSAT PSPC observation of the Pleiades (Rosner 1992) are shown. The latter clearly illustrates the dependence of PRF with position in the image. In each case a single detect cell size was used (12″ for the HRI and 30″ for the PSPC), and all sources with SNR$\geq$3.0 are displayed. For the PSPC observation a number of real sources remain undetected and illustrate the need for a larger detect cell size. *Einstein* experience suggests that cell size d should be chosen so that $d \simeq 3.6\sigma_{PRF}$ (Harnden et al. 1984). For instruments such as the PSPC, this requires the use of multiple cell sizes to cover the entire image to comparable sensitivity.

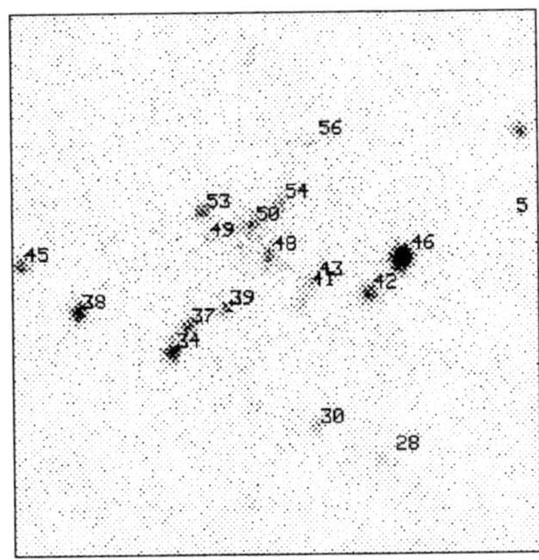

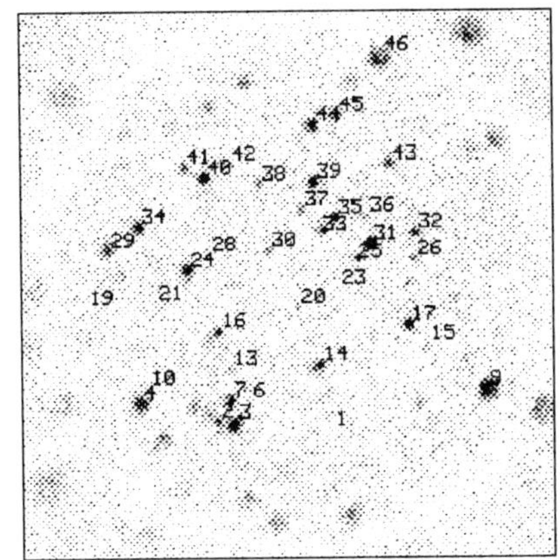

Figure 2. ROSAT HRI image of M31 (left) and PSPC image of the Pleiades (right), with Local Detect sources labeled.

6. Conclusion

Although by no means sophisticated, the Local Detect technique provides reasonable results in regions in which the point response function and exposure change only slowly. Since source significance is determined locally, the technique can also provide more reliable results in regions of extended x-ray emission than techniques which rely on model background maps. Local Detect may also be used as a first step in more elaborate source detection packages.

Acknowledgments. This work is partially supported by NASA contracts to the ROSAT Science Data Center (NAS5-30934) and Einstein (NAS8-30751).

References

Harnden, F.R., Fabricant, D.G., Harris, D.E., & Schwarz, J. 1984, Smithsonian Astrophysical Observatory Special Report 393

Primini, F.A., Forman, W., & Jones, C. 1993, submitted to ApJ

Rosner, R. 1992, private communication

Stetson, P.B. 1987, PASP, 99, 191

Tody, D. 1986, in Instrumentation in Astronomy VI, SPIE, 627, part 2

Worrall, D.M., Conroy, M., DePonte, J., Harnden, Jr., F.R., Mandel, E., Murray, S.S, Trinchieri, G., VanHilst, M., & Wilkes, B.J. 1992, in Data Analysis in Astronomy IV, eds. V. Di Gesu et al. (Plenum Press), 145

Astronomical Data Analysis Software and Systems II
ASP Conference Series, Vol. 52, 1993
R. J. Hanisch, R. J. V. Brissenden, and J. Barnes, eds.

Spatial Region Filtering in IRAF/PROS

E. Mandel, J. Roll, D. Schmidt

Harvard-Smithsonian Center for Astrophysics, Cambridge, MA 02138

M. VanHilst

University of Washington, Seattle, WA 98125

R. Burg

Johns Hopkins University, Baltimore, MD 21218

Abstract. We describe a spatial region filtering scheme in *IRAF* that allows users to specify one or more geometric shapes to be included or excluded when masking spatial data. These shapes can be combined using Boolean algebra to create the complex masks required for X-ray analysis.

1. Introduction

The analysis of X-ray data almost always requires the extraction of source and background events from a data set. Typically, the region of extraction is described in terms of geometric shapes or the combination of such shapes, but irregular shapes are also important. For example, it is not unusual to extract events from a circular region centered on a source position, and to extract background from an annulus whose inner ring coincides with the boundary of the source region. At the same time, it might be necessary to exclude a nearby region from the source or background in question. The excluded region might be described in terms of an ellipse or a rectangle or even an irregular polygon. The ability to filter event data spatially using a variety of mask specifiers, from simple geometric shapes to complex combinations of shapes to arbitrary polygons, is a fundamental requirement for X-ray analysis.

To meet this challenge in the *IRAF/PROS* X-ray analysis system, we have developed a spatial filtering scheme called *regions* that allows users to build up complex image masks using simple *ASCII* descriptions of geometric shapes (such as circles, ellipses, annuli, boxes, polygons, etc., and boolean combinations of shapes. These masks are used by *PROS* tasks to filter image data and photon event data.

2. The ASCII Region Descriptor

From the point of view of the user, the heart of the regions interface is the *region descriptor*. This is an *ASCII* string that is used by the region creation routines to build an image mask. A region descriptor contains one or more simple geometric shape specifications. A shape specification consists of a shape name (circle, box, polygon, etc., along with the shape's position, dimensions, and orientation. Alternatively, a region descriptor can be a file containing one or more of these geometric shape specifiers.

The following table lists the shapes supported by the regions interface, and their parameters:

shape	arguments
annulus	xcenter ycenter innerradius outerradius
box	xcenter ycenter xwidth yheight (optional angle)
circle	xcenter ycenter radius
ellipse	xcenter ycenter xwidth yheight angle
field	none
pie	xcenter ycenter angle1 angle2
point	x1 y1 x2 y2 ... xn yn
polygon	x1 y1 x2 y2 ... xn yn
rotbox	xcenter ycenter xwidth yheight angle

The syntax for describing region descriptors is very flexible. Shapes can be specified using *command* syntax:

shape arg1 arg2 ...

or using *subroutine* syntax:

shape(arg1, arg2, ...)

or using any combination of the two. Shape specifiers can be separated by semicolons or by new-lines. Shape names are case insensitive and all shapes can be specified by their unique abbreviations. The numeric parameter arguments can be specified either as integers or as floating point values (they are always converted to floating point values). The following region descriptor illustrates some of these options:

annulus 100 100 65 85; box(100,100,70,70)
circ(70, 100.0, 20); ellipse 130.0 100, 20, 30.0 45

The regions system converts a region descriptor such as the example above into a spatial mask by parsing the region string and compiling a set of *software CPU instructions* to build the mask. When parsing is complete, the software CPU executes the instructions. This results in the creation of a mask in which each pixel is assigned a positive integer denoting a region (with a value of zero indicating that the pixel is not in any region). If a pixel is covered by two different regions, it is given the value of the second region, i.e., successive regions overwrite previous regions in the mask. Thus, no pixel is covered by more than one region. The four regions of the example region descriptor above are shown in Figure 1.

Figure 1. A display of a mask containing an annulus, a box, a circle, and an ellipse.

3. Excluding Regions

It is often necessary to exclude regions from a mask. This can be done by placing a minus sign before the geometric shape name to specify an *exclude* region. Thus, if the circle and ellipse regions are excluded from the previous example, the region descriptor would be as follows:

annulus 100 100 65 85; –circ(70, 100.0, 20)
box(100,100,70,70); –ellipse 130.0 100, 20, 30.0 45

The resulting mask is shown in Figure 2. Note that exclude regions are always processed by the software CPU *after* the include regions, so that they unconditionally unset the pixels in the mask.

4. Combining Shapes Using Boolean Algebra

Each spatial region in a region descriptor can consist of a single geometric shape, as in the examples above, or it can consist of two or more shapes combined using the rules of boolean algebra. The following boolean operations are supported between geometric shapes in a region:

symbol	operator	associativity
!	not	right to left
&	and	left to right
∧	exclusive or	left to right
|	inclusive or	left to right

Parentheses can be used to change the associative order in which the boolean operations are carried out.

Figure 2. A display of a mask containing an annulus and a box that excludes a circle and an ellipse.

Figure 3. A display of a 2-region mask containing boolean operations on the four geometric shapes.

For example, consider the following region descriptor:

annulus 100 100 65 85 | circ(70, 100.0, 20)
box(100,100,70,70) ^ ellipse 130.0 100, 20, 30.0 45

Two regions are now defined. The first consists of the *inclusive or* of an annulus and a circle, i.e., it consists of all pixels that are either in the annulus or in the circle. The second region consists of the *exclusive or* of a box and an ellipse, i.e., it consists of all pixels that are in the box or in the ellipse, but not in both. The resulting mask is shown in Figure 3.

Note that one can use the & and **!** symbols together as an *and-complement* operator to clear a given region. The behavior of *and-complement*, however, differs from that of an *exclude* region, in that the cleared pixels of the former can be reset by subsequent region expressions, whereas the latter unconditionally unsets the pixels in the region.

5. Celestial Coordinate Systems

In all of the above examples, the units of position and length have been image pixels. It is also possible to specify these parameters in celestial units, by adding special symbols to the values, as illustrated in the following table:

symbol	units
2	two image pixels
2″	two arc seconds
2′	two arc minutes
2d	two degrees
2r	two radians
2:3:4	2 hours, 3 minutes, 4 seconds

These units are interpreted relative to the sky system of the reference image specified when the mask is created. Use of this default reference system can be overridden by giving one of the following commands in a region specification: **B[equinox]** (Besselian equinox)) **J[equinox]** (Julian equinox), **ecliptic**, **galactic**, or **supergalactic**.

Thus, for example, a region specification such as:

B1950;C 12:13:14 20.123d 10′

describes a ten arc-minute circle centered on the specified position, in 1950 FK4 coordinates.

6. Conclusions

The region filtering scheme described above provides great flexibility in specifying spatial masks for image and photon event data. It has proven to be a valuable part of the *IRAF/PROS* analysis environment.

Acknowledgments. This work was supported under NASA contract to the ROSAT Science Data Center (NAS5-30934).

Part 5. Data Analysis Applications

Section B. Spectral Algorithms and Techniques

Astronomical Data Analysis Software and Systems II
ASP Conference Series, Vol. 52, 1993
R. J. Hanisch, R. J. V. Brissenden, and J. Barnes, eds.

ASpect: A New Spectrum and Line Analysis Package

S. J. Hulbert, J. D. Eisenhamer, Z. G. Levay, and R. A. Shaw

Space Telescope Science Institute, 3700 San Martin Drive, Baltimore, MD 21218

Abstract. During the next two years we will write a new spectral analysis package, *ASpect*, that will incorporate analysis techniques for astronomical spectra in all wavelength domains. *ASpect* will: operate on spectra from a wide variety of ground-based and space-based instruments, spanning wavelengths from radio to gamma rays; accommodate non-linear dispersion relations; provide a variety of functions, individually or in combination, with which to fit spectral features and the continuum; mask known bad data; and propagate uncertainties throughout the calculations in order for astronomers to evaluate the reliability of results. Most importantly, this new package will provide a powerful, intuitive user interface to handle the burden of data input/output (I/O), on-line "help", selection of relevant features for analysis, plotting and graphical interaction, and data base management, all in a comprehensible environment.

1. Survey of Analysis Software for Spectroscopic Data

We compared the capabilities of several of the software packages which are currently used at the Space Telescope Science Institute (STScI). These packages provide reduction and analysis capabilities for data spanning the electromagnetic spectrum from radio to X-ray. While this survey is not exhaustive, it includes software packages widely available in the astronomical community. Most of the packages emphasize data reduction over data analysis. All of the reviewed packages have the capability of fitting and removing a baseline or continuum and the ability to fit at least one function (usually a Gaussian) to a spectral feature. Some, but not all, support the use of non-linear wavelength dispersions, while some required resampling the spectra to a linear wavelength scale. Some packages support the propagation of errors using an error vector. None of the packages reviewed, however, allows for the exclusion of data points determined to be of "bad" quality and which should properly be masked during the processing. None provides any facility for the straightforward manipulation of spectral data from a variety of wavelength regions. Only one of the packages can read a variety of data formats in a straightforward and simple manner. All of the packages reviewed use a text-oriented command-driven interface, i.e., keystrokes or text commands are used to communicate with the application. Not surprisingly, each package has its own "language" of commands, even for similar tasks. While each reviewed software package provides valuable reduction capabilities for specific kinds of data, none is comprehensive and not even the ensemble of packages provides for straightforward analysis of multi-wavelength spectroscopic data.

2. The *ASpect* Environment

We have chosen the IRAF environment in which to develop *ASpect* as IRAF provides resources (Tody 1986) that we believe will maximize the usefulness of this new software to the scientific community. By virtue of being developed within the IRAF system the code will be transportable to many existing hardware environments and it will adapt to new systems. The IRAF environment contains a rich set of existing tasks that are suitable for analyzing astronomical data, and supports a wide variety of peripheral hardware, including hardcopy devices and image displays. Code is easy to maintain and the accompanying documentation is easy to generate and available on-line. Finally, the IRAF system is freely available to the astronomical community and is already widely used.

The user interface to any software determines its convenience and in the long run is the single most important factor in its user acceptance. In particular, *interactive* software depends fundamentally on its user interface. We are committed to develop a highly interactive and intuitive user interface for the *ASpect* package. We will develop this interface outside of the existing IRAF graphics environment; specifically, using the sophisticated capabilities that are available with a modern graphics user interface (GUI), such as X-Windows. The X-Windows GUI will provide direct access to objects such as buttons, menus, and file requesters. The development of this interface outside of the IRAF environment means, of course, that the code defining the user interface will not necessarily run on all of the devices supported by IRAF, but rather will only be available on those display devices that are also running the appropriate window manager. We feel strongly that access to these GUI features is critical to the *ASpect* package functionality despite the fact that will we be violating the established IRAF interface. We choose to model our new interface after the successful image displays, *imtool* and *SAOimage*. Our plan is to provide a variety of analysis tools in the IRAF environment that are accessible from outside IRAF using the new user interface *or* accessible from within the IRAF environment in a noninteractive mode. Figures 1–3 demonstrate the planned "look-and-feel" of the *ASpect* graphical user interface. Figure 1 shows the main working window. Figures 2 and 3 show two of the planned optional menu windows that would be available for user input to the tasks.

3. Functionality of *ASpect*

To support the manipulation of spectra from different wavelength regions *ASpect* will be able to transparently read a variety of data formats. At a minimum, FITS files and all supported IRAF image formats will be accessible within *ASpect*. *ASpect* will be able to plot any combination of spectra on the same plot with user selected plot parameters. The wavelength or energy scale will be chosen by the user and *ASpect* will fully support a variety of non-linear wavelength scales, including polynomial, piecewise linear, echelle 2-dimensional format, uneven energy bins, and independent wavelength vectors. *ASpect* will also allow

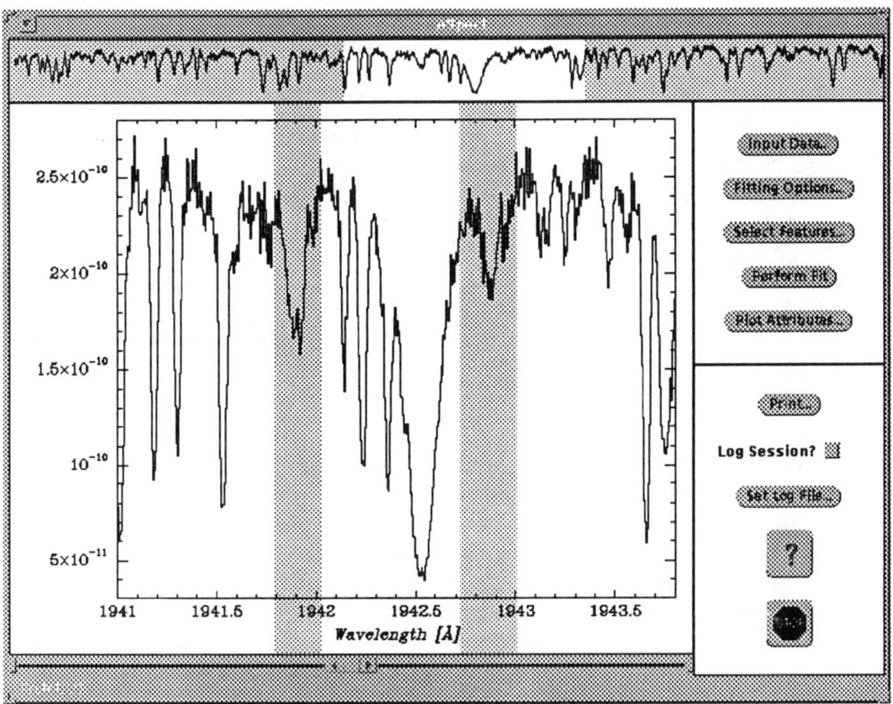

Figure 1. *ASpect* Main Window

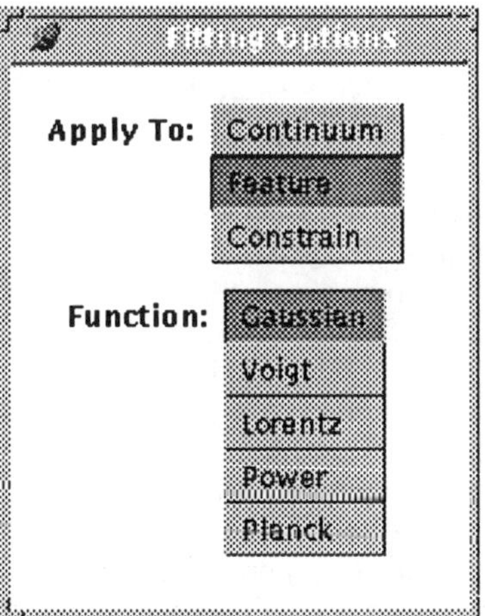

Figure 2. *ASpect* Fitting Menu

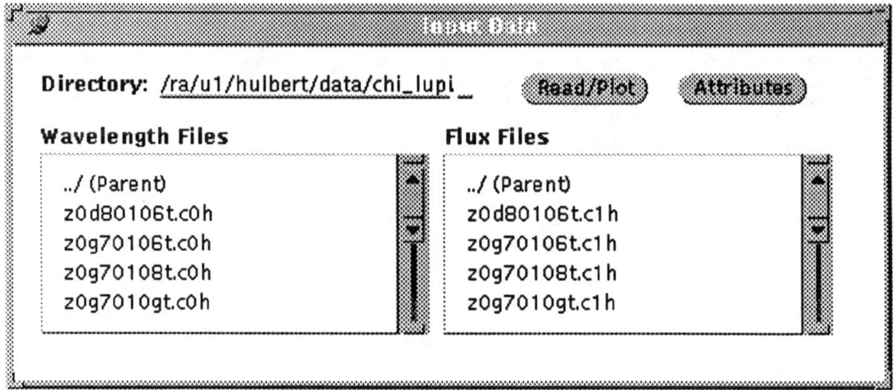

Figure 3. *ASpect* File Menu

conversion between independent variables and the user will be able to interactively merge or combine overlapping spectra.

A fundamental design consideration of the *ASpect* package is the internal data structure or data base used to store the spectra once they have been read, to store intermediate results, and to store the final results. The selection of this data format must accommodate the variety of spectral data. *ASpect* will adopt the STSDAS table as its internal data format. *ASpect* will be able to read "foreign" data files and store information in an STSDAS table. *ASpect* will be able to read and write STSDAS tables.

ASpect will perform error propagation during analysis, wherever possible. The propagation of errors is made more difficult if good uncertainties for the raw—or even calibrated—data are unavailable. *ASpect* will make use of error vectors when such data are available. One of two methods will be used to calculate errors: least squares and, in the case of model fitting, maximum likelihood or M-estimates. *ASpect* will be able to exclude data from the model fits when the data are known to be of low or bad quality. Rather than interpolating over such data points, *ASpect* will properly exclude them from the relevant calculations. The exclusion of flagged data from calculations will be supported using the current implementation of IRAF bad pixel lists.

ASpect will be able to fit both continua and spectral features with a variety of functions and sums of functions including: continuous functions, such as, Gaussian, Lorentzian, Voigt, rotational, polynomial, black-body, power law; piecewise functions, such as spline; as well as a facility for a user-defined or empirical functions. *ASpect* will permit fitting to features that simultaneously appear in different spectra without requiring the user to resample or merge the individual spectra. *ASpect* will permit the user to fit all relevant parameters, to interactively modify them, and to impose a constraint between certain parameters in the fit, such as specifying a fixed ratio between two parameters.

As a general facility, *ASpect* will able to perform arithmetic on spectra. This will include: addition, subtraction, multiplication, and division by constant values, other spectra, standard functions (i.e., polynomials), and special functions, such as correcting for extinction. Additionally, *ASpect* will provide the ability to filter spectra using mean, median, polynomial, and optimal filters. This facility will permit simple or even reasonably sophisticated series of computations to be performed on data held in *ASpect*'s buffers.

Acknowledgments. The *ASpect* project is funded under a contract from the National Aeronautics and Space Administration Astrophysics Data Program.

References

Tody, D. 1986, in Proc. SPIE Instrumentation in Astronomy VI, ed. D.L. Crawford, 627, 733

Astronomical Data Analysis Software and Systems II
ASP Conference Series, Vol. 52, 1993
R. J. Hanisch, R. J. V. Brissenden, and J. Barnes, eds.

Adaptive Filtering of Echelle Spectra of Distant Quasars

A. Priebe, D.-E. Liebscher, H. Lorenz, and G.-M. Richter

Astrophysical Institute, An der Sternwarte 16, O–1590 Potsdam, Germany

Abstract. We present an adaptive filtering technique combined with a special procedure for eliminating cosmic ray events in the original Echelle CCD-image. The procedure is applied to the two-dimensional frames before extracting the spectra. A similar procedure based on adaptive filtering was earlier applied to other tasks in astronomical image processing.

1. Introduction

The study of the $Ly\ \alpha$ – forest of distant quasars ($z > 3$) is an important tool in obtaining a more detailed picture of the distribution of matter along the line of sight and thus of the general distribution of matter in the Universe and is therefore of important cosmological significance. Obviously this is one of the tasks, where spectral resolution plays an important role. On the other hand these objects are faint and the observations are always limited by the signal to noise ratio.

Noise is a fundamental problem in image processing generally; and of particular importance in astronomy, because usually the faintest signals are of most interest. A well known tool to tackle noise is filtering. However, in image processing the conventional (stationary and linear) filters provide usually unsatisfactory results. Therefore non-linear filters, particularly median filters, are often applied. But the use of *linear* filters has several advantages:

1. Linear filters are bias-free. Nonlinear filters may be biased and it is difficult to prove in any particular case if the bias affects the accuracy of the result.

2. Linear estimators are generally the most efficient (i.e., have the smallest possible uncertainty) compared to nonlinear ones.

These features are very important if one has to process the filtered image further, e.g., to calibrate and to measure magnitudes or intensities of any kind. The aim of the present paper is to show the limitations of non-linear filtering and to introduce adaptive filtering.

We developed an Echelle reduction procedure on the basis of a space variable filter described by Richter (1978) (see also Richter et al. 1991) which recognizes the local resolution in the presence of noise and adapts to it. The data reduction is described in section 2. In section 3 it will be shown that this technique leads to an improvement in resolution by a factor of 2 with respect to standard procedures and a quasar spectrum will be used to illustrate these capabilities.

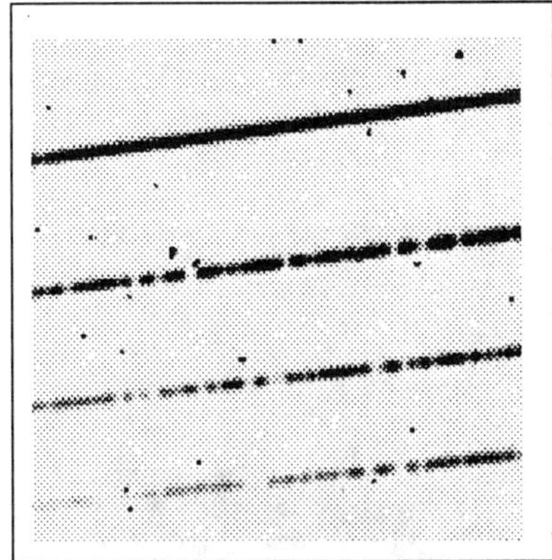

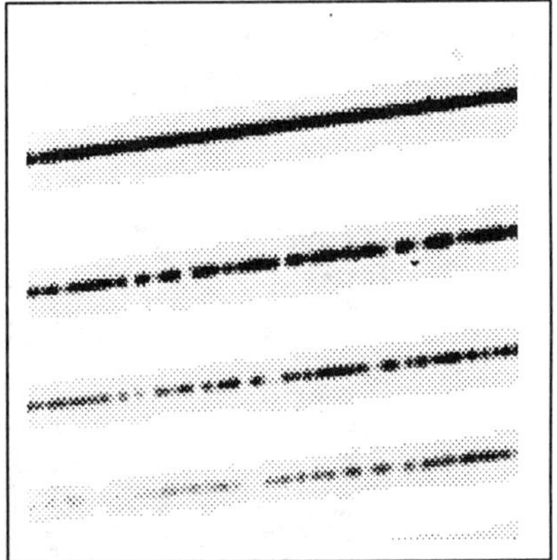

Figure 1. Enlarged part of the original (left) and the cosmic ray removed and background subtracted (right) Echelle spectrum of the Quasar 0420–388

2. Data and Data Processing

2.1. Removal of Cosmic Ray Events

The spectra of the quasars were obtained at the ESO EFOSC spectrograph in the Echelle mode on the 3.6 m telescope in December 1991. Applying the standard Echelle procedure to the data reduction for that instrument, as it is implemented in the MIDAS-package, one uses stationary median filters for noise and cosmic particle event reduction in the 2-dimensional Echelle image. These filters are useful if the spatial spectrum of the noise reaches essentially higher frequencies than the highest resolution features in the image. Otherwise the resolution in the data will be degraded and the spectral lines smoothed. However, in the Echelle spectra the highest resolution is already in the range of one to a few pixels and therefore stationary filtering always means a loss of resolution. The spectra that were obtained as a result of this procedure are presented in section 3 and compared with our reduction. Before analysing the differences in detail let us first describe our technique to remove the cosmics.

The removal of cosmic ray events is essential for the reduction of images with long integration times. The special problem for Echelle spectra is the existence of sharp signal features which are not cosmic ray events, i.e., the night sky lines.

To create a mask for the removal of the cosmics we performed separate Laplacian filtering in the x and y axes and thresholded them. The resulting masks contain the cosmic ray events, the night-sky lines and the spectra. To discriminate between the "objects" found we performed a logical OR on both masks and selected the cosmics by a size criterium. The masked cosmics are then removed from the frame by an interpolation algorithm.

Figure 1 illustrates the cleaning of the images and displays an enlarged part of the original and the background subtracted frame.

2.2. Filtering

The standard Echelle reduction procedure uses a median filtering scheme for the elimination of the cosmic ray events in the CCD exposure and the background extraction as well as for a proposed noise reduction in the frame.

Figure 2. Filtering of the spectrum Q 0420–388. *Upper left:* enlarged portion of the original frame, containing only one order of the echelle spectrum. *Upper right:* the adaptive filtered image. *Lower left:* difference between raw image and standard MIDAS-filtering. *Lower right:* difference between raw image and adaptive filtered frame. It is evident, that the parts of the image containing the spectral information, are not degraded at all in the case of adaptive filtering. The result of the removal of the cosmics and the reduction of the noise in the background is evident, whereas in the case of median filtering strong artifacts in the spectrum are seen.

In Figure 2 the results of these filterings are illustrated. Displayed is a very enlarged portion of a frame, containing only one order of the echelle spectrum. The upper left subimage shows the original. On the right hand side the adaptive filtered image is placed. It is evident, that the parts of the image containing the spectral information, are not degraded at all. But the result of the removal of the cosmics and the reduction of the noise in the background is clearly seen.

In order to illustrate both reduction procedures in detail, the differences between the input and the processed images are shown in the lower part of Figure 2. As one can easily see, the standard MIDAS-filtering (left subimage) highly degrades the original.

3. The Spectra

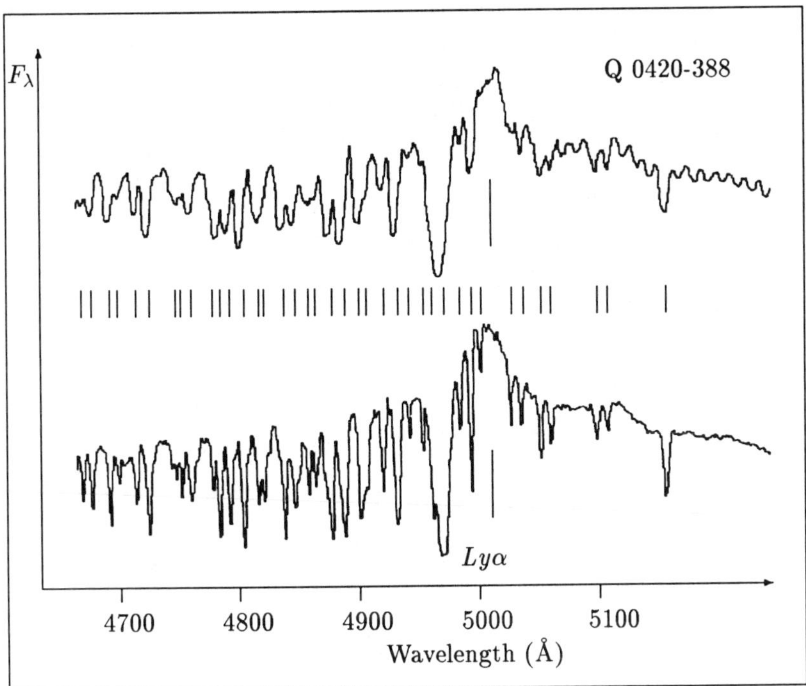

Figure 3. Spectrum of the quasar Q 0420+003. *Upper panel:* result of the standard MIDAS filtering. *Lower panel:* adaptive filtered spectrum; displayed are the absorption systems given by Atwood et al. (1985). It is evident that the spectrum shows a significantly higher resolution then the median filtered frame and resembles the marked absorption systems.

The cleaned and noise reduced images as they resulted from the procedure described in the last section were used as input for the order detection, spectral extraction and merging. Figure 3 shows the spectrum of the Quasar 0420+388. The upper panel contains the result of the median filtering, the lower one that of the adaptive filter. As it can be easily seen, the adaptive filtered spectrum shows a significantly higher resolution and more features. On the other hand, the median filter introduces considerable artifacts, seen as fringes at the right end of the spectrum.

4. Conclusions

The technique presented here can significantly improve the quality of Echelle spectra, compared to standard reduction and use of non-adaptive filters.

References

Atwood, B., Baldwin, J.A., & Carswell, R.F. 1985, ApJ, 292, 58

Lorenz, H., Richter, G.M., & Capaccioli, M. 1992, Adaptive filtering in astronomical image processing, in preparation

Richter, G.M. 1978, Astron. Nachr., 299, 282

Richter, G.M., Böhm, P., Lorenz, H., Priebe, A., & Capaccioli, M. 1991, Astron. Nachr., 312, 345

Astronomical Data Analysis Software and Systems II
ASP Conference Series, Vol. 52, 1993
R. J. Hanisch, R. J. V. Brissenden, and J. Barnes, eds.

The IRAF Fabry-Perot Analysis Package: The Incomplete Phase Surface

Patrick L. Shopbell, Jonathan Bland-Hawthorn

Rice University, Houston, TX 77251-1892

Gerald Cecil

University of North Carolina, Chapel Hill, NC 27514

1. Introduction

As introduced at ADASS I (Bland-Hawthorn et al. 1992), a Fabry-Perot analysis package for IRAF is currently under development as a joint effort between Rice University and Frank Valdes of the IRAF group. This package will extend the spectrophotometric analysis capabilities of IRAF, currently found in `onedspec` and `twodspec`, to the three dimensional data structures of imaging Fabry-Perot interferometry. In certain cases, this extension to a third dimension requires only small enhancements to the current IRAF code. The fitting of incomplete calibration rings is an example of an integral stage of Fabry-Perot data analysis for which there is currently no IRAF functionality.

Although additional portions of the package have also been implemented, we report here on the development of a robust task for fitting sections of calibration rings that arise in Fabry-Perot interferometry. The radial equation for an ellipse is fit to the shape of the rings, providing information on ring center, ellipticity, and position angle. Such parameters provide valuable information on the wavelength response of the etalon and the geometric stability of the system. In addition, fits to the rings of a calibration "cube" provide a simple means of parameterizing the three-dimensional Airy phase surface. Appropriate statistical weighting is applied to the pixels to account for increasing numbers with radius, the Lorentzian cross-section, and uneven illumination. The major problems of incomplete, non-uniform, and multiple rings have been addressed with the final task capable of fitting rings regardless of center, cross-section, or completeness. The task requires minimal user intervention in order to establish reasonable initial conditions for the ring fitting.

2. Motivation

As outlined in Shopbell et al. (1992), the instrumental response of a Fabry-Perot etalon to a monochromatic source is given by the Airy function:

$$I(x,y,z) = \frac{I_0}{1+\alpha^2 \sin^2(\beta z \cos(\gamma\sqrt{(x-x_0)^2+\epsilon^2(y-y_0)^2})+\delta)}, \qquad (1)$$

where I_0 is the intensity of the input beam, (x_0, y_0) defines the optical axis, and $\alpha, \beta, \gamma, \delta$, and ϵ are properties of the optical system. All of these constants are measured directly from the observed three-dimensional phase surface within the calibration cube. Due to the large number of unknowns, Bland-Hawthorn et al. (1992) suggest that each constant be fit separately and these values used as initial conditions for the full fit.

One of the primary motivations behind a ring fitting task is to determine initial values for some of the system constants. Accurate estimates of ϵ, x_0, and y_0 in Eq. 1 can be made, in preparation for a complete fit of the three-dimensional Airy function. Furthermore, changes in the system parameters over the course of an observing run can be monitored. The most crucial parameter in this respect is the ring radius, as this depends directly upon the wavelength response of the etalon (Bland & Tully 1989). Such variations usually imply fluctuations in the optical gap of the etalon, μl, due to changes in pressure and humidity. Variations in other parameters imply geometrical instabilities such as flexure in the observing system. In addition, the task can be used to estimate the instrumental resolution of the etalon across the field and to map asymmetric ghost reflections.

Practically, there are several complications that arise in fitting Fabry-Perot rings. Normally the etalon is tilted with respect to the optical axis, in order to reflect primary ghost reflections out of the field of view. The net result of this is that the phase surface will be about 50% complete. The ability to fit partial rings is integral to any Fabry-Perot ring fitting software. Other complications include the possible presence of multiple rings in a frame, uneven illumination across the frame, noise, and vignetting. The nature of the rings suggests that several types of weighting should be performed by the fitting task, including radial weighting (i.e., the number of pixels per annulus decreases with radius) and flux weighting. As with most Fabry-Perot software, the enormous amount of data requires that the task be able to fit a large number of rings quickly with little user intervention. All of these complications are addressed by the `ringfit` task described below.

3. Operation

The `ringfit` task implements the fitting of Fabry-Perot calibration rings as a two stage process: finding the rings in a frame and fitting to determine the ring parameters. In order to automate `ringfit` as much as possible, a ring finding algorithm has been implemented to locate and enumerate the rings in an image, returning the approximate center and estimates of the ring radii. Two primary user inputs are required by both stages of the task: an estimate of the average ring width and the approximate background pixel value. Three other parameters can be adjusted to modify the behavior of the ring finding algorithm itself, including the desired accuracy of the results and the maximum number of iterations for convergence. An option is provided to specify that rings only be found and not fit. As the former is much more rapidly executed, the user is able to quickly determine values for the ring-specific parameters before fits are attempted. Since the same parameters are usually viable for all the rings in a data set, this process need only be completed once.

Due to its complexity, the ring fitting stage has several parameters beyond those required for ring finding, all of which modify the behavior of the ring fitting algorithm. As for the ring finding stage, there are parameters that specify the fitting tolerance and the maximum number of iterations. Several parameters are provided for the automatic rejection of outlying data values. This is especially useful in the case of very noisy data. There are switches to enable azimuthal binning, azimuthal and radial flux weighting, and interactive fitting. Azimuthal binning of the data and flux weighting of the fit are extremely useful in the cases of noisy or unevenly illuminated data. The interactive fitting option provides the user with graphical information for improving the fit manually (i.e., deleting points, increasing the number of iterations, etc.).

The **`ringfit`** task has been specifically designed for fitting large numbers of Fabry-Perot calibration rings in an automated manner. As such, the task will find and fit multiple rings in multiple image frames with only a brief interactive setup period. Typical operation of the task consists of several quick runs on a sample ring, with the fitting stage disabled, allowing the user to set the ring width and background level. Following this, the ring fitting stage is enabled, and the entire set of rings may be fit non-interactively. Depending on the complexity of the data, a few rings may be fit interactively to better estimate the fitting tolerance and maximum number of iterations. Radial weighting of the data is performed automatically, and flux weighting is highly recommended. In cases of severe vignetting, the task allows for both azimuthal and radial flux weighting.

4. Algorithms

The ring finding algorithm employs the `center1d` routine from IRAF's `xtools` package to locate peaks in the image. Once a peak is found, an attempt is made to trace a ring azimuthally. If a ring has truly been located, the tracing algorithm will converge to an approximate center for the ring. Note that the ring center need not be within the frame, as is often the case for partial rings. With an approximate center, the algorithm then enumerates the number of rings present by searching for peaks radially outward. The results returned include the approximate center and the estimated radii of all rings detected. The ring width and background level parameters are crucial in this stage, as is the parameter which specifies the search radius for the `center1d` routine. The ring finding stage is sufficiently robust to handle multiple partial rings having any orientation within the frame. The only assumption made concerning multiple rings is that of concentricity, the standard condition for Fabry-Perot data.

The ring fitting stage is independently applied to each ring found in a frame. Due to the complexity of fitting a general ellipse to data, the optimization is performed in two steps. First, the radial equation of a circle of radius r_0, centered at (α, h) is fit to a radially-interpreted azimuthally-binned image:

$$r^2 - r_0^2 + h^2 \quad 2rh\cos(\theta - \alpha) = 0. \tag{2}$$

The known estimates of the ring center and radius are used as initial values for this fit. Two parameters of the final fit are determined: x_0 and y_0, the accurate center of the ring. The rings encountered in Fabry-Perot analysis are extremely close to circular and are sufficiently symmetric that fitting the above circle results in an accurate estimate of the ring center. Indeed, most Fabry-

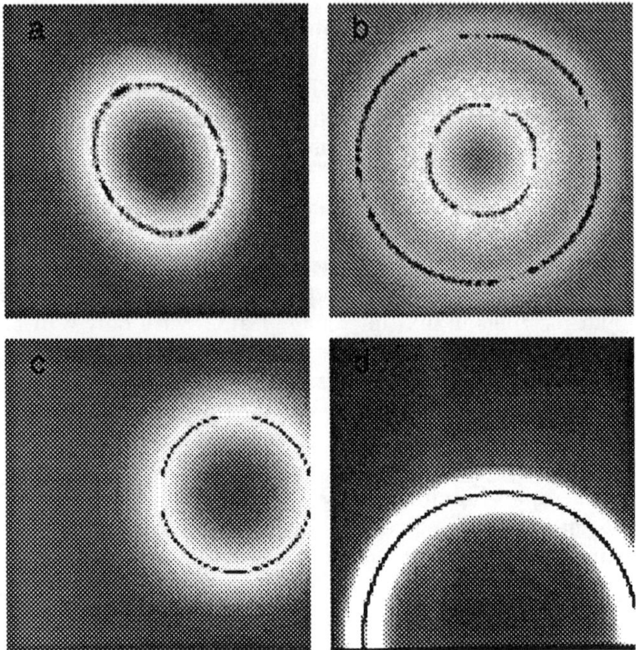

Figure 1. Four sample Fabry-Perot calibration rings and their fits, as determined by the `ringfit` task. Rings *a*, *b*, and *c* consist of artificially constructed data and serve merely to demonstrate the capabilities of the task. Frame *d* is an actual Fabry-Perot calibration ring and is quite typical in its partiality, noise level, and shape.

Perot ring fitting algorithms implemented prior to this work assume circularity and stop the optimization at this point.

The second step in the ring fitting stage is to fit the radial equation of an ellipse with axes a and b, position angle α, and center at the origin to a radially-interpreted azimuthally-binned image:

$$\frac{\cos^2(\theta-\alpha)}{a^2}+\frac{\sin^2(\theta-\alpha)}{b^2}=\frac{1}{r^2}. \tag{3}$$

The image is reinterpolated for this fit using the more accurate ring center found by the circle fit. This interpolation in effect performs the radial weighting. Prior to this fit, the data can also be weighted by the azimuthal or radial flux within the ring. Both the circle and ellipse fits are performed using the IRAF `nlfit` nonlinear fitting routines (Davis 1991), although the ellipse fit can be performed interactively using `inlfit`. The final three parameters of the complete fit, the ring ellipticity, ϵ, and the ring axes, a and b, are determined from the ellipse fit. All of the results and status messages are logged and can be used for further studies of the entire three dimensional phase surface.

Figure 1 illustrates the results of four runs of the **ringfit** task. Out to extreme ellipticities and poor levels of completeness, the task gives excellent results on both artificial and real Fabry-Perot calibration rings. Typically we find that it takes a couple of minutes to locate and fit an 800×800 calibration frame with several rings on a Sun 4/260 workstation. Much of the running time is spent not in fitting but in image I/O, and can be traced to the line-oriented image access routines in IRAF. Execution speeds will improve both as processors increase in speed and as machine memories become large enough to keep significant portions of the image resident in memory.

5. Conclusions

We have reported herein on the development of a robust automated ring fitting algorithm for Fabry-Perot analysis. This algorithm has been implemented as an IRAF task, **ringfit**, which is part of the Fabry-Perot analysis package currently under development. The task will rapidly fit multiple incomplete elliptical rings on multiple image frames in an automated manner. The results can be used to monitor the stability of system parameters over the course of an observing run, or as rough inputs to a parameterization of the full three-dimensional Airy phase surface. Further enhancements under study include improved ring finding using a minimization-pruning algorithm (Mihovilovic & Samadani 1992), graphical support for ring parameter specification, image masking of bad pixels, and ring profile fitting. Current work on the Fabry-Perot package as a whole includes parameterization of the three-dimensional Airy surface and phase correction, flux calibration, and three-dimensional drift correction. The unfinished package can be provided at any time to interested parties for testing and suggestions, while the final package will serve the astronomical community as a complete and robust Fabry-Perot analysis environment within IRAF.

The authors would like to express their thanks to the National Optical Astronomy Observatories, and the IRAF group in particular, for their continued support in this project. P.L.S. and J.B.H. are partially funded through NSF grant AST 88-18900.

References

Bland, J., & Tully, R.B. 1989, AJ, 98, 723

Bland-Hawthorn, J., Shopbell, P.L., & Cecil, G. 1992, in Astronomical Data Analysis Software and Systems I, A.S.P Conf. Ser., Vol. 25, eds. D.M. Worrall, C. Biemesderfer & J. Barnes, 393

Davis, L. 1991, NLFIT/INLFIT README files, IRAF distribution

Mihovilovic, D.A., & Samadani, R. 1992, in SPIE Vol. 1657 Image Processing Algorithms and Techniques III, 546

Shopbell, P.L., Bland-Hawthorn, J., & Cecil, G. 1992, in Astronomical Data Analysis Software and Systems I, A.S.P Conf. Ser., Vol. 25, eds. D.M. Worrall, C. Biemesderfer & J. Barnes, 442

Astronomical Data Analysis Software and Systems II
ASP Conference Series, Vol. 52, 1993
R. J. Hanisch, R. J. V. Brissenden, and J. Barnes, eds.

SPECFOCUS: An IRAF Task for Focusing Spectrographs

Francisco Valdes

IRAF Group, NOAO[1]*, PO Box 26732, Tucson, AZ 85726*

Abstract. An IRAF task for measuring the point-spread function width along the dispersion and wavelength shifts across the dispersion in two dimensional arc spectra is described.

1. Introduction

The IRAF[2] task **specfocus** estimates the dispersion width of spectral lines in sequences of arc spectra taken at different focus (or other parameter which affects the spectral line widths) settings. The widths can be measured at different spatial and dispersion positions, called *samples*, on the detector. The width estimates are recorded and displayed graphically in order to investigate dependencies and determine appropriate settings for the spectrograph setup. The task may also measure dispersion shifts when multiple spectral samples are specified. This task does not measure the focus point-spread function width across the dispersion.

The input images need not be bias corrected or flat fielded since the intention is to operate directly on the raw CCD images as they are obtained at the telescope. The images are specified with an image list which may consist of image names, wildcard templates, and @-files.

A *focus* value is associated with each image. This may be any numeric quantity. The focus values may be specified in several ways. If no value is given then integer numbers are assigned to the images in the order in which they appear in the image list. A range list may be specified which consists of individual values, ranges of values, a starting value and a step, and a range with a step. For example a range list could be "`500-250x-50, 225, 200-150x-10`". A long list, such as a list of individual focus values, may be placed in a file and specified with the IRAF @-file convention. Finally, a parameter in the image header may be used for the focus values by simply specifying the parameter name.

Two dimensional long slit images are summed into one or more one dimensional spectra across the dispersion. The dispersion axis is defined either by the

---

[1]National Optical Astronomy Observatories, operated by the Association of Universities for Research in Astronomy, Inc. (AURA) under cooperative agreement with the National Science Foundation

[2]Image Reduction and Analysis Facility, distributed by the National Optical Astronomy Observatories

image header parameter DISPAXIS or a task parameter with the image header parameter having precedence. The range of lines or columns across the dispersion may be specified to define the limits of the slit, otherwise the full width of the image is used. This range is then divided into a number of specified spectra. Use of more than one spectrum across the dispersion allows investigation of variations along the slit. In addition, if desired, the spectrum nearest the center may be used as a reference against which shifts in the dispersion positions of the features in the other spectra are determined by crosscorrelation.

The conversion of two dimensional spectra to one dimensional spectra may also be performed separately using the tasks in the IRAF **apextract** package. This would be done typically for multifiber or echelle format spectra. If the two dimensional spectra have been extracted to one dimensional spectra then the individual spectra are used without any summation. Measuring relative shifts between spectra may also be done and makes sense for multifiber spectra but not for echelle spectra.

In addition to dividing the spatial axis into a number of spectra the dispersion axis may also be divided into a set of subspectra. This applies to both long slit and 1D extracted spectra. When the dispersion axis is divided into more than one sample, the dependence of the dispersion widths and shifts along the dispersion may be investigated.

After computing the correlation profiles, the profile widths and shifts, and the best focus values, an interactive graphics mode is entered. Upon exiting the interactive graphics the results are written to the terminal and to a logfile if one is specified. A sample execution and output is shown in Figure 1.

Figure 1. Example command and output

```
cl> specfocus @imlist focus=400x50 slit1=50 slit2=130 nspec=3 ndisp=3 shifts-
<Interactive graphics which is exited with the 'q' key>
SPECFOCUS: NOAO/IRAF V2.10EXPORT valdes Thu 19:41:41 17-Sep-92
  Best avg focus at 206.6584 with avg width of 2.91 at 50% of peak

  -- Average Over All Samples

                                      Image  Focus  Width
                                jdv010.imh   150.    3.28
                                jdv009.imh   200.    2.95
                                jdv008.imh   250.    3.17
                                jdv007.imh   300.    3.41

  -- Image jdv009.imh at Focus 200. --

        Width at 50% of Peak:

                         Columns
                          50-76       77-103     104-130
           Lines  +---------------------------------
            2-267  |    2.93        2.58         2.74
          268-533  |    3.17        2.76         2.89
          534-799  |    3.77        2.23         3.50
```

2. Algorithms

Each spectral sample has a low order continuum subtracted using a noninteractive iterative rejection algorithm to exclude the spectral lines. This technique is the same as that commonly used in other IRAF applications such as *continuum.* The continuum subtracted spectrum is then tapered with a cosine bell function and autocorrelated. The length of the taper and the range of shifts for the correlation is set by another task parameter. This parameter should be set only slightly larger than the expected feature widths to prevent correlations between different spectral lines. The correlation profile is offset to zero at the edges of the profile and normalized to unity at the profile center. The profiles may be viewed as described below.

If there is more than one spatial sample the central spectrum is also crosscorrelated against the other spectra at the same dispersion sample. The crosscorrelation is computed in exactly the same way as the autocorrelation. The crosscorrelation profiles are only used for determining shifts between the two samples and are not used in the width determinations.

A cubic spline interpolator is fit to the profiles and this interpolation function is used to determined the profile width and center. The width is measured at a point given by a *level* parameter relative to the profile peak. The default value selects the full width at half maximum. The autocorrelation width is divided by the square root of two to yield an estimate of the width of the spectral features in the spectrum in units of pixels.

Having computed the width and shift for each input image at each sample, the *best focus* values (focus, width, and shift) are estimated for each sample. As mentioned later, it is possible to exclude some samples from this calculation by deleting them graphically. First the images with the smallest measured width at each distinct focus are selected since it is possible to input more than one image at the same focus. The selected images are sorted by focus value and the image with the smallest width is found. If that image has the lowest or highest focus (which will always be the case if there are only one or two images) then the best focus, width, and shift are those measured for that image. If there are three or more focus values and the minimum width focus image is not an endpoint then parabolic interpolation is used to find the minimum width. The focus at this minimum width is the *best focus.* The dispersion shift is the parabolic interpolation of the shifts at the best focus. The *average best focus* values are then the average of the best focus values over all samples.

3. Interactive Graphics

There are several interesting aspects of this program dealing with interactive graphics. First was the development of various informative plot formats to allow visualizing the multiparameter space. Another is that the plots had to be sensitive to the amount of data available. For instance if there is very little data some plots are inappropriate while at the other extreme of large amounts of data the plots have to be abbreviated in some ways for legibility.

The last feature to mention is that the graphics use a point-and-type interface to select data and plot formats. In all plots there is a concept of the

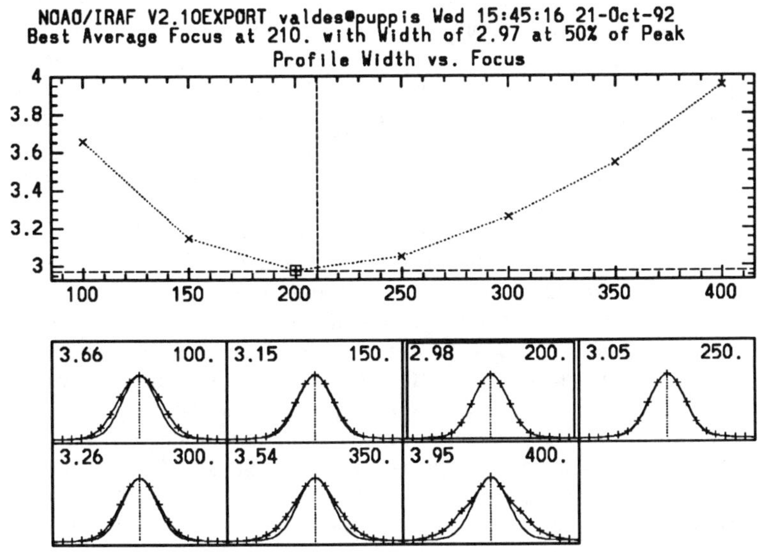

Figure 2. Width/profile plot with multiple spectra and one sample

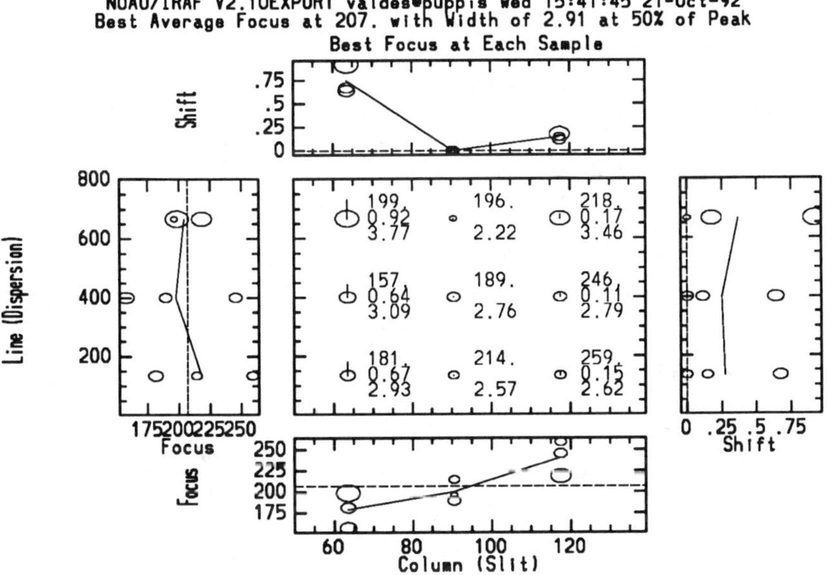

Figure 3. Best focus plot with multiple spectra and samples

current image and the current sample. In general there is an indication, usually a box, of which image and sample is the current one. The current image and sample are changed by pointing at a particular point, box, circle, or symbol for that image and sample and typing a key. For example to zoom on a particular subsample from a particular image one points to the appropriate symbol in the current plot and types 'z'.

There are five types of plot formats. The *width* format produces a graph showing the sample widths as a function of focus value. This is the default plot if there is only one sample over a set of images at different focus values. The top graph is a symbol plot of width verses focus. The lower portion of this format are either graphs of the autocorrelation profiles (described below) if there is only one sample per image or graphs showing the widths as circles with size proportional to the width and position corresponding to the spatial position of the samples in the image. Figure 2 shows an example of a width format plot.

The *best focus* format shows summary graphs of the best focus values (as described above) at each sample position. This is the default plot when there is sufficient data. The central graph represents the best focus (smallest) width at each sample by circles of size proportional to the width. The position of the circle indicates the central line and column of the sample and the relative shifts, if calculated, are represented by little vectors. In addition, there may be graphs along the line or column axes which, again, show the widths as circles but one axis is either the line or column and the other axis is either the best focus value or the shift. This identifies best focus trends along and across the dispersion. Figure 3 shows an example of this type of plot.

The *profile* format produces graphs of the autocorrelation profiles. This requires more than one image and more than one sample. The top graph shows the profiles of all images at a particular sample and the bottom graph shows the profiles of all samples at a particular image.

The profiles are drawn with a solid line using the interpolator function and the actual pixel lags are indicated with pluses. The profiles are drawn shifted by the amount computed from the crosscorrelation. Note that the shift is added to the autocorrelation profile and the crosscorrelation profile is not what is plotted. The zero shift position is indicated by a vertical line.

The *spectrum* format is similar to the *profile* format but shows the spectra rather than the profiles. The top graphs are the spectra of each image at a particular sample and the bottom graphs are the spectra of each sample for a particular image.

The *zoom* format graphs the autocorrelation profile and the spectrum of a single sample. This graph provides scales which are not provided with the *profile* and *spectrum* graphs. If there is only one image and one sample then this is the only plot available.

It is possible to exclude some of the images and samples from the calculation of the best focus and best average focus values. This is done with the *delete* command. There is also an *undelete* command to recall deleted data. When the task exits the printed and logged results will have the deleted data excluded.

The remaining commands give a command help, redraws the current plot, prints information about the sample nearest the cursor, and exits the task.

Astronomical Data Analysis Software and Systems II
ASP Conference Series, Vol. 52, 1993
R. J. Hanisch, R. J. V. Brissenden, and J. Barnes, eds.

Interactive Spectral Analysis And Computation (ISAAC)

D. M. Lytle

National Optical Astronomy Observatories, Tucson, AZ 85719

Abstract. *Isaac* is a task in the NSO external package for IRAF. A descendant of a FORTRAN program written to analyze data from a Fourier transform spectrometer, the current implementation has been generalized sufficiently to make it useful for general spectral analysis and other one dimensional data analysis tasks. The user interface for *Isaac* is implemented as an interpreted mini-language containing a powerful, programmable vector calculator. Built-in commands provide much of the functionality needed to produce accurate line lists from input spectra. These built-in functions include automated spectral line finding, least squares fitting of Voigt profiles to spectral lines including equality constraints, various filters including an optimal filter construction tool, continuum fitting, and various I/O functions.

1. Introduction

Isaac is a software system designed for the analysis of one dimensional spectra. It runs as a task under IRAF. The user interface for *Isaac* is implemented as an interpreted mini-language containing a programmable vector calculator.

Isaac's design philosophy is based on the toolbox metaphor. The built-in functions are quite general and can be combined by the user into scripts and aliases for particular applications.

1.1. Background

Isaac is a distant relative of a FORTRAN program called DECOMP which in turn was a descendant of another FORTRAN program called REDUCER, both written at NOAO by James Brault to analyze Fourier transform spectrometer (FTS) data.

1.2. Foreground

As with all IRAF programs, *Isaac* inherits the task structure and much of its graphical capability from this parent system. *Isaac* is the first program of its kind in IRAF so much of the user interface is new. The mini-language paradigm evolved from the parent DECOMP program which has a command interpreter interface containing many built-in functions. *Isaac's* mini-language is a small programming language containing *if, while, print, declaration,* and *assignment* statements.

The major built-in algorithms in *Isaac* are the continuum correction algorithms, the filter algorithms, the line finding algorithm, and the algorithm used to fit Voigt functions to spectral line profiles.

Isaac can read IRAF image format as well as the old DECOMP disk format. Spectra are considered to have two parts, the actual data and the associated spectral line list. *Isaac's* I/O functions automatically read and write these associated line lists when they are present.

2. Mini-Language

The mini-language used in Isaac is loosely based on the calculator language "hoc" developed as an example in the book *The UNIX Programming Environment* by Kernighan and Pike. The parser is constructed using a variation of YACC, the lexical analyzer was written from scratch.

2.1. Data Types and Variables

There are four datatypes in *Isaac: real scalar, real vector, line,* and *linelist.* The variable names can be any alphanumeric sequence starting with a letter (up to 100 characters). Scaler variables do not have to be declared, all the others must be declared before use.

Vector variables not only contain a 1-D array of real numbers but also a number of attributes, that is, a vector has a header. This header contains the length, flags for wavelength/wavenumber and emission/absorption, the continuum level, the reference wavenumber and the dispersion, and a number of others. A particularly important header entry is the pointer to the associated linelist. Each vector has associated with it in memory a list of spectral lines. This list is initially null and is only created when lines are found by "find" or inserted by the user. There are a number of ways the user can modify this header.

The other two data types are *line* and *linelist.* A *line* refers to a spectral line and has attributes such as position, width, amplitude, and damping as well as various flags and an identification string. A *linelist* is just an array containing lines. Linelists can be modified independently of any particular vector and may be associated and disassociated with any particular vector.

2.2. Syntax of Expressions

Arithmetic expressions in *Isaac* are very similar to those found in other procedural languages such as C and FORTRAN. Infix notation is used and subexpressions are grouped by parenthesis. *Isaac* has an unusually large number of unary operators. Besides the usual "–" for negation and "!" for logical not, *Isaac* has operators for *norm, derivative, second derivative, forward FFT, reverse FFT, modulus,* and *power spectrum.* The binary operations are the usual ones including arithmetic operations and logical operations. All operators are overloaded to work with all data types, however many combinations are not allowed (for instance, the Fourier transform of a linelist!).

2.3. Declarations, Assignment, and Print

Vectors are declared by using the keyword *vector* on a line followed by the vector name and then its initial length. Lines and linelists are declared similarly except no length is given. Declarations must be given in a script before conditional or loop structures are called since the declarations happen when the script is compiled into intermediate code, not at run time.

The assignment statement is very similar to assignment statements in other languages with a few important exceptions concerning the *Isaac* data types. For example, if the variable on the left hand side of the assignment statement is a vector and the value of the expression on the right hand side has a scalar value, every data entry in the vector will be set equal to the value of the right hand side.

The *print* statement can print strings, scalars, vectors, lines, and linelists. Care must be taken when printing very long vectors as this can take quite a while! There is no user adjustable format for the *print* statement.

2.4. Conditionals and Loops

The *Isaac* mini-language contains the *if* and *if-else* constructs. Braces are used to enclose the conditional execution block and/or the alternate conditional execution block unless said blocks only contain a single command.

The only loop construct provided in *Isaac* is the *while* loop. This statement consists of the *while* keyword, a conditional statement enclosed in parenthesis, and a conditional execution block enclosed in braces.

3. Preprocessor

The *Isaac* preprocessor intercepts certain commands before they get to the lexical analyzer and thus to the parser. Features of this preprocessor are a simple history mechanism, an alias function, mini-language escape, and the *gofile* mechanism. The *gofile* command is followed by a filename and instructs the preprocessor to begin reading commands from that filename instead of from the keyboard. Gofiles can call gofiles and this nesting can be 20 levels deep.

The preprocessor also captures calls to the online help facility. Online help is available for all built-in functions and for other general things like syntax.

For complex scripts *Isaac* has a *program* mode which requires a change in input syntax but which allows the user to easily nest built-in functions. The default *interactive* mode does not require the user to delimit argument lists to built-in functions, *program* mode requires that argument lists be enclosed in parenthesis and that arguments be separated by commas. This allows built-in functions that return values to be passed as arguments to other built-in functions. The user switches between these modes by typing *program* or *interactive* at the *Isaac* prompt.

4. Input/Output

In most case, it is expected that data input to *Isaac* will be via IRAF image format and output will consist of IRAF images and linelists. Linelists are stored on disk as text files.

There are various commands for reading input from and writing output to IRAF images. Support is provided for both one and two dimensional images. The user can assign file descriptors to individual input and output files and can thus have more than one input and/or output file open simultaneously. The important *Isaac* commands for I/O are *open* and *close*, *read* and *write*, *readline, readcol,* and *write2d.*

For compatibility, support is also provided for reading from the old DECOMP format files. The important commands in this case are *openp, readp,* and *pair* .

Linelists can also be read and written in isolation, that is without reference to any particular vector data. These independent linelists can also be associated and disassociated with *Isaac* vectors using the *llswap* command.

Unformatted I/O is also available to text files in case the user wants to read and write data values as text or perhaps create a personalized log file of program execution.

5. Vector Manipulation

Isaac contains a number of built-in functions for manipulating vectors beyond the syntactical manipulation. These include functions to cut and paste vectors, generate specified special functions like sinc, shah, Voigt, and noise, and other things like shift, reverse, convolution, etc.

Other built-in functions are the trigonometric functions, log, apodization, and functions for extracting the amplitude and phase from Fourier transforms.

Isaac contains various data filters including boxcar, Gaussian, and an approximation to a Voigt optimum filter. There is also an interactive optimal filter construction tool that allows the user to construct an optimal Fourier filter by examining the data power spectrum and setting the noise level and then fitting a parabola to the data power. Once this filter has been constructed, the user can save it in a filter vector and apply it to other data vectors at will.

Of course, *Isaac* also includes a plot routine to allow the user to examine data, look at filters, look at residuals, etc.

6. Finding Lines

This section and the last section describe two of the most fundamental built-in functions in *Isaac.* The command *find* is used to automatically locate spectral lines in a spectrum. In general this function just locates maxima or minima in one dimensional data. *Find* works by convolving the target vector with various kernels and examining the results. These kernels are first and second derivatives of the sinc function. Convolution with the derivative of the sinc function returns the derivative of the target vector. Extrema are found by looking at zero crossings of the second derivative and their positions can be determined by finding the point halfway between two zero crossings or by looking for a zero crossing of the first derivative in that region.

Filtered convolution kernels are also provided which help with noisy data or with data with very narrow spectral lines that sometimes cause ringing in FTS data (for example).

7. Lsqfit and Related Commands

Lsqfit allows the user to fit Voigt profiles to spectral lines. The Voigt function is the convolution of a Gaussian and a Lorentzian (both of which are special cases of the Voigt function). The algorithm is a least squares method and iterates to minimize a figure of merit function. Lsqfit can fit any number of profiles simultaneously and allows equality constraints to be specified between profile parameters.

Acknowledgments. I am grateful to James Brault of NSO/Tucson for many discussions of strategy and algorithms. I am also grateful to Bill Jefferies of UT/Austin for his suggestions concerning the constraints in the least squares algorithm. *Isaac* has been improved though suggestions of a couple users in the last year. Gregg Kopp of NSO/Tucson and Karin Muglach of the Karl Franzens-Universitaet/Graz have both used *Isaac* for data reduction and have made useful bug reports and suggestions.

Astronomical Data Analysis Software and Systems II
ASP Conference Series, Vol. 52, 1993
R. J. Hanisch, R. J. V. Brissenden, and J. Barnes, eds.

Factor Analysis as a Tool for Spectral Line Component Separation

L. Viktor Tóth[1], Kalevi Mattila, Lauri Haikala

Helsinki University Observatory, Tähtitorninmäki, SF-00130, Finland

Lajos G. Balázs

Konkoly Observatory, Budapest, P.O. Box 67., H-1525, Hungary

Abstract. The spectra of the 21 cm HI radiation from the direction of L1780, a small high-galactic latitude dark/molecular cloud, have been analysed by multivariate methods. Factor analysis has been performed on the HI spectra in order to separate the different components responsible for the spectral features.

Our method has been found to be effective in separating small spectral features, and was able to differentiate among very similarly looking input spectra.

The rotated, orthogonal factors explain the spectra as a sum of radiation (1) from the background (an extended HI emission layer with galactic latitude dependent density distribution), and (2) from the L1780 dark cloud.

Our statistically derived "background" and "cloud" spectral line profiles, as well as the spatial distribution of the HI halo emission have been compared to the results of a previous study which used conventional methods analysing nearly the same data set.

1. Introduction

In analysing radio spectral line data (especially HI 21 cm spectra) one has to separate two or more components responsible for the observed line profiles. Having no previous knowledge on the number, velocity and shape of the components only a multidimensional statistical approach (see Murtagh and Heck 1987) may provide an unbiased result. In our previous study (Tóth et al. 1992a) we tested the principal component analysis (PCA) on HI 21 cm data of a large region. This paper reports the results of a different method carried out on HI spectra of a small interstellar cloud.

The dark nebula L1778/1780 (the name L1780 will be used in the following) is especially suitable for studies of 21 cm HI emission associated with dark clouds: (1) L1780 is located at a high galactic latitude ($b = 37°$) where the confusion

[1]Eötvös University, Department of Astronomy, Budapest, Ludovika tér 2., H-1083, Hungary. E-mail: EASD101@ HUECO.Uni-Wien.AC.AT or TOTH@ phcu.Helsinki.FI

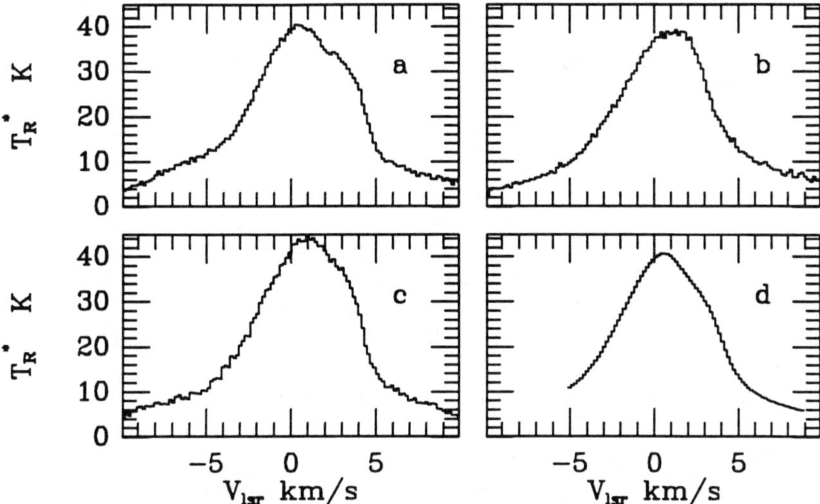

Figure 1. Input data: a, b, c, HI 21 cm sample spectra in the direction of L1780; d, Average of the 209 input spectra

by foreground/background HI is minimized; (2) L1780 has a size of ca. 30 arc min which is small enough so that the background emission does not change too much over the cloud area; (3) the size is on the other hand large enough for mapping of the 21 cm line with a large single dish radio telescope.

In a previous paper Mattila and Sandell (1979) reported on the detection of a 21 cm excess emission component in L1780. Their method of separating the emission from the cloud itself was based on the estimation of an expected background emission profile at each position of the cloud. It was derived from six positions outside the (optical) cloud boundaries.

In the present study we have extended the observational material especially in the surroundings of the cloud to get a better idea of the background emission. In addition, we have applied a new method, based on rotated factor analysis, to separate the background emission and the emission of the cloud.

2. HI 21 cm Input Data

Observations were made with the Effelsberg 100-m telescope in 1975 and 1979. The central part of the cloud has was observed with half beam (4.5) spacing. The velocity resolution was 0.17 km/s. Linear baselines were subtracted (fitted in the velocity interval −43, −23 km/s). The antenna temperatures were scaled according to Williams (1973) using S6 as reference area. The maximum internal error in the data (RMS noise) is less than 5%.

In Figure 1 we present the baseline-fitted HI 21 cm input spectra at three positions (see Figure 3d for locations). The velocity and antenna temperature (T_R^*) range of the spectra are from −10 km/s to 10 km/s and from 0 K to 45 K, respectively.

209 spectra were used in the analysis with velocity range from –5 km/s to +9 km/s corresponding to 82 spectrometer channels. In the 209×82 data matrix the columns stand for the channels so that one row contains one input spectrum as one row vector.

3. Data Analysis

Our task was to reduce the dimensionality of the R^{82} data space resulting from the 82 radiation temperatures as 'variables' given for the 209 positions ('objects'). Factor analysis has been performed on the correlation matrix obtained from the data matrix of the HI radiation temperatures. In this way spectral features have been searched for as deviations from the average spectrum (see Figure 1d).

The factor analysis with varimax rotation (which converged in 17 iterations) resulted in seven main factors (i.e., the corresponding eigenvalues were greater than unity). The first 8 factors have been extracted allowing only eigenvalues greater than 0.5. With these 8 factors the explained cumulative percentage of variance was 90%, i.e., they explain 90% of the 'deviations' from the average. The maximum deviations for T_R^* at the different channels were about 20–40% of the average value at the given channel. Thus with the first 8 factors the input data is explained down to the noise level (ca 5% rms).

Our factors may be considered as artificial spectral line components. The factors are (1×82) vectors consisting of 82 factor loading values corresponding to the 82 channels. The higher the i-th factor loading is in a given factor the stronger this factor contributes to the description of the deviations in T_R^* in the i-th channel. In Figure 2 we show the 1st, 2nd, 5th and 6th normalized factors as spectra.

The 1st factor as a spectral line has a central velocity of about $v_{LSR} = -1$km/s, the 2nd, 5th and 6th factors have central velocities of about 2.5, 4 and 5.5 km/s, respectively.

Factor scores have been calculated for the 8 extracted factors. These act as coefficients measuring the weight of each factor to the excess spectrum (linear combination of the factors) at a given position.

4. Discussion and Results

Performance of the Method The above mentioned computations may be done using standard procedures of statistical software packages available for mini and mainframe computers, and small routines for appropriate data handling. The present method works fast, and is sensitive for small peculiarities. The sensitivity seems to be determined only by the noise in the input data.

The factor analysis performed on the correlation matrix provides more factors with large eigenvalues than the analysis performed on the 'sums of squares and crossproducts' matrix. For radio spectral line type data the data matrix may contain the spectra in its columns; then we perform the analysis on the 'dual space' of attributes. This method was followed by Tóth et al. (1992a) for analysing HI 21 cm spectra. The deviations in this kind of analysis will be

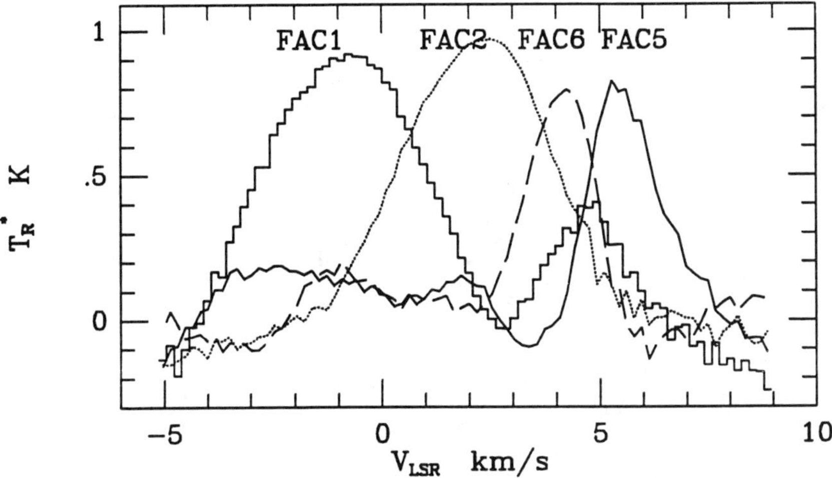

Figure 2. Normalized factors presented as spectra, and features corresponding to their velocities: 1st—background HI main component, 2nd, 5th and 6th—excess emission of L1780

large, and one may get only the major spectral features out of the statistics, as we found with the present data set.

We also tried to apply the non-rotated factor analysis (principal component analysis) on the covariance matrix. The resulting factors were not as Gaussian-looking as they are after rotations.

Distribution of HI towards L1780 The central velocities of the factor spectra and the spatial distribution of the factor scores may be used to identify the explained features with known objects in the studied region.

Both its velocity and its factor score distribution show that the 1st factor describes the general HI background emission, or more precisely, its changing part depending on the galactic latitude (decreasing towards higher b). Some clumpiness is also seen in this background.

The "velocity" of the 2nd, 5th, and 6th components correspond to the velocity of the molecular gas of the L1780 cloud. The measured v_{LSR} for OH, CH and CO lines was about 3.5–3.7 km/s (Mattila and Sandell 1979, Mattila 1986, Tóth et al. 1992b). Besides, only for these factors do the factor scores have large values in the region where the dark cloud is seen. We may conclude that the excess emission of L1780 is described by these factors.

The excess emission spectra for L1780 have been created from linear combinations of the 2nd, 5th and 6th factors with their scores (as weights). The resulting spectral lines for three different positions are shown in Figure 3.

We notice that the average HI spectrum (see Figure 1d) has a contribution also from the cloud. Thus, at the background position we see the cloud in "absorption" in the differential spectrum, and the two spectra from the direction of the cloud do not show the full amplitude of the excess emission line.

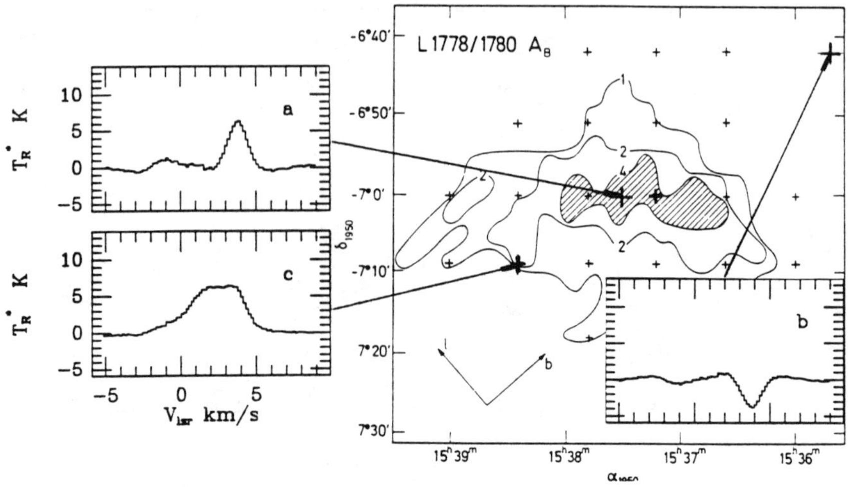

Figure 3. a, b, c: HI excess emission spectra of L1780 (created as linear combination of factors 2, 5 and 6) as they appear at different regions shown on d: A_B contours 1, 2, 4 mag of Mattila (1986).

Comparison with a Previous Study The result of Mattila and Sandell (1979) was quite similar to the present one in the respect that the HI excess peak was offset southwards from the extinction maximum. The shape of the excess emission line was also similar: almost symmetrical at the cloud centre and wider and asymmetrical (with blue shifted wing) at the excess maximum.

Our method provides a more reliable (unbiased) background, and the HI excess obtained is slightly larger than the one derived in the previous study.

Acknowledgments. L. V. Tóth acknowledges the research grants from the MHB *For the Hungarian Science* Foundation and from the NSRC of Finland, and a travel grant of the Peregrinatio ELTE Foundation.

References

Mattila, K., & Sandell, G. 1979, A&A, 78, 264

Mattila, K. 1986, A&A, 160, 157

Murtagh, F., & Heck, A. 1987, Multivariate Data Analysis, Ap. Sp. Sc. Lib. (Dordrecht: Reidel), 13

Tóth, L.V., Balázs, L.G., & Ábrahám, P. 1992, in Astronomical Data Analysis Software and Systems I, A.S.P. Conf. Ser., Vol. 25, eds. D.M. Worrall, C. Biemesderfer & J. Barnes, 251

Tóth, L.V., Haikala, L., Liljeström, T., & Mattila, K. 1992, in prep.

Williams, D.R.W. 1973, A&ASuppl., 8, 505

Astronomical Data Analysis Software and Systems II
ASP Conference Series, Vol. 52, 1993
R. J. Hanisch, R. J. V. Brissenden, and J. Barnes, eds.

The IRAF/NOAO Spectral World Coordinate Systems

Francisco Valdes

IRAF Group, NOAO[1]*, PO Box 26732, Tucson, AZ 85726*

Abstract. The IRAF (Version 2.10) world coordinate systems for dispersion calibrated spectra are presented. In particular the FITS keywords describing the coordinates are defined.

1. Introduction

Within the IRAF[2] environment users need not concern themselves with the details of the world coordinate systems (WCS) for spectra. However, to export the spectral coordinate systems to other software or to import dispersion calibrated spectra into IRAF requires understanding the external representation of the spectral coordinate systems. By this we mean a FITS image header. Because the current IRAF image formats use the same syntax as FITS to store image header information, such as the WCS, this description also applies to IRAF spectral images. The purpose of this paper is to describe the spectral WCS FITS representations in sufficient detail to allow someone to interpret them and possibly implement software outside of IRAF to access them. This description applies to version 2.10.3 though much of it also applies to earlier versions of 2.10.

Due to page limitations and the complexity of the full coordinate systems implementation only the simplest case in which the spectral images have not been modified by any image operators, such as image sections, block averages, etc., is described. A more detailed paper about these coordinate systems may be obtained from the IRAF anonymous FTP account (iraf.noao.edu, iraf/docs/specwcs.ps.Z). Also some relevant material on the IRAF *Mini World Coordinate Systems*, or MWCS, is not covered.

2. Types of Spectral Data

The spectra are stored as images. Images may be one, two, or three dimensional with one axis being the dispersion axis. There are two types of spectral image formats. One type has spatial axes for the other dimensions and the dispersion axis may be along any of the image axes. Typically this type of format is used for

[1]National Optical Astronomy Observatories, operated by the Association of Universities for Research in Astronomy, Inc. (AURA) under cooperative agreement with the National Science Foundation

[2]Image Reduction and Analysis Facility, distributed by the National Optical Astronomy Observatories

long slit (two dimensional) and Fabry-Perot (three dimensional) spectra. This format will be referred to in this paper as *spatial* format.

The second type of spectral image format consists of multiple one dimensional spectra stored in a higher dimensional image with the first image axis being the dispersion axis. This format allows associating many spectra and related parameters into a single data object. This format is called *multispec* format. A special case of this is when all spectra have the same linear dispersion relation in which case a simpler WCS representation is used and the format is called *equispec*. These formats are important since maintaining large numbers of one dimensional spectra as individual images is very unwieldy for the user and inefficient for the software.

Examples of equispec/multispec format are the extracted spectra from a multifiber or multiaperture spectrograph or the extracted orders from an echelle spectrum. The second axis is some arbitrary indexing of the spectra, called *apertures* in IRAF tasks, and the third dimension is used for associated quantities such as a sigma spectrum.

One dimensional spectra, whether from a multispec/equispec image or implicitly extracted from spatial spectra, have several associated quantities which may appear in the image header as part of the coordinate system description. The primary identification of a spectrum is an integer aperture number. A secondary identification is an integer beam number. Since most 1D spectra are derived from an the integration over one or more spatial axes, two additional aperture parameters recorded are the aperture limits in the original data format.

An additional WCS parameter which appears in the description below is a doppler factor. The equispec WCS records this value and also folds it into the WCS coefficients. In contrast, the multispec WCS does not modify the WCS coefficients but applies it separately whenever a wavelength is evaluated. The spatial format WCS does not include a doppler factor.

In addition to individual FITS keywords the WCS also stores information in keyword identified *attribute* strings. These attribute strings are stored as FITS keywords by collecting them into one very long string with the individual attributes beginning with the keyword, followed by an equal sign, and then the attribute string. If the attribute string contains whitespace it is quoted. The long string of attributes is then stored as a series of indexed FITS cards by breaking the string at the end of each card. This provides a maximally efficient storage of the arbitrarily long character string attributes. The FITS card keywords begin with the characters `WAT` followed by an axis number (0 applying to all axes), an underscore, and a sequence number.

Accessing the attributes from the FITS representation in an non-IRAF program can be a challenge since the strings may be arbitrarily long and may occur at any point in the set of keywords. Extracting a single attribute string by name from an OIF (original IRAF image format) in C or FORTRAN programs may be accomplished using an IRAF IMFORT routine available from the author. The user program must still parse the attribute string based on the information given below.

Some common attribute strings are the system name and the axes coordinate types, labels, units, and formats. Axis attributes may occur for each axis in the image. The axes coordinate types also appear in the FITS `CTYPE` keywords.

3. Linear Spectral World Coordinate Systems

When there is a linear or log-linear relation between pixels and dispersion coordinates which is the same for all spectra, the external representation used is simple linear FITS. This applies to one, two, and three dimensional data. The higher dimensional data may have either linear spatial axes such as long slit or Fabry-Perot spectra or the equispec format where each one dimensional spectrum has the same dispersion.

The FITS image header keywords describing the spectral world coordinates are **CTYPEi**, **CRPIXi**, **CRVALi**, **CDELTi**, and **CDi_i** where i is the axis number. The coordinate type has the value **LINEAR**. Equations 1 and 2 define a wavelength in terms of these parameters for axis i and pixel coordinate p. The keyword **DC-FLAG** identifies the dispersion type by a value of 0 for linear or 1 for log-linear sampling. Note that though the log-linear coefficients are defined in log space the IRAF tasks convert to non-log wavelength. For spatial spectra there should also be a **DISPAXIS** parameter identifying the image axis, that is i, along which the dispersion runs. For equispec format the dispersion axis i is always 1.

$$\lambda = \mathtt{CRVAL}i + \mathtt{CD}i\_i \cdot (p - \mathtt{CRPIX}i) \quad (1)$$

$$\lambda = 10^{\mathtt{CRVAL}i + \mathtt{CD}i\_i \cdot (p - \mathtt{CRPIX}i)} \quad (2)$$

4. Multispec Spectral World Coordinate System

The multispec spectral world coordinate systems apply only to one dimensional spectra; i.e., there is no analog for the spatial type spectra. They are used either when there are multiple 1D spectra with differing dispersion functions in a single image or when the dispersion functions are nonlinear.

The multispec coordinate system is always two dimensional though there may be an independent third axis. The two axes are coupled and they both have axis type **multispec**. When the image is one dimensional the line is specified by the dimensional reduction keyword **WAXMAP01** (see the longer paper for more on dimensional reduction). The second, line axis, has world coordinates of aperture number.

The dispersion functions are specified by attribute strings with the identifier *specN* where *N* is the image line. The attribute strings contain a series of numeric fields. The fields are indicated symbolically in (3) and (4).

$$specN = ap\ beam\ dtype\ \lambda_1\ d\lambda\ n_\lambda\ z\ aplow\ aphigh\ [functions] \quad (3)$$

$$function_i = w_i\ \Delta\lambda_i\ ftype\ [parameters]\ [coefficients] \quad (4)$$

The first nine fields in the attribute are common to all the dispersion functions. The first field of the WCS attribute is the aperture number, the second field is the beam number, the seventh is a doppler factor, and the eighth and ninth fields are the aperture limits. These parameters were discussed earlier. The third field, *dtype*, is an integer code with the same function as **DC-FLAG**. A value of −1 indicates the coordinates are not dispersion coordinates (the spectrum is not dispersion calibrated), a value of 0 indicates linear dispersion, a value of 1 indicates log-linear dispersion, and a value of 2 indicates a nonlinear dispersion.

The next two fields are the dispersion coordinate of the first pixel and the average dispersion interval per pixel. For linear and log-linear dispersion types the dispersion is exact while for the nonlinear dispersion functions it is approximate. The next field is the number of valid pixels. It is possible to have spectra with varying lengths in the same image. In that case the image is as big as the biggest spectrum and the number of pixels selects the actual data in each image line.

Following these fields are zero or more function descriptions. For linear or log-linear dispersion coordinate systems there are no function fields. For the nonlinear dispersion systems the function fields specify a weight, a wavelength offset, the type of dispersion function, and the parameters and coefficients describing it. The function type codes, *ftype*, are 1 for a Chebyshev polynomial, 2 for a Legendre polynomial, 3 for a cubic spline, 4 for a linear spline, 5 for a pixel coordinate array, and 6 for a sampled coordinate array. The number of fields before the next function and the number of functions are determined from the parameters of the preceding function until the end of the attribute is reached.

Equation 5 shows how the final wavelength is computed based on the $nfunc$ individual dispersion functions $\Lambda(p)$. Note that this is completely general in that different functions types may be combined. However, in practice when multiple functions are used they are generally of the same type and represent a calibration before and after the actual object observation with the weights based on the relative time difference between the calibration dispersion functions and the object observation.

$$\lambda = \sum_{i=1}^{nfunc} w_i(\Delta\lambda_i + \Lambda_i(p))/(1+z) \tag{5}$$

The multispec coordinate systems define a transformation between pixel, p, and world coordinates, λ. Generally there is an intermediate coordinate system used. The following equations define these coordinates. The polynomial functions are defined in terms of a normalized coordinate, n, as shown in equation 6. The normalized coordinates run between −1 and 1 over the range of pixel coordinates, $p_{\rm min}$ and $p_{\rm max}$ which are parameters of the function, upon which the coefficients were defined. The spline functions map the range into an index over the number of evenly divided spline pieces, *npieces*, which is a parameter of the function. This mapping is shown in equations 7 and 8 where s is the continuous spline coordinate and j is the nearest integer less than or equal to s.

$$\begin{aligned} n &= (p - p_{\rm middle})/(2 * p_{\rm range}) \\ &= (p - (p_{\rm max} + p_{\rm min})/2)/(2 * (p_{\rm max} - p_{\rm min})) && (6) \\ s &= (p - p_{\rm min})/(p_{\rm max} - p_{\rm min}) * npieces && (7) \\ j &= {\rm int}(s) && (8) \end{aligned}$$

4.1. Linear and Log Linear Dispersion Functions

The linear and log-linear dispersion functions are described by a wavelength at the first pixel and a wavelength increment per pixel. A doppler correction may also be applied. Equations 9 and 10 show the two forms. Note that the coordinates returned are always wavelength even though the internal representation and the coefficient values may be log-linear.

$$\lambda = (\lambda_1 + d\lambda \cdot (p-1))/(1+z) \tag{9}$$

$$\lambda = 10^{(\lambda_1 + d\lambda \cdot (p-1))}/(1+z) \tag{10}$$

4.2. Chebyshev and Legendre Polynomial Dispersion Functions

The parameters for the Chebyshev and Legendre polynomial dispersion functions are the *order* (number of coefficients) and the normalizing range of pixel coordinates, $p_{\rm min}$ and $p_{\rm max}$, over which the functions are defined and which are used to compute n. Following the parameters are the *order* coefficients, c_i. Equation 11 shows how to evaluate the function where the x_i are defined iteratively. For the Chebyshev function $x_1 = 1$, $x_2 = n$, and $x_i = 2nx_{i-1} - x_{i-2}$. For the Legendre function $x_1 = 1$, $x_2 = n$, and $x_i = ((2i-3)nx_{i-1} - (i-2)x_{i-2})/(i-1)$.

$$\Lambda = \sum_{i=1}^{order} c_i x_i \tag{11}$$

4.3. Linear and Cubic Spline Dispersion Functions

The parameters for the linear and cubic spline dispersion functions are the number of spline pieces, *npieces*, and the range of pixel coordinates, $p_{\rm min}$ and $p_{\rm max}$, over which the functions are defined and which are used to compute the spline coordinate s. Following the parameters are the *npieces* + 1 (linear spline) or *npieces* + 3 (cubic spline) coefficients, c_i. The coefficients used are selected based on the spline coordinate. The fractions of the interval between the nearest integer spline knots are given by a and b, $a = (j+1) - s$, $b = s - j$. The x_i for the linear spline are $x_0 = a$, and $x_1 = b$ and for the cubic spline are $x_0 = a^3$, $x_1 = (1 + 3a(1+ab))$, $x_2 = (1 + 3b(1+ab))$, and $x_3 = b^3$.

$$\Lambda = \sum_{i=0}^{1|3} c_{i+j} x_i \tag{12}$$

4.4. Pixel and Sampled Array Dispersion Functions

The parameters for the pixel and sampled array dispersion function consists of just the number of coordinates *ncoords*. For the pixel array there are then the wavelengths at the integer pixel coordinates. The sampled array starts with a dummy field and then coordinate and wavelength pairs in increasing order of pixel coordinate. To evaluate the functions the nearest integer pixel coordinates are determined and a linear interpolation between the wavelengths is computed.

Astronomical Data Analysis Software and Systems II
ASP Conference Series, Vol. 52, 1993
R. J. Hanisch, R. J. V. Brissenden, and J. Barnes, eds.

The IRAF Radial Velocity Analysis Package

Michael J. Fitzpatrick

IRAF Group, NOAO[1], PO Box 26732, Tucson, AZ 85726

Abstract. The IRAF Radial Velocity Analysis package is described and future plans are presented. A discussion of the current strengths and weaknesses of the package is given and future plans for new tasks and algorithms are also described. An overview of the cross-correlation task and it's many features is presented, along with a simple test of the package comparing the various fitting options.

1. Introduction

The RV package was first released as part of the IRAF[2] V2.10 system earlier this year. Prior to that it was released, in two separate forms, as an external package while the algorithms were being refined. Presently the package consists of a Fourier cross correlation task and several parameter sets to support the various features of the package.

Like other radial velocity software in use today, the cross correlation algorithm, which is implemented as the **fxcor** task, is that of Tonry and Davis (Tonry & Davis 1979) and is well understood. As part of the widely used IRAF package, the **fxcor** task has many advantages as a general purpose radial velocity task:

- Ability to process arbitrary lists of one or two dimensional spectra.
- Fit blended correlation peaks when doing multiple star work.
- A choice of six fitting functions. Some functions permit a weighted fit to the selected points.
- The ability specify a search window for velocity peaks.
- A choice of six output format options, including any combination of short or long tabular data, verbose (human readable) output, and a graphical summary of the correlation.

---

[1]National Optical Astronomy Observatories, operated by the Association of Universities for Research in Astronomy, Inc. (AURA) under cooperative agreement with the National Science Foundation

[2]Image Reduction and Analysis Facility, distributed by the National Optical Astronomy Observatories

- Interactive editing of task parameters. Being able to set parameters for a specific data set on the fly and then processing the list with those parameters.
- FFT data filtering. Four filter functions are supported.
- Review of input spectra both before and after filtering.
- Full access to IRAF MWCS image headers.
- Multiple, independent correlation sample regions.
- Automatic (or interactive) continuum subtraction.

While the current package may not meet the needs of all researchers, favorable results have thus far been obtained for a wide variety of applications. The generality and portability of the package allow researchers to use the same software in many areas of radial velocity work.

2. Overview

The **fxcor** task was designed with the idea that not everyone knows the best parameters for a given dataset until they've had a chance to experiment. The many commands in this task allow the user to "tweak" various options until they provide the best results for the given data. The user is allowed to save any changes or unlearn them all and start from the defaults, and once satisfied can process the remainder of the input lists in batch mode with those parameters.

The **fxcor** task is fully interactive and allows the user to view the data in many different ways. At task startup the user is presented with a plot of the cross-correlation function (CCF) and the fit based on the initial parameters (see Figure 1). A search window (specified in velocity units if dispersion header keywords are present) may be given to limit the range of acceptable velocities or pixel shifts. It is possible to change the fit parameters or even select a new peak to be fit by selecting it from a plot of the whole CCF with the cursor. The user may move through lists of spectra (either images or echelle/multispec apertures) quickly with a simple keystroke or randomly using colon commands.

Results may be saved automatically, however only the last correlation for the pair is saved to eliminate multiple entries in the output files. A text summary of the correlation containing additional information on the fit quality and velocity parameters may be viewed before saving the results to the output logs. The task may also be used for correlating images with no dispersion information in the image headers, the output will simply reflect the pixel shift of the data. A package parameter may be set to specify a threshold used to determine whether the output velocities are in units of km/s, or the more useful z value for high redshift objects.

Three independent sub-modes of operation are also supported: The *Spectrum Mode* allows the user to view the raw, prepared or filtered spectra as well as interactively set the correlation sample regions. The *Fourier Mode* lets the user select filtering parameters and view various properties of the FFT. Finally, a *Continuum Mode* exists to permit interactive continuum fitting and subtraction.

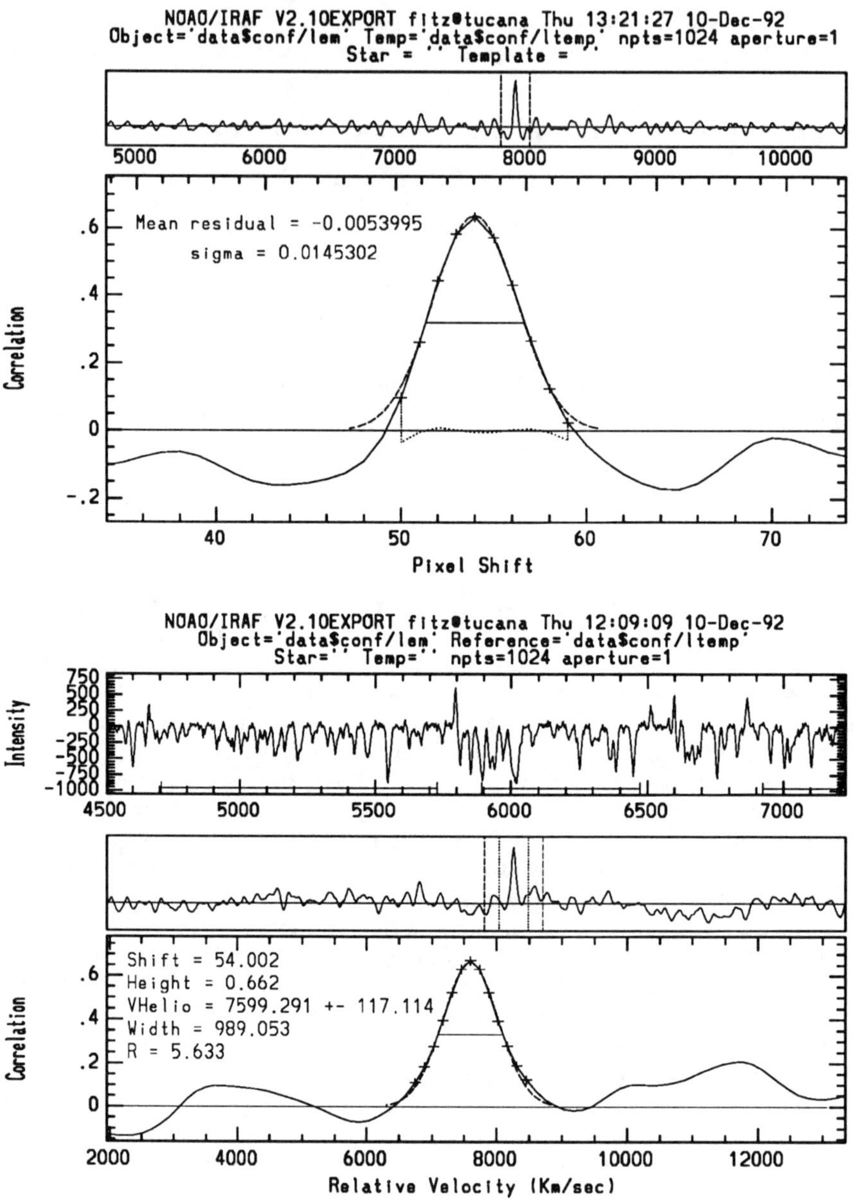

Figure 1. The top figure shows initial correlation plot and fit to the peak. The user may select a new peak from the top plot of the whole CCF, or select new fit points from the bottom plot. The optional residual plot of the fit is also shown. The bottom figure shows the (optional) summary plot that may be generated for each correlation. The bars on the spectrum plot indicate the sample regions used.

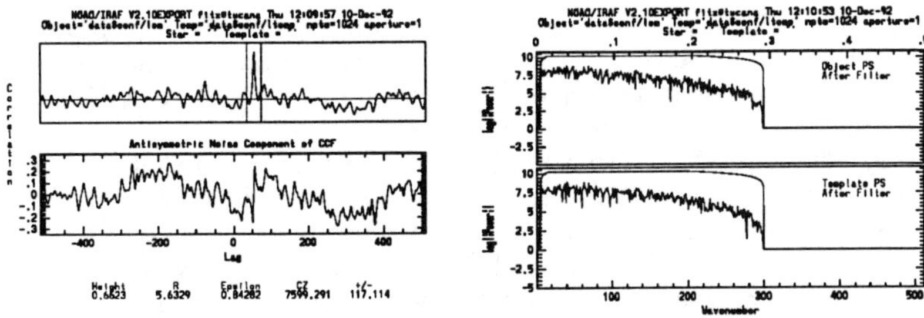

Figure 2. The left hand plot shows the antisymmetric noise component plot which can be used to judge the correlation quality. The right hand plot is one of several available in the Fourier Mode. In this case we see the filtered object and template power spectra.

3. Plotting Capabilities

One of the design goals of the **RV** package was a fully interactive task. To this end there exist a wide variety of plots with which the user may view the data and the correlation. The task makes use of multiple plots where possible to convey as much information to the user in the most intuitive manner.

The dual plots in the Fourier and Spectrum mode allow the user to see the same plot for both the object and template without the confusion of overlaying vectors. Cursor commands, such as those for specifying a sample region or finding the period in an FFT spectrum, are sensitive to whether the user is referring to the top or bottom plot and will affect the object or template parameter accordingly. Even within a command mode, the command is checked to see whether it is appropriate for the current screen. Figures 1 and 2 show some of the plots currently available.

4. An Example of Testing Accuracies

Since the results obtained for a given correlation depend on a half dozen fitting parameters, continuum and fitting parameters, dispersion and noise characteristics of the data, etc., a full analysis of the package is a difficult undertaking. As an example we present the results obtained using 30 noiseless spectra with velocities ranging from 1000 km/s to 1290 km/s. The **fxcor** task parameters were all left at their defaults. The wavelength range of the data was fixed, the dispersion was changed in the four cases to alter the sampling of the CCF peak.

Fitting Function	Δ_V=130.91 km/s npts=512	Δ_V=87.22 km/s npts=768	Δ_V=65.39 km/s npts=1024	Δ_V=52.22 km/s npts=1282
Gaussian	1.315916 0.010052	3.417086 0.039177	2.055306 0.031431	1.262360 0.024173
Lorentzian	1.260320 0.009627	3.453826 0.039599	2.061913 0.031532	1.263450 0.024194
Parabola	1.527330 0.011667	3.328400 0.038160	2.043076 0.031244	1.259486 0.024118
Center1d	1.676123 0.012803	3.592980 0.041194	2.121523 0.032444	1.268426 0.024290
Sinc	1.298756 0.009920	3.830740 0.043920	3.436960 0.052560	6.393450 0.122432

The numbers in each column represent the mean difference between the known velocity and the computed velocity for each fitting function over the 30 spectra. The number on the left of each column is the velocity difference, the number on the right is the difference in pixels.

While these results are encouraging, it should be noted that *no* effort was made to tune the parameters to the data. Doing so could have improved these results some (reproducible accuracies of one one-hundredth of a pixel have been reported), the wrong parameters could have just as easily have worsened them. The user is encouraged to experiment with various parameter settings before deciding on which is correct for his/her data.

5. Future Plans

Much work still needs to be done to provide a package that meets *all* of the needs of astronomers doing radial velocity work. It is hoped that in the future development of this package some of the following features may be added:

- A Fourier Quotient (or Fourier Difference) task.
- Improved support for fitting blended lines (binary star velocities).
- Computation of rotational velocities.
- Tracing of emission lines for velocity information in Longslit data.
- A graphical interface for specifying FFT filter parameters.
- Computation of dispersion velocities within galaxies.
- A tool for automatically making catalogue from output tables.
- A Wiener filter option and phase plot.

References

Press, W.H., et al. 1986, Numerical Recipes (Cambridge Univ. Press, Cambridge), Ch. 12

Rabiner, L.R., & Gold, B. 1975, Theory and Application of Digital Signal Processing (Prentice Hall, Englewood Cliffs), Ch. 3

Tonry, J., & Davis, M. 1979, AJ, 84, 1511

Weiss, W.W., et al. 1978, A&A, 63, 247

Willmarth, D.W., & Abt, H.A. 1985, "Radial Velocities From CCD Detectors" in IAU Coll. No. 88, Stellar Radial Velocities, 99

Wyatt, W.F. 1985, "The CfA System for Digital Correlations" in IAU Coll. No 88, Stellar Radial Velocities, 123

Part 5. Data Analysis Applications

Section C. Other Algorithms and Techniques

Astronomical Data Analysis Software and Systems II
ASP Conference Series, Vol. 52, 1993
R. J. Hanisch, R. J. V. Brissenden, and J. Barnes, eds.

PHOTCAL: The IRAF Photometric Calibration Package

L. E. Davis

NOAO/IRAF Group, Tucson, AZ 85726

P. Gigoux

NOAO/CTIO, La Serena, Chile

1. Introduction

PHOTCAL is an IRAF package designed to derive and evaluate the transformation from the instrumental photometric system to the standard photometric system. PHOTCAL contains tasks for: 1) creating and/or editing the input standard star catalog and observations files, 2) creating, checking and editing the input configuration or setup file, 3) solving the transformation equations interactively or non-interactively using non-linear least-squares techniques, and 4) applying the transformation equations to the observations.

2. The Standard Star Catalog and Observations Files

The standard photometric indices for a set of standard stars are stored in a catalog file, which must contain only one entry per star. The instrumental photometric indices of a set of standard and/or program stars are stored in an observations file, which may contain any number of entries per star. Catalog and observations files are multi-column text files whose columns are delimited by whitespace. The first column is reserved for the star name, which is used to match the standard star catalog and observations file entries, but the remaining columns may contain data of any type, e.g., positions, magnitudes, colors, errors, airmass, or time. Portions of simple catalog and observations files are shown in Tables 1 and 2 respectively.

PHOTCAL maintains a library of standard star catalogs in a default standard star catalog directory, which is always searched by the PHOTCAL tasks prior to the current directory. The catalog directory may be reconfigured to point to a local site or personal catalog directory. PHOTCAL catalog and observations files may be typed in by hand, prepared automatically from IRAF/APPHOT, IRAF/DAOPHOT, or outside package output catalogs using PHOTCAL tasks and/or other IRAF list processing tools, or since their format is so simple, written with a user computer program.

3. The PHOTCAL Configuration File

The PHOTCAL configuration file is a text file written by the user in the PHOTCAL mini-language. The configuration file is divided into three sections: 1)

Table 1. Sample Catalog File

# ID	V	error(V)	BV	error(BV)	UB	error(UB)
105-405	8.309	0.004	1.521	0.001	1.905	0.007
105-411	10.620	0.014	0.950	0.010	0.620	0.008
105-257	9.140	0.003	0.490	0.013	0.020	0.008
...	...	...	...	...	...	...

Table 2. Sample Observations File

# ID	FILTER	AIRMASS	XCENTER	YCENTER	MAG	MERR
105-405	u	1.276	481.39	357.19	18.683	0.009
*	b	1.270	480.57	360.07	14.919	0.005
*	v	1.265	477.07	358.62	13.212	0.002
105-411	u	1.276	507.69	128.53	19.144	0.014
*	b	1.270	507.06	131.44	16.612	0.020
*	v	1.265	503.42	130.29	15.487	0.008
105-257	u	1.305	470.72	393.68	16.675	0.005
*	b	1.315	469.71	396.22	14.743	0.004
*	v	1.320	466.58	397.27	14.030	0.004
...	.	...	...	...	...	...

Table 3. Sample Configuration File

```
catalog

V            2   # V magnitude
error(V)     3   # error in V magnitude
BV           4   # B-V color
error(BV)    5   # error in B-V color
UB           6   # U-B color
error(UB)    7   # error in U-B color

observations

Xu           3   # airmass in filter u
mu           6   # instrumental magnitude in filter u
error(mu)    7   # magnitude error in filter u

Xb           9   # airmass in filter b
mb          12   # instrumental magnitude in filter b
error(mb)   13   # magnitude error in filter b

Xv          15   # airmass in filter v
mv          18   # instrumental magnitude in filter v
error(mv)   19   # magnitude error in filter v

transformation

fit u1 = 0.0, u2 = -.07, u3 = 0.70
fit b1 = 0.0, b2 = -.06, b3 = 0.30
fit v1 = 0.0, v2 = 0.05, v3 = 0.20
UFIT: mu = V + BV + UB + u1 + u2 * UB + u3 * Xu
BFIT: mb = V + BV + b1 + b2 * BV + b3 * Xb
VFIT: mv = V + v1 + v2 * BV + v3 * Xv
```

the catalog section which assigns variable names and associated errors to the columns of the catalog file, 2) the observations section which assigns variable names and associated errors to the columns of the observations file, and 3) the transformation section which assigns names to and defines the transformation equations, assigns names and initial values to the parameters to be fit or held constant, and, optionally, assigns derivative, weight, error and/or default plot expressions to each equation to be fit. A simple configuration file for calibrating UBV CCD photometry is shown in Table 3.

4. The PHOTCAL Mini-language

The configuration file consists of a series of instructions written by the user in the PHOTCAL mini-language and parsed by the PHOTCAL parser. The PHOTCAL parser was generated using a yacc grammer and parser generator. The PHOTCAL mini-language elements are numerical constants, identifiers, arithmetic operators, arithmetic expressions, and comment statements.

Numerical constants are decimal integers or floating point numbers. Double precision and complex numbers are not presently supported. The INDEF constant is not supported, although it is permitted in the input data.

An identifier (keyword, name, label, function) is a word consisting of an upper or lower case letter, followed by zero or more upper or lower case letters or digits. Keywords are identifiers with special meaning to the PHOTCAL parser. For example the three identifiers "catalog", "observations", and "transformation" are used to declare the beginning of the catalog, observations, and transformation sections of the configuration file respectively. There are currently eleven reserved keywords in the PHOTCAL mini-language: "catalog", "constant", "delta", "derivative", "error", "fit", "observations", "plot", "set", "transformation", and "weight". Names are variables that have been declared in the catalog or observation sections, or declared as parameters or temporary variables in the transformation section of the configuration file. Labels are identifiers which name an equation. Labels are used to tell the parser which transformation equation, the optional derivative, weight, error or plot expressions belong to. Functions are built-in mathematical functions that can be used in expressions. The supported functions are: "abs", "acos", "asin", "atan", "cos", "exp", "log", "log10", "sin", "sqrt", and "tan".

Expressions are arithmetic FORTRAN expressions containing variables, parameters, and one or more of the the operators +, −, *, /, **.

Comments are statements preceded by a #, and may occur anywhere in the configuration file.

5. Equation Fitting and Evaluation

The photometric transformation equations are solved using non-linear least-squares techniques. Any number of photometric transformation equations containing any number of variables may be solved. The photometric transformation equations may defined, fit, and evaluated in the forward sense (the standard indices are a function of the instrumental indices) or the inverse sense (the instrumental indices are a function of the standard indices). The left and right sides

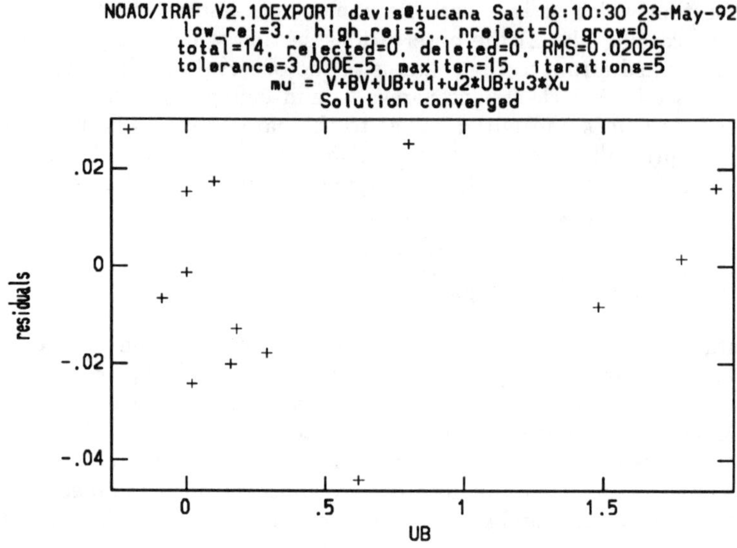

Figure 1. The fit residuals plotted versus U–B color.

of the transformation equations may be any FORTRAN arithmetic expression containing variables, parameters, constants, the operators +, −, /, *, and **, or the built-in functions listed in the previous section. Optional expressions for computing the derivatives with respect to the parameters, weights, and errors may be associated with each transformation equation.

The photometric transformation equations can be fit either interactively or non-interactively. After computing the initial fit for an equation in interactive mode, the user may examine and/or interact with the fit by, examining the various default graphs, reprogramming the default graph keys, editing the default convergence or data rejection parameters, deleting and undeleting points, altering which parameters in the fitting function are to be fit and which are to be held constant, and refitting the data. A sample PHOTCAL plot of the fit residuals versus U–B for equation UFIT in Table 3 is shown in Figure 1.

6. Using Photcal for General Function Fitting

Although PHOTCAL is optimized for solving photometric transformation equations, the PHOTCAL tasks can also be used to fit general non-linear functions of n variables, with or without standard catalog matching. Figure 2 shows a plot of a list of x and y values, which have been fit with a gaussian function, using the PHOTCAL fitting task.

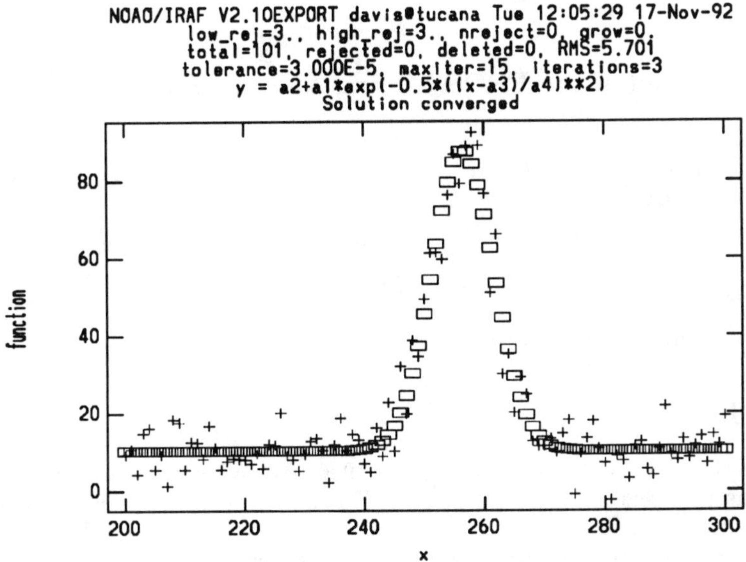

Figure 2. Plot of the observed and fitted gaussian function.

7. Conclusions

Future plans for the package include increasing the flexibility and programmability of the interactive plotting code, and providing support for combining, and fitting simultaneously, data taken on more than one night.

PHOTCAL is installed in IRAF 2.10. An on-line tutorial document is provided and there is on-line help for all the package tasks. More detailed information on using the package, especially in the context of doing APPHOT and DAOPHOT reductions, can be found in, "A User's Guide to Stellar CCD Photometry with IRAF", by P. Massey and L. E. Davis, available from the IRAF network archive.

Acknowledgments. One of us (LED) wishes to thank P. Massey and M. Fitzpatrick for testing various versions of PHOTCAL, and P. Massey for critical comments which greatly improved the package.

Astronomical Data Analysis Software and Systems II
ASP Conference Series, Vol. 52, 1993
R. J. Hanisch, R. J. V. Brissenden, and J. Barnes, eds.

Verification of the PROS Timing Analysis Package

K. R. Manning, M. A. Conroy, J. DePonte, J. F. Moran, F. A. Primini, F. D. Seward

Smithsonian Astrophysical Observatory, Cambridge, MA 02138

B. Aschenbach

MPE, Garching bei München, Germany

1. Introduction

The **IRAF/PROS** analysis package (Tody 1986; Worrall et al. 1992) contains two tasks used to search for periodicities in X-ray data. One is the **FFT** (Fast Fourier Transform) and the other is the epoch folding task **PERIOD**. These tasks were originally tested with a short *Einstein* observation of the supernova remnant CTB 109. This remnant contains a bright central X-ray source which is pulsating with a period of 7 s. The power is almost totally in the second harmonic, so the apparent period is 3.5 s. A continuous data sequence of 1,800 s was used to check the task. The period was easily detected, and the time required to run the tasks was short. The **FFT** required only 2^{11} bins (bin length = 0.88 s) to search for periods longer than 1.8 s.

Many pulsars of astrophysical interest have very short periods, extending down to about 1 ms. Furthermore, the detection of a weak source requires a long observation so that enough source counts are present to make analysis possible. The combination of a short period and a weak source requires a more demanding analysis than the pulsar in CTB 109.

A greater precision in the timing of data events is also necessary. Short periods require corrections to the satellite clock time for the motion of the satellite and Earth. ROSAT observations of two known fast X-ray pulsars have been used to test the **PROS** timing tasks. These results demonstrate the tasks' capabilities for both a very strong and a weak signal.

2. Analysis of the Pulsar in the Crab Nebula

A 4-day ROSAT observation of the Crab Nebula contains the central 33 ms (30 Hz), bright X-ray pulsar. A subset of the data was created by isolating the region immediately surrounding the pulsar from the rest of the nebula. The resulting data set contains a signal which is 90% pulsed with a strength of 15 cts/s. Figure 1 shows an **FFT** power spectrum for a 160 s interval of the data set. Seven harmonics appear above the noise. Most of the power is in the second harmonic because this pulsar produces two pulses per cycle. The non-sinusoidal pulse shape also causes significant power to be distributed to higher harmonics. The seventh harmonic is actually above the Nyquist frequency of 205 Hz, but an alias appears at 200 Hz.

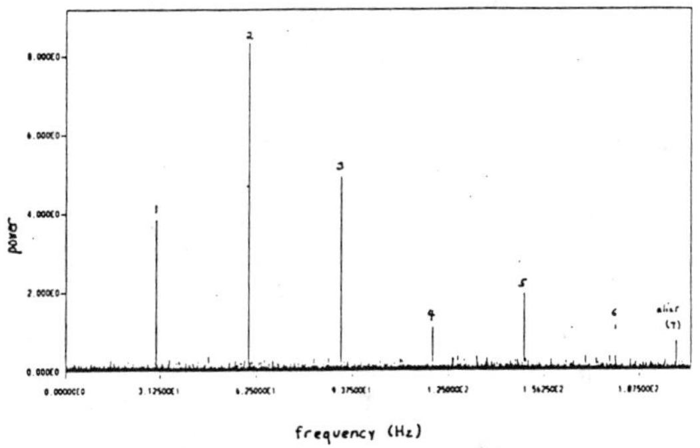

Figure 1.
FFT Power Spectrum of Crab Pulsar.

The Crab Pulsar data tests the validity of the ROSAT data itself and demonstrates the timing tasks' response to a strong signal. The data set was divided into six intervals, and the epoch folding program was used to find the value of the period in each interval. This object is an isolated pulsar and the period is slowly increasing with time as the neutron star loses rotational energy. Although the increase in the period, $\dot{P}$, is very small, the expected change over this four-day span was easily observed. Folding was performed on each of the six intervals using the best value of the period for that interval. The resulting pulsar light curves had exactly the pulse shape observed in previous missions.

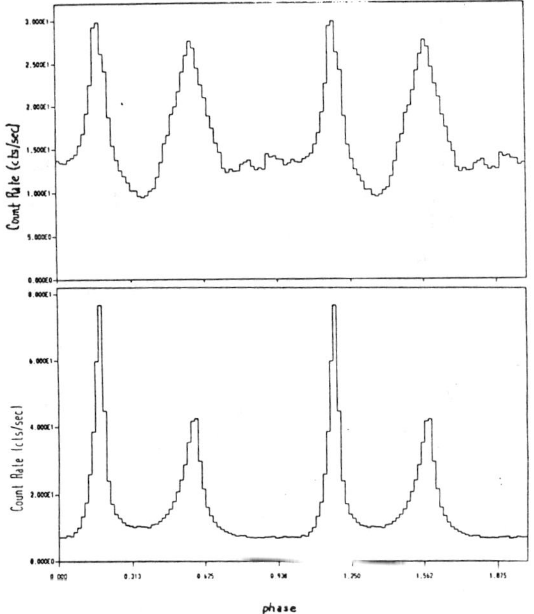

Figure 2.
Crab Pulsar Light Curve.
$\dot{P} = 0$ and
$\dot{P} = 4.22 \times 10^{-13}$.

Event timing of this observation is accurate to 0.3 ms and there is no evidence for erroneous timing in any events. The complete data set was folded to determine the period precisely. These results are very sensitive to the value of $\dot{P}$ used. Figure 2 shows the epoch folding result with $\dot{P}$*=0*, and the result using the correct value of $\dot{P}$. Using this four-day ROSAT observation, the period of the Crab Pulsar can be determined to one part in 10^8 and the value of $\dot{P}$ to one part in 100.

3. Analysis of the Pulsar PSR 0540-69

A distant supernova remnant in the Large Magellanic Cloud contains a weak 50 ms X-ray pulsar, PSR 0540-69. This object was observed by ROSAT in June of 1990. Figure 3 shows the sequence of good data intervals and gaps in the 32-hour data sequence. This data set represents a typical ROSAT observation, and provides a more practical test of the timing tasks. The count rate is only 0.15 cts/s, and the sequence is dominated by large gaps separating short data intervals. The entire data set must be used for the analysis. A single Fourier transform of this interval, sensitive to periods as short as 50 ms, requires 2^{22} bins (bin length = 0.027 s).

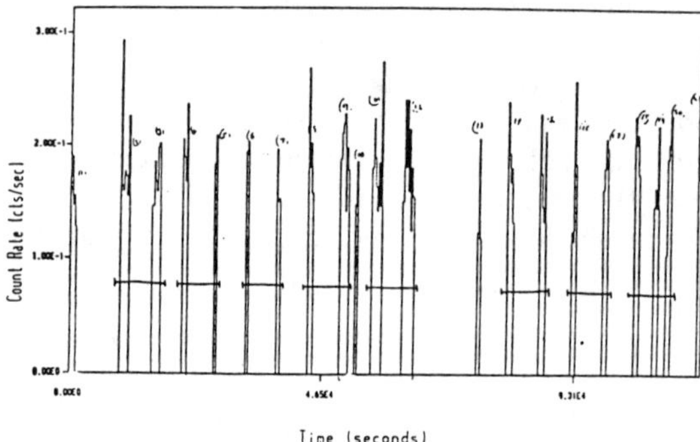

Figure 3.
PSR 0540-69, good-data intervals and gaps.

In the usual timing analysis, the **FFT** is used for the initial search of periodicities in the data set. The epoch folding program is then used to investigate small frequency ranges where the **FFT** has found large power coefficients. It is, of course, possible to search for a period using the **PERIOD** task, but this technique is time consuming. If the period of a pulsar is well known, the tasks are easily used to measure the pulsar's signal. This has been done for PSR 0540-69. If the period is not known, it is much more difficult to find and detect the pulsar.

The weak signal from PSR 0540-69 puts this object at the limit of detection. This pulsar is also slowing with time and Figure 4 shows the improvement when the correct value of $\dot{P}$ (which is well known) is used. The pulsar light curve is sinusoidal.

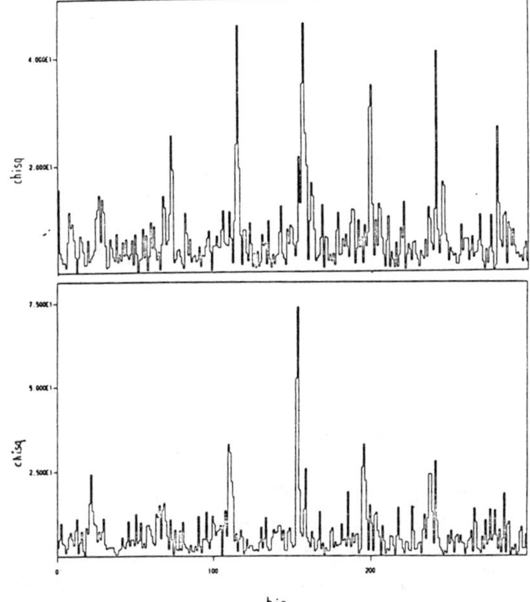

Figure 4.
PSR 0540-69 Epoch folding with $\dot{P} = 0$ and $\dot{P} = 4.79 \times 10^{-13}$.

A single, coherent, 2^{22} bin **FFT** transform was performed on the data stream. A small fraction (10^{-4}) of the resulting power spectrum is shown in Figure 5. The signal can be found at the location of the period. However, even in this small data set, there are frequencies with higher power coefficients than the pulsar period. If the period were unknown, it could not be identified by this technique.

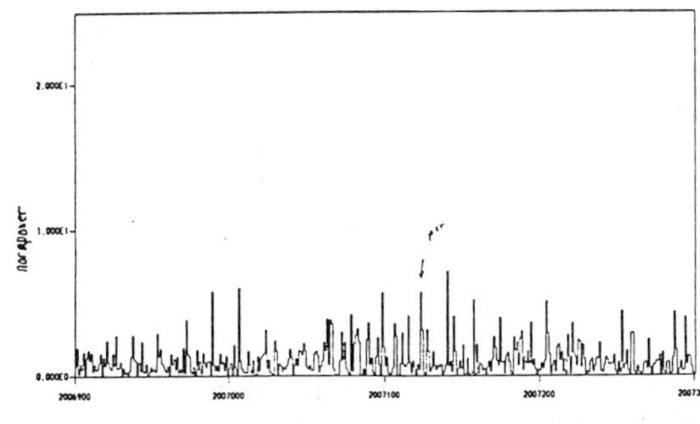

Figure 5.
Part of coherent FFT power spectrum PSR 0540-69.

One reason for the small power coefficient in the pulsar signal is that the frequency changes during the observation. To avoid this difficulty, the data set was divided into eight intervals, and an **FFT** performed on each interval. The individual power spectra were then added. Figure 6 again shows a small fraction (10^{-2}) of the summed **FFT** power spectrum resulting from this analysis. The signal to noise ratio is much better, but outside this illustration, there are a few higher power coefficients which are noise. The pulsar could be located and identified by using first a Summed **FFT** and then the epoch folding **PERIOD** to investigate the highest power coefficients.

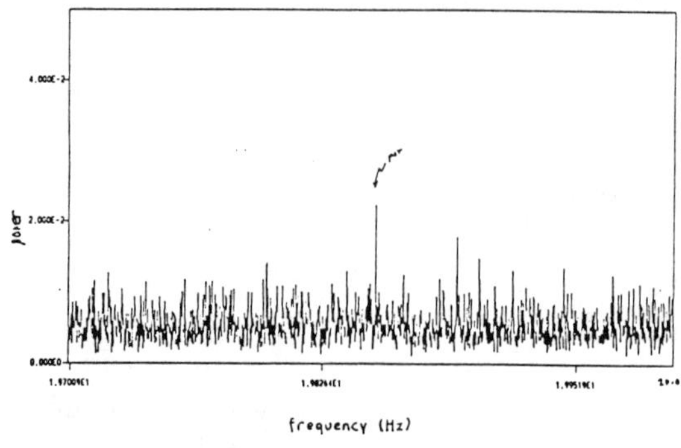

Figure 6.
Part of summed FFT power spectrum PSR 0540-69.

4. Conclusion

The timing analysis results do support the integrity of the ROSAT data, as well as the functionality of the **PROS** timing tasks. The analysis of the two ROSAT observations provided a more rigorous test of the **PROS** timing package than the original 3.5 s pulsar data, and prompted the modification of some of the tasks. The analysis of the fast and weak pulsars still requires a significant amount of temporary disk space for intermediate files, and is not fully optimized. However, the number of bins (and hence the brevity of the period) that can be analyzed by the **FFT** task is limited only by disk space.

Although the **PROS** timing tasks were developed for the analysis of *Einstein* and ROSAT data, many of the tasks can be used on data from other sources. The **FFT** task can be run either on an **IRAF/QPOE** event list, or on a simple light curve table. Any ASCII light curve table can be easily converted into **IRAF** table format for input to the **FFT** task. The remaining timing tasks can run on any **IRAF/QPOE** file that has a 'time' attribute defined. Only the **TIMCOR** package, which calculates the time corrections, is mission-specific.

Acknowledgments. **PROS** is partially supported by NASA contracts NAS5-30934 (RSDC) and NAS5-30751 (*Einstein*).

References

Tody, D. 1986, in Instrumentation in Astronomy VI, SPIE, 627, part 2

Worrall, D.M., Conroy, M., DePonte, J., Harnden, Jr., F.R., Mandel, E., Murray, S.S, Trinchieri, G., VanHilst, M., Wilkes, B.J., et al. 1991, in Data Analysis in Astronomy IV, eds. V. Di Gesu et al., Plenum Press, 1992, 145

Astronomical Data Analysis Software and Systems II
ASP Conference Series, Vol. 52, 1993
R. J. Hanisch, R. J. V. Brissenden, and J. Barnes, eds.

Constraining Galactic Structure Parameters from Multivariate Density Estimation

B. Chen, M. Crézé

Observatoire de Strasbourg, F-67000,strasbourg, France (cdsxb2::chen, cdsxb2::creze)

A. C. Robin, O. Bienaymé

Observatoire de Besançon, BP1615,F-25010 Besançon Cedex, France (obsbea::annie, obsbea::olivier)

Abstract. Galactic structure studies involve the manipulation of multivariate catalogues and their comparison with sophisticated synthesis models. Model parameter fitting must deal with multivariate data analysis while most existing methods use a binning procedure applied to one or two parameters only to extract informations. In this paper, we describe how to use multivariate density estimation to constrain the physical model parameters without binning, to detect systematic shifts between data and model and to accurately estimate the density of each population. This method can also be applied to a wide variety of astrophysical problems involving multivariate data.

1. Introduction

Over the last two decades, complete samples of stars have been obtained with a good accuracy in photometry, astrometry and radial velocities over sufficiently large areas of sky that random errors due to counting statistics are unimportant. On the other hand, the main features of Galaxy properties have been identified. It has been therefore possible to design synthetic models of galactic stellar populations.

However, in the field of fitting star counts, most authors compare the model and the observations by binning the data to a histogram of counts as a function of the variables of interest (e.g., B-V), in a limited area of the sky. Such a blind fitting by visual method is oversimplified and uses a small part of the information available. We need to perform a global analysis to the multivariate star count samples including all available observable simultaneously.

2. Methodology

2.1. Multivariate Density Definition

Presently available star count data provide properly calibrated measurement including magnitude, colour and proper motions for extended deep probes and so may strengthen the constraints in Galaxy models. Multivariate statistical analysis, such as cluster analysis, discriminant analysis, principal components analysis, and nonparameter density estimation have been successfully used (Crézé et al. 1991; Robin & Chen 1992) to search for meaningful features in the five-dimensional space of observables between observed and simulated samples generated from a synthetic approach of galaxy modeling (Robin & Crézé 1986; Bienaymé et al. 1987). In this paper, we describe how to use multivariate density estimations to constrain the model parameters.

In a multidimensional space of observables, the density distribution at a point $\vec{x}$ can be estimated using a spherical gaussian filter (Hand 1981):

$$\hat{P}(\vec{x}) = \frac{1}{n}\sum_{i=1}^{n}\frac{1}{h(2\pi)^{d/2}}e^{-\frac{1}{2}(\frac{\vec{x}-\vec{x}_i}{h})^2} \tag{1}$$

where n is the number of stars in the sample, $\vec{x}_1$, ... $\vec{x}_n$, is the given multivariate data set whose underlying density is to be estimated, h is the window width and d represents the number of observable, therefore the dimension of $\vec{x}$, let $\hat{\theta}(\vec{x})$ be the multivariate density in a local neighbourhood L of $\vec{x}$, then we have $\hat{P}(\vec{x}) = \hat{\theta}(\vec{x})/\mathrm{n}$.

2.2. Fitting Procedure

Galactic structure studies involve the manipulation of multivariate catalogues and their comparison with sophisticated synthesis models. Model simulated catalogues have been created according to the same selected criteria. Intrinsic as well as observed properties of each simulated star are known. At first, we classify simulated stars in several groups according to their physical properties. For example groups can be formed with different populations or age groups (disc, halo,...) and/or luminosity class (dwarfs, giants). One can define the density of each physical group ω_m in the simulated catalogue by $\hat{\theta}(\vec{x}|\omega_m)$ and the total observed density $\hat{\theta}(\vec{x})$ at each observed point $\vec{x}$ by equation 1. To fit the model parameters to the data we suppose that the observed density of stars in each point of the 5-D space can be fitted by a linear combination of the densities of the model groups. Let α_{ω_m} be the coefficient to apply to the group ω_m in order to fit the data:

$$\Sigma\alpha_{\omega_m}\hat{\theta}(\vec{x}|\omega_m) = \hat{\theta}(\vec{x}) \tag{2}$$

This equation must be valid at each point $\vec{x}$ leading to a system of n equations if n is the number of stars in the observed catalogue.

Suppose $\hat{P}(\omega_m)$ is a prior probability density of ω_m, by Bayes rules, the joint probability density can be expressed as :

$$\hat{P}(\omega_m|\vec{x}) = C \times \hat{P}(\vec{x}|\omega_m)\hat{P}(\omega_m)$$

where C is a normalization. These probabilities are functions of $\vec{x}$ and have to be normalized to unity:

$$\Sigma_{\omega_m=1}^{K}\hat{P}(\omega_m|\vec{x}) = 1 \tag{3}$$

We derive weighted means and the convariance matrix for each group as follows:

$$\overline{X_l^{\omega_m}} = \frac{\sum_{i=1}^{n} \hat{P}(\omega_m|\vec{x_i})x_{il}}{\sum_{i=1}^{n} \hat{P}(\omega_m|x_i)} \tag{4}$$

$$\sigma_{lj}^{\omega_m 2} = \frac{\sum_{i=1}^{n} \hat{P}(\omega_m|\vec{x_i})(x_{il} - \overline{x}_l)(x_{ij} - \overline{x}_j)}{\sum_{i=1}^{n} \hat{P}(\omega_m|\vec{x_i})} \tag{5}$$

where the weight is the posterior probability ($\hat{P}(\omega_m|\vec{x})$), l,j are two axes of the multi-dimensional space. Equations 4 and 5 describe the statistical properties of observed stars in each group. We compare them with the one of the model catalogues, any disagreement between the model and the data being interpreted as erroneous physical properties of the model groups.

3. Monte Carlo Simulations

In order to test the reliability of the method it has been applied to simulated 2-dimensional data containing three physical groups, in which the observables have multivariate normal distributions. We also produce a set of simulated data that will be used as "observed" data to compare with. (Hereafter the term "observed" refers to this second set of simulations). They differ from the model data by the number of stars, the mean and standard deviation of each physical group. Our goal is to detect these differences by the multivariate density estimation method.

The process has been applied in two steps:

1. Consider the two-dimensional case (X,Y). In Table 1, the input parameters and the derived parameters are shown on the left side and the right side respectively. C is the mean, σ the standard deviation, and α the ratio of number of stars between the model and the "observed" data in each group. Some differences (italic) between the model data and the "observed" data have been assumed in the input parameters. Both the model data and "observed" data have been generated one hundred times by Monte Carlo simulation. For the model the derived parameters are computed by averaging the 100 samples, while for the "observed" data they are computed from the multivariate density fitting method (equations 2, 4 and 5).

   In the derived parameters in Table 1, we notice that the systematic errors introduced in the simulation have been mostly recovered while the proportions α are not well recovered in groups 2 and 3. This is mainly due to the fact that the groups are not well centered because of the shift introduced.

Table 1.

		input parameters						derived parameters					
		group 1		group 2		group 3		group 1		group 2		group 3	
		X	Y	X	Y	X	Y	X	Y	X	Y	X	Y
Model	C	1	6	*3*	*8*	*5*	*10*	1.0	6.0	3.0	8.0	*5.0*	*10.0*
	σ	1	1	1	1	1	1	1.0	1.0	1.0	1.0	1.0	1.0
Data	C	1	6	*4*	*9*	*8*	*13*	0.9	5.9	3.0	7.9	*7.6*	*12.8*
	σ	1	1	1	1	1	1	0.9	1.0	1.1	1.3	1.5	1.5
α		1		0.5		2.0		0.95		0.26		0.51	

2. In a second step, we give to the model the shifted mean derived from the data in the first step (i.e., 7.6 and 12.8 in X and Y in group 3) and we compute as before the derived parameters (Table 2). This step is a simulation of a correction applied to a model in order to fit the data.

   In Table 2 we see that, once the shifts (hence the systematic errors) in group 3 have been corrected, the correct values of the mean and sigma in group 2 are well recovered (i.e., 3.5 in place of 4, 8.5 in place of 9) as well as the values of α for each group (0.4 in place of 0.5 in group 2 and 1.94 in place of 2. in group 3).

Table 2.

		input parameters						derived parameters					
		group 1		group 2		group 3		group 1		group 2		group 3	
		X	Y	X	Y	X	Y	X	Y	X	Y	X	Y
Model	C	1	6	*3*	*8*	7.6	12.8	1.0	6.0	*3.0*	*8.0*	7.6	12.8
	σ	1	1	1	1	1	1	1.0	1.0	1.0	1.0	1.0	1.0
Data	C	1	6	*4*	*9*	8	13	0.9	5.9	*3.5*	*8.5*	8.0	13.0
	σ	1	1	1	1	1	1	0.9	1.2	1.1	1.0	1.3	1.1
α		1		0.5		2.0		0.91		0.40		1.94	

4. Conclusions

We have tested the possibility of using the multivariate density estimation method to analyse multivariate catalogues and compare them with model simulations. The method has been applied using a stepwise approach. A first comparison of the multivariate densities of the model (defined by a sum of populations or physical groups) and the data allows us to detect systematic shifts between the group characteristics and the observed distributions. Then, applying these shifts to the model, the same computation allows us to measure the relative density of each physical group present in the data.

Moreover, in comparison with other analysis methods, this one allows us to compare model predictions and data in a whole survey of several galactic directions at the same time.

Because the addressed problem is universal, this methodology may be applied to other astronomical fields, involving multivariate data. Though the software has been written and programmed in a general way.

References

Bienaymé, O., Robin, A.C., & Crézé, M. 1987, A&A, 180, 94

Bienaymé, O., Mohan, V., Crézé, M., Considère, S., & Robin, A.C. 1992. A&A, 253, 389

Crézé, M., Chen, B., Robin, A.C., & Bienaymé, O. 1991, The Stellar Populations of Galaxies, IAU Symp. 149

Hand, D.J. 1981, Discrimination and classification, (John Wiley & Sons Ltd.)

Robin, A.C., & Crézé, M. 1986, A&A, 157, 71

Robin, A.C., & Chen B. 1992, Back to the Galaxy, 3rd Annual October Astrophysics Conference in Maryland, College Park

Astronomical Data Analysis Software and Systems II
ASP Conference Series, Vol. 52, 1993
R. J. Hanisch, R. J. V. Brissenden, and J. Barnes, eds.

Tests of a Simple Data Merging Algorithm for the GONG Project

W. Williams, F. Hill, and C. Toner

National Solar Observatory, NOAO[1], *P.O. Box 26732, Tucson, Arizona 85726*

Abstract. The GONG (Global Oscillation Network Group) project proposes to reduce the impact of diurnal variations on helioseismic measurements by using the data from six sites placed around the globe. The data from the sites must be combined into a single time series in order to determine mode frequencies, amplitudes and line widths. Here, we report on tests of a simple (all weights = 1) average merging algorithm emphasizing the results in the p-mode frequency band around 3 mHz.

1. Data Description

In order to test merging strategies, an artificial data set was created. The data set consists of a set of "perfect" solar disk velocity and intensity images which have no atmospheric effects and one set of "observed" solar disk images for each of the six observing sites in the network. The perfect data set is described in more detail in Bogart et al. (1992). With oscillation frequencies, rotational splitting, and mode amplitudes and widths as input parameters, artificial power spectra and time series were computed using E. Anderson's GRTOOLS package in the GONG project's GRASP software system. Spherical harmonics were generated from code written by T. Duvall using intensity-weighted integrals of the Legendre functions. A background steady-flow velocity field was generated from code originally written by D. Hathaway (1988), and modified by R. Toussaint and F. Hill. These flows include differential rotation, meridional flows, giant cells, supergranulation, and the limb effect.

A total of 1440 "perfect" solar disk images at one-minute cadence were produced. The solar disk images have a resolution of about 8 arc seconds per pixel. A total of 4364 oscillation modes for all m values for l=20 and 100 and all solar oscillation frequencies between 1.5 and 5 mHz were included. The mode amplitudes were rescaled so that the incoherent sum of the modes was about 300 m/s. There are three known deficiencies in the data: the steady flow velocity field is rather aliased at the limb; the spherical harmonics have constant spatial phase; and the phase shift due to rotation was incorrectly applied resulting in a "double" image of the ridges in the spectrum.

---

[1]The National Optical Astronomy Observatories are operated by the Association of Universities for Research in Astronomy, Inc. (AURA) under cooperative agreement with the National Science Foundation

To test the merging algorithm, the "perfect" data was distorted in various ways to simulate the actual observations which might be taken at the various sites, around the globe. This distortion was time and site dependent. Earth-based observer's motion was applied to the velocities. Atmospheric factors included exponential scattering, Gaussian seeing, clouds (stratus, cumulus and cirrus), and atmospheric distortion due to differential refraction. Instrument factors applied to the data include $\text{sinc}^2(\text{aperture}) \times$ quadratic (defocus) instrumental modulation transfer function and CCD read, shot, and photon noise. Also included were occasional bizarre events, such as striped CCD noise patterns, etc. Further information can be found in Williams et al. (1992).

2. Data Analysis

In this work, data analysis for the helioseismic measurements used the three major steps commonly accepted by workers in the field (Hill et al. 1991). In the first step, the solar images are remapped on a heliographic coordinate grid and filtered to remove low temporal frequency changes ($\nu < 800\mu$Hz). In the second step, the spherical harmonic mode coefficients for each image are found and the Fourier transform of the time series of each spatial oscillation mode are computed. In the third step, the power in the spherical harmonic modes is plotted as a function of mode degree and frequency (an $l - \nu$ diagram), and a precise determination of the temporal frequencies of the oscillation modes is made. All of the analysis from solar disk images through time series was done with W. Williams' SUNTRANS package from the GRASP software system. All of the analysis from power series through mode frequency identification was done with E. Anderson's GRTOOLS package from the GRASP software system.

The solar disk images from each site and the "perfect" solar disk images were analysed in as nearly an identical fashion as possible. The analysis of the velocity images from the six sites was done with identical remapping and temporal filtering parameters to those for the "perfect" data. This step minimized problems due to changing observer motion and variations in image scale from site to site. In order to obtain a single $l - \nu$ diagram which contained data from all six sites, it was necessary to combine the data at some point before the generation of the power series. Potential types of data to combine were the solar disk images, the remapped images, the temporally-filtered remapped images and the mode coefficient time series.

The constraints due to image scale, atmospheric and instrument parameters and velocity offset (due to observer motion) dictated a choice between merging images which had been remapped and temporally filtered and merging time series of mode coefficients. A merging scheme using each type of data was tried. Both of these merging schemes of data yielded the same $l - \nu$ diagrams to four or more decimal places for the simple, equally-weighted average.

3. Evaluation of the Merge

The results of the merging schemes using a simple equally-weighted average were evaluated by comparison to results at the same point in the data processing for the "perfect" data. Comparison of movies of the perfect images and the merged

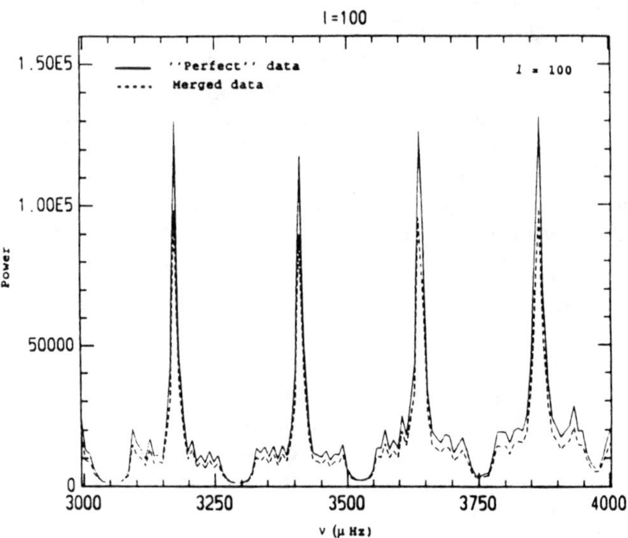

Figure 1. A comparison of the m-averaged, rotation corrected $l - \nu$ diagram for frequencies between 3 and 4 mHz at oscillation mode l=100. The reduced amplitude in the merged spectrum at l=100 is present at all frequencies.

images showed no overt jerks or sudden changes in level. Overplotting the same spot on the image for the perfect data and the merged data shows differences in heliographic remapping between the merged and "perfect" data of less than a pixel.

The overall magnitude of the mode coefficients for the merged time series was less than the overall magnitude of the mode coefficients for the perfect time series. This indicates that the distortions put into the site data decreased the signal power. The amplitudes of the mode coefficients for the merged data are less than those of the perfect data by 2.5% in the mean at l=20 and by 10% at l=100. The amplitude of the differences between the merged and perfect time series was not found to be dependent on the number of sites observing or the number of changes in the number of sites on line for either l=20 or l=100.

The reduced amplitudes are consistent with our estimate of reduction in mode coefficient magnitude using the Modulation Transfer Function, i.e., the Fourier transform of the point spread function from atmospheric and instrumental distortion. The Modulation Transfer Function was obtained by dividing the transform of the median image radial profile by the transform of a realistic limb darkening profile constructed using parameters estimated from the image. Further details on this new method may be found in Toner and Jefferies (1992).

Regions of the m-averaged, rotation-corrected $l - \nu$ diagrams for the merged and perfect data were compared. The merged data shows some anomalous power at very low l. The peaks are in very good agreement in both amplitude and

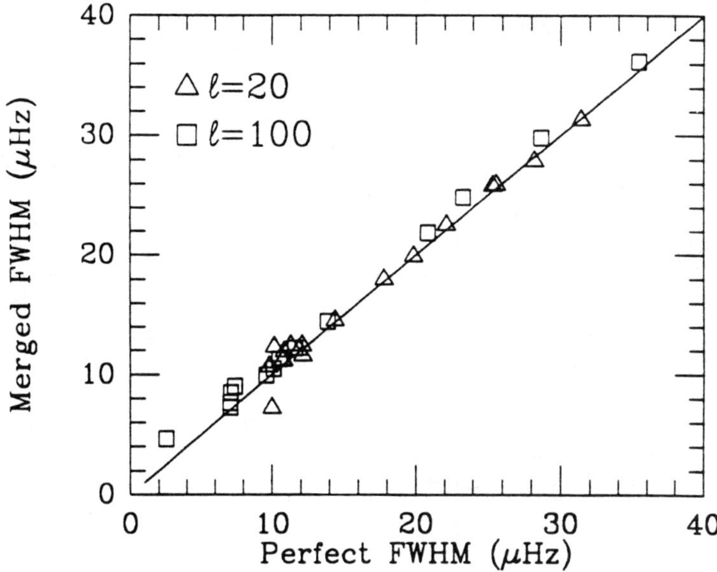

Figure 2. Comparison of measured Full Width at Half Maximum obtained from the merged and perfect spectra for all of the available solar oscillation modes. The merged peaks are broadened by about 2 μHz.

frequency at l=20. Reduced amplitude at l=100 is present at all frequencies, as indicated in Figure 1. Examination of the rest of the spectrum shows that the background power at both high and low frequencies has been increased. This suggests that the merging process has redistributed power out of the peaks and into the background. At high frequencies, the background enhancement is greater at l=100 than at l=20. The background enhancement appears to be independent of l at low frequencies.

We measured the oscillation line parameters in both the merged and perfect spectra using E. Anderson's PEAKFIND package in GRASP. For frequency, we found that the RMS deviation for all modes was 0.31 μHz, about an order of magnitude less than the 2 μHz random error expected from the underlying stochastic nature of the oscillations for one day of data (Anderson, Duvall & Jefferies 1990). Figure 2 shows a comparison of the Full Width at Half Maximum of the spectral lines measured from the merged and perfect spectra. The RMS average deviation was measured to be 0.9 μHz at l=100, 0.4 μHz at l=20, and 0.6 μHz for all modes. This systematic error is probably caused by the redistribution of power in the merged spectrum which results in lower amplitude peaks and higher backgrounds. However, the doubled ridges may also contribute to this systematic effect.

We also noticed that the peaks in the merged spectrum were shifted slightly toward lower l values. The data indicates a shift of approximately 2 in l. The shift in l value may be due to varying registration across the disk due to seeing and other atmospheric phenomena, or to the incorrectly-applied rotation in the m values, which is possibly smeared into lower l values.

4. Conclusions

The results presented here demonstrate that the simple average merge is probably acceptable for p modes. For a one-day time series, the noise from the merge is much less than the inherent noise in the central frequency determination. The merged peaks are systematically wider by 0.6 μHz RMS and have a RMS deviation of 0.3 μHz between the central frequencies of the merged and perfect data. The merged peaks are shifted to lower l at l=100. The power is redistributed out of the peaks into the background in the merged power spectrum. These differences are due to residual misregistration and/or the imperfections in the "perfect" data.

In the future we will be investigating more sophisticated weighted averages for use in merging the data from the GONG sites. Weights will be computed from the Modulation Transfer Function and other measures of instrument performance and sky conditions.

References

Anderson, E.R., Duvall, T.L., Jr., & Jefferies, S.M. 1990, ApJ, 364, 699

Bogart, R.S., Hill, F., Toussaint, R., Hathaway, D.H., & Duvall, T.L., Jr. 1992, in GONG 1992: Seismic Investigation of the Sun and Stars, A.S.P. Conf. Ser., ed. T.M. Brown, in press

Hathaway, D.H. 1988, Solar Phys., 117, 329

Hill, F., Deubner, F.-L., & Isaak, G. 1991, in Solar Interior and Atmosphere, eds. A.N. Cox, W.C. Livingston & M.S. Mathews (University of Arizona Press), 329

Toner, C.G., & Jefferies, S.M. 1992, ApJ, submitted

Williams, W., Hill, F., Toner, C., & Brown, T.M. 1992, in GONG 1992: Seismic Investigation of the Sun and Stars, A.S.P. Conf. Ser., ed. T.M. Brown, in press

Discussion

Fowler: The simple merge works for 24 hours. Will it work over the 3 year data set?

W. Williams: We can't generate an artificial data set long enough to answer that question in advance. We are aware of the possibility of long-term, subtle diurnal variations which may complicate the merge for the real data.

Astronomical Data Analysis Software and Systems II
ASP Conference Series, Vol. 52, 1993
R. J. Hanisch, R. J. V. Brissenden, and J. Barnes, eds.

SKYMAP: Exploring the Universe in Software

Douglas J. Mink

Smithsonian Astrophysical Observatory, 60 Garden St., Cambridge, MA 02138

Abstract. SKYMAP is a computer program which produces maps of arbitrary portions of the sky in a variety of projections and coordinate systems. Over the past 10 years it has been used to produce finder charts for occultations by planets, display scan and image data from the Spacelab 2 Infrared Telescope, and make maps of fields for astronomical observations at X-ray, optical, infrared, and radio wavelengths. It can display multiple source catalogs, including the HST Guide Star Catalog, as well as solar system objects with astrometric accuracy. SKYMAP can be tuned to a specific task using an ASCII parameter file which controls how information is displayed on any Tektronix-compatible graphics display or hardcopy device. The program contains a variety of interactive graphic and image processing features and has been ported to a variety of computer systems.

1. Introduction

SKYMAP is an interactive astronomical display program which projects a map of the sky in any of several coordinate systems in any of several projections. This map can be displayed on any Tektronix-compatible graphics terminal, such as xterm, and printed to any hard-copy device which supports Tektronix graphics code or for which a translator exists. Star catalogs, planet positions, images, or telescope scan directions can be plotted over the coordinate grid. SKYMAP can access the IRAS, SAO, PPM, and HST Guide Star catalogs, as well as planetary positions from 1950 to 2000. Additional catalogs (and ephemerides) can be accommodated as well. Images can be displayed in halftones or contours with any of the other information overplotted.

2. Uses

Creation of finder charts is the most common use of SKYMAP. The ability to plot at a particular plate scale, such as that of the Palomar Sky Survey, is particularly useful. The map can be centered on a specific catalog object and include up to two catalogs plus HST Guide Stars. Planet positions can be plotted over time to aid in planning observations. All-sky maps can also be produced to show large structures or the position of a catalog in the big picture. Catalogs can also be plotted over contour or halftone images.

3. Parameters

Almost everything in the program is parameterized; the map format can be modified to fit the user's requirements, and there are many possibilities for plotting information over images or other spatial information. Parameters which define a SKYMAP session are stored in an ASCII parameter file, which is modified by all of the interaction modes. A default parameter file, "skymap.par", is moved into the current working directory when SKYMAP is invoked there for the first time.

An ASCII parameter file format has been designed based on the FITS image file header format. The parameter file structure is clear enough that such a file can be edited by an ASCII file editor such as vi or Emacs. In addition, parameters can be modified on the command line or values returned, so the Unix shell can be used for batch operation of programs. Although this is a fairly rudimentary system, it has been in use for several years and is fairly well debugged.

The parameter file is very simple. Each parameter gets one line of the form "keyword = value / comment" where "/ comment" is optional, and each line is terminated with a linefeed. As in FITS, the final line of the file is simply "END". Character strings get no special treatment. The comment "/" is differentiated from any "/"s which may occur legitimately in character strings by the fact that character string values are surrounded by quotes. Quotes, which may be single or double as long as they're matched, are only needed if "/"s are in the strings, but they may be used to delimit any character string. While the order of the elements of the line is important, the parsing subroutines will omit leading and trailing spaces and tabs from each element, so they may be positioned for clarity.

4. Mapping

SKYMAP knows about Aitoff, Mercator, Lambert cylindrical, gnomonic, orthographic, polar, and linear projections and can plot a grid or a set of tick-marked box axes for any of them. All of the IRAS map projections are included so objects can be plotted against contoured IR sky flux. Equatorial (J2000 or B1950), galactic, and ecliptic coordinates may be specified, as well as orbital, where an arbitrary pole position in right ascension and declination is specified.

The map may be labeled using terminal or computer-generated characters. Computer drawn characters use the Hershey font set and are, in decreasing detail, Complex, Italic, Duplex, and Simplex. Four standard font sizes, corresponding to the four Tektronix sizes, plus arbitrary sizes in Tektronix coordinates (height=10 to 4095) are available. The font size may be set separately for headings, tickmark labels, axis labels, time tag, plate scale, and magnitude scale table. Font type is the same for heading, tick, and axis labels.

Portions of the displayed map may be enlarged using the Zoom menu or cursor command. The original display parameters are remembered and may be recovered using the Unzoom commands. If the default automatic grid spacing has been specified, an optimum spacing is program-selected when zooming. If plotting to a specific plate scale, the size of a zoomed map can be specified by plate scale. By zooming, portions of images can be selected in sky coordinates rather than in rows and columns.

5. Plotting Objects on the Sky

SKYMAP can plot sources from the HST Guide Star Catalog plus up to two other catalogs and any number of solar system objects on a map. The SAO Catalog, the Yale Bright Star Catalog, the IRAS Point Source Catalog, and the PPM Catalog are available in a rapid-plotting binary format. Catalogs also may be plotted from the simple ASCII format:

<catalog name>

<one-line catalog description>

<number><RA as hh.mmsssss><Dec as dd.mmssss><magnitude><name>

...

A program, STARCAT, exists to turn files in this format into binary files for faster plotting in SKYMAP.

Magnitude and flux limits may be set, and the size of the plotted source scaled by its flux. Foreground objects may be labeled and plotted as open crosses or half-crosses as well as circles. Sources may be identified using the cursor. A table of source size vs. magnitude may be plotted in an arbitrary location on the display. Sources may be labeled with their number, or if plotting from an ASCII file, their name. The location of this label is specified in display (Tektronix) coordinates, where coordinates less than 100 are used as offsets from the plotted source position. The font types and sizes of the star labels may also be specified. Optional log files (one per catalog) may be kept which list positions and magnitudes of all of the stars plotted in the current region.

If the Hubble Space Telescope Guide Star Catalog CD-ROM set is mounted on the computer as a Unix file system, SKYMAP can plot sources directly from the CD-ROMs. If only one CD-ROM drive is available, only the hemisphere of interest need be mounted. As with other catalogs, magnitude limits may be set. In addition, objects may be selected by whether or not they have been flagged as non-stellar. Guide Stars may be plotted in the same way as the other catalogs. If the stars are labeled, the number of the region is omitted. An optional log file listing all Guide Stars with region and source number may be kept. Another program, RGSC, is a standalone version of this searching software.

Positions of many solar system objects, including the sun, the moon, the other eight planets, and the satellites of Uranus, Neptune, and Pluto, may be plotted for any time between 1980 and 2000. The positions are from the current JPL DE-130 ephemeris and are the most accurate available. More than one object can be displayed, and variation in position over time can be plotted. Rings may be plotted by selecting the appropriate data base entry. Additional objects may be added to the planet database if ephemeris files in a simple ASCII format exist:

<Julian date><right ascension><declination><distance>.

In this file, <right ascension> is in fractional hours, <declination> is in fractional degrees, and <distance> is the distance from Earth in Astronomical Units. Planet positions and distances for any object in the planet database may be printed using a separate program called PLANET.

Time series of pointing directions, such as those of the Spacelab 2 Infrared Telescope can be plotted on the sky.

6. User Interface

Of the four ways to interact with SKYMAP, parameter file editing, command line, menu/prompt, and cursor, the menu system is the most user-friendly. It is a fairly bomb-proof interface perfected over the past ten years with several levels of menus. The top level handles action commands, parameter file actions, and selection of submenus. The submenus handle information concerning source catalogs, planets, mapping, image characteristics, IRT scan data, display format, and image operations.

A single keyboard character chooses the menu selection; the main menu is displayed if a space or "?" is entered in response to the top level prompt, "SKYMAP?". Lists of submenu commands can be obtained by responding with a space or "?" after the submenu is selected. An "=" in response to a submenu prompt lists the values of all of the parameters which can be set within that submenu. Out of range numbers and illegal characters are trapped, and the prompt is repeated without crashing the program.

Menu interaction can be bypassed by setting the parameter MENU=F; SKYMAP can then be run directly from the Unix shell. In this mode, parameter values can be checked and modified from the Unix command line. SKYMAP can thus be run in a Unix shell script, and selected parameters, which are returned on standard output, can be tested. Parameter names are included if the parameter PARVAL is "N." If PARVAL=Y, "skymap cra cdec" returns the single line " 19.3936 –1.". Typing "skymap cra cdec" returns the following if PARVAL=N:

CRA = 19.3936 [Center right ascension in hh.mmssss]

CDEC = –1. [Center declination in dd.mmsss]

Comments from the parameter file are included if they are present. Parameter values may be changed on the command line:

skymap <keyword1>=<value1> <keyword2> + <value2> ...

where parameter <keyword1> will be set to <value1> and <keyword2> will be incremented by <value2>. Because SKYMAP's command line parser allows arithmetic on parameters, Unix shell scripts can be written to generate reams of finding charts. SKYMAP may also be run with an alternate parameter file by typing

skymap –par <parameter file name>.

The most commonly used parameters are listed with their current values when you type "skymap help". To set parameters without plotting anything, or to simply install the default parameter file in the current working directory, add "stop" to the command line, and SKYMAP will process the preceding parameter entries on the command line and stop without attempting to plot anything.

The cursor command set is based on the Tektronix GIN mode which is available to a wide variety of graphics terminals. Cursor mode is entered from menu mode after a plot has been made or automatically after the plot is complete if the CURSOR parameter is set. When the cursor is moved to the appropriate position using a mouse or other pointing device, a single keyboard character selects the use to which the cursor coordinates will be put. Scaling, zooming, coordinates, image values, and source identifications are available in this way. A menu of cursor commands appears when a space or "?" is entered.

7. Image Display

SKYMAP reads FITS images from disk files which are specified in two parts: the image file name and the image directory pathname. When plotting different images from the same directory, only the image name needs to be retyped. Mapping parameters may be read from the FITS image header or set using the Mapping submenu. Image pixels are assumed to be square. If an image is displayed, it is plotted before axes or other information. Only 2-byte integer (Fortran Integer*2, C short) images are supported; SHRINK, a separate program, reduces 4-byte integer images to 2-byte images.

Two modes of image display are currently implemented, halftone and contour, on any Tektronix-emulating terminal. Contour intervals and gray levels for halftones are separately maintained. Halftones may be scaled linearly, logarithmicly, or by a power law. Two halftone modes differ in the way they fill the display. One plots dots randomly, the other repeats patterns. The image is centered within a defined display area on the terminal, as is any map that is drawn. The image is redisplayed whenever the range of values or scaling technique is changed as these affect the mapping of image pixels to display pixels.

8. Hardcopy Output

Hardcopy output is available for line graphs by dumping a log file whose name is specified in the PLOTFILE parameter. It may also be specified by the ">" menu command. While the default file is skymap.plot in the current working directory, it is safer to send complicated plots to a file in /tmp to avoid quota or disk overflows. This log file contains all Tektronix commands and character strings which were sent to the display between the time the "P" command was issued and the "SKYMAP?" prompt or cursor appeared. It may be displayed on a Tektronix-compatible terminal or printed on any Tektronix-compatible output device such as a plotter or a laser printer. A program, IMTEK, has been written to print this log file to Imagen laser printers using the best resolution possible. The public domain program TEK2PS can be used to produce Postscript files for most laser printers or publication. Separate log files containing lists of plotted stars and their positions or responses to cursor commands may also be kept and printed.

9. Future Features

It would be nice to reimplement the color image display functions written for AED color graphics terminals for the X Window System, though this could require restructuring of the whole program. The graphic image display, color map graph, and linear cross section could all be plotted in separate windows from the command window, the latter two being popped up only when needed. If a color graphics terminal emulator were written for X, porting would be much easier as the image refresh function would be taken care of by the terminal emulator rather than the image display program.

Astronomical Data Analysis Software and Systems II
ASP Conference Series, Vol. 52, 1993
R. J. Hanisch, R. J. V. Brissenden, and J. Barnes, eds.

Guide Star Catalog Data Retrieval Software II

O. M. Smirnov and O. Yu. Malkov

Institute of Astronomy of the Russian Academy of Sciences, 48 Pyatnitskaya st., Moscow 109017 Russia

1. Introduction

The Guide Star Catalog (GSC) was initially created at STScI to support the operational requirements of the Hubble Space Telescope (Lasker et al. 1990; Russel et al. 1990; Jenkner et al. 1990), specifically, an all-sky set of reference coordinates for pointing and tracking. A new version of the catalog, GSC 1.1, includes bright stars from the HIPPARCOS INCA database.

The GSC is widely used by the astronomical community for many different applications, such as:

- statistical studies of sky regions;
- searches for candidate optical counterparts of astronomical objects found at other wavelengths;
- mapping of object vicinities (in particular, finder charts for variable stars).

The catalog's distribution format (a set of two CD-ROMs) requires minimum hardware and is well suited for all sorts of conditions, especially observations.

Unfortunately, the actual data in the catalog is not easily accessible. It takes the form of FITS tables, and the coordinates are given in one standard system (J2000.0). The included software (Priou 1990) doesn't help much—it can only search the GSC for objects from a rectangular region specified by coordinates in J2000.0, and list the objects on the screen, without any possibility of storing them in a file for later use. Thus, even the generation of a simple finder chart is no trivial undertaking.

To help PC users solve this problem, we have created the Guide Star Catalog Data Retrieval Software, or GUIDARES (Malkov et al. 1992). This is a user-friendly program which allows users to easily produce text samplings of the catalog and sky maps in projected celestial coordinates. This paper describes GUIDARES version 1.2.

GUIDARES requires a PC with a CD-ROM drive and CGA, EGA or VGA graphics (for plotting sky maps only). It will work without a CD-ROM if the necessary GSC files are copied onto the hard disk, preserving the GSC's directory structure.

The main function of GUIDARES is to produce an ASCII table of objects from a specified region, and optionally, a graphical sky map of the region. It can handle rectangular and circular regions in different coordinate systems.

2. User Interface

From the user's viewpoint, GUIDARES looks like a self-explanatory menu with automatic help that is used to enter the following information:

- The logical drive where the GSC files are located.
- The coordinate system in which the region is specified: equatorial, ecliptic, galactic or supergalactic. For equatorial and ecliptic coordinates, the equinox may also be set (the default is 2000).
- The shape of the region: rectangular or circular.
- The actual coordinates of the region (minimal and maximal or central and radius).

The software performs a lot of error control along the way. Illegal values of coordinates are impossible to enter, the menu option to start retrieval is inaccessible until all the necessary data have been specified, inverted coordinates are swapped if necessary, etc.

3. Data Retrieval

During retrieval, the evolution of the process is shown on the screen. In addition, if the region is suitable (i.e., small enough—less than 16 square degrees in area), a sky map in projected celestial coordinates is simultaneously plotted on the screen. Sky maps of bigger regions are not practical because the objects are too dense for a meaningful plot.

Internally, GUIDARES must perform the following steps:

- Convert the region coordinates to the standard GSC system.
- Determine which GSC files overlap the user's region. This is done algorithmically. In contrast, Priou's software uses a lookup table, and for this reason it is twice the size of GUIDARES (340 Kb vs. 150 Kb), while providing much less functionality.
- Scan the GSC files for objects that fall within the user's region and convert the objects' coordinates back to the user's system.

The GSC contains a lot of so-called few entry objects (FEOs), i.e., objects that appear on several plates, and therefore have several corresponding GSC entries. For FEOs, GUIDARES computes mean weighted values and errors of coordinates and magnitude. Some FEOs are classified both as stellar and non-stellar objects in different entries, here we'll call them "mixed objects".

We were able to improve the retrieval time somewhat in comparison with Priou's program by employing a buffering scheme. However, because both programs spend most of their time accessing the slow CD-ROM, this improvement was not very significant—80 vs. 90 seconds for 4 GSC files. If the user accesses the same region many times, he can copy the relevant GSC files to his hard disk (GUIDARES makes this possible by displaying the names of the files as it accesses them) and speed up the process drastically.

If errors occur, or if the CD-ROM must be changed, GUIDARES notifies the user via a message box with standard "Abort, Retry, Ignore" or "Continue, Cancel" options.

4. Output and Sky Maps

For every region, GUIDARES creates two output files. One of them (called the GUIDARES sampling file) is an ASCII table with the following entries for each object:

- GUIDARES ID of object, obtained by appending the number of the object in its GSC file to the GSC file number
- position and position error
- magnitude and magnitude error
- classification

The other file is a script for the MONGO graphics package. MONGO can generate a hard copy of a sky map, otherwise it is not much use because it only supports rectangular coordinates, while the GUIDARES SkyMap facility features projected celestial ones.

A GUIDARES SkyMap can be produced both during retrieval and afterwards, by reading the sampling file. A coordinate grid with labels is displayed, and objects are plotted over it. Stellar objects are represented by disks, non-stellar ones by circles, and mixed objects by semi-filled circles; the radius of the disks and circles is proportional to the objects' magnitude.

GUIDARES automatically selects the highest resolution available and takes into account the screen's aspect ratio, so that square regions look truly square in any resolution.

5. Distribution

To avoid the hassle of separate "Read Me" files, all the documentation has been built into the executable module. By running GUIDARES with a command-line option (`guidares/d`) the user can produce a documentation file whenever he needs. Therefore, the whole package is distributed as a single file.

GUIDARES is a shareware product. An unregistered copy of the latest version can be picked up by anonymous ftp from `iraf.noao.edu` (`140.252.1.1`), directory `iraf/contrib`.

If you have any questions, comments, or even nasty criticisms, we can be contacted by e-mail either at `oms@airas.msk.su` (Oleg Smirnov), or at `iaas@adonis.ias.msk.su` (Oleg Malkov).

Acknowledgments. Financial support from the Smithsonian Astrophysical Observatory is gratefully acknowledged. Also, many thanks to Jeannette Barnes for helping make GUIDARES available by ftp.

References

Jenkner, H., Lasker, B.M., Sturch, C.R., McLean, B.J., Shara, M.M., & Russell, J.L. 1990, AJ, 99, 2081

Lasker, B.M., Sturch, C.R., McLean, B.J., Russell, J.L., Jenkner, H., & Shara, M.M. 1990, AJ, 99, 2019

Priou, D. 1990, Guide Star Catalog Listing Program, Institute Géographique National

Malkov, O.Yu., Kulkova, L.I., & Smirnov, O.M. 1992, in Astronomical Data Analysis Software and Systems I, A.S.P. Conf. Ser., Vol. 25, eds. D.M. Worrall, C. Biemesderfer & J. Barnes, 79

Russell, J.L., Lasker, B.M., McLean, B.J., Sturch, C.R., & Jenkner, H. 1990, AJ, 99, 2059

Astronomical Data Analysis Software and Systems II
ASP Conference Series, Vol. 52, 1993
R. J. Hanisch, R. J. V. Brissenden, and J. Barnes, eds.

Enhancements to IRAF/STSDAS Graphics

J. D. Eisenhamer and Z. G. Levay

Space Telescope Science Institute, 3700 San Martin Dr., Baltimore, MD 21218

Abstract. The IRAF graphics kernel, *psikern*, is a true encapsulated PostScript implementation, an improvement over the former SGI-based PostScript output available from IRAF. The *psikern* kernel implements many more capabilities of gio/gki such as cell arrays (grayscale images), color, filled-area patterns and true PostScript fonts. Several of the general-purpose graphics tasks in STSDAS such as *igi*, *sgraph*, *skymap*, *newcont* and *wcslab* have been modified to use these capabilities explicitly. Other graphics tasks not enhanced explicitly can also make use of new capabilities such as PostScript font support. We present an overview of *psikern* and several examples of output created by the enhanced STSDAS tasks.

1. The PostScript IRAF Graphics Kernel, psikern

A major improvement to graphics produced using IRAF is now possible with the development of *psikern*, an IRAF graphics kernel implementing true encapsulated PostScript capabilities. The existing IRAF PostScript interface implements only vector drawing, even including its own vector-based font set for drawing text.

An IRAF graphics kernel is a task that renders (draws) graphics on a device, translating internal IRAF graphics metacode into a form recognizable by the hardware. Normally, a kernel is executed transparently by the IRAF cl, but it may be executed explicitly as well, similar to any other IRAF task.

The *psikern* kernel is a task in the STSDAS *stplot* package. It implements all fundamental graphics capabilities now available in the IRAF gio/gki applications software interface. No existing kernel, *stdgraph*, *stdplot*, *sgikern*, or *imdkern*, implements all of them. Therefore, many existing graphics tasks do not use these capabilities such as color, filled-area patterns, or raster image display. Using new graphcap device descriptions, *psikern* may be installed to be used as the default kernel for drawing graphics on PostScript printers.

Some capabilities enhance existing tasks without modifying code. For instance, *psikern* uses true PostScript fonts installed in the printer. Any text written by an IRAF graphics task will be written using these fonts with *psikern*. It is possible to specify the font to use for the various fonts selectable via gio (Figure 1):

- Default (Roman): Times-Roman, Helvetica
- Bold: Times-Bold, Helvetica-Bold

- Italic: Times-Italic, Helvetica-Oblique
- Greek: Symbol

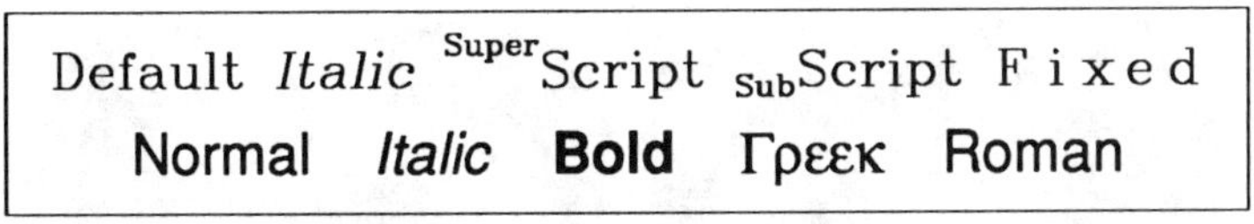

Default *Italic* ^{Super}Script $_{\text{Sub}}$Script F i x e d

Normal *Italic* **Bold** Γρεεκ Roman

Figure 1. Examples of text drawn by *igi* showing *igi* (software) fonts above, and gio (hardware) fonts below rendered using Helvetica in PostScript.

Other capabilities have not been available with any existing kernel. The *psikern* kernel permits selecting color from task by setting a gio parameter, as shown in Figure 2. It is also possible to draw filled-area patterns or textures by selecting a gio pattern type. Figure 3 shows the available textures.

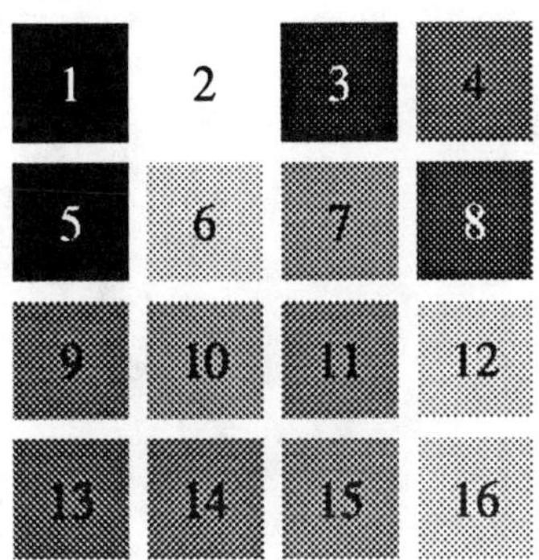

Figure 2. Colors selectable in *igi* and rendered by *psikern*. Here, they are rendered in black and white as halftones by PostScript. The numerals are the color index used in igi to obtain the rendered color.

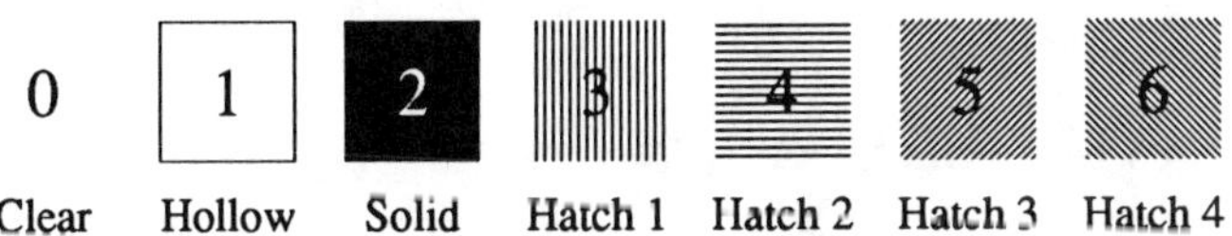

Figure 3. Fill patterns selectable in *igi* and rendered by *psikern*.

Another major addition is the ability to display images (cell-arrays in gio terminology) from an IRAF graphics program with the result printed on a PostScript device. The images are spatially scaled to a viewport on the page, conveniently specified from gio via world coordinates. Two ways of scaling images are shown in Figures 4 and 5. Optionally, colors may be assigned to pixel values via a separate palette from that used to draw other graphic elements.

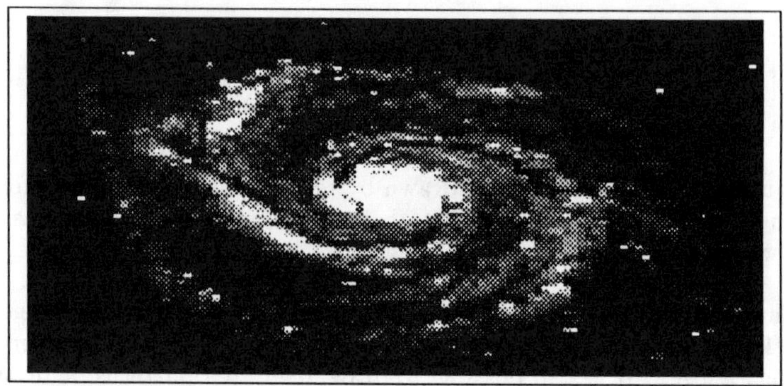

Figure 4. An image drawn with *igi* and rendered by *psikern*. The image is rendered using the default spatial and intensity scaling and is therefore stretched to fill the specified viewport (drawing area).

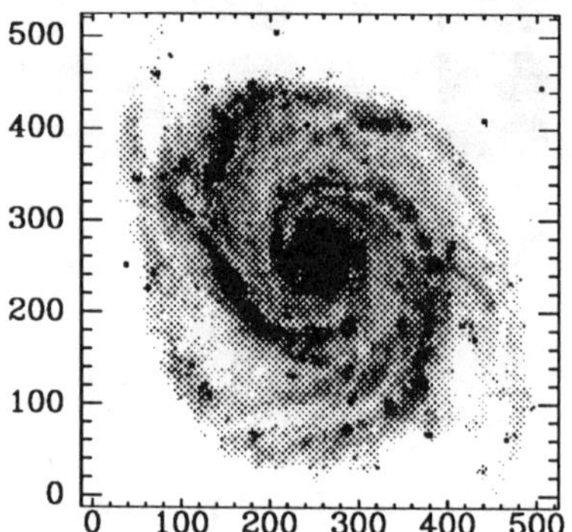

Figure 5. The same image as in Figure 4, but drawn with negative intensity scaling and scaled to preserve the natural aspect ratio.

2. Changes to STSDAS Graphics Tasks

Several tasks in the STSDAS *stplot* package have been modified to take advantage of gio capabilities now available with *psikern*. The most common change was to include the selection of colors from the available palette.

Several commands in the *igi* graphics interpreter task were modified and others added to provide additional capability:

- `COLOR` — Select color from palette

- `FILLPAT` — Select fill pattern by index
- `ZSECTION` — Read and display images
- `FITPIX`, `ZRANGE` — Scale image data
- `SAOCMAP` — Apply an image colormap
- `FONTSET` — Select *igi* or gio (PostScript) fonts.
- `BARGRAPH`, `POINTS` — Existing commands modified to use area patterns

The *skymap* task may now draw various plot features such as the catalog objects, coordinate axesand annotation, and chart key, in different colors. Object symbols may be drawn all in the same color or a catalog table column may be used as a color index to select a color from the palette. The symbols may also be drawn using filled areas with a clear edge, to permit distinguishing overlapping symbols.

Other modified *stplot* tasks include *newcont*, *sgraph*, and *wcslab*, which all now include drawing in color. In *newcont*, the user may now select the colors of the contour lines. In *sgraph*, curves and axes with annotations may be drawn in color. It is also possible to cycle colors and/or line styles for multiple curves on the same plot. In *wcslab*, colors may be selected for drawing the coordinate axes and labels.

3. Existing IRAF and STSDAS Graphics Tasks

All existing tasks producing graphics from IRAF can make use of true PostScript fonts with no change. Simply using *psikern* to render the plot will do this automatically. In addition, any task that uses extended gio capabilities such as filled-area patterns or cell-arrays will benefit from the PostScript kernel. A greater benefit, however, comes from newly developed software, which can take advantage of features already implemented in gio but not available with other kernels.

4. Software Interface to psikern

In addition to the usual gio capabilities for drawing and setting attributes, *psikern* makes use of the escape function of gio for capabilities specific to PostScript. These permit specifying the palette for selecting the "pen" color, the color map used for rendering raster images, and the PostScript fonts to use for rendering text.

5. Graphcap Considerations

The graphcap is the file IRAF uses to determine capabilities of graphics devices. Fully implementing *psikern* requires additions to the graphcap as delivered by IRAF. The following is an example of a representative psikern graphcap entry defining a "device" named **`psi_def`**, with the resulting PostScript saved in a file:

```
psi_def|Postscript Kernel Default 8.5x11in 300dpi Landscape:\
        :kf=stsdas$bin/x_{\em psikern}.e:tn={\em psikern}:\
        :xs#0.2594:ys#0.1959:X0#0.01:Y0#0.01:\
        :cw#.02:ch#.02644:ar#0.7552:\
        :fs#6:li#24:pl:pm:se:tq#0:xr#3063:yr#2313:\
        :IF=stsdas$lib/psikern_prolog.ps:\
        :FR=Times-Roman:FI=Times-Italic:FB=Times-Bold:FG=Symbol\
        :PW#.00011:PI#1.:TD#.01221:TP#.001221:TS#.006104:MF#10:BO:\
        :DD=psi_def,tmp$psk,!{ echo $F; }&:
```

For more information including additional sample graphcap entries, contact the STSDAS group at ST ScI.

Part 5. Data Analysis Applications

Section D. Image Restoration

Astronomical Data Analysis Software and Systems II
ASP Conference Series, Vol. 52, 1993
R. J. Hanisch, R. J. V. Brissenden, and J. Barnes, eds.

An IDL Based Image Deconvolution Software Package

F. Városi and W. B. Landsman

Hughes STX Co., Code 685, NASA/GSFC, Greenbelt, MD 20771

Abstract. Using the Interactive Data Language (IDL), we have implemented some of the basic iterative algorithms for deconvolution of blurred images in a unified environment called DeConv_Tool. Most algorithms have as their goal the optimization of some goodness of fit statistic, possibly combined with a smoothing criterion. All the statistics for all the implemented algorithms are computed by DeConv_Tool for the purpose of determining the performance and convergence of any particular algorithm and comparing the different methods. DeConv_Tool allows interactive monitoring of the statistics and the deconvolved image during computation. Results are stored in a structure convenient for making comparisons between methods and reviewing the deconvolution process. The DeConv_Tool package can be acquired by anonymous FTP from the IDL Astronomy User's Library (Landsman 1993).

1. Introduction

The standard model of image formation involves first knowing or estimating the point spread function (PSF), which is the response of the instrument to a point source of photons (e.g., a star). Effects such as atmospheric seeing and optical abberations are then presumably described by the PSF, and for convenience, assume that the PSF is spatially invariant over the imaging detector. Observed images are the result of convolution of the photon signal I with H, the PSF :

$$D = I * H + N\,, \tag{1}$$

where D is the observed image (the data), and N is the noise added during data acquisition, which we assume to follow either a Gaussian or Poisson distribution. Convolution is defined as

$$(I * H)(i,j) = \sum_{k,l} I(k,l)H(i-k, j-l)\,, \tag{2}$$

where (i,j) are image pixel indices and H is assumed to be centered at $(0,0)$. Similarly, the correlation operator, used below, is defined as

$$(I \otimes H)(i,j) = \sum_{k,l} I(k+i, l+j)H(k,l)\,, \tag{3}$$

where now (i,j) are referred to as the "lag" indices. The objective of deconvolution is to reconstruct the image I given the data D, the PSF $= H$, and knowledge of the noise N. Equation (1) is part of what is referred to as the imaging model M. In reality the deconvolution problem has more unknown than known quantities because of noise and instrument limitations. For this reason a unique solution does not exist, so it is more realistic to consider the probabilities of solutions.

The statistical formulation of the deconvolution problem involves the use of Bayes' theorem to derive (see, e.g., Weir 1991, Piña & Puetter 1992)

$$p(I|D,M) = \frac{p(D|I,M)p(I|M)}{p(D|M)} \tag{4}$$

where $p(I|D,M)$ = probability of the restored image given the data and model, $p(D|I,M)$ = probability of the data for a given restored image and model. Assuming the model is fixed, the term $p(D|M)$ is independent of I so it is constant, and $p(I|M)$ is the prior probability of the image given the model. A solution to the deconvolution problem is obtained by maximizing the probability of the restored image, and this can be accomplished by maximizing $p(D|I,M)$, the "goodness of fit," leading to "maximum likelihood" methods. Also, the product $p(D|I,M)p(I|M)$ could be maximized, with $p(I|M) = e^S$ and $S \propto$ "entropy" of I, then leading to "maximum entropy" methods. Hence the Bayesian formulation leads to a variety of algorithms for deconvolution, each having its own statistic to be optimized in order to reconstruct the image.

2. Algorithms Implemented

The deconvolution algorithms implemented in DeConv_Tool are: the Maximum Likelihood method for Poisson or Gaussian noise (MLP and MLG resp.), and the Maximum Residual Likelihood (MRL) method for Gaussian noise. An implementation of the Maximum Entropy (MEM) method is in progress, and other algorithms which are iterative in nature can be integrated into DeConv_Tool. The convolution and correlation operations are computed via the Fast Fourier Transform (FFT) by default, and this assumes periodic boundary conditions. Convolutions can be performed directly if desired, but in any case the edge regions of observed image should not contain significant data.

The Maximum Likelihood algorithms are derived by noting that when $p(D|I,M)$ is at a maximum

$$\nabla_I[\ln p(D|I,M)] = \frac{\partial}{\partial I_j}[\ln p(D|I,M)] = 0\,. \tag{5}$$

Applying this condition leads to iterative algorithms which converge toward a fixed point of the iteration (see, e.g., Meinel 1986, Shepp & Vardi 1982).

For the case of Poisson noise in the imaging model

$$p(D|I,M) = \prod_{k=1}^{n} \frac{(I*H)_k{}^{D_k}}{D_k!} \exp[-(I*H)_k] \tag{6}$$

where the product is over all pixel indices k. Applying the condition (5) and manipulating the result gives the following iterative algorithm:

$$I_{\rm new} = I_{\rm old} \left[\left(\frac{D}{I_{\rm old} * H} \right) \otimes H \right] , \tag{7}$$

which is the Richardson-Lucy algorithm (Richardson 1972, Lucy 1974). The relevant statistic for the above MLP algorithm is the log-likelihood $\ln p(D|I,M)$, and note that the restored image is always positive (if $D \geq 0$).

The case of Gaussian noise in the imaging model gives

$$p(D|I,M) = \prod_{k=1}^{n} \exp\left(- \frac{[D_k - (I * H)_k]^2}{2\sigma_N^2} \right) \tag{8}$$

where the product is over all pixel indices k and σ_N^2 is the constant variance of the noise. Applying the condition (5) results in the iterative algorithm

$$I_{\rm new} = I_{\rm old} + [D - (I_{\rm old} * H)] \otimes H , \tag{9}$$

called MLG, which is similar in concept to MLP, but can result in negative restored image values, and so must be forced positive. Defining the residual image as

$$R = D - (I * H) , \tag{10}$$

the relevant MLG statistic is the ratio of standard deviations, σ_R/σ_N, which should approach unity (note that $n\sigma_R^2$ = Variance$[\ln p(D|I,M)]$). Gaussian noise is the limit of Poisson noise at high intensities, as in the case of high background imaging. We have found that the MLP algorithm, equation (7), can also achieve good image restoration results for the case of Gaussian noise.

The residual image R should approach the noise N not only in mean and variance but in all properties. If the additive noise is Gaussian *white noise*, then there should be no correlation between pixels. Recalling the definition of autocorrelation, $A_N = N \otimes N$, the "white" property of the noise gives

$$\langle A_N(i,j) \rangle = \begin{cases} n\,\sigma_N^2 & \text{if } i = j = 0 \\ 0 & \text{otherwise} , \end{cases}$$

where $\langle\,\rangle$ denotes expectation and n equals the total number of pixels in the image. Since there is no spatial information in the white noise there should be no spatial information left in the residual image after deconvolution. This leads to a new figure of merit for the goodness of fit, the Maximum Residual Likelihood (MRL) statistic, which explicitly incorporates spatial information into the deconvolution.

The MRL statistic, developed by Piña and Puetter (1992), is defined as

$$\chi^2(A_R) = \sum_{i,j} \left(\frac{A_R - \langle A_N \rangle}{\sigma(A_N)} \right)^2 \tag{11}$$

where the sum is over nonredundant lags (i,j), since $A_R(i,j) = A_R(-i,-j)$. The range of practical lags is determined by the FWHM of the PSF. This MRL statistic follows the standard χ^2 distribution with f degrees of freedom, where

f is number of terms in the sum of equation (11). Another likelihood statistic is then defined by setting $p(D|I,M) = p(\chi^2(A_R))$, which is maximized when $\chi^2(A_R) = f-2$. Then the autocorrelation of the residuals is consistent with the white noise model and the residuals have minimal spatial structure. However, the assumption of white noise is sometimes unrealistic since detectors can exhibit noise which is slightly correlated. In such a case the correlation usually occurs over adjacent pixels, so the MRL statistic can still be applied by leaving out the single pixel lags from the sum in equation (11).

The goal of maximizing $p(\chi^2(A_R))$, or equivalently, minimizing the difference between $\chi^2(A_R)$ and its most probable value $f-2$, is accomplished by employing the "conjugate-gradient" method for n-dimensional function minimization. The resulting iterative MRL algorithm is computation intensive, since, at each iteration it requires computing $\nabla_I[\chi^2(A_R)]$. However, the MRL statistic proves to be a useful measure for the goodness of fit when using other deconvolution algorithms. We find that as iteration of a deconvolution algorithm proceeds, the point at which the MRL statistic reaches a minimum, or when it stops decreasing at a steady rate, indicates a reasonable cutoff point for the deconvolution iterations.

3. Software Features and Options

DeConv_Tool provides options for the setup of the observed image and PSF, and the user input is via mouse-menu choices and/or direct keyboard entry. A more advanced widget interface for setup and control will be implemented soon. Starting with an image containing the PSF, any subregion of the image can be selected and extracted, and the result is automatically centered and normalized for use in convolutions. Optionally, the observed PSF can be modeled analytically, where model parameters are estimated by fitting x-y profiles. If the PSF is not supplied, an analytic PSF will be used by inquiring the user for model profile parameters. The PSF model is either a Gaussian functional form or a modified Lorentzian form (Diego 1985), defined as

$$\begin{aligned} H &= \frac{C}{1+\alpha^{P(1+\beta)}} \qquad (12) \\ \alpha &= \sqrt{\left(\frac{x-x_0}{r_x}\right)^2 + \left(\frac{y-y_0}{r_y}\right)^2} \\ \beta &= \sqrt{\rho_x(x-x_0)^2 + \rho_y(y-y_0)^2} \ , \end{aligned}$$

where (x_0, y_0) is the center pixel, (r_x, r_y) specifies the radius at half-max of the PSF x-y profiles, and parameters P, ρ_x, ρ_y control the steepness of the profiles.

The standard deviation, σ_N, of Gaussian noise in the observed image is determined automatically by examining histograms of the local image variances around each pixel and finding the most probable variance. Background (mean sky) level is then determined by fitting a Gaussian having the determined σ_N to the histogram of image data values. The user can adjust the sky level, and specify an alternate value for σ_N if desired. Any subregion of the observed image can be extracted and/or a larger region can be defined for the subsequent computations, to make, for example, the size a power of two for optimal FFT convolutions. Empty pixels can then be substituted with appropriate Gaussian noise or a constant value.

Progress of the deconvolution is monitored automatically at every b^m iterations ($m = 1, 2, 3, ...$), where the base b can be specified to give more or less frequent monitoring, and default base $b = 2$ (setting $b = 1$ causes every iteration to be monitored). In addition, the user can press a key at any time to get the current status, or press "P" to pause with options to save results, adjust colors, or create a zoom window. During the monitoring events DeConv_Tool displays: the current deconvolution result (I) alongside the data (D), the residual image ($R = D - I * H$), and the residual autocorrelation ($A_R = R \otimes R$). Also plotted is the histogram of the residual image, with comparison to a Gaussian curve of same variance. For the case of Poisson noise distribution, since the variance equals the mean, the residuals are first divided by $I * H$ so that all variances are normalized. Each monitoring event also prints a line of statistics giving: the mean and standard deviation of residuals (Gaussian noise case), the log-likelihood and RMS difference (Poisson noise case), and the ratio of the residual autocorrelation MRL statistic to the most probable value: $\chi^2(A_R)/\chi^2(A_N)$. These statistics can be monitored to determine convergence and stopping iterations. Also printed are the total flux conservation ratio $\sum I / \sum D$, the current maximum of the restored image, and FWHM of x-y profiles.

All deconvolution statistics are stored in IDL structures and are automatically saved every b^m iterations ($m = 1, 2, 3, ...$). The observed image, PSF, and deconvolved images can also be saved, if requested. By "save" we mean using the IDL **save** procedure which writes to a file, using External Data Representation (XDR), the designated variables with IDL descriptors so that they can later be easily and completely "restored" into memory. Thus at a later time, the computations can be restored and replayed using the procedure DeConv_Review, giving the same output as DeConv_Tool, and further, the statistics for all iterations are then concatenated to form a structured array for convenient examination and graphical display.

In summary, DeConv_Tool is a modular IDL software package which provides a convenient and powerful environment for applying deconvolution algorithms. The modules (procedures and functions) comprising it can also be utilized separately for special image analysis tasks, and of course, any stage of the computations can be manipulated and examined using IDL.

References

Diego, F. 1985, PASP, 97, 1209

Landsman, W.B. 1993, this volume

Lucy, L.B. 1974, AJ, 79, 745

Meinel, E.S. 1986, J. Opt. Soc. Am. A, 3, 787

Piña, R.K., & Puetter, R.C. 1992, PASP, 104, 1096

Richardson, B.H. 1972, J. Opt. Soc. Am., 62, 55

Shepp, L.A., & Vardi, Y. 1982, IEEE Trans. Med. Imag., MI-1, 113

Weir, N. 1991, in Proceedings of the 3rd ESO/ST-ECF Data Analysis Workshop

Astronomical Data Analysis Software and Systems II
ASP Conference Series, Vol. 52, 1993
R. J. Hanisch, R. J. V. Brissenden, and J. Barnes, eds.

MEM Package for Image Restoration in IRAF

Nailong Wu

SCARS/STSDAS, STScI, 3700 San Martin Drive, Baltimore, MD 21218

CCS, NOAO, 950 N. Cherry Ave., P.O. Box 26732, Tucson, AZ 85726

Beijing Astronomical Observatory, Chinese Academy of Sciences, Beijing 100080, China

Abstract. The MEM package called `mem0` for image restoration written in the SPP language in IRAF is described. Emphasis is put on the task `irme0b` for image restoration. Its underlying algorithm, programming, and usage are introduced in some detail. Its merits and limitations are discussed. Finally, the future development of this package is proposed.

1. Introduction

The Maximum Entropy Method (MEM) is a well-known technique for image restoration. It has interested many researchers. However, the use of MEM programs is not as widely spread as it could be mainly because they are not gratis to the public.

Now a MEM package in IRAF is available. The first version of this package, called `mem0`, was released in May 1992 . The package has been updated to version b, and is now in the IRAF archive ready for delivery by `ftp`. An accompanying *Technical Report No. 2* is also available.

The package `mem0 v.b` consists of five tasks.

1. `irme0b`: performs 2-D deconvolution for an image using MEM.
2. `immakeb`: generates an image having an object of Gaussian type.
3. `imconvb`: convolves an image with a point spread function.
4. `irfftesb`: tests the 2-D FFT procedure especially for its speed.
5. `foolfactor`: factorizes a natural number to its prime factors.

The most important and complex task is `irme0b`. We will describe this task in some detail in this paper. The functions of other tasks can be understood from the above brief description. Their usage, etc., can be found in the IRAF on-line help.

2. Algorithm

The task **irme0b** is for image restoration using MEM. The entropy of the image (in the 1-D notation)

$$H2 = -\sum_j b_j \log(b_j/em_j) \tag{1}$$

is maximized subject to the data (mis)fit constraint in the image domain

$$E = \chi^2 = \sum_j |\sum_k p_{jk}b_k - d_j|^2/\mathrm{var}(j) \leq E_c \tag{2}$$

and the total power constraint

$$F = \sum_j b_j = F_{ob} \tag{3}$$

where b_j represent the image to be restored, m_j the prior estimate of the image, d_j the degraded image, var(j) the noise variance; p_{jk} is the p.s.f.; the subscript c stands for critical value, and ob for observational value.

Form the objective function

$$J = H2 - \alpha E - \beta F \tag{4}$$

then use the Newton method to find an ME solution, i.e., maximize J for particular values of the Lagrange multipliers α and β. The desired values of E and F are achieved by choosing appropriate α and β.

The Newton method is efficient in optimization. However, its exact implementation in image restoration needs inversion of matrices of large size in order to calculate the changes in the iteration. This cannot be done in practice. So some kind of approximation is inevitable. In the underlying algorithm for this task, the solution is to simply ignore the non-diagonal elements of the p.s.f. (p_{jk}) and increase the diagonal ones in compensation so that the matrix $\nabla\nabla J$ becomes a diagonal one. In this way the inversion of the matrix becomes a simple operation (for details, see Cornwell & Evans 1985).

3. Programming

The SPP (Subset Preprocessor Language) in IRAF is a reasonably high level language for image restoration programming using methods like MEM. The `clio` package provides the user with a flexible means to enter parameters. The dynamic memory allocation mechanism makes it possible to use the core memory efficiently. However, one is required to follow rather strict standards and conventions for a uniform programming style and readability, and to eliminate avoidable errors.

The source code `t_irme0b.x` consists of 15 procedures of its own having a bit more than 1000 lines. It also shares 8 procedures with other tasks.

4. Usage

Running a MEM program is intrinsically difficult in that quite a few parameters should be entered by the user. In order to choose parameters properly, the user needs to understand MEM to some extent and to have some experience. Much effort has been made in writing the program to set reasonable default values for some parameters, and to automatically calculate some of the others. The user must carefully read the on-line help or technical report before the first attempt to run the task. The output messages displayed while running the task are useful for monitoring the iteration process.

A detailed description of the usage of the task `irme0b` is beyond the scope of this paper. The following are some suggestions and comments.

(1) *Start with simple images.* If you have not run any MEM program before, you should run the task on a synthetic image of simple configuration first. Use the tasks `immakeb` and `imconvb` to make a degraded image having 2 or 3 peaks and a point spread function. In this case you know all the parameters related to the images. You can run the task repeatedly with different values of other parameters, and analyze the results. The experience you gained in this way will be helpful in setting parameters for processing more complex images.

When you process real images, you should also begin with images having simple configuration. If you attempt to run the task on a complex real image before doing any "warm up" exercise, it is likely that you will get nothing useful but two-dimensional junk.

(2) This task works with input images of arbitrary size. However, the user still needs to spend some time preparing the input images. Six real and two complex arrays are needed in the deconvolution procedure. The array size is determined by the maximum of the degraded image and p.s.f. file sizes. So, for example, to process a 512×512 image, the required core memory is somewhat more than $1.0 \times 10 = 10$ MB.

As a general guideline, keep the p.s.f., prior image and ICF to the smallest sizes possible. Perform deconvolution only on the degraded image's area of interest. The output image will have the same size as this area. In this way not only can the memory requirement be kept to a minimum, but also the CPU time can be reduced. However, the user should be aware of the aliasing problem due to the convolution performed by the FFT technique.

From the point of view of FFT speed, the array size should be a power of two if possible. If this is impractical, then the user should avoid using an array size (usually equal to the input degraded image area size) having a large prime factor. As a good example, on a Sun 4/370 a 128×128 FFT takes 0.42 second, a 127×127 FFT takes 6.9 seconds, and a 512×512 FFT takes 8.7 seconds.

In order to assist the user in choosing the array size, two tasks are provided: `foolfactor` used to factorize a natural number, and `irfftesb` used to determine the FFT speed. They are easy to use.

5. Merits and Limitations

Generally speaking, MEM is very good in resolution enhancement and noise suppression. However, because of its high degree of nonlinearity, its mathematical treatment is complex, and its programming is rather difficult.

From the user's point of view, the most serious problem is that it is hard to use the MEM images for photometry. MEM tends to suppress lower peaks. The difficulty in choosing proper parameters may be another problem. During an iteration quite a few arrays are needed to hold the input and intermediate images, and a large number of 2-D FFTs are performed. With the arbitrary FFT size and usually an advanced computer system at an IRAF site, the problems concerning the core memory and computational time are no longer as serious as they used to be.

Regarding the task `irme0b`, it is flexible and reasonably easy to use. The input images can have arbitrary sizes. Some basic parameters can be automatically calculated. Gaussian and Poisson noises can be handled. The results are satisfactory in most cases. However, the zeroth-order approximate Newton method used in this level 0 package may result in slow convergence, especially when the number of pixels, i.e., volume of the p.s.f. is large. The development of a better task based on the first-order approximate Newton method is under consideration.

6. Concluding Remarks

Programming MEM for image restoration is a challenging task. A considerable amount of work has been done to complete the original and updated versions of the `mem0` package.

The function of this package will be enhanced in its future development. The revisions may include:

1. Use subpixelization (overresampling) technique.
2. Use Poisson distribution exclusively in noise model (no χ^2-statistic).
3. Multichannel deconvolution.
4. MEM for photometry.
5. Space-variant restoration.
6. Use the first-order approximate Newton algorithm (level 1 package).

This MEM task should be compared with other MEM programs and other image restoration programs.

It is now the time to encourage IRAF users to run the package, find bugs, make requests, or even revise the source codes. Given feedback, the programs can be revised accordingly. We hope that the package will evolve from a simple to more sophisticated package with the author's and users' efforts.

Acknowledgments. The author is most grateful to the IRAF group at NOAO and STSDAS group at STScI for their assistance in developing the package.

References

Cornwell, J., & Evans, K. 1985, A&A, 143, 77

Astronomical Data Analysis Software and Systems II
ASP Conference Series, Vol. 52, 1993
R. J. Hanisch, R. J. V. Brissenden, and J. Barnes, eds.

HST Image Restoration: Current Results and Post-Servicing Mission Prospects

Robert J. Hanisch and Jinger Mo

Space Telescope Science Institute, 3700 San Martin Drive, Baltimore, MD 21218

Abstract. The spherical aberration in the primary mirror of the Hubble Space Telescope causes more than 80% of the light from a point source to be spread into a halo of radius of 2–3 arcsec. The point spread function (PSF) is both time variant (resulting from spacecraft jitter and desorption of the secondary mirror support structure) and space variant (owing to the Cassegrain repeater optics in the Wide Field / Planetary Camera). A variety of image restoration algorithms have been utilized on HST data with some success, although optimal restorations require better modeling of the PSF and the development of efficient restoration algorithms that accommodate a space-variant PSF. The first HST servicing mission (December 1993) will deploy a corrective optics system for the Faint Object Camera and the two spectrographs and a second generation WF/PC with internal corrective optics. As simulations demonstrate, however, the restoration algorithms developed now for aberrated images will be very useful for removing the remaining diffraction features and optimizing dynamic range in post-servicing mission data.

1. The HST Imaging Problem

The Hubble Space Telescope (HST) suffers from spherical aberration in the primary mirror (see, for example, Burrows et al. 1991). The aberration has been measured as −0.43 waves rms at λ633 nm, which leads to a 4.6 μm optical path length error for marginal rays at paraxial focus. The error in the surface scales as $(\text{radius})^4$. Marginal focus is 43 mm from paraxial focus. The aberration causes the point-spread function (PSF) to cover a region over 5 arcsec in diameter, with only about 15% of the light in a sharp core with radius $0\farcs1$. Had the telescope performed at the design specification 70–80% of the light would have been concentrated within the core. The core of the PSF is formed by the inner region of the mirror, and the halo (comprised of diffraction rings and tendrils) originates from aberrated rays from the outer region of the mirror and from obscurations in the optical telescope assembly (OTA).

There are a number of effects that need to be considered in doing HST image restoration work:

• The PSF for the WF/PC is strongly space-variant. At the edge of the field the internal Cassegrain repeater causes vignetting of the OTA pupil. Even well within the field, however, the WF/PC internal obscurations are offset from the

OTA obscurations, and thus the PSF is formed by rays traveling over differing paths and having differing phase errors as a function of field position.

• The PSF for all instruments is time dependent. Spacecraft motion is induced by the solar array panels, especially as the telescope enters and exits from the earth's shadow. The resulting jitter (with typical frequency of about 0.1 Hz) degrades resolution. The secondary mirror support system is known to be shrinking with time from desorption, giving rise to changes in focus on time scales of months. There is also some evidence for focus changes on time scales comparable to the orbital period, suggesting that the secondary support system may be subject to thermal instabilities.

• PSFs are wavelength dependent. Images taken in broad-band filters will show different PSFs for red and blue objects.

• Both the FOC and WF/PC produce data with low dynamic range. WF/PC is limited by its 12-bit analog/digital converter and the associated read-out noise (13.5 e-) to a dynamic range of about 2000. The FOC response becomes strongly non-linear at count rates exceeding 1 $\sec^{-1}$pixel^{-1} (512×512 pixel format, $f/96$) yielding an optimum dynamic range of about 1700. In practice these values are rarely achieved because of the overlapping halos of bright objects and uncertainties in the PSF models.

• Some restoration algorithms require that bad pixel values (cosmic ray hits, data drop-outs, etc.), be repaired prior to deconvolution.

• WF/PC data suffers from undersampling in the spatial domain, especially in the UV part of the spectrum. Restoration techniques based on Fourier transforms have potential problems with aliasing.

• The modulation transfer function (Fourier transform of the PSF) may have points with zero amplitude. Restoration techniques will have difficulty restoring structures at the corresponding spatial frequencies.

• Noise in the data makes the restoration problem "ill-posed," and one of the major objectives in image restoration is to find techniques that restore the object structure without excessive noise amplification. Since noise in the PSF adds further instabilities to the restoration processing, it is important to be able to compute PSFs based on knowledge of the telescope optics. Computed PSFs are also required for wide-field restorations since observing time is generally not available to map the spatial dependence of the PSF adequately.

• Aside from simply improving HST images qualitatively, astronomers are concerned with the photometric integrity of the data. Restoration techniques must conserve flux and the linearity of response, at least over some well-defined intensity range, or well-calibrated correction procedures must be available.

2. Current Studies

2.1. Jitter

Thermal expansion and contraction of HST's solar panels cause the entire spacecraft to wobble as it moves out of or into the earth's shadow. The spacecraft's guidance system attempts to compensate for the motion but cannot always react to the large magnitude of the resulting pointing error, resulting in a loss of lock on the guide stars. Revisions to the onboard software installed earlier this year have greatly improved HST's tracking performance from jitter with a magnitude of 20–50 milli-arcsec, a level which significantly blurs the PSF, to 7 milli-arcsec rms or less more than 90% of the time, where it has virtually no affect on the PSF or on image restoration. Restoration of archival data for which the pointing stability was poor, or for the small fraction of data that continues to be troubled with poor pointing, will need to utilize jitter functions generated from the engineering data. These can be convolved with model PSFs in order to generate a PSF suitable for restoration processing.

2.2. Secondary Mirror Motion

The desorption curve for the secondary mirror support structure has been monitored since launch. The position of the secondary mirror is known to within ± 5 μm given the best fit to the desorption measurements, although discrepancies as large as 11 μm have been seen. We have done numerical tests on simulated data in order to understand the effects on restoration of using a PSF model computed for an incorrect focus position. For focus errors of 5 μm or less the effects are negligible, whereas errors at the level of 11 μm introduce systematic errors in the profiles of restored stellar images at a level of about 1 part in 3000. Thus, focus errors can in general be ignored except when striving for maximum dynamic range.

2.3. Photometric Linearity

While several restoration techniques (such as Fourier inversion and Wiener filtering) are intrinsically linear, the techniques that are best suited to HST image restoration are inherently non-linear. This does not mean, however, that linearity is *necessarily* corrupted. Our tests of the Lucy-Richardson algorithm (Lucy 1974, Richardson 1972, Snyder 1990) indicate that the photometric response is linear over a dynamic range of at least 6 stellar magnitudes. The linearity of other restoration techniques has been investigated (Busko 1992), and modifications to the maximum entropy method that preserve linearity have been suggested (Weir 1991). In general, MEM techniques are notorious for corrupting photometry and must be used with caution.

2.4. PSF Modeling via Phase Retrieval

PSF modeling software such as TIM (Telescope Image Modeling software; Krist, Hasan, and Burrows 1992) and TinyTIM (Krist 1993) has, to date, been only partially successful in computing satisfactory PSFs for image restoration purposes (e.g., Krist & Hasan 1993). One of the primary problems with the PSF models is that they require a full specification of the telescope optical system, usually given in terms of the Zernike polynomial coefficients and a residual phase error map. This specification is a best overall fit to observed PSFs for a variety of wavelengths and filters, but frequently there are substantial mismatches between the models and real data. Our current efforts in this area, led by Rick White (ST ScI), utilize a non-parametric approach to solving for the residual phase errors. An initial guess for the phase distribution over the aperture which incorporates the known telescope optical characteristics is used as a starting point for an iterative method derived from the maximum likelihood equation for Poisson statistics. The resulting algorithm is similar to the Richardson-Lucy

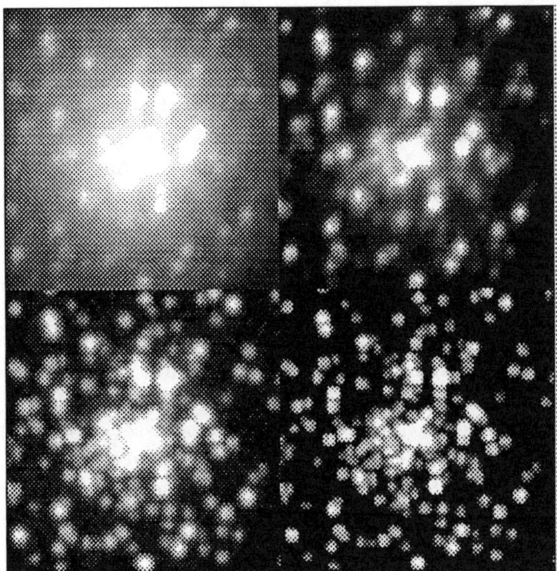

Figure 1. HST simulated images of a star cluster. *Upper left:* current telescope performance. The model is convolved with the current PSF. Poisson noise and detector read-out noise have been added. The gray scale runs from 0 to 10% of peak flux. *Upper right:* Lucy restoration of frame at upper left. Gray scale for this and subsequent frames runs from 0–5% of peak flux. *Lower left:* WFPC-II performance. The model is convolved with the WFPC-II PSF assuming perfect correction for the aberration in the primary mirror, and Poisson noise and detector read-out noise have been added. Note added sensitivity to fainter stars. *Lower right:* WFPC-II after restoration. Resolution is improved by approximately 20–25% after restoration. Actual results from in-orbit data will depend on pointing behavior and the performance of the corrective optics in WFPC-II.

iteration: at each step of the iteration the phase map is adjusted by comparing the observed PSF with the computed PSF. After about 20 iterations a PSF is computed that is in much better agreement with the data. This approach works well when an approximate solution for the large-scale phase errors is already known, but tends not to converge properly if the initial guess is poor. We do not yet have enough experience with this new technique to determine how useful it will be in *predicting* PSFs as a function of wavelength or camera position, as opposed to simply *fitting* PSFs for a particular image.

A second potential problem with the approach used by TIM and TinyTIM is that they model the obscurations in the telescope as if they all occur in the same diffraction plane. For the WF/PC in particular, which uses a Cassegrain repeater system that introduces obscurations that are field-angle dependent, this assumption is incorrect and may lead to significant errors in the PSF model. This problem is now being investigated further.

3. The HST Servicing Mission and Image Restoration

The first servicing mission for HST is currently scheduled for December 1993. The plan for this mission includes replacement of the solar panels, removal of the High-Speed Photometer and its replacement by COSTAR (Corrective Optics Space Telescope Axial Replacement instrument), removal of WF/PC and its replacement by WF/PC II (which has internal corrective optics), and repair of several other minor problems. COSTAR will correct the spherical aberration for the two spectrographs and the Faint Object Camera.

We have done simulations to compare current and post-servicing mission images and the benefits of applying the same restoration techniques to aberration-free data. Typical results are shown in Figure 1. The simulations assume perfect correction for spherical aberration and perfect knowledge of the PSF, and thus represent a best-case scenario.

Application of restoration algorithms to post-servicing mission data will continue to have a major benefit. Substantial improvements in dynamic range and spatial resolution are achievable even in the ideal case we have simulated, and these techniques can help correct for any residual optical errors resulting from an incomplete prescription in COSTAR or WF/PC II. The corrective optics are necessary, of course, in order to recover absolute sensitivity and dynamic range, but restorations of high signal-to-noise data now achieve virtually the same spatial resolution as will be obtained for raw data after the servicing mission.

Acknowledgments. Collaborators in this work include Rick White and Nailong Wu (ST ScI).

References

Burrows, C.J., et al. 1991, ApJ 369, L21

Busko, I. 1992, submitted to PASP

Krist, J.E., Hasan, H., & Burrows, C.J. 1992, in Astronomical Data Analysis Software and Systems I, A.S.P. Conf. Ser., Vol.25, eds. D.M. Worrall, C. Biemesderfer, & J. Barnes, 223

Krist, J. 1993, this volume

Krist, J., & Hasan, H. 1993, this volume

Lucy, L.B. 1974, AJ 79, 745

Richardson, W.H. 1972, J. Opt. Soc. Am. 62, 55

Snyder, D. 1990, in The Restoration of HST Images and Spectra, eds. R.L. White & R.J. Allen, Space Telescope Science Institute, p. 56

Weir, N. 1991, in Proceedings, Third ESO/ST-ECF Data Analysis Workshop, eds. P.J. Grosbøl & R.H. Warmels, European Southern Observatory Conference and Workshop Proceedings No. 38, p. 115

Discussion

Misra: Did you try MEM for image restoration?

Hanisch: Yes, we use MEM extensively, but not without some caveats. Most MEM implementations introduce flux non-linearities, and these may be of concern for certain applications. For morphological work and especially for difficult brightness distributions (such as a strong point source superimposed on an extended object) MEM techniques are very valuable and constitute a major part of the image restoration toolkit we provide to HST observers. ST ScI provides a public domain MEM package (see Wu, this volume) and can provide the MEMSYS 5 package to users working solely on HST data.

Astronomical Data Analysis Software and Systems II
ASP Conference Series, Vol. 52, 1993
R. J. Hanisch, R. J. V. Brissenden, and J. Barnes, eds.

Deconvolution of HST WFPC Images using Simulated PSFs

J. Krist and H. Hasan

STScI, 3700 San Martin Drive, Baltimore, MD 21218

Abstract. Given the difficulty of obtaining high quality point spread functions for deconvolution of Hubble Space Telescope images, some have turned to computer generated models. We have briefly studied the use of simulated PSFs in deconvolving Wide Field/Planetary Camera (WFPC) images.

1. Introduction

HST image deconvolution requires a linear, unsaturated PSF with a high signal-to-noise ratio, no cosmic rays, and taken through the correct filter at the proper focus position. In addition, the WFPC has a field dependent PSF, due to its additional set of obscurations. Obtaining PSFs which match these constraints is difficult, so using models generated for a particular set of conditions is an enticing possibility.

However, because our knowledge of the HST optics is not complete, the models do not perfectly match the real PSFs, especially in the shorter wavelengths. The lack of an accurate map of the mirror zonal errors and uncertainties about the camera obscurations lead to errors in the simulations. Given these limitations, it is still worthwhile to determine up to which point models are usable in deconvolutions.

2. Generating PSFs

Currently, the only widely available HST PSF generators are TIM and Tiny Tim. TIM (Telescope Image Modeling) was written by Chris Burrows and Hasan at STScI for the analysis of the HST optics, and was later used for generating PSFs for deconvolution. It is a large and complex program with a steep learning curve. However, it can be used for other optical systems and has a large number of options, including dust and microroughness simulation. It runs only on VAX VMS systems and is distributed as executables.

Tiny Tim was written by Krist to provide simple and fast PSF generation capabilities. It is written and distributed in C and has been used on many computer platforms. It can only model HST cameras.

These programs can model the position dependence of the WFPC PSF, as well as produce PSFs for a given focus position. Both systems are available via anonymous ftp to stsci.edu and are located in the software/tinytim and software/tim directories.

3. Deconvolutions

In all of our tests, we used forty to eighty iterations of the Lucy-Richardson algorithm.

Our first test was on a single observed Planetary Camera (PC) PSF taken through the filter F439W. At this wavelength, the mirror maps have a significant affect on the rings, highlighting some and attenuating others. We used forty iterations of the Lucy-Richardson algorithm to deconvolve the star with itself, with the same star at a slightly different focus, and with a model. The resulting encircled energy curves are shown in Figure 1. In all cases the deconvolutions significantly increase contrast of the core to the wings. The defocused, observed PSF did a little better in the wings, where the errors in the models tend to be greater. The pixel-to-pixel residuals in the model's deconvolution were higher than those for the others, but not by a great amount. From this test it seems that the models are useful, at least for a first look, for deconvolving PC data down to about 4000 Angstroms. The residuals would increase as one goes to shorter wavelengths because of the mirror zonal errors.

The next test was of the field dependent effects in the Wide Field Camera (WFC). A grid of PSFs at different positions on a single WFC chip was obtained by Roberto Gilmozzi at STScI. From this grid, we extracted two PSFs, one from near the center of the chip and another from the edge, where the vignetting from the WFC entrance aperture causes severe attenuation of the inner-facing wings.

The central PSF was deconvolved with a model generated for that position and with an observed, nearby (100 pixels distant) PSF. The results highlighted an additional attenuation of one side of the wings which occurs uniformly over the field. Since this effect is not modeled by the current software, the simulated PSFs tend to remove too much flux from this portion of the wings. This can be remedied, however, by adding another vignetting rim at the edge of the pupil. What is causing this attenuation is unknown. The result using the observed PSF was better in terms of residuals in the wings than the one using the model, even though the PSF was taken at a different location. However, deconvolution with the model produces a sharper core, which is consistent with the fact that the simulations are more accurate in the core than in the wings.

The results for deconvolving the PSF from the edge of the chip were quite different. The PSF there is extremely position dependent, due to the entrance aperture vignetting and the WFC secondary mirror shadow. When deconvolved with an observed PSF taken from the center of the field (thus lacking the same vignetting in the wings), the star was left with a large "pit" where the flux was removed. On the other hand, the deconvolution using a model generated for that star's position provided a much better result. Thus, if an object to be deconvolved is at the field edge and an observed PSF is not available from very near the same location, one is better off using a model.

Our last tests were on extended objects in the WFC, namely Saturn and the Eta Carinae nebulosity. In both cases, the results agreed well with those from deconvolutions using observed PSFs. However, near the edges of Saturn, a faint, dark line could be seen, caused by the wing attenuation mentioned above. With an additional obscuration, this will likely go away in the models.

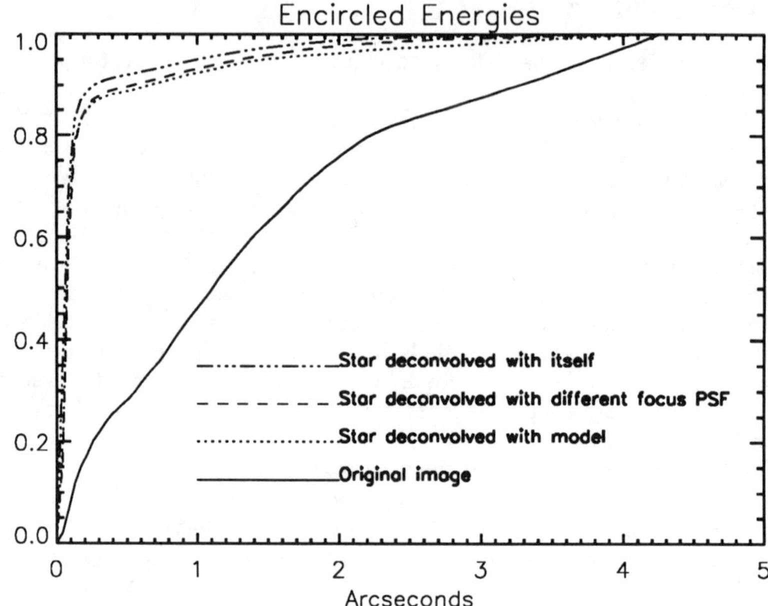

Figure 1. These are encircled energy curves for a Planetary Camera PSF deconvolved with itself, an observed PSF with a slighty different focus, and a model PSF. Forty iterations of the Lucy-Richardson method were used. In all cases, a significant improvement in the flux distribution was obtained. Note that the original specification for HST was to put about 70% of the light into a 0.1 arcsecond radius.

4. Conclusions

Until we better understand the HST telescope and camera optics, high quality observed PSFs will remain the best solution for deconvolution. However, when a good, observed PSF is not available, models can be successfully used. When deconvolving objects at the edges of WFC chips, or when the only available PSF is at a significantly different focus, then models will likely do a better job. If better maps of the mirror zonal errors can be produced, then the quality of simulations will substantially increase.

Astronomical Data Analysis Software and Systems II
ASP Conference Series, Vol. 52, 1993
R. J. Hanisch, R. J. V. Brissenden, and J. Barnes, eds.

Telescope Image Modelling Software in STSDAS

P. E. Hodge, J. D. Eisenhamer, R. A. Shaw, and R. L. Williamson II

Space Telescope Science Institute

Abstract. A software package for creating model point spread functions is being modified to run under IRAF.

1. Introduction

The Telescope Image Modelling (TIM) system written by Chris Burrows and Hashima Hasan at the Space Telescope Science Institute creates model point spread functions (PSFs) for optical systems based on ray-trace information. Aside from its usefulness as an analysis tool, TIM is used for generating PSFs for deconvolving images taken with the Hubble Space Telescope (HST). Model PSFs are also being explored as a means to improve the accuracy of digital photometry software. The original TIM runs only on VAX/VMS systems, and it calls subroutines in the Numerical Algorithms Group (NAG) and IMSL packages. We are modifying a copy of TIM to run under IRAF in the STSDAS package in order to make TIM portable, and therefore more widely available. Initially the changes will be restricted to the user interface and replacing VAX-specific code. Soon thereafter the NAG and IMSL subroutines will be replaced by public-domain software.

TIM should not be confused with Tiny Tim, an image modelling program written by John Krist, also at STScI.

2. Overview of TIM

While our emphasis will be on generating PSFs which model HST PSFs, TIM can also create modulation transfer functions and ideal PSFs. The effects of spacecraft jitter, dust and surface roughness can be included. The basic approach is to create a complex image of the telescope aperture; then the PSF is the squared amplitude of the complex Fourier transform of the aperture image. The aperture image has an amplitude of unity where the aperture is clear and an amplitude of zero outside the aperture and within obstructions such as the spider and the secondary mirror. The phase of the aperture image at a pixel is the wavefront error of the optical system for a ray that enters the optical system at the location of that pixel. For example, for perfect optics the phase would be zero, and for a focus error the phase would be parabolic with vertex at the aperture center.

The entire process is done in several steps, called "stages." Stage I computes the Fourier transform of the aperture image. Stage II then creates a PSF or modulation transfer function. The pixel spacing of the images output by the first two stages is set by the software to be critically sampled. Integration of the PSF to the user-specified pixel size is then performed by Stage III, which is the most time-consuming step. In order to simulate a PSF through a filter with finite bandwidth, multiple wavelengths may be specified when running the first three stages, which results in a PSF for each wavelength. Stage IV then integrates these PSFs over wavelength and offers the option of including other detector effects as well. Stage V, which we will not implement in IRAF initially, creates a realistic simulation of a CCD image by adding noise, cosmic rays, saturation effects, etc.

The user interface is well developed but somewhat daunting for the novice user. Each stage gets some information from one or two ASCII files and prompts the user for a number of parameters. The ASCII files contain such information as polynomial coefficients that describe the optical aberrations and a specification of the aperture obscurations. The user is prompted for the size of the image to create and what options to include. There are typically two dozen such parameters per stage. The images created by each stage are STSDAS format images.

3. Changes for IRAF Version

Each stage in VMS TIM is implemented as a task in the IRAF version. The ASCII files are read the same way in both versions so that the files written by one version can be read by the other. Instead of explicitly prompting for parameters, however, the parameters are fetched from IRAF parameter files which can be set up using `eparam` before running the task. VMS TIM includes plotting options in each stage. These were removed for the IRAF version because IRAF has extensive plotting and image display capability.

Many modifications to the Fortran code were required for IRAF compatibility. These changes are probably typical of what would need to be done when converting Fortran programs to IRAF. The authors used mostly standard Fortran, but we did find a few VAX-specific features. The most common of these were the use of in-line comments and `DO-ENDDO` constructs. A frequently used construct that was more difficult to fix was a variable format expression such as `I<NCHAR>`, where `NCHAR` is an integer variable. These were converted to run-time formats, where the value of `NCHAR` would be written into the format string. The use of tabs and lower case text was not changed, as these appear to be acceptable extensions to any Fortran compiler used by IRAF.

Some features of standard Fortran also cannot be used in IRAF tasks. Data statements for initializing variables can cause problems if the values of the variables are subsequently changed, since variables are not reinitialized each time the task is run, unless the process cache is flushed. We replaced block data and some other data statements with direct assignment of values. Fortran I/O is not permitted in IRAF tasks because the process main is not written in Fortran. TIM does a lot of formatted I/O with ASCII files and the terminal. These were changed to internal reads and writes with character-string buffers; Fortran-to-SPP interface routines in the STSDAS IRAF77 library were then called to read or write the character strings to ASCII files.

Two non-standard constructs were only discovered after attempting to compile the IRAF version on a Cray. While `READ (iounit,*)` and `WRITE (iounit,*)` are standard Fortran if `iounit` is a logical unit number, they are non-standard for internal reads and writes to a character string. We encountered this problem when converting Fortran I/O to internal reads and writes. The other problem was that generic functions such as `MAX` were used in `PARAMETER` statements for declaring the sizes of buffers. This makes good sense, as it shows the dependence between parameters, and it protects against changing one value and forgetting to change another, but it is not standard Fortran.

4. Status of IRAF TIM

IRAF TIM is well along in its development, but some work still remains. Stages I, II and III have been converted and are running under IRAF. Several NAG and IMSL routines are still used, however, so we must modify the code to use public-domain routines instead. In addition, only minimal testing has been done so far, so we expect that some further debugging will be necessary.

Nearly all the code changes for Stage IV have been done, but Stage IV links with XCAL subroutines. XCAL is a system written by Keith Horne at STScI for simulating HST photometry and spectra. A stellar spectrum, the mirror reflectivities, filter throughputs, and detector sensitivities are all combined to predict the count rate as a function of wavelength. Stage IV needs this information in order to combine PSFs at different wavelengths to form a single PSF corresponding to a particular instrument and filter. We will change this to use the STSDAS SYNPHOT library instead.

Burrows and Hasan are continually improving VMS TIM. When they release a new version, they also give us a copy, and we update the IRAF version. This is fairly straightforward for two reasons. We make only the minimal changes necessary for IRAF compatibility, and their changes tend to be in the algorithms, which are easier to update than changes to the I/O statements.

Astronomical Data Analysis Software and Systems II
ASP Conference Series, Vol. 52, 1993
R. J. Hanisch, R. J. V. Brissenden, and J. Barnes, eds.

Tiny Tim : An HST PSF Simulator

J. Krist

STScI, 3700 San Martin Drive, Baltimore, MD 21218

Abstract. Tiny Tim is a portable program for generating synthetic point spread functions for the Hubble Space Telescope cameras. PSFs generated by it have been used for deconvolution, optical analysis, and photometric studies.

1. Introduction

The Tiny Tim PSF generator was designed to provide synthetic PSFs for use in deconvolving Hubble Space Telescope images. It is written and distributed in C, and has been used on a wide variety of systems. Ease of use, speed, and accuracy (given our current knowledge of the telescope) are the key features of the software.

Tiny Tim consists of two programs. The first, **tiny1**, asks the user a few questions and generates a parameter file. This file is the input to the second stage, **tiny2**.

2. Inputs

The user is asked for the following:

- The camera being modeled (WFPC, FOC, WFPC II, COSTARed FOC)
- The position on the detector (if WFPC or WFPC II)
- The date of observation (for the first generation cameras)
- The filter (or wavelength)
- The size of the PSF in arcseconds
- Whether or not to include jitter, and if so, how much
- Subsampling factor (if camera undersamples the PSF)

The program contains important parameters for each of the cameras (focal ratio, pixel size, rotations).

If the WFPC or WFPC II camera is being modeled, the user is asked for the position of the PSF on the detector, since the field dependent variability of the PSF is taken into account. A list of positions can also be entered.

Since the telescope has been refocused at various times, the date of observation is needed in order to determine the proper focus parameter. Tiny Tim contains a table of the secondary mirror moves. Between mirror moves, the affect of desorption is accounted for by an additional focus offset, using equations based on studies by Chris Burrows and Hashima Hasan at STScI.

To reasonably model the wavelength dependence of the HST PSF, three to five monochromatic PSFs are generated and summed with weighting to account for the bandpass of a given filter. Tiny Tim contains wavelengths and weights for each WFPC and FOC filter for a selected range of object colors.

The size and wavelength values will determine the necessary grid sizes for computing the PSF. The shorter the wavelength, the larger a grid size is needed for computing the critically sampled PSF. Thus, it will take longer to generate a blue PSF than a red one.

Jitter is modeled by convolving the PSF with an appropriate bidimensional Gaussian function. If the camera undersamples the diffraction limited PSF in the given wavelength range, then the user is allowed the option of computing a subsampled PSF, which is useful for enhanced deconvolution.

All of the above parameters are stored in a file. The aberration values are given as coefficients to Zernike polynomials, and can be changed by editing the file.

3. Computing the PSF

The program **tiny2** reads in the parameter file and does the actual PSF generation. The optical path differences function (OPD) is computed as the sum of the Zernike polynomials for each aberration, multiplied by their respective coefficients (from the parameter file). A map of the zonal errors in the primary and secondary mirrors is added into the OPD. The aperture function, which is simply a mask containing the obscurations, is generated based on the known camera and telescope components. The OPD and aperture function form the pupil function. By taking the Fourier transform of the pupil function, the amplitude spread function (ASF) is obtained. Taking the modulus squared of the ASF produces the PSF.

The above PSF is generated to be critically spaced (Nyquist spaced) at the required wavelength. To integrate it over detector pixels, it is first FFTed to the optical transfer function (OTF) and multipled by the analytical expression for the integrated OTF of a detector pixel. While still in OTF space, the function is multipled by a Gaussian to simulate jitter. The result is FFTed back, producing a critically spaced PSF with detector sized integrated pixels. By using sinc interpolation, a PSF with detector sized and spaced pixels is obtained.

The above process is repeated for a series of wavelengths. The PSFs are added together with weights to form a polychromatic PSF which approximates the affect of a specified filter's bandpass. This is written out as either a FITS or STSDAS file.

Here is a list of execution times on a Sparcstation 2 for polychromatic, six arcsec (268 by 268 pixels) FOC f96 PSFs:

Filter Name	Component Wavelengths (nm)	Critical Grid Sizes (n by n)	Execution Time
F120M	115 120 125	1280 1280 1280	20 min
F190M	180 195 210	1024 768 768	12 min
F346M	320 340 370	512 512 512	6 min
F630M	600 630 650	256 256 256	2 min

4. The Use and Limits of Tiny Tim PSFs

While the Tiny Tim software produces high quality simulations, our limited knowledge of the HST optics creates a disagreement between theory and reality. The most significant problem is the need for a better map of the mirror zonal errors (left by the polishing tool), since these dominate the structure in the UV. At longer wavelengths, the models agree better with the observed PSFs since structure is instead dominated by the obscurations. Since the Faint Object Camera operates primarily in the shorter wavelengths, models for it are currently poor.

The ability to produce PSFs at any location, through any filter, and at a given sampling is an important feature. This allows the WFPC simulations to be fairly accurate, though the obscuration positions and aberrations need to be refined.

The software is continuously updated with the latest information on focus positions and desorption. It is available via anonymous ftp from stsci.edu in the directory software/tinytim.

Part 5. Data Analysis Applications

Section E. Data Formats

Astronomical Data Analysis Software and Systems II
ASP Conference Series, Vol. 52, 1993
R. J. Hanisch, R. J. V. Brissenden, and J. Barnes, eds.

FTOOLS—A FITS Utility Package for Multiple Environments

W. Pence, J. K. Blackburn, and E. Greene

HEASARC, Code 668, NASA/GSFC, Greenbelt, Md, 20771

Abstract. FTOOLS is a collection of general analysis programs to create, examine, or modify data files in FITS format. These programs use the FITSIO subroutine interface to access the FITS files. Each task is modular and performs one type of operation which can be strung together in command files to perform more elaborate functions. Many of these tasks are designed to operate on FITS tables (either ASCII TABLE or BINTABLE extensions) but FITS IMAGE extensions and primary arrays are also supported by many of the tasks. The FTOOLS package is very portable and can be run either as a set of IRAF tasks or as standalone executable programs.

A set of programs to create, examine, or modify data files in FITS format has been developed by the High Energy Astrophysics Science Archive Research Center (HEASARC) at Goddard Space Flight Center. This package is called FTOOLS and currently contains more than 40 individual tasks. A few of the tasks are specifically designed for high energy astrophysics applications, but most of them are very general and may be used on any type of FITS file. For example, there are tasks to dump the header or data portions of FITS files to ASCII format, to select rows out of a FITS table based on an arbitrary boolean expression, to extract or append FITS extensions from one FITS file to another, and to compute statistics or generate histograms of the values in a FITS table.

This package was developed because of the need to create and maintain a large variety of data files in a heterogeneous computer environment. FITS was the natural format to adopt because it is widely used throughout astronomy and because the format is machine independent and hence the FITS files may simply be copied from machine to machine without having to be concerned about the native file formats and number representations on each machine. After deciding to adopt FITS formats, we then needed a set of software to be able to create and manipulate files in this format.

The FTOOLS tasks are designed to be very portable. Most of the code is written in standard Fortran-77, and all interfaces to the data or to the user are isolated through standard subroutine calls. All of the data I/O is channeled through calls to the portable FITSIO subroutine interface which is available for most machines. The user interaction is also isolated through a standard set of subroutine calls which read or write a parameter of a specified data type. This interface is implemented using the same subroutines and the same physical parameter file format as is used in IRAF tasks. (More specifically, it is the same interface used in the F77 interface to IRAF that was developed at the STScI). Because we use the IRAF conventions, it it very easy to link the FTOOLS software as an IRAF package which can be run on any machine which has IRAF already installed.

For non-IRAF users, a similar parameter interface library has been developed at SAO which can be used to link the FTOOLS subroutines into standalone executable programs which can be run directly from the host operating system. This 'Host' interface is currently available for Sun workstations and DECstations, and a VMS port is expected to be available in early 1993.

The FTOOLS package is freely available from the HEASARC in the anonymous FTP account on the legacy.gsfc.nasa.gov machine. Interested users should contact the HEASARC for the latest information about this package.

Astronomical Data Analysis Software and Systems II
ASP Conference Series, Vol. 52, 1993
R. J. Hanisch, R. J. V. Brissenden, and J. Barnes, eds.

FITS Data Conversion Efforts at the Compton Observatory Science Support Center

D. G. Jennings, J. M. Jordan, T. A. McGlynn, N. G. Ruggiero, and T. A. Serlemitsos

COSSC/NASA-GSFC/CSC,Code 668.1,NASA/GSFC, Greenbelt, MD, 20771

Abstract. The Compton Gamma Ray Observatory (CGRO) is an active satellite consisting of four gamma ray telescopes. Each telescope is maintained by an independent team of investigators, and each team has devised separate data formats to handle the needs of their particular instrument.

As mandated by NASA, the Compton Observatory Science Support Center (COSSC) intends to archive and distribute data from all four CGRO instruments in FITS. This paper describes the problems encountered in transcribing large amounts of data into a standard FITS form, the capabilities of the COSSC-built conversion software designed to perform the transformations and the ToFU conversion tools on which this software is based.

1. Introduction

As the name suggests, the Astronomical Data Analysis and Software Systems (ADASS) conference acts primarily as a forum for the astronomical community to exchange ideas about the data analysis software tools currently under development. Implicit in the design of these tools is the assumption that the data will be provided to the community in a form that the software can understand; therefore, facilities which produce data products must take care to assure that they do not invalidate this assumption.

The software development group of the Compton Observatory Science Support Center (COSSC) has been putting a considerable fraction of its efforts into converting Compton Gamma Ray Observatory data from its raw, native format to a standard format that is usable by the astronomical community at large. The intent of this paper is to report to the community the problems encountered, strategies developed and solutions implemented by our group in this endeavor.

2. Background on the Compton Observatory and COSSC

The Compton Gamma Ray Observatory (CGRO) is a low earth-orbiting (450 km) satellite launched in April of 1991 from STS-37. The satellite itself provides an operating environment for four gamma ray telescopes:

- BATSE: Burst And Transient Source Experiment – 8 detectors that provide continuous full sky monitoring from .03 to 2 MeV;
- COMPTEL: COMPton TELescope – an imaging instrument that uses the Compton scattering effect to detect photons from 1 to 30 MeV;
- EGRET: Energetic Gamma Ray Experimental Telescope – a "spark chamber" imaging instrument that detects photons from 20 MeV to 30 GeV; and,
- OSSE: Oriented Scintillation Spectrometer Experiment – 4 movable scintillator/PMT spectrometers that detect photons in the .1 to 10 MeV energy range.

Each instrument aboard CGRO was built by, and is managed by, its own Principal Investigator (PI) team. The PI teams produce data products and analysis software which are tailored to their own needs and the particularities of their instrument.

The COSSC was established by NASA to provide the general astronomical community with an interface to Compton Observatory data, analysis software, and observational opportunities. COSSC's responsibilities include: supporting CGRO guest investigators, providing CGRO data analysis software and analysis environments, building and maintaining the CGRO data archive and converting the PI team supplied data products into a "standard" format for distribution and archival storage.

3. Data Conversion Software Design Considerations

The software system developed to perform the conversion of native PI team data into standard formats needed to account for several constraining factors in its design. Some of these constraints were imposed by NASA policy, some by a fundamental lack of staff, and others by the very nature of the PI team data.

The most significant constraint placed upon the design by NASA was that CGRO data had to be made available to the astronomical community in FITS (Flexible Image Transport System) format. Therefore, at least one of the "standard" formats that the software had to support was FITS.

Personnel constraints provided another set of software design considerations. With approximately one FTE (Full Time Equivalent) to build and maintain the data conversion system, the scope of the design and implementation had to remain simple. The project used a rapid prototyping approach to the problem where software design, testing and implementation were done in parallel. The design also had to provide for a system that was as automated as possible, utilizing schemes that minimize human operator intervention and perform automatic data conversion verification.

The PI-supplied data products created the last set of design constraints. PI data products come in several hundred different formats and the conversion system had to assimilate them all into the data conversion stream. The native data formats are subject to modification by the PI teams; therefore, the conversion software had to allow for fast, error-free updates to the program code and FITS structures. The conversion system's throughput rate needed to accommodate a large amount of data, since the combined PI data product output is anticipated to be 150 GB per year. Finally, the conversion software was required to run on several different machine platforms to span the various computer architectures on which the native data is produced.

In general, computing power and disk space was NOT a constraining factor on the software system design. High performance Unix-based workstations and "kilobuck per gigabyte" disk drives alleviated most of the potential hardware shortfalls. Although, when use of VMS systems is necessary, CPU limitations have been significant.

4. Data Conversion Software Implementation Tools

In implementing a data conversion system with the above-mentioned design constraints, we found it efficient to write a set of software tools that performed most of the housekeeping chores associated with the FITS conversion process. This way, format-specific conversion code could be written quickly in a modular, machine independent fashion and with a minimum of FITS expertise required by the programmer.

The resulting product of this effort is ToFU, which stands for To-FITS Utilities. ToFU consists of a set of fortran callable routines that overlay the FITSIO library (Pence 1992) and run on DECstation, SUN and VAX/VMS platforms. ToFU routines perform three main functions: create and manipulate FITS headers, write generic byte streams into FITS table extensions, and read FITS table extension elements out to a generic byte stream. Error and value checks are done on table keyword values, with incorrect values being corrected if possible.

One of the most useful aspects of ToFU is its template files. Template files are ASCII text files which contain information to build or modify a FITS header. As with X11 resource files, template files allow the conversion software's actions to be changed without code modification or recompilation. Programs written with ToFU parse each line of the specified template file(s) and create FITS headers and tables according to the directions given. Template file syntax is a super-set of FITS header syntax; hence, any valid FITS header constitutes a valid ToFU template file.

ToFU software and documentation is available to anyone who wishes it via anonymous ftp on enemy.gsfc.nasa.gov in the /pub/TOFU directory. While ToFU has had extensive use within COSSC, it is only distributed to others on a "without warranty" basis.

5. Data Conversion Software Implementation Philosophy

To facilitate the building of an automated, error-free conversion system the following implementation philosophy was adopted: FITS data formats should follow the native PI data format structure whenever possible, and all data conversions must be invertible.

By building FITS formats that mimic the original native data formats, a minimum of data manipulation takes place during the conversion process and,

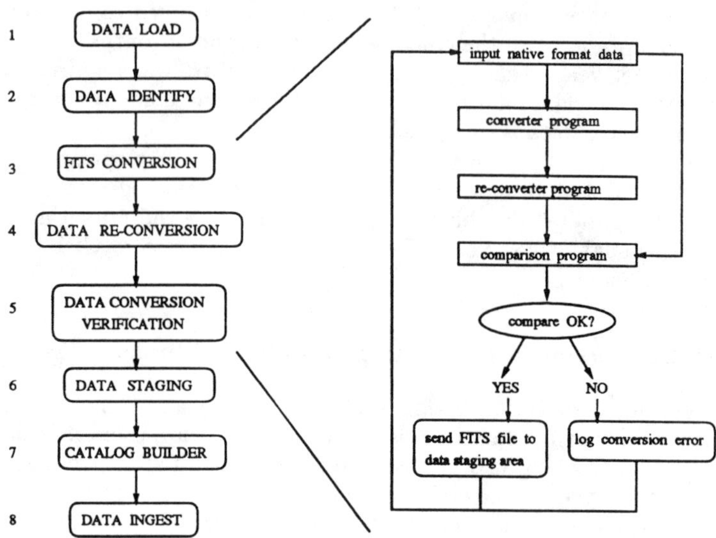

Figure 1. The CGRO data pipeline

therefore, the introduction of error into the data stream is minimized. After this "keep it simple, stupid" (KISS) conversion is made, it is possible to massage the FITS formatted data into other FITS formats that conform to more general conventions, such as those being developed by the HEASARC at NASA-Goddard.

By demanding that all data conversions be invertible, the system can automate data verification; thus, the condition that

$$native\_data \rightarrow FITS\_formatted\_data \rightarrow reconverted\_data == native\_data$$

can be imposed upon the system. In practice, this requires programs be written in FITS converter and FITS "anti-converter" pairs. Then, each time a data file is transformed into FITS it is immediately re-transformed into its original format and a byte-by-byte comparison is made between the original and re-transformed data files. This scheme does not preclude the possibility of compensating errors in the conversion/reconversion process masking data corruption. Human intervention is still required to spot check the native data → FITS conversion.

6. Data Conversion Software System Operation

The left-hand side of Figure 1 shows the entire data pipeline for the Compton Observatory data archive system. Steps 3, 4 and 5 are concerned with the conversion of native PI team data into FITS, and are expanded on the right-hand side of Figure 1. The operator identifies the type of CGRO data entered into the pipeline, and then calls the appropriate data conversion script. The conversion script applies the appropriate FITS data converter to the native data file, reconverts the data converter's output with its anti-converter twin, and compares the original and reconverted data files byte by byte. If the two files match then the script copies the FITS file to a data staging area for ingestion into the CGRO archive, otherwise it logs the conversion error and continues on to the next native data file. Human operator intervention is required only for the data load and data identify steps (also for error conversion errors if they occur), and these steps may also be automated in the future.

7. Concluding Remarks

The Compton Observatory Science Support Center has shown, by example, that a relatively small organization with scant resources can undertake a significant data conversion task for a major NASA mission. The ingredients that make this possible are:

- use of the unconventional, albeit slightly risky, rapid prototyping software design and implementation strategy, as opposed the traditional and structured software lifecycle strategy;
- the commercial availability of "cheap," high powered workstations;
- a set of modular software tools to facilitate the building of the data converter programs (ToFU); and
- the use of the KISS philosophy for FITS format design and data conversion verification.

If there are any other organizations facing similar problems, with data conversion or data archiving, we at COSSC would be happy to hear from you. Please direct all e-mail to jennings@enemy.gsfc.nasa.gov.

References

Pence, W. 1992, FITSIO - A Subroutine Interface to FITS Format Files, HEASARC/GSFC

Discussion

Blum: Our project has 11 instruments each with a different data format. The Data Center has already agreed that all data will be archived in FITS. Two years are being spent on defining keywords/headers.

Jennings: Glad you are planning ahead.

Percival: Why did you choose to deviate from the standard FITS header rules when defining your header template files?

Jennings: The template files are intended for human data entry, so a relaxation of the rules seemed desirable.

Barg: What are the aspects of the GRO format that makes them difficult to convert to FITS?

Jennings: Each instruments' data format is built around or for different hard architectures.

Conroy: What considerations do you make when defining what your FITS file should look like? Were the standard formats being developed by HEASRC for FITS files in the High Energy Astrophysics considered?

Jennings: The most important consideration used by COSSC when defining FITS formats for Compton Observatory data is that the format must completely encapsulate the original instrument team data structure. Every byte of information present in the native data file must go into the FITS file. We also attempt to devise FITS formats which mirror the native data formats as closely as possible in order to minimize manipulation of the native data as it is transformed into FITS format.

COSSC has strong scientific and administrative ties to HEASARC, and is often consulted by HEASARC to represent the gamma ray community's interests in the development of high energy FITS format standards. However, these FITS standards are still in development and are only for highly processed data products. Data products currently produced by the Compton Observatory instrument teams are of a lower level of complexity than those being considered by HEASARC for standardization.

Astronomical Data Analysis Software and Systems II
ASP Conference Series, Vol. 52, 1993
R. J. Hanisch, R. J. V. Brissenden, and J. Barnes, eds.

The ROSAT Implementation of a Proposed Multi-Mission X-ray Data Format

M. F. Corcoran, W. Pence

Code 668 GSFC, Greenbelt, MD 20771

R. White

Code 932 GSFC, Greenbelt, MD 20771

M. Conroy

HEA-CFA, MS-3, 60 Garden St., Cambridge, MA 02138

1. Introduction

The use of mission-specific data formats hampers the analysis of archival data, and makes difficult the comparing of observations obtained by different instruments. An extra burden is placed on the archive user who must

- waste time learning arcane data file structures
- waste time writing software to interpret these file structures
- waste time reformatting data into files which are amenable to analysis by the user's favorite software package

2. Development of Rationalized Files

The High Energy Astrophysics Science Archive Research Center (HEASARC) located at GSFC proposed that data from high energy astronomy missions be distributed and archived in RrationalizedS data files which share a common data format.

HEASARC in conjunction with SAO has proposed that these "Rationalized Files"

- Use standard FITS and ASCII, binary table and image extensions to ensure portability; preferentially use binary table extensions for efficiency
- Use a common set of header keywords
- Contain detailed COMMENT and HISTORY cards in the header to define contents of files/extensions and indicate data processing history
- Use same data structures for common information (e.g., photon lists, orbit information, aspect information)

- Isolate instrument-specific information in separate files/extensions (housekeeping information, calibration data, derived products)
- keep related information together so user's don't have to search for it

3. ROSAT Implementation of Rationalized File Format

ROSAT is the first active mission to make use of the suggested rationalized format. The general scheme which has been adopted is to divide all data output from standard ROSAT data processing into 1 of 4 data types. Each data type consists of 1 or more FITS files. Each FITS file consists of a primary header + image (which may be NULL) and 0 or more extensions.

3.1. The BASIC file

The Basic file contains "bare-bones" science data: photon lists, good times, instrument statuses. The format of this file is the same for the HRI and PSPC except where noted. The file consists of

- a NULL primary array
- a BINTABLE giving start and stop times for the standard good time intervals in spacecraft clock seconds, along with corresponding observation interval number
- a BINTABLE giving times (in spacecraft clock seconds) when the instrument was collecting data (regardless of whether the data were "good"), along with corresponding observation interval number
- a BINTABLE giving the time of arrival for the photon (in s.c. clock seconds), the corrected and uncorrected photon energy amplitude (corrected amplitude not available for the HRI), the position of the photon in raw telemetered coordinates (unavailable for the PSPC), linearized detector coordinates and sky coordinates
- a BINTABLE giving the temporal status history of instrument parameters during the observation (parameters listed are instrument dependent)

3.2. The ANCILLARY file

The Ancillary file contains information useful for analysis of processed science data but not crucial for standard analyses: orbit, aspect, housekeeping. This file consists of:

- a NULL primary array
- a BINTABLE giving satellite orbit information
- a BINTABLE giving satellite aspect information
- a BINTABLE consisting of housekeeping information
- a BINTABLE giving rates of accepted and rejected events, particle event rates, etc., as a function of time. Some of this information is detector-specific.

3.3. The CALIBRATION File

The Calibration file contains information needed to calibrate results of standard analyses of processed science data. Because it's the most instrument-specific file, in the ROSAT implementation there is the widest difference between the HRI and PSPC versions of this data type.

The PSPC calibration file includes:

- A NULL primary array
- A BINTABLE extension containing the redistribution matrix in the HEASARC standard form
- A BINTABLE containing an the filter transmission in the HEASARC ancillary response file format
- An IMAGE extension which gives the instrument map
- A BINTABLE which gives the change in effective area with energy and off-axis angle
- A BINTABLE which gives the results of the fit to observations of the radioactive calibration source
- A BINTABLE which gives solar activity indicators

The HRI Calibration File includes:

- A NULL primary array
- an IMAGE extension which gives the quantum efficiency map
- a BINTABLE giving the gap map
- a BINTABLE giving the location of hotspots
- an IMAGE extension giving the bright earth background map
- an IMAGE extension giving the detector background map
- an IMAGE extension giving the charged particle background map

3.4. The DERIVED File

The Derived file contains products derived by standard processing system such as images, spectra, lightcurves, etc. Derived files are subdivided into 3 categories:

- Image Files: These files consist of an image as primary array and (optionally) one or more BINTABLE extensions.
- Source File: This file contains a NULL primary array and a X-ray source list giving characteristics of each detected source. Also included is a BINTABLE list of optical counterparts in the field, a BINTABLE giving lightcurve and/or timing analysis results and (for the PSPC) a BINTABLE extension giving source spectra (one per source) and a BINTABLE giving results of standard spectral analysis.

- Processing History File: This file consists of a NULL primary array and ASCII table extensions which list the processing system parameters used to process the data.

4. Further Information

Sample files are available via anonymous ftp from legacy.gsfc.nasa.gov. The files are kept in the directories

- DATA/rosat/pspc/sample_files
- DATA/rosat/hri/sample_files

Or one can contact Dr. Michael F. Corcoran (corcoran@heasrc.gsfc.nasa.gov).

Astronomical Data Analysis Software and Systems II
ASP Conference Series, Vol. 52, 1993
R. J. Hanisch, R. J. V. Brissenden, and J. Barnes, eds.

A Self-Defining Hierarchical Data System

Jeremy Bailey

Anglo-Australian Observatory, PO Box 296, Epping, NSW 2121, Australia

Abstract. The Self-Defining Data System (SDS) allows the creation of self-defining hierarchical data structures in a form which allows the data to be moved between different machine architectures. Because the structures are self-defining they can be used for communication between independent modules in a distributed system. The data structures are dynamic, allowing components to be added or deleted, or the size of arrays to be changed.

1. Introduction

SDS allows the building of essentially the same sort of data structures that can be built in most programming languages, for example in C using the struct keyword. A C struct groups together a number of items, each of which may be a simple variable, or may itself be another struct (thus giving rise to the hierarchical nature of the structures). However, SDS provides a number of additional features as follows:

1.1. Portable Structures

Differences between machine architectures may be encountered in the byte order of numeric items, in the representation of floating point numbers, and in the alignment requirements for structure components. The DEC VAX and SUN Sparc architectures, for example, differ in all these respects. SDS is designed to look after all these architectural dependencies, and enable structures to be moved between machines, while guaranteeing to provide data in the correct format for the local machine.

1.2. Dynamic Structures

A C struct is a static structure, fixed at compile time. It is, of course, possible to create dynamic structures in C using pointers, linked lists, etc., but such a structure is not then easily accessible as a single localized entity which can be written to a file or moved between machines. SDS allows structures to be manipulated dynamically, while retaining the ability to move a structure as a single entity between machines.

1.3. Self-Defining Structures

If a data structure is to be passed between two communicating programs, then both need to have an identical copy of the structure definition to ensure they interpret the structure identically. In the case of a C struct this definition is the C source code declaring the structure. For two tightly linked programs, it is not too difficult to ensure that both are always using the same copy of the definition. For data moved around a widely distributed system, this can be much more difficult to accomplish, and it becomes very difficult to safely make any changes to a data structure without undesirable effects on other programs.

SDS solves this problem by including the definition of the structure with the data. This definition contains the name, type and dimensionality of each data item, and enables an item to be accessed by name, without the program necessarily having to know everything about the contents of the structure. It is thus much easier to develop communicating programs independently, since a new item may be added to a structure without requiring any changes in another program which reads that structure.

2. SDS Structures

An example of a the sort of structure which can be built using SDS is given below. This example uses Starlink's NDF protocol normally used in conjunction with the HDS disk based hierarchical data format, but equally applicable to SDS. In this listing the first word on each line is the object name followed by its type, dimensions (if an array) and value. This particular example represents a spectropolarimetry dataset.

```
EXAMPLE                   Struct
  DATA_ARRAY                Struct
    DATA                      Float  [1024] { 0.0767828, 0.38353,...}
    ORIGIN                    Int      [1] {1}
  LABEL                     Char    [64] "Flux                        ..."
  UNITS                     Char    [64] "mJy                         ..."
  AXIS                      Struct [1]
    AXIS[1]                   Struct
      DATA_ARRAY                Struct
        DATA                      Float  [1024] { 4442.9, 4445.54,...}
        ORIGIN                    Int      [1] {1}
      LABEL                     Char    [64] "Wavelength              ..."
      UNITS                     Char    [64] "Angstroms               ..."
  MORE                      Struct
    POLARIMETRY               Struct
      STOKES_V                  Struct
        DATA_ARRAY                Struct
          DATA                      Float  [1024] { 0.0244155,...}
          ORIGIN                    Int      [1] {1}
        VARIANCE                  Struct
          DATA                      Float  [1024] { 0.0254768,...}
          ORIGIN                    Int      [1] {1}
```

An SDS structure is built out of three types of objects.

Primitive items - A primitive item is an item which can contain some data and may be either a scalar or an n dimensional array (where $n \leq 7$). In the example the items called DATA and LABEL are examples of primitive items.

Structures - A structure is a list of named items each of which may be a primitive item, a structure or a structure array. The EXAMPLE structure contains components, DATA_ARRAY, LABEL, UNITS, AXIS and MORE some of which are themselves structures.

Structure Arrays - A structure array is an n-dimensional array (where $n \leq 7$) of items, each of which is a structure. AXIS in the example is a structure array, in this case with only one component (more components would be used with multi-dimensional data arrays).

3. Applications of SDS

Some possible applications of SDS include the following:

3.1. Message Format in a Distributed System

The instrument control system for the AAO 2 degree field will consist of a number of VME machines running the VxWorks operating system coordinated by a control task running on a VAX/VMS machine. The VAX sends messages via TCP/IP to request actions in the VME software. By encoding the messages using SDS we automatically handle the different data representations on the two machine types, while the self-defining nature means we can easily add new components to structures without affecting existing software.

3.2. Providing Dynamic Data Structures in Fortran

Fortran 77 has very poor capabilities for supporting data structures compared with languages such as Pascal and C and provides no support for dynamic data structures. The Fortran interface to SDS enables this capability to be provided for Fortran programmers.

3.3. Interchange Format for Astronomical Data

Traditionally FITS has been the standard interchange format for astronomical data, but while it copes well with simple image formats, it cannot easily deal with some of the more complex types of data which are now being encountered. A general hierarchical data format might be a suitable replacement for FITS. However previous systems such as HDS have been unsuitable for this role because of the lack of a defined data format (only the callable interface to the software is defined) and lack of portability of the software. SDS might be a suitable candidate.

It is important to understand that such a format only provides part of the solution to the data interchange problem since SDS provides no specification of how its structure components are to be used. We need in addition a specification of how actual astronomical data items are to be represented in terms of such general hierarchical structures. A possible solution to this part of the problem is the Starlink NDF (extensible N-dimensional Data Format) which provides great flexibility for representing data, while providing support for the addition of extensions to add any capabilities not provided in the standard.

4. SDS Software

The SDS system consists of the following items of software:

4.1. The SDS Kernel

The SDS kernel is a package of C functions which provide the following basic set of capabilities:

- Creation of SDS structures and structure components.
- Navigation of structures to find items by name or index.
- Reading and Writing of data from primitive items and structures.
- Import and Export of structures.
- Editing Structures (adding new components, deleting components, renaming or resizing objects).

The SDS Kernel operates completely in memory. This distinguishes SDS from disk based hierarchical systems such as Starlink's HDS. Structures are created as internal objects in memory managed by SDS itself, but may be exported into a memory buffer supplied by the caller. This external structure can then be moved between machines by writing it to a file or sending it as a message and imported back into SDS on the remote machine.

4.2. The SDS Utility Package

The SDS utility package provides C functions which read and write SDS structures to and from files as well as a function which lists the contents of an SDS structure.

4.3. The Arg Package

The Arg package is a set of functions which provide a simpler interface to SDS for the creation of simple structures consisting of only scalar, and character string components. Arg calls can be interspersed with SDS calls if necessary.

4.4. The SDS Compiler

The SDS compiler automatically generates the sequence of SDS calls to build an SDS structure from an equivalent C struct definition. It can also be used to generate the SDS structure itself. It is available as a standalone program or as a callable function.

4.5. Fortran Interface

A Fortran callable interface to the SDS kernel, utilities and Arg package is available on some machines. This is implemented using Starlink's CNF/F77 package which provides a portable Fortran/C interface.

4.6. Programs

Programs to test the SDS system, and to list the contents of SDS structures are provided. On machines for which the Fortran interface is supported programs to copy HDS structures to SDS and vice versa are provided.

5. SDS Format

SDS external structures are stored in a defined format which specifies how a complex hierarchical structure can be encoded into a sequence of bytes. The format definition is described in the SDS documentation.

6. Availability

SDS can be obtained by anonymous ftp from aaoepp.oz.au. SDS releases are available for the following systems (SDS should be easily portable to many other machines).

- VAX/VMS.
- SUN Sparcstation.
- DECstation.
- Apple Macintosh (using MPW).
- VxWorks real time operating system.

Part 5. Data Analysis Applications

Section F. Hardware Issues

Astronomical Data Analysis Software and Systems II
ASP Conference Series, Vol. 52, 1993
R. J. Hanisch, R. J. V. Brissenden, and J. Barnes, eds.

IRAF Port to the Dec Alpha Machine

Nelson Zarate

Space Telescope Science Institute, Baltimore, MD 21218

1. Introduction

The Alpha architecture is a high performance RISC architecture that increases CPU performance through the following features:

- a simplified instruction set
- multi-instruction issue
- a load/store operation model
- elimination of microcode
- a processor that allows parallel instruction execution and out-of-order completion of instructions

Porting IRAF can be divided into three groups: code that is portable across different computer architectures and requires no changes, i.e., the SPP code; code that can be modified so that it becomes portable across VMS platforms; and the code that must be specific to either VAX or Alpha systems.

The process of porting IRAF in the OpenVMS environment took two levels of effort that were dictated by the resources available at the Institute at any given time. First, we had only the Migration Kit on the VAX, which is a set of tools and cross-compilers, to determine how easy or difficult is the process of compilation or, translate a VAX image into an Alpha image when there is no source available or is time critical to have it run under the new architecture. VEST is the tool to translate a VAX binary image file into a native Alpha image. Secondly, after we installed the native C, MACRO and FORTRAN compilers on the machine, a recompilation of most of the IRAF host system interface was necessary.

2. Larger Page Size in Alpha

In general an application may contain page-size dependencies if it calls system services or run-time library routines to perform memory management functions such as:

- allocating virtual memory
- mapping sections in your application's virtual address space

- locking memory into your working set
- protecting segments of your virtual address space

The Alpha VMS architecture supports an 8 K, 16 K, 32 K, or 64 K page size, depending on the implementation. Whenever possible the Alpha system attempts to preserve compatibility with its VAX counterpart; for example, the Alpha version of the routines that accepts page-count values still interprets this in 512-byte quantities, called "pagelets" to distinguish them from the CPU-specific pages.

3. Data Types

The data-type compatibility between VAX and Alpha is important, because VAX and Alpha systems can be mixed in a single VMS cluster. Data on disk placed there by a VAX will be accessible by Alpha and vice versa. However, you may need to make special considerations for data on disk in the D_FLOAT format, which is the default format for double precision on the VAX. The default for the Alpha is G_FLOAT.

Because the Alpha instruction set does not support D_floating computations, D_floating data is converted to G_floating format for arithmetic computations and then converted back to D_floating format. Thus, using D_floating data is slower than G_floating.

4. Data Alignment

A VAX implementation almost always performs memory reads and writes better when the data are naturally aligned by size. VAX hardware implementations are built to handle unaligned data without software intervention and the performance penalty is small.

The Alpha architecture on the other hand can perform memory reads and writes only in 32-bit longword boundaries. If a memory read or write is done on an unaligned memory address, an exception occurs. This exception will read or write correctly but it may execute up to 100 times slower under these circumstances.

So, it is important to insure that misaligned memory operations do not occur. Fortunately, the C and Fortran compiler will take care of data alignment for you at a cost of extra instructions.

5. The IRAF Port

5.1. Routines in IRAF/VMS/OS

The IRAF system relies on the so called Z-routines to communicate with the host operating system. These routines are mostly written in C language compiled with the VAX C compiler. There are a number of MACRO assembler routines written to speed up processing or as in the case of error handling have to be in assembler.

5.2. IRAF C Language Routines

The OpenVMS DEC C compiler is in full ANSI-C compliance with facilities to compile under a VAX C mode (CC/Standard=vaxc) or ANSI plus DEC extensions facilities (CC/Standard=relaxed). When the Z-routines were first written for VAX VMS, only VAXC was available. Now with DEC C these routines can be compiled under more than one standard and then we can be as strict as we can afford.

At first I started to compile with the **relaxed** standard and got errors like variables declared as **int** that actually needed to be **short** or vice versa, lots of warnings and errors about trying to assign a **struct** to an **int**. These needed to be fixed by changing variable type and/or adding a cast before assignment. In the end I fixed only a couple of Z-routines and proceeded by changing the compilation standard to VAXC.

The VAX-Specific hardware exceptions like SS$_FLTDIV_F (Float divide by zero) are no longer recognized and most of them are grouped under the new Alpha VMS SS$_HPARITH (High Performance arithmetic exception). The **zxwhen.c** routine requires these kinds of changes.

The routines in the BOOT directory needed to be compiled with the CC/defined=NOVOS since at this stage of the port there are no IRAF libraries available to get file name mapping for example. The NOVOS macro allows compilation of code in the BOOTLIB directory to make crude IRAF file name mapping to VMS file names; this is necessary for the XC and MKPKG programs to work so they can create the IRAF system libraries. The most important routine for this purpose is bootlib/vfn2osfn.c and it needed modifications and enhancements to produce the correct VMS pathnames.

The more important changes to the IRAF C routines relate to the external linkages storage class specifiers and modifiers that are specific to DEC C for VMS systems. These are **globaldef**, **globalref**, **globalvalue** and **extern** among others. The **#pragma extern_model** directive controls how the compiler interprets these objects; for example in "hlib$libc/stdio.h" the declaration:

```
extern int FIOCOM[];          /* the FIO common */
```

needs to be controlled with the following pragma.

```
#pragma extern_model comon_block
extern int FIOCOM[];          /* the FIO common */
```

The absence of the pragma results in an Undefined Symbol at linking time. Also to avoid linking warning messages like "Conflicting Attributes for Psect" that appear after every MKPKG command, the C compiler qualifier Share_globals is required.

5.3. IRAF Assembly Routines

The most famous assembly routines in IRAF are the pair ZSVJMP/ ZDOJMP to handle non_local GOTO's. The creation of these routines for new hardware has always been a challenge for any person attempting to port IRAF. The Alpha machine with its modern RISC architecture puts many constraints on the handling of VMS calling sequences, PC and FP manipulations. The assembler instructions are a lot simpler than its predecessor the VAX MACRO. Also the treatment of the Frame Pointer to save the calling environment necessary in the ZSVJMP routine is not available in here.

A first attempt to solve this problem was to write these in the same way as in the UNIX environment; i.e., using the C RTL setjmp/longjmp protocol. So `zsvjmp.s` calls setjmp and is written in Macro64 and the `zdojmp.c` is written in C and does a straight call to longjmp.

The problem that still remains with this approach is the zdojmp target statement that is one more than the target statement obtained in the VAX environment when running an application under the CL. In cl/main/execute() under Alpha, the statement:

```
if (setjmp(jumpcom))
   onerr();
```

is reached by the longjmp routine at the onerr() statement rather than the `if(setjmp(jumpcom))` as in the VAX case. In both cases the error flag is cleared since the error has already been posted before reaching this part of the code.

Since the C compiler is very efficient in producing good optimized code, the routines written in VAX assembler, `os/str.s` which comprises `_strcpy`, `_strupk` and `_strpak` have been rewritten in the C language. All the rest of the Assembler routines have been dropped with the exception of IEEER and IEEED.

6. IEEE Support

The Alpha OpenVMS system supports among other things the IEEE data types for single and double floating point representation for internal computation, although the default types are S_FLOAT and G_FLOAT. In order to be as closely compatible as possible with the VAX VMS environment I decided to take the default floating point representation and rewrite the IEEER and IEEED routines needed for the IEEE conversion application.

The Alpha OpenVMS RTL library (LIB$) has a set of routines to convert from native format to IEEE and vice versa. I wrote ieeer.c and ieeed.c using the CVT$CONVERT_FLOAT library call.

7. IRAF Shareable Image S_IRAF.EXE

In order to save disk space, linking time, and start up time, a shared image approach is more convenient when we have hundreds of applications that call the same basic libraries; as in the case of IRAF, the libsys, libex, libvops and libos. The shared image creation procedure is essentially the same and only some extra requirements are necessary. The UNIVERSAL symbol is now SYMBOL_VECTOR to indicate that a transfer vector approach is included; COMMON names need to be declared as PSECT with the same SYMBOL_VECTOR approach since they are no longer included automatically in the global symbol table as in the VAX/VMS case.

Now it is possible to change the content of some of the routines in the shared image without the need to relink the applications that use it. It is important however to conserve the order in which the routines are declared the very first time one builds the shared image.

8. Size of Object Files and Executables

The size of the compiled object code scales with the simpler instruction set. Here is a random list that compares the sizes of one object file and an image file between the Alpha and the VAX.

Filename	Alpha size	VAX size (Kbytes)
t_wfits.obj	16.1	4.8
x_dataio.exe	1372	710 (No S_IRAF)
x_dataio.exe	209	124 (With S_IRAF)

9. Efficiency

In rough terms, since there are no official numbers yet, the Alpha machine under the OpenVMS operating system performs better than my best expectations. Making a subjective comparison with a VAX 4000/500 the DCL operations take at least half the time on the Alpha machine; the start up time of an EDIT session has improved remarkably as well as compilation and linking time. Under IRAF/CL a mkpkg takes no time in comparison with the long wait on a VAX. Relinking is surprisingly faster when using the IRAF shared image.

10. Performance

When running the LUCY deconvolution task in the STSDAS/RESTORE package the following numbers were obtained. Remember that the figures for the Alpha machine were obtained on a machine that was not tuned up for best performance.

```
LUCY Deconvolution Benchmark  (800x800 32bits/pixel)
     (50 iterations)
```

Machine	User-time (minutes)	system-time (minutes)
CRAY-YMP/gsfc	1.4	0:03.8
DEC Alpha	9.0	8:05
SGI-personal	34.7	42:08
SUN sparc2	65.7	90:00

Astronomical Data Analysis Software and Systems II
ASP Conference Series, Vol. 52, 1993
R. J. Hanisch, R. J. V. Brissenden, and J. Barnes, eds.

Seeing the Forest for the Trees: Networked Workstations as a Parallel Processing Computer

J. O. Breen and D. M. Meleedy

Smithsonian Astrophysical Observatory 60 Garden Street, Cambridge, MA 02138

Abstract. Unlike traditional computers in which one central processing unit performs one instruction at a time, parallel processing computers contain several processing units, thereby allowing them to perform several instructions at once. Many of today's fastest supercomputers achieve their speed by employing thousands of processing elements working in parallel. While few institutions can afford these state-of-the-art parallel processors, many already have the makings of a modest parallel processing system. Workstations on existing high-speed networks can be harnessed as nodes in a parallel processing environment, bringing the benefits of parallel processing to many. While such a system can not rival the industry's latest machines, many common tasks can be accelerated greatly by distributing the processing burden and exploiting idle network resources. We have used p4, a freely-available parallel processing environment, as the foundation for our investigation, and our preliminary results have been very encouraging. With the ever-growing volume of observational data and the ever-increasing complexity of theoretical calculations, it is essential to utilize our computing resources fully.

1. Introduction

There is a trend towards the decentralization of computing resources in research institutions, due in large part to the growing popularity of high-speed RISC-based engineering and scientific workstations. For many users, gone are the days when central computation facilities billed users and departments for computer time spent running calculations, sorting data, or reading e-mail. Instead, it is now common to find workstations as powerful as yesterday's supercomputers on the desks of researchers. But some efficiency was inevitably traded for the freedom and flexibility of personal workstations. Some workstations may sit idle while their owners are away due to travel, vacation, or perhaps just lunch, while others may be taxed to the limit by particularly demanding applications. Better management of distributed resources can lead to higher performance for all users.

One way to manage workstations is as processing nodes in a parallel processing system. The idea behind parallel processing is a very simple one. If one processor can execute one instruction at a time, then multiple processors should execute multiple instructions concurrently. As a machine running in a vacuum is of no use, communication facilities are also necessary for a parallel processing

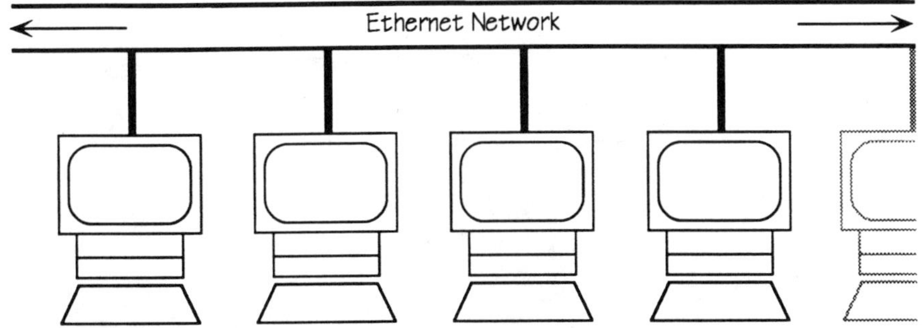

Figure 1. The physical topology of workstations via Ethernet. The workstations may be phyically separated by feet, miles, or continents.

system. There are several software packages which provide the requisite computational and communication facilities. We have chosen to work with p4 from Argonne National Labs as described by Boyle et al. (1987). It is essentially a library of routines, able to be called from either C and Fortran, which handles the creation and management of processes on machines across the network. It has been ported to more than a dozen architectures, from Sun to Cray to Thinking Machines, and even handles data conversion between the machine-specific formats.

2. Basic Implementation

The p4 system itself uses a very simple mechanism to manage processes on different machines. For each p4 program to be run, the user creates a "procgroup" file, which contains a list of machines on which to run processes. One can run multiple processes on a given machine simply by including multiple instances of that machine in the procgroup file. The system is built around the master-slave computing paradigm. The user's machine (the "local" machine) runs the one and only master process, which is responsible for creating and destroying the slave processes. Slave processes may be run on any machine, including the local machine. Figure 1 shows a typical physical topology for networked workstations. The p4 program sees not the machines, however, but only the processes (see Figure 2), thereby insulating the program from any machine-dependencies.

3. Opportunity for Customization

While the p4 system requires a procgroup file in order to be fully operational, creation of the procgroup file is left entirely to the user's discretion. For some installations, where there are just a handful of workstations, merely a list of the hosts would do. Other installations may have one very fast computer and would prefer to run several processes on it at once. Our site is different still. We have access to over 80 Sun SPARCstations, with a range of performance greater than a factor of four. Some machines are heavily used, others lightly, and most

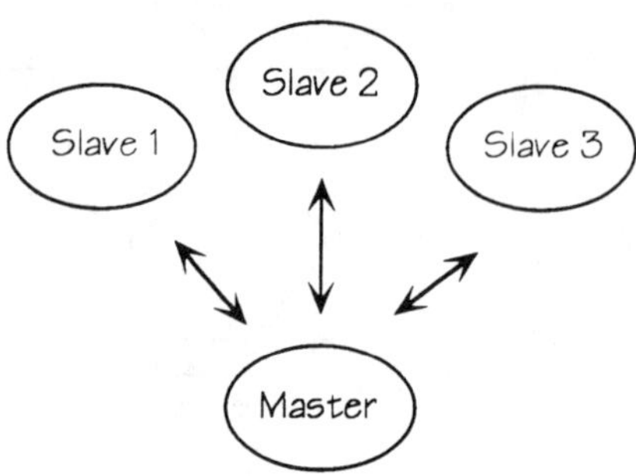

Figure 2. The master process is insulated from the details of the hardware running each process. All of these processes could be running on one machine, for instance.

intermittently. We have developed some tools to help choose which machines are the best to use at any given time, and to create an appropriate procgroup file.

First, we estimate the intrinsic performance of each workstation, in the absence of any other users or processes. To this end, we have run a simple floating-point arithmetic benchmark program called flops.c (Aburto 1992). Flops uses the same mix of floating point instructions as does the popular Dhrystone benchmark (Weicker 1984) to measure floating point performance in million floating point instructions per second (MFLOPS). As these results (should) never change, we store them in a file, and add new hosts as is needed. Then, we query each system for its five-minute Unix load average. This is a number which indicates the number of tasks the processor is trying to perform at any one time. As we do not wish to overburden any machines, a strict cutoff at a load average of 1.0 is applied. This load average is indicative of a program running without a pause, such as (alas) xlock. (It would be simple to make exceptions for xlock or another such program, however) We have determined an empirical relationship between load average and performance, with which we apply a correction to the intrinsic MFLOPS rating, and store these predictions in a "performance" file. If a user needs five machines, a procgroup file is created containing the first five hosts from the performance file. These performance estimators may be run as often or as seldom as is appropriate or desired. Some applications are well-suited to include load-management features themselves. A program in which slaves are given more work if they complete tasks early will benefit by having more of its work performed by the faster slaves.

4. Preliminary Results

In our preliminary tests, we have found the greatest bottleneck to be interprocess communication. The worst performance was exhibited by a simple program to compare observed radial profiles with various empirical models stored in files. It took almost as much time for the master to load and send each model as it did for the slave to compute the quality of the fit—there was no observable increase in performance past two slaves! This is to be expected. Massively parallel machines like the Connection Machines have dedicated high-speed networks specifically for interprocess communication. Our network, on the other hand, is servicing all of the needs of all of the machines in our division, from X-Windows servers to telnet sessions, from NFS access to mail and news. The applications which gain the most benefit in such an environment are computationally-intensive tasks, such as modeling and image processing which either generate their own data or access large sections of data only occasionally. The performance of these applications approaches the correlation with the number of processors one would expect, at least until other bottlenecks interfere.

5. Software and Sources

The p4 Programming System is available via anonymous FTP from info.mcs.anl.gov in the directory /pub/p4/. The distribution includes sources, documentation, and example programs. Our performance evaluation utilities and sample applications will be made available via anonymous FTP when they are documented and ready for general release. It is our hope to maintain a collection of p4-based applications of interest to astronomers and space scientists Please feel free to contact either author for more information.

References

Aburto, A. 1992, notes to flops.c computer program available via anonymous FTP from marlin.nosc.mil in /pub/aburto/

Boyle, J., Butler, R., Disz, T., Glickfeld, B., Lusk, E., Overbeek, R., Patterson, J., & Stevens, R. 1987, Portable Programs for Parallel Processors (Holt, Rinehart & Winston)

Weicker, R. 1984, Communications of the ACM, 27, 1013

Astronomical Data Analysis Software and Systems II
ASP Conference Series, Vol. 52, 1993
R. J. Hanisch, R. J. V. Brissenden, and J. Barnes, eds.

A Low-Cost Vector Processor for Speeding-Up Compute-Intensive Image Processing

Hans-Martin Adorf

Space Telescope–European Coordinating Facility, European Southern Observatory, Karl-Schwarzschild-Str. 2, D-8046 Garching b. München, FRG. (Internet: adorf@eso.org—SPAN: ESO::ADORF)

Abstract. The advent of affordable add-on vector processing boards for standard workstations, complemented by mathematical/statistical libraries, is beginning to impact compute-intensive tasks such as image processing. A case in point is the restoration of distorted images from the Hubble Space Telescope. A low-cost vector processor implementation of the standard Tarasko-Richardson-Lucy (TRL) restoration algorithm is presented which is seamlessly interfaced to a commercial, memory-based image processing system.

There should not be a contradiction between
ease of writing correct programs and
ease of writing fast programs.
The first is an advantage,
while the second is a requirement.
— *Jim Allard, 1992*

1. Background

The spherical aberration present in the optics of the Hubble Space Telescope (HST) has triggered the development and exploitation of image restoration methods on a scale hitherto not seen in optical astronomy (Adorf 1992a).

The restoration problems for the two cameras on board HST, the Faint Object Camera (FOC) and the Wide Field and Planetary Camera (WF/PC), are quite distinct. The FOC currently shows an almost *space-invariant* PSF (SI-PSF). The associated restoration task is compute-intensive, but appears manageable on a 1992 state-of-the-art workstation. For SI-PSFs the standard Tarasko-Richardson-Lucy (TRL) algorithm (Tarasko 1969; Richardson 1972; Lucy 1974) is expressible on the abstraction level of image frames (Adorf 1990) and thus can readily be vectorized. This opens the possibility of speeding-up the restoration process with the help of a vector processor (VP).

The WF/PC restoration problem, on the other hand, is characterized by a *space-variant* PSF (SV-PSF) which inhibits efficient global restorations. However, the "sectioned restoration" method pioneered by Trussel & Hunt (1978) would equally benefit from the availability of VP-hardware.

The processing of space-borne imagery is not the only computationally demanding data analysis task. Ground-based optical astronomy of the later 1990s is likely to produce data volumes 10 to 100 times higher than those usually seen today and at rates that are also factors higher. A case in point is the image dissecting spectrograph, proposed for one of the later ESO-VLT units, which is supposed to deliver data frames of 128 Megabyte each. These data volumes, each equivalent to about one sixth of a typical digitized Schmidt plate, will have to be analyzed automatically with fast, robust and reliable algorithms.

2. Vector Processing Hard- and Software

After a survey of contemporary parallel computing options (Adorf 1991), an initial assessment of a particular VP-board, based on Intel's i860 64-bit RISC-microprocessor, was carried out (Adorf & Oldfield 1992; Adorf 1992b). Subsequently, in order to enable ST-ECF staff to provide guidance to European astronomers interested in vector processing, it was decided to acquire a VP-board for further experiments. The selected board, the "MemSys i860 Accelerator" from Cambridge (UK)-based MaxEnt Solutions Ltd. (MESL), equipped with 8 MB of fast, dynamic random access memory (DRAM), arrived in April 1992 and was installed in the S-bus slot of a standard Sun workstation.

The i860-board was accompanied by the *MESL Hostmode Library* (HML, MESL 1992) comprising pre-compiled procedures for program and data management, basic numerical arithmetic, special function computation, etc., roughly on the completeness level of Fortran-77 (F77). All image processing operators are vectorized and can be called from F77- or C-programs.

3. The Vectorized Tarasko-Richardson-Lucy Algorithm

Backed by the 1991-experience, the re-implementation of the standard SI-PSF TRL-algorithm proved straightforward. In order to facilitate the use of the i860-VP, it was considered necessary to seamlessly interface it to one of the standard astronomical data analysis systems. Since much of the image restoration development work at the ST-ECF is carried out within IDL (RSI 1991), it appeared natural to interface the VP to that particular system (Figure 1).

Implementation of the main C-program invoking the HML-routines and the required IDL-to-C "glue" procedures was facilitated by program templates kindly provided by MESL and Research Systems, Inc. (RSI), respectively. The resulting C-code consists of about 130 lines of code, of which at most 35% can be considered essential, whereas the other 65% is overhead. Note that, on the appropriate mathematical abstraction level, the TRL-algorithm can be described in essentially *one line* via two co-recursive equations (Adorf et al. 1992).

Consistent with previous performance estimates, the vectorized TRL-algorithm carries out a 40-iteration restoration on a standard 512x512 frame in 85 seconds (wallclock time). The VP-solution achieves a considerable (factor 8) reduction of overall turn-around time by raw compute power. As such it has to compete with recent more "intelligent" developments to accelerate the convergence of the TRL-algorithm in software (Adorf et al. 1992, and refs. therein).

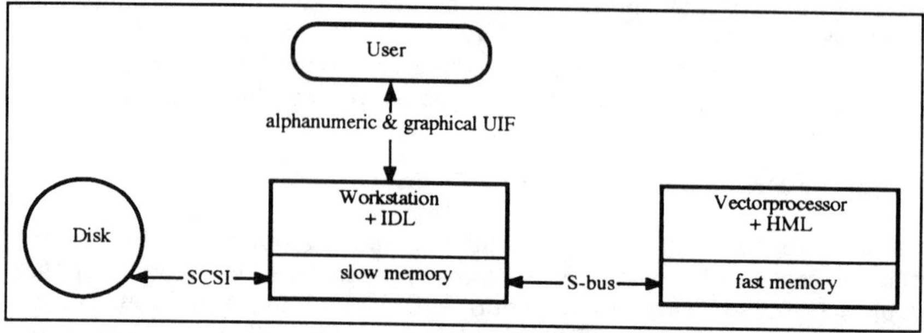

Figure 1. Hard- and software architecture for vector processing. Under the control of the user, the image processing system, executing on the scalar host, reads data from an external disk into host memory. (Alternatively, data may already be resident in host-memory from a previous processing step.) Data are subsequently transferred to the vector processor board, where processing (restoration) takes place. Results are transferred back to the host for further processing and/or storage onto an external disk.

4. Accelerating General Image Processing Operators by the VP

Naturally the question arises whether one could speed-up image processing operators of a finer granularity (i.e., lower abstraction) level than a complete iterative algorithm. Decomposing the standard TRL-algorithm for SI-PSFs into lower-level image operators, leads to multiplication and division, convolution and correlation as building blocks (Adorf et al. 1992). Convolution, for instance, can be further decomposed into a multiplication and two or three FFTs.

An efficient VP-based implementation of individual operators would permit a combination of the flexibility of the image processing system's high-level language with the VP's computational power. Also, a given algorithm would have to exist only in a single form, namely a high-level implementation. Thus the problem of developing and maintaining two implementations of an evolving algorithm (one for a scalar, one for a vector architecture) would be alleviated.

An experiment was therefore carried out to compare the efficiency of individual VP-accelerated FFTs with those computed by a scalar workstation. Neglecting different data organizations, the efficiency of real-to-complex fast Fourier transform `RFT2D` found in the HML was compared to IDL's built-in complex-to-complex FFT executing on either a SPARCstation-1 or a SPARCstation-2. Performance results are given in Table 1. In terms of elapsed wall-clock time, the i860-VP wins marginally on 512x512 pixel arrays, and considerably on 1024x1024 pixel arrays.

Array dims	SPARC-1 & IDL	SPARC-2 & IDL	i860 & HML FFT only	i860 & HML FFT + overh.
256x256	1.7 sec	0.9 sec	0.1 sec	1.8 sec
512x512	7.5 sec	3.8 sec	0.5 sec	2.4 sec
1024x1024	52.3 sec	16.5 sec	2.0 sec	5.5 sec

Table 1. Results of timing tests comparing IDL's FFT on a scalar host with one of the *MESL Hostmode Library* FFTs executing on an i860 vector processor. Times on the SPARCstations are best times obtained.

5. The Costs of Vector Processing

It would be futile only to discuss computational performance of a VP-board and neglect costs. Our initial investment into hardware and indispensable software amounted to ca. 15,000DM and 1,500 DM, respectively. From the restricted perspective of "MFLOPS returned per currency-unit invested", the i860-VP is certainly a bargain compared to a standard workstation, though scalar machines are catching up.

However, the hardware investment is just one, obvious cost factor of many. Less obvious is the cost absorbed by software development in a general sense. An attempt was made to retrospectively identify the steps taken within what might be called the "ST-ECF parallel processing initiative" together with tentatively estimated associated costs. The conclusions: Firstly, the pure coding phase absorbed only a relatively minor fraction of the overall costs—a generally observed fact in software engineering. Secondly, at an assumed overall cost of about 500 DM per person-day, the secondary software cost of the VP-exercise at the ST-ECF has already matched, if not exceeded, the primary hardware cost. Surely, some of the start-up investment has to be borne once only, and European astronomical sites could save software development cost by relying on the experience and code now available at the ST-ECF.

6. Summary and Outlook

The standard Tarasko-Richardson-Lucy (TRL) restoration algorithm, implemented on an Intel i860-based vector processor (VP) board, outperforms the equivalent scalar algorithm on a Sun SPARCstation-2 by a factor of about 8.

When seamlessly controlled via a high-level, vectorized image processing language, the VP is easy to use. Indeed, user response so far has been quite positive, since the speed of the VP-board re-introduces interactivity into restoration work. Calling sequences have deliberately been kept the same, thus allowing the continued use of pre-existing data analysis "notepads" with little or no modification.

As expected, programming the VP is more involved and—considering the primitive programming environment used—less flexible than programming a scalar workstation. In particular, the *MESL Hostmode Library* lacks a conditional construct analogous to the WHERE-statement of IDL, Fortran-90 or CM-Fortran, a deficiency which makes the VP less suitable for sophisticated algorithms, e.g., those involving regions of interest.

For the future, experiments are being considered to investigate the feasibility of detaching i860-board *initialization* from board *use*, which would allow a speed-up of the execution of individual VP-supported image processing-operators. Experiments are also envisaged to control the board with "VPL", a vectorized programming language currently implemented on top of Common Lisp. Thus VP-implementations of a statistical classifier and a trainable neural network should become feasible. However, the speed of progress is seriously limited by available human resources.

References

Adorf, H.-M. 1990, ST-ECF Newslett., 14, 8

Adorf, H.-M. 1991, ST-ECF Newslett., 15, 8

Adorf, H.-M. 1992a, in Proc. Conf. Science with the Hubble Space Telescope, eds. P. Benvenuti & E. Schreier (ESO, Garching, FRG), in press

Adorf, H.-M. 1992b, ST-ECF Newslett., 17, 9

Adorf, H.-M., Hook, R.N., Lucy, L.B., & Murtagh, F.D. 1992, in Proc. 4th ESO/ST-ECF Data Analysis Workshop, eds. P. Grosbøl & R.C.E. de Ruijsscher (ESO, Garching, FRG), 99

Adorf, H.-M., & Oldfield, M.J. 1992, in Astronomical Data Analysis Software and Systems I, A.S.P. Conf. Ser., Vol. 25, eds. D.M. Worrall, C. Biemesderfer & J. Barnes, 215

Lucy, L.B. 1974, AJ, 79, 745

MESL, 1992, MemSys i860 Accelerator Host Mode Library User Manual, MaxEnt Solutions Ltd., Cambridge, England

RSI, 1991, IDL User's Guide — Interactive Data Language, Version 2.2, August 1991, Research Systems Inc., 777 29th Street, Suite 302, Boulder, CO 80303

Richardson, B.H. 1972, J. Opt. Soc. America, 62, 55

Tarasko, M.Z. 1969, FEI-156 (1969), Obninsk, preprint (in Russian)

Trussel, H.J., & Hunt, B.R. 1978, IEEE Trans. Acoust. Speech Signal Process., ASSP-26, No. 2, 157

Author Index

Subject Index

Colophon

These Proceedings were prepared with LaTeX using a style file customized for the PASP series. Authors submitted manuscripts electronically, either by electronic mail or by depositing files in an designated anonymous FTP area. All of the 104 papers in this volume were submitted with usable LaTeX markup generated by the authors themselves.

Final proof pages for the entire book (from page v on) were produced at NOAO on an Apple LaserWriter IIntx. Tomas Rokicki's *dvips* program was used to translate the device-independent output from LaTeX. This combination of PostScript printer and driver program enabled us to take advantage of the PostScript language to merge figures into the pages as they were printed.

49 of the papers submitted were accompanied by 109 figures. Of these papers, 41 authors submitted their figures as Encapsulated PostScript (EPS) files. There were 90 figures submitted as EPS files, and we were able to incorporate 73 of them directly into the manuscripts as part of the printing process. 30 papers with figures were generated completely electronically requiring no cutting and pasting. 4 papers included EPS files for some figures but not all. The automatic inclusion of EPS graphics is advantageous and produces a superior product, and it is our hope that graphics-producing software will evolve toward a common implementation of the EPS standard. Already, we see a vast improvement in the production of EPS figures over the past year when we compare the results of this Proceedings with that of last year.

Most of the front and back matter was generated mechanically from the material submitted by the authors. A database of pertinent information about each paper was maintained, and the table of contents, author index, etc. were derived from it. The overall pagination of the volume was determined by software after the papers were edited and ordered, and the running heads and folio numbers were applied by LaTeX using standard markup commands. The software that was developed for these chores is generalized so that it can be used for other projects, and it is available to other editors through the PASP Conference Series Office.